Platelet Membrane Glycoproteins

Platelet Membrane Glycoproteins

Edited by

James N. George

University of Texas Health Science Center at San Antonio
San Antonio, Texas

Alan T. Nurden

Unité 150 INSERM
Hôpital Lariboisière
Paris, France

and

David R. Phillips

Gladstone Foundation Laboratories for Cardiovascular Disease
University of California, San Francisco
San Francisco, California

PLENUM PRESS · NEW YORK AND LONDON

Library of Congress Cataloging in Publication Data

Main entry under title:

Platelet membrane glycoproteins.

   Includes bibliographies and index.
   1. Blood platelets. 2. Membrane proteins. 3. Glycoproteins. I. George, James N. II.
Nurden, Alan T. III. Phillips, David R., 1942–   . [DNLM: 1. Blood Platelets—
physiology. 2. Glycoproteins—physiology.3. Membrane Proteins—physiology. QU 55
P716]
QP97.P55   1985   599′.0113   85-3498
ISBN 0-306-41857-6

©1985 Plenum Press, New York
A Division of Plenum Publishing Corporation
233 Spring Street, New York, N.Y. 10013

Printed in the United States of America

Contributors

Ralph M. Albrecht, M.D. • School of Pharmacy, University of Wisconsin, Madison, Wisconsin 53706

Joel S. Bennett, M.D. • Hematology-Oncology Section, Department of Medicine, University of Pennsylvania School of Medicine, Philadelphia, Pennsylvania 19104

Kenneth J. Clemetson, Ph.D. • Theodor Kocher Institute, University of Berne, CH-3000 Berne 9, Switzerland

Isaac Cohen, Ph.D. • Atherosclerosis Program, Rehabilitation Institute of Chicago, and Department of Molecular Biology, Northwestern University Medical School, Chicago, Illinois 60611

Barry S. Coller, M.D. • Division of Hematology, Departments of Medicine and Pathology, State University of New York at Stony Brook, Stony Brook, New York 11794

Neville Crawford, Ph.D. • Department of Biochemistry, Institute of Basic Medical Sciences, Royal College of Surgeons of England, London WC2A 3PN, England

Joan E. B. Fox, Ph.D. • The Gladstone Foundation Laboratories for Cardiovascular Disease, Cardiovascular Research Institute, University of California, San Francisco, California 94140

T. Kent Gartner, Ph.D. • Department of Biology, Memphis State University, Memphis, Tennessee 38152

James N. George, M.D. • Division of Hematology, Department of Medicine, University of Texas Health Science Center at San Antonio, San Antonio, Texas 78284

Inger Hagen, Ph.D. • Research Institute for Internal Medicine, Section on Hemostasis and Thrombosis, Rikshospitalet, University of Oslo, Oslo 1, Norway

Thomas J. Kunicki, Ph.D. • The Blood Center of Southeastern Wisconsin, Inc., Milwaukee, Wisconsin 53233

Lawrence L. K. Leung, M. D. • Division of Hematology-Oncology, Department of Medicine, The New York Hospital-Cornell Medical Center, New York, New York 10021

Joseph C. Loftus • School of Pharmacy, University of Wisconsin, Madison, Wisconsin 53706

Ernst F. Lüscher, Ph.D. • Theodor Kocher Institute, University of Berne, CH-3000 Berne 9, Switzerland

Deane F. Mosher, M.D. • Department of Medicine, University of Wisconsin, Madison, Wisconsin 53706

Ralph L. Nachman, M.D. • Division of Hematology-Oncology, Department of Medicine, The New York Hospital-Cornell Medical Center, New York, New York 10021

Alan T. Nurden, Ph.D. • Unité 150 INSERM, Hôpital Lariboisière, Paris ‚Cedex 10, France

Donna M. Pesciotta • Department of Medicine, University of Wisconsin, Madison, Wisconsin 53706

Sharron L. Pfueller, Ph.D. • Department of Medicine, Monash University Medical School, Alfred Hospital, Prahran 3181, Victoria, Australia

David R. Phillips, Ph.D. • Gladstone Foundation Laboratories for Cardiovascular Disease, Cardiovascular Research Institute, and Department of Pathology, University of California, San Francisco, San Francisco, California 94140

Margaret J. Polley, Ph.D. • Division of Hematology-Oncology, Department of Medicine, The New York Hospital-Cornell Medical Center, New York, New York 10021

Avner Rotman, Ph.D. • Department of Membrane Research, Weizmann Institute of Science, Rehovot, Israel

Nils Olav Solum, Ph.D. • Research Institute for Internal Medicine, Section on Hemostasis and Thrombosis, Rikshospitalet, University of Oslo, Oslo 1, Norway

Preface

It was just about ten years ago that platelet membrane glycoproteins were first characterized and their abnormalities in congenital bleeding disorders first recognized. During this decade there has been a remarkable growth in our understanding of the
structure and membrane organization of the platelet surface glycoproteins, their interactions with external ligands during the process of hemostasis, and their defects
causing hemorrhagic disease. These studies have advanced the knowledge of platelet
involvement in hemostasis from a cellular to a molecular level, and they have provided
a model for contact interactions among other cell types. This seemed a proper time to
ask those who contributed major observations and insights during these past years to
review their progress and to assess the state of our present knowledge. We have
planned this volume to begin with the biochemistry of platelet membrane glycoproteins
themselves and proceed through their involvement in platelet function to the final
considerations of the platelet's role in maintaining the integrity of the vascular system.
Our aim was an integrated presentation on the blood platelet from the perspective of its
highly specialized and reactive cell surface.

James N. George
Alan T. Nurden
David R. Phillips

Contents

4. Organization of Glycoproteins within the Platelet Plasma Membrane
Thomas J. Kunicki

5. Structure and Function of Platelet Membrane Glycoproteins as Studied by Crossed Immunoelectrophoresis
Inger Hagen and Nils Olav Solum

6. Platelet Membrane Electrical Potential: Its Regulation and Relationship to Platelet Activation
Avner Rotman

III. INTERACTION OF PLATELET MEMBRANE GLYCOPROTEINS WITH THE EXTRACELLULAR ENVIRONMENT

7. Receptors for Platelet Agonists
David R. Phillips

8. *Secreted Alpha Granule Proteins: The Race for Receptors*

Deane F. Mosher, Donna M. Pesciotta, Joseph C. Loftus, and Ralph M. Albrecht

9. *The Platelet–Fibrinogen Interaction*

Joel S. Bennett

10. *Platelet–von Willebrand Factor Interactions*

Barry S. Coller

11. Molecular Mechanisms of Platelet Adhesion and Platelet Aggregation

Ralph L. Nachman, Lawrence L. K. Leung, and Margaret J. Polley

12. Lectin–Carbohydrate Binding as a Model for Platelet Contact Interactions

T. Kent Gartner

IV. INTERACTIONS OF PLATELET MEMBRANE GLYCOPROTEINS WITH THE INTRACELLULAR CYTOSKELETON

13. The Organization of Platelet Contractile Proteins

Joan E. B. Fox

14.   *The Mechanism of Clot Retraction*
      Isaac Cohen

V.   *PLATELET MEMBRANE GLYCOPROTEIN IMMUNOLOGY AND ABNORMALITIES*

15.   *Immunology of the Platelet Surface*
      Sharron L. Pfueller

16. Glycoprotein Defects Responsible for Abnormal Platelet Function in Inherited Platelet Disorders
Alan T. Nurden

VI. CONCLUSION

17. The Role of Membrane Glycoproteins in Platelet Formation, Circulation, and Senescence: Review and Hypotheses

James N. George

1

Introduction

Plasma Membrane Receptors and Platelet Response

Ernst F. Lüscher

1. INTRODUCTION

It is slightly more than a hundred years ago that Bizzozero (1882) published his remarkable article on the blood platelets. He and his contemporaries were particularly impressed by the ease with which platelets adhered to vascular lesions and to each other. There is good reason to believe that the formation of a "white thrombus" and of a hemostatic plug by virtue of the formation of tight platelet aggregates must have been common knowledge before the end of the last century. It is rather amazing that the following 50 years were characterized by an almost total disregard of the blood platelets and the revival of interest in them in fact was mainly derived, not from their role in hemostasis and thrombus formation, but from their essential contribution to the intrinsic blood-clotting system. Today we know that platelets, besides their participation in blood coagulation, have a variety of functions including participation in the upkeep of vascular integrity and the provision of a remarkable diversity of materials that are released during their activation and that range from prostaglandins and thromboxane to the contents of their storage organelles with platelet-derived growth factor as a particularly prominent representative. Nonetheless, there is certainly agreement that the formation of the hemostatic plug and of its pathological counterpart, the "white" thrombus, is the most spectacular and, at least at first sight, most important manifestation of platelet activity.

The adhesion of the circulating platelet to a vascular lesion is an instantaneous event; obviously such "resting" platelets must have sites on their surface that recognize deviations from the normal structure of the vascular endothelium and react, most likely without further prerequisites, with components of the altered vascular wall. What follows is a most remarkable phenomenon consisting of the induction in the

---

Ernst F. Lüscher • Theodor Kocher Institute, University of Berne, CH-3000 Berne 9, Switzerland.

adhering platelet of a subtle and extremely fast alteration of its surface properties in the course of which a selective stickiness toward other, obviously "resting" platelets develops. Circulating platelets thus trapped will immediately suffer the same transformation, thereby becoming in turn attractive for the circulating species. In this way large aggregates are built up in an amazingly short time. As a consequence of complex secondary reactions, they solidify to form an effective hemostatic plug or an occluding white thrombus. Already Bizzozero has stressed that the primary aggregate which forms under in vivo conditions is composed of morphologically unchanged platelets and we certainly have to revise the widespread view that shape change precedes aggregation, which is derived from an artefact originating from the study of platelet aggregation under laminar flow conditions in an aggregometer. Since everybody will agree that the appearance of the "shape change peak" in the aggregometer is a remarkably fast event, it is evident that the alterations of the platelet surface that are the prerequisite for the formation of aggregates must be still faster. It is obvious that the elucidation of the mechanisms involved in the transformation of the platelet surface from the resting to an adhesive state is as important as it is difficult. The clue to the understanding of these vitally important reactions lies in a better knowledge of membrane reactivity, which in turn must be based on accurate information on the structure and function of the constituents that form the plasma membrane. The topic of this book, as highly specialized as it may appear, therefore touches on one of the very basic phenomena of living cells in very general terms.

It has already been mentioned that the rediscovery of the platelet as something worth serious research effort dates back not much more than 35 years. It is therefore not unexpected that the identification of specific membrane constituents that are directly related to functional parameters is of even younger date. The discovery by Nurden and Caen (1975) that the platelets from patients with certain bleeding disorders are characterized by a deficiency of specific membrane glycoproteins marks the beginning of an ever increasing effort to characterize membrane constituents, in particular the proteins and glycoproteins, with the aim of finding clues to a better insight into platelet function. In the following chapters the present state of knowledge will be summarized and it seems therefore more appropriate to point here to some of the problems, which invariably will require attention whenever it is attempted to relate the accumulated knowledge on single membrane constituents to functional parameters.

2. PLATELET RECEPTOR FUNCTIONS

The proteins, carbohydrates, and lipid constituents of the membrane determine the structure and properties of the platelet surface. They are involved in the adhesion of platelets to other structures and to each other as well as in the manifestation of procoagulant activity. Furthermore, it is generally assumed that the proteins and glycoproteins in particular act as receptors for the wide variety of agonists that are capable of activating the platelet. This is understandable since the complexity of protein and glycoprotein structure accounts best for the high degree of specificity that generally characterizes such receptors.

2.1. Receptors Involved in Platelet Adhesion and in Cell–Cell Contact

There can be no doubt that these types of membrane receptors are of vital importance, since in their absence platelets will be unable to adhere either to the site of a vascular injury or to each other.

Subendothelial structures capable of binding platelets are well known. In the foreground of interest stand collagen fibers, which *in vitro* not only bind, but also activate platelets without a requirement for particular cofactors. Furthermore, microfibrils may also contribute to platelet adhesion to denuded endothelium (Birembaut *et al.*, 1982). A platelet receptor for collagen has been described and partly characterized (Chiang and Kang, 1982). The situation becomes complicated due to the fact that with high shear rates, as encountered under *in vivo* conditions, platelet adhesion to subendothelium requires the presence of von Willebrand factor (VWF), which in a human system binds to platelets only if they are activated (Fujimoto *et al.*, 1982). This would imply that VWF-mediated adhesion cannot be an immediate process that is barely compatible with the instantaneous platelet deposition observed, e.g., in the course of the formation of a hemostatic plug. Bolhuis *et al.* (1979) have shown that human VWF binds to subendothelium and that platelets will then adhere to it in a flow chamber without any requirement for stimulation or cofactors such as ristocetin. This must be interpreted to mean that VWF, after combination with collagen, undergoes a structural alteration that enables it to bind to a platelet receptor, which otherwise has only a negligible affinity for it. This is an interesting situation in view of the fact that other examples are known, where the interaction of the platelet with agonists depends on the latter's quaternary structure. Thus, neither tropocollagen nor monomeric IgG exert any effect on platelets, whereas fibrillar collagen and aggregated IgG are potent activators. A repetitive structure, capable of combining at the same time with more than one specific receptor molecule, is required in order to establish a firm binding that in turn initiates sequential reactions. Von Willebrand factor is a flexible, linear polymer of rather large subunits. It is conceivable that by binding to the rigid collagen fiber, a spacial arrangement of the platelet-binding sites is provoked that favors a multisite combination with glycoprotein Ib (GPIb), the platelet VWF receptor molecule. Olson *et al.* (1983) have recently provided further evidence for the likelihood of such a hypothesis by showing that VWF immobilized on another solid-phase carrier (a dialysis membrane) will also bind platelets without a requirement for activation or cofactors. It should be noted that the dramatic increase in binding affinities on the aggregation of a potential inducer of cellular activities is also well established for other cells, particularly of the lymphoid series.

Another receptor that requires a stimulation of the platelet for activity is the GPIIb/IIIa complex, which binds fibrinogen in the presence of Ca^{2+} ions and ADP (Marguerie *et al.*, 1980). It is generally assumed that fibrinogen, due to its symmetric structure, will bind with one site to one platelet and with the other to the next one, thus explaining aggregation by a bridge formation. The location of the platelet-binding sites within the fibrinogen molecule are known (Kloczewiak *et al.*, 1982) and are indeed on the opposite ends of the molecule. However, it is to be expected that the chance of an activated platelet combining with fibrinogen is much higher than that of platelet-bound fibrinogen finding a free receptor on the next platelet. All this implies that a fibrinogen–fibrinogen interaction would have to be postulated in order to explain the mutual adhesion of fibrinogen-coated platelets. Fibrinogen molecules have little tendency to interact with each other, quite different from fibrin monomers, which polymerize spontaneously. It is tempting to speculate that the GPIIb–IIIa–fibrinogen interaction is linked to the establishment, in the fibrinogen molecule, of a structure, which lends itself to a "fibrin monomerlike" polymerization, thus explaining why such platelets stick to each other. Whether again a multiple receptor combination is required remains an open question; it is interesting, however, that the binding affinities for fibrinogen appear to increase with time (Marguerie *et al.*, 1980), suggesting that an increased number of receptors become involved. In this context it is of interest that ADP-stimulated platelets will become fully activated only if allowed to aggregate (Zucker and Peterson, 1967). Polymerizing fibrin monomer is an activator of platelets (Brosstad *et al.*, 1980) and it is tempting to speculate that "activation by cell–cell contact" in fact is due to a fibrinogen polymerization phenomenon resulting from the special type of attachment of the molecule to its receptor(s).

2.2. Receptors Involved in Platelet Activation

Platelets can be activated by a wide variety of agents and as a rule always the same response is evoked, consisting of aggregation, morphological and metabolic alterations, the synthesis of prostaglandins, thromboxane, and platelet activating factor, and the initiation of contractile activity and of the release reaction. This implies that the agonist–receptor combination is linked to the generation of a signal as a consequence of which a more or less uniform reaction sequence is started.

The question then arises: What particular properties of the agonist–receptor complex are responsible for the initiation of such a chain of reactions? The answer offered most often is that the receptor molecule undergoes a conformational change as a consequence of which certain reactions, mostly of an enzymatic nature, become possible and mark the beginning of the activation process. Among the very early events in a stimulated membrane is the activation of phosphatidylinositol metabolism that leads to the formation of diacylglycerol, which in turn is linked to protein phosphorylation via the activation of protein kinase C (Kawahara *et al.*, 1980). These processes and several others, such as the cleavage of arachidonic acid from phospholipids (for prostaglandin synthesis) or the activation of the contractile system, are Ca^{2+} dependent. One therefore might argue that the key event consists in the mobilization of Ca^{2+} ions. Billah and Lapetina (1982) have recently claimed that it is the decay, by cleavage of the calcium-chelating phosphate groups of phosphatidylinositol $4'$, $5'$ biphosphate, that is responsible for the early availability of this cation. This again would imply that a phosphatase somehow begins to manifest itself, either due to an activation step or because the substrate becomes accessible. It is of particular interest that, although there must be many agonist-specific receptor proteins, the consequences of their stimulation are generally the same. It therefore is unlikely that it is a particular property of one given receptor molecule that is responsible for the initiation of activation, but much more a uniform type of reactivity that is shared by all receptors. What could such a reactivity be?

The vast majority of the membrane proteins are in fact lipoproteins and occur in a noncovalent association with phospholipids (compare review by Marsh, 1983). There are several examples known which show that such complexes exist not only based on hydrophobic interaction with the fatty acid chains, but also influenced by the more polar regions of the phospholipid molecule. Thus, the protein kinase C complex requires exclusively phosphatidylserine (Kawahara *et al.*, 1980) and isolated erythrocyte membrane proteins have been shown in recombination experiments to display a certain selectivity for the type of phospholipid with which they normally are associated. Agonists combine with the surface-oriented part of the receptor molecule and it appears not unlikely that this is linked to a disturbance of the protein–lipid interaction. This could either result from a displacement of the more polar moiety of the phospholipid, in the induction of a conformational change that leads to the disruption of the hydrophobic interaction between protein and the phospholipid hydrocarbon chains, or, finally, from a superposition of both these effects. This idea is not at all new; thus Taraschi *et al.* (1982) propose that changes in the protein–lipid interaction are responsible for "signal production" in stimulated membranes and Epand (1983) has summarized evidence pointing in the same direction for the activation of cells by peptide hormones.

It is to be expected that the disruption of lipid–protein interaction should have consequences for the physicochemical properties of the plasma membrane. However, it is obvious that the effects due to an event which is restricted to the stimulation of one type of receptor molecule will barely be discernible. Thus, the available reports obtained with markers of the lipid phase on altered fluorescence polarization (Berlin, 1980; Nathan *et al.*, 1979) should be interpreted with caution, since such measurements may already reflect secondary processes, such as the appearance, due to the availability of Ca^{2+} ions, of products of phospholipid metabolism. Among these are substances such as diacylglycerol, phosphatidic acid, and lysophosphatides which in turn exert an effect on membrane structure. Thus, Imai and Nozawa (1981) have found that the appearance of such metabolites in fact goes in parallel with an increase of membrane fluidity as detected by electron spin resonance. The same deviation in the time scale from the original event may also enter into play when methods based on the assessment of protein–protein interactions are applied. These procedures have nonetheless established for several types of cells the rapid formation of clusters of proteins within the stimulated membrane (compare Metzger and Ishizaka, 1982; Hoebeke, 1981; see also review by Schlessinger, 1980). There is increasing evidence that the same process is also intimately linked to platelet activation (compare Ganguly and Gould, 1979; Steiner and Lüscher, 1984). In fact, the hypothesis that protein–protein interaction, culminating in cluster formation, is a crucial event in cell activation, is quite attractive indeed. The displacement of lipids between intrinsic protein molecules would be expected to create hydrophilic zones, allowing the movement of hydrated ions and the establishment of enzyme–substrate complexes, which otherwise could not exist. One is then further led to deduce that agonists that at the same time create conditions for intramembrane protein–protein interaction and force the receptors into an optimal position for such an interaction should be particularly good activators. This should be the case for all those agents that possess a repetitive pattern of binding sites, such as collagen or aggregated IgG. Von Willebrand factor in principle should also fulfill this requirement and it has indeed recently been shown that its combination with platelets is linked to the inhibition of adenylate cyclase (Kao *et al.,* 1983). It will be interesting to see whether solid-phase–bound VWF is not more active in this respect. It is obvious that the majority of the many substances that can activate platelets have no repetitive structure and act on individual receptor molecules, which will undergo a protein–protein interaction not influenced by the sterical properties of the agonist.

3. CONCLUSION

How membranes react on stimulation of a cell with an adequate agonist is one of the major problems of cell biology and biochemistry today. A picture slowly starts to emerge; however, it may be, as is often the case, incomplete and oversimplified. Experimental approaches are rather difficult, particularly if it turns out to be correct that not only the structure of the individual partners in this system, but also their sterical relationships play a crucial role. Furthermore, these reactions must be extremely fast. In this context it is worthwhile to come back to the striking phenomenon of the formation of a hemostatic plug, where inert platelets stick to already adhering ones and in an amazingly short time become themselves adhesive. Within this very short period the fibrinogen receptor must be mobilized (whereby it is quite unlikely that it is the direct target of ADP or any other agonist) and fibrinogen must bind to it. It is rather obvious that equally fast techniques will be required in order to follow these early events.

Practically all cells can be stimulated by external agents or by cell–cell contact into some type of activity that reflects their specialization. Some cells react slowly, because their reaction involves complex processes such as protein synthesis, whereas others react quickly. The platelet that is obviously built for immediate action is undoubtedly a particularly valuable model for studying activation processes: its answers to stimulation are both clear and rapid and, as the following chapters show, a wealth of information on particular aspects of its reactivity and on its constituents is already available.

REFERENCES

Berlin, E., 1980, Platelet membrane fluidity and aggregation of rabbit platelets, *Fed. Proc.* **39**:1894.

Billah, M. M., and Lapetina, E. G., 1982, Rapid decrease of phosphatidylinositol 4,5-bisphosphate in thrombin-stimulated platelets, *J. Biol. Chem.* **257**:12705–12708.

Birembaut, P., Legrand, Y. J., Bariety, J., Bretton, R., Fauvel, F., Belair, M. F., Pignaud, G., and Caen, J. P., 1982, Histochemical and ultrastructural characterization of subendothelial glycoprotein microfibrils interacting with platelets, *J. Histochem. Cytochem.* **30**:75–80.

Bizzozero, J., 1882, Ueber einen neuen Formbestandteil des Blutes und dessen Rolle bei der Thrombose und der Blutgerinnung, *Virchows Arch. Pathol. Anat. Physiol. Klin. Med.* **90**:261–332.

Bolhuis, P. A., Sakariassen, K. S., and Sixma, J. J, 1979, Adhesion of blood platelets to human arterial subendothelium: Role of factor VIII - von Willebrand factor, *Haemostasis* **8**:312–323.

Brosstad, F., Rygh, M., Kierulf, P., Godal, H. C., and Eika, C., 1980, Interaction of platelets with soluble, polymerizing and polymerized des-AA and des-AABB fibrin in human plasma, *Thromb. Res.* **18**:123–130.

Chiang, T. M., and Kang, A. H., 1982, Isolation and purification of collagen αl (I) receptor from human platelet membrane, *J. Biol. Chem.* **257:**7581–7586.

Epand, R. M., 1983, The amphipathic helix: Its possible role in the interaction of glucagon and other peptide hormones with membrane receptor sites, *Trends Biochem. Sci.* **8:**205–207.

Fujimoto, T., Ohara, S., and Hawiger, J., 1982, Thrombin induced exposure and prostacyclin inhibition of the receptor for factor VIII/von Willebrand factor on human platelets, *J. Clin. Invest.* **69:**1212–1222.

Ganguly, P., and Gould, M. L., 1979, Receptor aggregation: A possible mechanism of platelet stimulation by thrombin, *Thromb. Res.* **15:**879–884.

Hoebeke, J., 1981, Resonance energy transfer as a tool for studying membrane receptor microclustering, *Period. Biol.* **83:**95–98.

Imai, A., and Nozawa, Y., 1981, Dynamic structure of platelet membrane, *Blood and Vessel* **12:**509–519.

Kao, K. J., Tsai, B.-S., McKee, P. A., Lefkowitz, R. J., and Pizzo, S. V., 1983, Inhibitory effect of ristocetin and factor VIII/von Willebrand factor protein on human platelet adenylate cyclase activity, *Thromb. Res.* **30:**301–308.

Kawahara, Y., Takai, Y, Minakuchi, R., Sano, K., and Nishizuka, Y., 1980, Phospholipid turnover as a possible transmembrane signal for protein phosphorylation during human platelet activation by thrombin, *Biochem. Biophys. Res. Commun.* **97:**309–317.

Kloczewiak, M., Timmons, S., and Hawiger, J., 1982, Localization of a site interacting with human platelet receptor on carboxyterminal segment of human fibrinogen γ chain, *Biochem. Biophys. Res. Commun.* **107:**181–187.

Marguerie, G. A., Edgington, T. S., and Plow, E. F., 1980, Interaction of fibrinogen with its platelet receptor as part of a multistep reaction in ADP-induced platelet aggregation, *J. Biol. Chem.* **255:**154–161.

Marsh, D., 1983, Spin-label answers to lipid-protein interactions, *Trends Biochem. Sci.* **8:**330–333.

Metzger, H., and Ishizaka, T., 1982, Transmembrane signaling by receptor aggregation: The mast cell receptor for IgE as a case study, *Fed. Proc.* **41:**7–34.

Nathan, I., Fleischer, G., Livne, A., Dvilansky, A., and Parola, A. H., 1979, Membrane microenvironmental changes during activation of human blood platelets by thrombin. A study with a fluorescent probe, *J. Biol. Chem.* **254:**9822–9828.

Nurden, A. T., and Caen, J. P., 1975, Specific roles for platelet surface glycoproteins in platelet function, *Nature (London)* **255:**720–722.

Olson, J. D., Moake, J. L., Collins, M. F., and Michael, B. S., 1983, Adhesion of human platelets to purified solid-phase von Willebrand factor: Studies of normal and Bernard-Soulier platelets, *Thromb. Res.* **32:**115–122.

Schlessinger, J., 1980, The mechanism and role of hormone-induced clustering of membrane receptors, *Trends Biochem. Sci.* **5:**210–214.

Steiner, M., and Lüscher, E. F., 1984, Fluorescence anisotropy changes in platelet membranes during activation, *Biochemistry* **23:**247–252.

Taraschi, T. F., van der Steen, A. T. M., de Kruijff, B., Tellier, C., and Verkleij, A., 1982, Lectin-receptor interactions in liposomes: Evidence that binding of wheat germ agglutinin to glycoprotein-phosphatidylethanolamine vesicles induces nonbilayer structures, *Biochemistry* **21:**5756–5764.

Zucker, M. B., and Peterson, J., 1967, Serotonin, platelet factor 3-activity and platelet aggregating agent released by ADP, *Blood* **30:**556–565.

Plasma Membrane and Membrane Glycoprotein Structure and Function

Structural and Molecular Properties of Platelet Membrane Systems

Neville Crawford

1. INTRODUCTION

Our understanding about the chemical composition and functional properties of platelet membrane systems has been advancing very rapidly at both the morphological and molecular levels over the last ten years or so. This is well emphasized by the fact that in certain areas of membrane biochemistry the use of the platelet as a model cell for probing the more general aspects of membrane structure and behavior, as also for studying drug transport and disease-related membrane defects, has resulted in a popularity rating for the platelet almost equal to that enjoyed earlier by the red cell. Although both the red cell and the platelet are anucleate cells, in some respects the platelet with its highly interactive surface membrane, its mitochondria, internal membranes, and lysosomes and storage organelles more reflects the behavioral and metabolic activities of other cells in the body than does the red cell with its high level of functional specialization. Since the platelet circulates as a poised and potentially highly reactive cell, responding rapidly to external stimuli, and is also most sensitive to even minor biochemical changes in the surrounding milieu, considerably more attention has been focused on the plasma membrane and its constituents and properties than on the various intracellular membrane systems that are equally well developed for functional needs. These latter membrane elements include not only endoplasmic-reticulumlike (ER) structures often referred to as dense tubular membrane system (DTS), but also include the boundary membranes of the many different granular organelles residing

Neville Crawford • Department of Biochemistry, Institute of Basic Medical Sciences, Royal College of Surgeons of England, London WC2A 3PN, England.

within the cytoplasmic matrix (mitochondria, lysosomes, α-granules, dense body granules, etc.). More recently, with advances in electron microscopy techniques, there has been a developing appreciation of the wide variety of membrane systems within the platelet and in consequence a growth in interest in these intracellular membranes. As will be discussed more extensively later, it now seems certain that the DTS complex of membrane channels may share both the enzymatic processes associated with a classical ER-type membrane and the transport activities of muscle sarcoplasmic reticulum, the specialized form of ER that sequesters calcium and is normally only well expressed in skeletal, smooth, or cardiac muscle cells.

Since the platelet lacks a nucleus, protein and glycoprotein synthesis is limited or nil, and therefore during the lifespan of the platelet the chemical composition of the various membranes is perhaps more stable than in other cells where membrane synthesis and recycling processes are constantly operating. Moreover, even though the platelet is a secreting cell, and fusion of granule membranes with the plasma membrane almost certainly occurs during exocytotic export of granule-stored constituents, this is probably a "one off event." Unlike other secreting cells, there is little, if any, two-way traffic of membrane materials or recycling of vesicles for the continuous synthesis and packaging of granule constituents. Nonetheless, in comparison with other cell membranes, the various platelet membranes show the same dynamic properties and structural features, which have essentially similar implications for the functional and metabolic competence of the platelet as for other cells. This is particularly so with respect to transport functions, recognition sites for cell–cell interactions, ectoenzymes, compartmentation of metabolites, mobilization of membrane constituents, and the specificity of the release of granule-stored constituents.

In this review, in order to integrate with and yet avoid unnecessary overlap with other contributors in this volume, I have chosen to focus primarily on three aspects of the human platelet membrane systems: (1) ultrastructural definition, (2) isolation procedures, and (3) characterization. Around ten or so years ago, most membrane biochemistry was focused almost entirely on the unit bilayer structure and its protein and lipid constituents. Today, however, considerations of membrane architecture and function have to be extended well beyond this region, outward to include the exterior coating of the cell and inward to take account of the variety of interacting cytoskeletal systems, both stable and transient, that lie in the regions of a cell subadjacent to the cytoplasmic face of the surface membrane. In fact, it is these integrative considerations that are, without doubt, some of the most important components of the platelet's structural and functional physiology, particularly as they concern shape maintenance, contractile activity, exocytotic release processes, control of membrane transport, receptor revelation, and signal transduction. All these are basic to the cell's important role in hemostatic processes and may be deficient or overemphasized in thrombotic events. Accordingly, I shall indicate in this chapter other contributions in this volume, together with some key references elsewhere to important platelet membrane papers and reviews, which complement and enlarge on the information I am presenting.

The principal difficulty, however, in preparing this review has been that in recent years the interest in platelet biology has been almost explosive and it is virtually impossible to do full justice to the many researchers working in this very vast field. Some selectivity has been inevitable and I therefore tender in advance apologies to any platelet membrane enthusiasts who may consider that my treatment of their work has been somewhat superficial or perhaps even worse, less than just!

2. ULTRASTRUCTURAL DEFINITIONS OF THE DIFFERENT PLATELET MEMBRANE SYSTEMS

2.1. Platelet Origin in Megakaryocytes

Platelets circulate as discoid or lentiform cells with diameters across the equatorial region of the disk varying between 2–4 μm and a thickness of about 1 μm. Their cellular precursor is the bone marrow megakaryocyte from which platelets are released after extensive cytoplasmic partitioning has taken place. This release process and the various controlling factors for it are poorly understood, and even though it is a unique phenomenon in biology there are immense difficulties in its study. To the present author's knowledge, no one has yet been able to exactly simulate this platelet release process *in vitro* either in megakaryocytes harvested from bone marrow or in culture. *In vivo*, the various morphological changes have been well documented by Pennington (1981) and his colleagues (see review by Pennington *et al.*, 1974) and they involve introversion of the megakaryocyte plasma membrane, coalescence of the intracellular membranes, and a spreading of these processes throughout the cytoplasm to the extent that between 2000 and 3000 individual platelets are formed from one megakaryocyte, each equipped with its complement of intracellular organelles, proteins, enzymes, and fuel stores to last its circulating lifespan, which has been varyingly estimated as between 7 and 11 days. The release of the platelets involves the whole of the cytoplasm and may proceed until the megakaryocyte nucleus is completely denuded of cytoplasm. At release the cytoplasmic fragments are not discoid and they lack the marginal microtubule bundles. Discoidicity and microtubule reorganization are believed to occur through mechanical massaging in the microcirculation. There is some controversy about whether this megakaryocyte release of platelets is entirely confined to bone marrow cells or whether megakaryocytes sequestered in the lung microvasculature may also be a major contributor, with platelet release triggered by ventricular pounding and/or other mechanisms (see Chapter 17). For many years it has been accepted almost as dogma that most of the circulating platelets arise by budding off from the pseudopodia and cytoplasmic fragmentation of megakaryocytes resident in the sinusoidal spaces of the bone marrow, but there is now convincing evidence in support of an alternative theory that most, if not all, of the circulating platelet pool originates from megakaryocytes that have migrated from the bone marrow and fragment elsewhere, predominantly in the pulmonary vascular tree (Martin *et al.*, 1982, 1983; Trowbridge *et al.*, 1982; Slater *et al.*, 1983).

2.2. Heterogeneity of Circulating Platelets

Whatever their true source, there is, however, a considerable heterogeneity in the circulating platelets in volume, density, and functional and metabolic competence. Some of the platelet heterogeneity may be related to changes during circulating lifespan and particularly membrane modifications and some to the maturation and ploidy pattern of the parent megakaryocytes at the time of platelet release (Paulus, 1975; Pennington and Streatfield, 1975). George and his colleagues (George et al., 1976; George and Lewis, 1978) have demonstrated with rabbits that circulating platelets may make random and reversible encounters with vessel wall cells losing surface components in the process. Such interactions, which are apparently unrelated to the platelet's preprogrammed senescence, could account for the many difficulties in interpreting platelet lifespan curves since they often show both linear and exponential characteristics. To date, little attention has been turned to the likely changes in the platelet surface membrane resulting from such random encounters. It would be of some interest to know, for example, if the platelet surface can become sufficiently compromised that more IgG binding sites are revealed that could become collaborative for macrophage removal. During their sojourn in the circulation, if they remain uninfluenced by external factors or by surface interactions with the vessel wall, the discocytes generally have a relatively smooth but extensively invaginated plasma membrane. Even if fixation is very rapid, with phlebotomy carried out under minimal stasis conditions directly into glutaraldehyde fixatives, a small proportion of the platelets (approximately 10–15%) as seen in scanning electron micrographs appears to be slightly activated. These cells, although remaining largely discoid, show tiny protuberances arising from the surface and it is difficult to know if this represents their true *in vivo* morphology or that a small circulating subpopulation is so delicately poised that even the slightest disturbance in removing them from the circulation triggers activation with these consequent surface protrusions.

2.3. Platelet Surface Membranes

Specific carbohydrate or glycoprotein staining techniques applied to electron microscopic preparations of fresh platelets generally reveal a "fuzzy" amorphous coating or "glycocalyx" extending outward from the cell as much as 20–30 nm beyond the surface membrane unit bilayer. The term "glycocalyx" was first coined by Bennett (1963) to describe the carbohydrate-rich "fuzz" found on the exterior of most cells. This material includes of course "intrinsic" glycoconjugate molecules strongly bound to the lipid bilayer as well as "extrinsic" components such as secretion products and plasma constituents that may be bound adventitiously or specifically to the intrinsic components of the membrane. This distinction between intrinsic and extrinsic constituents of the glycocalyx is an important one in the context of platelet interactions with foreign and natural surfaces, as also in the choice of platelet isolation procedures, where ideally one wishes to remove them from their plasmatic milieu with minimum disturbance to the organization of the surface coat. The invaginated domains of the surface membrane that appear to be contiguous with the truly externally oriented plasma membrane are now generally referred to in ultrastructural studies as the open canalicular membrane system (OCS), and within these regions the glycocalyx appears to be less well expressed than at the platelet membrane/plasma boundary layer. The extent of these invaginations, which in some platelets can give the cell an almost spongelike appearance (White, 1972a), is believed by Pennington et al. (1974) to be also correlated with the ploidy class of the parent megakaryocyte from which they originate. Particles such as small latex beads are readily held within the channels of the invaginated OCS (White, 1972b) and this has led to some difficulty in ascertaining if platelets have an inherent phagocytic activity, capable of internalizing foreign particles and large molecular complexes. Platelets do contain lysosomelike organelles within the cytoplasm and their plasma membrane is certainly capable of participating in membrane fusion events during the release reaction so in these respects they have some of the aptitudes necessary for endocytosis, but phagocytosis has by no means been established to be part of the armamentarium of normal platelet function.

2.4. The Dense Tubular System and Internal Organelles

Within the interior of the platelet, the next most prominent membrane system is the dense tubular membrane (DTS). These are interconnecting membrane channels coursing throughout the cytoplasm and are often swollen at intervals into vacuoles that may contain some amorphous material, the character of which is not known. This membrane system is an ER-like complex of channels and in certain areas it appears to be in close juxtaposition with the cytoplasmic face of the open canalicular membrane, a feature not unlike the relationship that exists between the sarcoplasmic reticulum of skeletal muscle and its closely associated transverse tubular membranes. More recently, the increase in our knowledge about the protein components of the platelet's miniature muscle systems and the control of contractility by Ca^{2+} has given this plasma membrane/intracellular membrane association, or structural apposition, more functional importance in platelet behavior than was perhaps considered when these complexes were first observed. Reference will be made to this point again in Sections 2.6.1. and 6.3. Very occasionally a small stack of tubular cisternae can be seen in some platelets and this appears to be a poorly developed Golgilike complex. This is a rare feature, however, and since the platelet is not involved in the synthesizing and packaging of its granule-stored proteins, one must presume such structures are a vestigial legacy of the megakaryocyte cytoplasm. The association of proteolytic activity with this Golgilike complex as in, for example, pancreatic cells would be particularly interesting if identifiable, since although the platelet has virtually no protein-synthesizing capacity, it may be involved in certain posttranslational modifications.

Within the cytoplasm of the platelet are numerous organelles, all of which are surrounded by well-defined boundary membranes. The mitochondria, which seem to have thinner and less well-developed cristae than those seen in other cells, have the usual double membrane, whereas the various storage organelles (α granules, lysosomelike granules, and the 5-HT-storing dense bodies) appear in general to have a single-unit membrane bilayer morphologically indistinguishable from the plasma membrane with which these storage organelle membranes fuse during release of their stored constituents. The outer cytoplasmic face of these organelle membranes lacks a glycocalyx but filamentous processes are occasionally seen arising from the membrane which may be actin, but these have not been definitively identified as such.

2.5. Membrane-Associated Components

In this context, reference should be made to the platelet's cytoskeletal equipment, since in some regions of the cell this is clearly associated in some way with both the plasma membrane and the intracellular membrane systems.

2.5.1. Marginal Microtubules

The most prominent cytoskeletal organization in the discoid-shaped platelet is the marginal bundle of microtubules. This polymeric complex is analagous in protein subunit composition to microtubules seen in brain, cilia, and flagella, but the organization and general morphology appear to be quite different in a number of important respects. This peripheral bundle of tubules lies subadjacent to the cytoplasmic face of the plasma membrane and subtantially confined to the equatorial plane of the discocyte (Behnke, 1965; Behnke, 1967, 1970; Behnke and Zelander, 1966; Hovig, 1968; White, 1971). In transverse sectional preparations of inactivated platelets, the tubules, which may number as many as 30 or more, are seen in cross section in the apical region of the disk. They are quite discrete from each other with an intertubule spacing that seems fairly constant, suggesting that they have outer coatings not readily visualized by the usual electron microscopic staining techniques. Filamentous processes are also occasionally seen connecting these tubules to the cytoplasmic face of the plasma membrane in the apices of the disk, but again these have not been definitively characterized. Their width suggests that some may be actin filaments, but the possibility that 10-nm filaments may be present in platelets cannot be discounted, and a number of groups have been searching to try to identify in platelets the subunit proteins characteristic of these kinds of structures in other cells. A recent report (Koteliansky *et al.*, 1983) that platelets contain vinculin suggests another possible anchor protein candidate for membranes and for microtubules. In other cells this has been considered to be the major-link protein between actin filaments and integral components of the membrane. It would of course be of some interest to know the polarity of the actinlike filaments and whether nucleating sites for F-actin polymerization are on the tubules or membranes. The subunit composition of these microtubules and the factors that control assembly and disassembly have been reviewed by Crawford *et al.* (1980) following earlier reports (Crawford, 1976; Castle and Crawford, 1977) of the isolation, characterization, and *in vitro* assembly of platelet microtubules. A concept has been presented (Crawford *et al.*, 1980) that the high-molecular-weight microtubule-associated proteins (or MAPs) that appear to be regularly disposed as a superhelical lattice array on the outer surface of the assembled platelet microtubule may be the interactive sites for connection with membranes and other intracellular features. In fact, Schliwa and van Blerkom (1981) have demonstrated in other cell systems numerous end-to-side contacts of actin filaments with microtubules. Such an association has been recently examined *in vitro* by Griffith and Pollard (1982) using muscle actin and brain microtubule proteins. Their findings suggest that the interaction specifically requires the presence of MAPs, which are a feature of all isolated microtubule protein preparations. One dynamic aspect of the microtubule ring that may relate to changes in surface membrane properties is that, immediately following a hemostatic signal to the plasma membrane, the microtubule bundle contracts toward the central zone of the cell, the rings reducing their diameter by as much as one third or more. This centripetal movement results in the intracellular granular organelles becoming localized in the middle of the platelet as the diameter of the bundle reduces. How this contraction event occurs is the subject of some controversy, but what seems certain is that preparatory to this radial shortening, dislocation of any connecting bridges with plasma membrane constituents would have to take place. Recent evidence indicates decrease in diameter of the marginal bundle of tubules during platelet shape change may be analogous to a "clock spring" effect in which the intact coils of microtubules progressively constrict (White and Sank, 1984). Following this microtubule contraction, the whole complex appears to fragment or totally disassemble since microtubule rings do not seem to be present in shape-changed and activated platelets. In the fully hemostatically active platelet, an entirely new reassembly of microtubules takes place extending along the full length of the pseudopodia from an area in the center of the platelet to the pseudo-pod tips. Presumably for this new assembly the platelet draws on the cytoplasmic microtubule subunit pool, but the direction of polymerization has not been determined. This new configuration of microtubules does not appear to be membrane associated and this phenomenon is therefore perhaps outside the scope of this present membrane review, although for favorable conditions for reassembly, Ca^{2+} ions must be low ($<$ 10^{-6} M) and the activity of the calcium-sequestering DTS and the plasma membrane Ca^{2+} extrusion pump may well be implicated in producing the right polymerizing conditions for reassembly.

2.5.2. Microfilaments

Microfilaments, which are considerably smaller in diameter than the micro-tubules, are seen frequently in the cytoplasm and they increase in number following hemostatic activation, particularly within the pseudopodia. Some of these filaments are close to, if not actually inserted into, the lipid bilayer at the cytoplasmic side and are identifiable as filaments of the contractile protein actin. They can be decorated with heavy meromyosin, showing the classical chevron or arrowhead array, and they also bind antiactin antibodies. The slightly larger filaments may be side-to-side associated F-actin bundles or small oligomeric assemblies of the other important muscle protein, myosin, which is known to be present in the platelet cytoplasm. These structures are more fugitive than the actin filaments and have not been characterized with certainty, but for an excellent review and catalogue of the platelet contractile and associated proteins the reader is referred to the paper of Harris (1981), and for a survey of the platelet cytoskeleton with emphasis on functional considerations to Chapter 13 in this present volume.

2.6. Characteristics of Internal Membranes

2.6.1. Dense Tubular System

The intracellular membrane complex, which is a prominent feature of the cytoplasm of the platelet, consists of a maze of interconnecting membrane channels. These channels widen at intervals and the presence of amorphous material inside gives them an appearance that is clearly distinguishable from the surface-connected OCS. As will be discussed more fully in Section 6, the physical properties, enzyme activities, and lipid and polypeptide composition of these intracellular membranes are very different from the plasma membrane. Briefly, some outstanding distinguishing features are the specific localization in the intracellular membranes and high activity of the ER marker enzyme NADH cytochrome-c-reductase, a low microviscosity that correlates well with a low sphingomyelin and cholesterol content, and the complete absence of any of the cytoskeleton proteins such as actin, myosin, and actin-binding protein that are such prominent features of the polypeptide profile of the plasma membrane. In electron micrographs of discoid-shaped platelets, there is some suggestion that the DTS may be differentially localized in the cytoplasm at the apices of the disk with less of these membrane elements in the waist regions of the cell. This regional disposition is striking in some electron micrographs, but has not been properly substantiated by morphometric analysis. One histochemical distinguishing feature of the DTS is the presence of peroxidase activities within the membrane channels that White and his colleagues (Gerrard et al., 1976, 1978a) consider may relate to enzymes concerned with prostanoid synthesis. Their view is that the DTS is a prostanoid-synthesizing site that has its origin in the smooth ER of the parent megakaryocyte. Whether the platelet possesses any rough ER-like structures has not been established and synthesis of proteins is, if not entirely absent, certainly limited. Some slight incorporation of radiolabeled amino acids into proteins can be demonstrated with very fresh platelets, although not consistently, so this seems to be due to the presence of some stable RNA and perhaps some residual fragments of ribosomal-rich ER or polysomes. In fact, as will be referred to later, the DTS also has many properties resembling those of the sarcoplasmic reticulum (SR) of skeletal and cardiac muscle cells and more recent evidence suggests its role in regulating cytosolic calcium concentrations may be analogous to the muscle SR system (Käser-Glanzmann et al., 1977, 1978). The morphological characteristics of the DTS and its relationship with other platelet membranes have been presented in a number of papers by White and his colleagues (White, 1971, 1972a; Gerrard et al., 1974, 1976).

2.6.2. Mitochondrial and Granule Membranes

Although platelet granular organelles are numerous, less attention has been paid to their boundary membranes than to the other platelet membrane systems. Granule fractions can be harvested reasonably free from contamination by surface and DTS membranes, but clean granule membranes are difficult to isolate free from adherent stored constituents. Some chemical and enzymatic studies have been made on the platelet mitochondria although the preparations have not been quite as pure as mitochondrial fractions isolated from other cells and tissues. With perhaps only quantitative differences, most of the enzyme activities associated with the mitochondrial membranes and "mitosol" (the intramitochondrial soluble phase) in other cells seem to be present in platelet mitochondria, and it is clear that coupled oxidative phosphorylation mechanisms generating ATP as an energy source are well expressed. The difficulty in clearly identifying mitochondria in platelet electron microscope sections, because they do not all display well-formed cristae, has led to the erroneous belief that they are a rare feature in platelets. Most do show a double membrane, however, and in fact, with respect to their number per platelet, the careful morphometric analyses performed by Mangenstern and Stark (1975) revealed that the percent of the platelet volume that mitochondria occupy in platelets (3–5%) is equal to their volume in monocytes and significantly exceeds their volume in polymorphonuclear leukocytes. The platelet is therefore well-equipped energetically to sustain its various ATP-requiring metabolic and functional activities, providing adequate sources of fuel are available for the various metabolic pathways involved. In this context, since platelets have glycogen granules in the cytoplasm and the various enzymes of the glycogenolytic cascade are also present, the ATP-generating metabolic activities should continue even if the availability of glucose to the cell has been compromised as in the case of a tightly organized platelet plug that apparently can still display energy-dependent contractile activity (Cohen and de Vries, 1973; Cohen and Lüscher, 1975). In addition to glycogen, seen as aggregates of tiny granules, small lipid inclusions have also been observed in the cytoplasm and also what appear to be a few scattered polysomes.

Concerning the membranes of the storage organelles (α granules and dense bodies) and lysosomelike granules, there has been no reported feature, morphological or chemical, capable of differentiating these granule boundary membranes either from each other or from the surface or ER-type membrane systems. Some granules may have a number of attached microfilaments, but this is not a prominent feature and does not appear to be confined to a distinct morphologically identifiable subpopulation. A more detailed investigation of these granule membranes could in fact be most revealing, since, for example, a comparison of their polypeptide composition with the distinctive polypeptide profiles of the surface and intracellular membranes (see Section 6.1) might provide some insight into their site of origin. A similarity with the polypeptides of the intracellular membranes could suggest that the stored proteins therein are synthesized and packaged within the megakaryocyte through a ribosomal ER/Golgi process, whereas a similarity of a granule membrane with a surface membrane polypeptide profile might suggest that the granule contents have been internalized in vesicles by some type of receptor-mediated endocytic process at the megakaryocyte level. Such studies will, however, require better separation techniques for the various storage granules and a more complete catalogue of their various contents than we have at present. To date we have very little knowledge about these granule membranes, of either their composition or site of origin. As referred to earlier, short actinlike filaments are occasionally seen arising from the granule boundary membranes and incubation of an intact granule preparation with platelet or muscle myosin raises the basal Mg ATPase activity of the myosin slightly, although significantly, suggestive of contractile actomyosin formation (N. Crawford, unpublished observations) and confirming the actinlike nature of the associated filaments. It is not known if these filaments have a role in the movement of the granules to the surface membrane preparatory to exocytotic release.

3. PLATELET SUBCELLULAR FRACTIONATION AND MEMBRANE ISOLATION PROCEDURES

3.1. Ideals and Practical Considerations

In isolating membranes from any type of cell, the choice of the procedure is conditioned by a number of desirable but rarely fully attainable objectives. First, the membrane fraction should be relatively free from contamination by other cell organelles. Second, the analytical and functional properties of the membrane are unaltered during the isolation. Third, the recoveries and yields should be high enough to ensure that the subfraction is reasonably representative and adequate for the application of the most precise and reliable analytical techniques available. Most procedures developed to date for platelet membrane isolation have generally been a compromise in which one or another of these goals has to some extent been sacrificed in the interests of the remainder. Nevertheless, some progress has been made in taking the platelet apart at the subcellular level to provide reasonably discrete identifiable elements, and with cautious interpretation of the current information available, a picture is beginning to emerge of the complexity of the various intracellular relationships between the different morphologically identifiable features concerned in the expression of platelet function. Some of the techniques that have been used to provide subcellular elements for biochemical studies will be discussed in the following section.

3.2. Platelet Isolation

Whatever the experimental intentions, one of the most important considerations in most *in vitro* platelet research requires the establishment of a satisfactory procedure for the initial isolation of the cells from whole blood. Such a procedure must be reasonably rapid, reproducible, and, most importantly, harvest the cells in a substantially inactivated state, free from contamination by other blood-formed elements. In most circumstances it is usual to prepare first a platelet-rich plasma (PRP) from the anticoagulated blood by a single differential centrifugation step (usually 130–200g for 15–20 min at room temperature). A temperature within the range 18–23 °C is suitable since it is now known from earlier studies that processing at 4 °C causes some disassembly of the coldlabile microtubular complexes. Even with the most carefully controlled conditions, a significant proportion (20–30%) of the platelets may remain in the red cell layer. Such losses would be acceptable if those trapped were fully representative of the circulating population, but dependent on the gravitational force used and the viscosity of the plasma, some selection has almost certainly been made with platelets of higher density and/or larger volume excluded from the PRP when it is removed. Cycles of resuspension of the red cell layer and further low-speed recentrifugation can significantly increase the recovery, in the best circumstances to > 90%. For most studies one such repeat cycle is an acceptable compromise given that it is also desirable to reduce handling procedures to the minimum. For optimum recovery it is more ideal if the red cell layer can be resuspended in homologous platelet-free plasma, thus maintaining a colloidal environment of similar viscosity and density and also fibrinogen content for the second and subsequent centrifugation stages. If 0.154 M saline or similar aqueous isotonic solutions are used for resuspension of the red cell layer, the "gravity ×
minute" value for the centrifugation must be reduced (say to 80–120g for 15 min) and this is best determined experimentally in each laboratory. Different separation procedures are necessary if instead of conventionally anticoagulated whole blood, buffy coat packs or platelet concentrates are used, as are supplied by some transfusion centers, and some guidance for dealing with these is offered in the papers by Baenziger and Majerus (1974) and Day *et al.* (1969).

The currently most popular procedures for isolating platelets from PRP are differential centrifugation, gel filtration, and albumin density gradient separation. Of the three, perhaps centrifugation is the most widely used method, but care must be exercised to avoid adventitious activation due to too-close packing that can seriously disturb the topography of surface membrane constituents (Feinberg, 1982). Reference should be made to the papers of Sixma and Lips (1978), Manucci (1972), and Rittenhouse-Simmons and Deykin (1976) for discussion of the various factors that affect platelet integrity during isolation and to some of their recommended precautionary measures. More recently, various additives to the anticoagulated blood, or incorporated within the anticoagulant cocktail, have been recommended such as adenosine, PGE_1, theophylline, apyrase, EDTA, and prostacyclin in order to try to maintain the platelets in an inactivated state during the initial handling procedures. One particularly effective method, suitable for volumes of blood up to about 50 ml, is the procedure of Lagarde *et al.* (1981), which depends on a pH adjustment to pH 6.4 by the addition of citric acid to the PRP. Although there is some slight contamination of the resuspended platelets with plasma proteins and resuspension in a calcium-free buffer is necessary to avoid fibrin formation, good ultrastructural integrity is achieved by this procedure, and if calcium ions are added before functional testing, responses to aggregation inducers such as thrombin, collagen, arachidonate, and PGH_2 are well sustained. In a study by George and his colleagues (George *et al.*, 1981) that was directed toward maximizing platelet recovery during washing procedures while at the same time maintaining surface membrane integrity with respect to glycoprotein content, it was found that the inclusion of albumin in the washing buffers was particularly advantageous. This protective effect of albumin resulted in approximately 90% recovery of platelets after two wash cycles with a significantly improved quantitative profile of PAS-positive surface membrane glycoproteins over more conventional washing procedures using protein-free buffers. All these platelet-stabilizing procedures have their enthusiasts, but in general, whatever approach is taken, some customization to the requirements for the subsequent studies is necessary. For example, should the intention be to study protein phosphorylation mechanisms, additives such as PGE_1 or theophylline, which may artifactually raise the platelet cAMP content and activate protein kinases, would be contraindicated. Similarly, if a study of membrane protein constituents is the goal, EDTA would be undesirable because of its known selective solubilizing effects on membrane proteins and removal of extrinsic membrane components. In the absence of a calcium chelator, as in heparinization, a protease inhibitor cocktail may have to be substituted to suppress the action of the platelet's very active Ca^{2+}-dependent proteases released when the platelets are ruptured. If straightforward membrane analytical and enzymatic characterization is the major aim, then a procedure acceptable for these assays might be entirely different from one in which well-sealed membrane vesicles are required for the study of transport functions, ligand receptor binding, or lipid asymmetry. These are but a few examples of why it is impossible to set down firm guidelines of universal applicability, and the starting point for the new researcher is still best taken from the contributions of others whose experimental interests coincide or are closely analogous.

3.3. Platelet Lysis and Fractionation of Subcellular Components

The next technical problem to be faced is that, in sharp contrast to many other cells, platelets are particularly resistant to mechanical shear and physical forces and are not disrupted by the relatively mild lytic or homogenization procedures suitable, for example, for hepatocytes or leukocytes. Sixma and Lips (1978) reviewed most of the homogenization procedures in use at that time for disrupting platelets. These included freezing and thawing, osmotic lysis, glycerol lysis, ultrasonics, blender homogenization, nitrogen decompression, and teflon pestle techniques. Two other techniques or modifications have been reported since their review—digitonin lysis (Akkerman *et al.,* 1980) and the electrostatic binding of whole platelets to polylysine beads followed by sonic disruption (Kinoshita *et al.,* 1979). Table 1 updates the review of Sixma and Lips (1978) with references and some details of all the techniques that have been applied to the problem of platelet disruption. The reference given is not necessarily the first description of the procedure, but chosen according to the best and most detailed description. Included in this table are details of some of the procedures for subsequent subfractionation of the homogenate, sonicate, or lysate and it will be seen that by far the most popular procedure involves density gradient centrifugation in either a continuous or discontinuous mode. Although this approach can certainly separate a membrane

Table 1. Procedures for the Isolation of Platelet Membranes

Method of cell disruption	Subsequent fractionation	Reference[a]
"Nil clearance" pestle	CDG[b] (30–60% sucrose)	Marcus *et al.,* 1966
Top drive blender	DDG (sucrose)	Siegel and Lüscher, 1967
Blender	CDG (1–2 M sucrose)	Minter and Crawford, 1967
Blender, pestle	DDG (27.5–55% sucrose)	Day *et al.,*1969[c]
Glycerol lysis	DC onto sucrose cushion (27%) followed by CDG (15–40% sucrose)	Barber and Jamieson, 1970
Nil clearance pestle, sonication, nitrogen decompression, osmotic lysis, glycerol lysis	CDG (5–80% sucrose) also DC followed by CDG	Barber *et al.,*1971[c]
Low clearance pestle	CDG (0.6–2.0 M sucrose)	Harris and Crawford, 1973
Freezing and thawing 5 ×, sonication, pestle, glycerol lysis	DDG (18, 25, 30, 35% sucrose) Also DDG (4, 12, 23% Ficol)	Kaulen and Gross, 1973[c]
Nitrogen cavitation	DC—fractions	Broekman *et al.,* 1974
Top drive blender	ZRC (30–55% sucrose)	Taylor and Crawford, 1974a
French pressure cell	DC—4 fractions	Salganicoff *et al.,* 1975
Glycerol lysis + pestle	CDG (20–30% diatrizoate)	Rittenhouse-Simmons and Deykins, 1976[c]
Polylysine beads and sonication	Cells attached to beads for lysis and centrifuged for membranes	Kinoshita *et al.,* 1979
Digitonin lysis	DC for cytosol and particulate fractions	Akkerman *et al.,* 1980
Sonication after neuramidase treatment at whole cell level	CDG (1–3.5 M sorbitol) and free-flow electrophoresis	Menashi *et al.,* 1981

[a]The reference given is usually the first description of a technique but in some cases the reference with the best description and evaluation has been chosen.
[b]Abbreviations: CDG = continuous density gradient, DDG = discontinuous density gradient, DC = differential centrifugation, ZRC = zonal rotor centrifugation.
[c]Indicates comparative studies.

fraction from the granular organelles with minimal cross contamination, there has been no claim that this is other than a mixed population of membrane vesicles of both surface and intracellular origin. One of the more recent methods described, that of Menashi *et al.* (1981), does, however, make such a claim, but their procedure involves further fractionation of the mixed membranes by free-flow electrophoresis that depends on differences in the surface electrical charge of the vesicles. This technique will be discussed in more detail in Section 5. With all these procedures and their many modifications, the severity of the initial mechanical, physical, or chemical insult applied to the platelets determines whether one achieves good membrane recoveries of low purity or highly purified membrane fractions of generally poor yield. The major likely contaminating components are of course the granular organelles, their boundary membranes, and their contents. Mitochondria and the 5-HT storage granules are perhaps marginally the most robust of the internal granular organelles and the likelihood of contamination of a membrane fraction by the contents of these is less than by the contents of the α granules and lysosomelike organelles. Generally, contamination is revealed by the presence of serotonin, granule proteins, and/or mitochondrial and lysosomal enzymes. The choice of the physical or chemical disruption procedure must rest with the investigator since it is largely dictated by his overall experimental intentions. For our own rather special purposes, which will be described in Section 5, we have elected to use repeated short periods of ultrasonication with intervals for recooling. In our hands, the advantages of this procedure has been its reproducibility, high shear forces in short time intervals (10 sec) that avoid heating artifacts, a reasonable maintenance of granule integrity, and uniformity of membrane vesicle size, allowing better resolution of membrane subfractions. In fact, we now firmly believe that one of the major factors in the nonreproducibility of most other mechanical disruption procedures, particularly where these have been combined with density gradient fractionation, is the variation in the size (or volume) of the membrane vesicles. Since most density gradient procedures applied to platelets are by rate sedimentation not isopyknic densities, both size and density of the vesicles are the operative parameters and discrete membrane zones of reproducible chemical and enzymatic characteristics simply cannot be harvested if the size range of the vesicles arising from any particular membrane domain is wide. Perhaps one of the most commonly used procedures for platelet lysis is the glycerol-loading technique first introduced by Barber and Jamieson in 1970. This involves the slow sedimentation of the cells through a glycerol gradient. The glycerol enters the platelet, enhancing their sensitivity to osmotic shock, which causes lysis when they are placed in a cold buffer solution. This lysate is applied to a sucrose density gradient and the membranes collected in the low-density region. Over 80% of the platelets are disrupted and this technique has been compared with low clearance pestle homogenization, nitrogen decompression, and sonication with favorable results (Barber *et al.*, 1971). Although the membrane vesicles from the glycerol loading separate into two bands on the sucrose gradient, electron microscopy, surface labels, analytical and enzymatic data all suggest they are both essentially mixed membrane fractions with plasma membrane elements enriched in each but varyingly contaminated with membranes of intracellular origin. This membrane subfractionation almost certainly reflects variations in vesicle size and density as mentioned above rather than differences in cellular origin. The procedure has, nevertheless, been a useful advance on some of the earlier fractionations and has enabled studies to be made of a mixed membrane fraction relatively uncontaminated with granular organelles. Cell disruption by digitonin lysis (Akkerman *et al.*, 1980), although a very rapid procedure, has limitations since it only yields crude particulate and cytosol fractions. Digitonin has a high affinity for cholesterol, forming stable complexes; thus its effect on the general distribution of membrane lipids may be quite profound, although marker enzymes do seem largely unaffected.

3.4. Identification of Subcellular Fractions

Almost all strategies applied to cell membrane isolation and characterization require the use of markers, both to follow the purification protocol and to assess the degree of enrichment of the membrane components in the final fraction or fractions. Ideally in platelet studies, as with other cells, if one strictly followed the tenets of de Duve (the pioneer of subcellular fractionation and Nobel Laureate) one should be seeking an endogenous or intrinsic membrane component whose presence or characteristics are unique to the membrane, has stability and properties that do not significantly change during isolation, and which can be readily measured with specificity and confidence. In practice, in most cellular studies single-location markers are the exception rather than the rule, many are bimodally distributed, and redistribution artifacts during homogenization and fractionation procedures add a further dimension of complexity. Furthermore, membrane marker enzymes that have been useful in other cells and tissues are just not well expressed in the platelet membranes, possibly because of their origin from the megakaryocyte cytoplasmic matrix. For example, alkaline phosphatase, 5′nucleotidase, $Mg^{2+}(Na^+, K^+)$ATPase, leucyl aminopeptidase, and bis-*p*-nitrophenyl phosphodiesterase, all good membrane marker enzymes for other cells, are low in activity in platelets, often bimodally distributed, or not sufficiently membrane specific to be useful. Table 2 lists a number of enzyme activities that have been reported to be associated with platelet membranes. Most of these have had some value in monitoring membrane fractions containing elements of both surface and intracellular origin, but enrichment values (relative specific activities) have been disappointingly low and generally less than tenfold increased with respect to homogenate values. The procedure shortly to be described for separating platelet surface membranes from intracellular membranes stimulated a re-examination of many of these alleged markers. Only adenylate cyclase (for the surface membrane) and NADH cytochrome-*c*-reductase and a nonspecific phosphatase that hydrolyzes glucose-6-phosphate (ER membrane enzymes) proved to have real marker value in our hands for the platelet membrane subfractions. These difficulties in defining appropriate enzyme markers have stimulated many platelet membrane researchers to explore the value of surface probes such as radioactive lectins and antibodies or chemical labeling by the covalent addition of ^{125}I or ^{3}H to groups contained in membrane proteins. These membrane-attached molecules have perhaps been of more value than enzymatic studies in following the enrichment of surface-oriented protein or glycoprotein constituents during fractionation procedures and, moreover, they can often reveal useful information about membrane orientation in the isolated vesicles or intramembrane associations revealed by

Table 2. Subcellular Localization of Human Platelet Membrane Enzyme Activities

Enzyme	Substrate	Comments	References
Adenylate cyclase	ATP	Exclusively SM[a]	Brodie *et al.*, 1972; Haslam, 1973; Menashi *et al.*, 1981; Smith and Limbird, 1982
5'Nucleotidase	AMP (or XMP)	SM/IM, IMP is the best substrate	Day *et al.*, 1969; Brodie *et al.*, 1972; Harris and Crawford, 1973; Menashi *et al.*, 1981
Acid phosphatase	*p*-Nitrophenyl phosphate	SM and cytosol. True physiological substrate unknown	Day *et al.*, 1969; Barber *et al.*, 1971; Harris and Crawford, 1973; Kaulen and Gross, 1973
Phosphodiesterase	Bis-(*p*-nitrophenyl)-phosphate diester	SM. Good marker for pig platelets, but poor marker for human platelets	Taylor and Crawford, 1974a,b
Mg ATPase	Mg ATP	SM. Basal Mg ATPase	Barber *et al.*, 1971; Saba *et al.*, 1969
Mg(Na+K)ATPase	Mg ATP	SM. Ouabain inhibited component. Not well expressed	French *et al.*, 1970; Kinoshita *et al.*, 1979; Moake *et al.*, 1970; Salganicoff *et al.*, 1975
Ca, Mg ATPase	Mg ATP	SM/IM. Associated with SM extrusion and IM uptake of Ca^{2+}	Barber *et al.*, 1971; Cutler *et al.*, 1980; French *et al.*, 1970; Käser-Glanzmann *et al.*, 1977; Menashi *et al.*, 1981
Ca ATPase (high μ)	Ca ATP	SM and cytosol. Myosin enzyme inhibited by low Mg^{2+}	Menashi *et al.*, 1981
Nucleoside diphosphokinase	ADP (or XDP)	SM. Ecto enzyme	Sixma and Lips, 1978
Collagen glycosyl-transferase	Collagen + UDP-glucose	SM. These enzymes at first thought to be bridge-forming molecules at membrane/collagen attachment sites. Theory now discounted (see Menashi *et al.*, 1976)	Barber and Jamieson, 1971a,b
Collagen galactosyl-transferase	Collagen + UDP-galactose		
Cholinesterase	Acetylcholine, iodine	SM/IM. Megakaryocyte legacy, not well expressed	Saba *et al.*, 1969

(continued)

Table 2. (Continued)

Enzyme	Substrate	Comments	References
cAMP-dependent protein kinase	ATP + protein	Dependence varies. Different substrates. May be associated with Ca^{2+} uptake	Käser-Glanzmann et al.,1978
Leucyl amino peptidase	L-leucyl-naphthylamide	MSU, IM	Harris and Crawford, 1973; Menashi et al.,1981
Phospholipase-A_2	Phospholipids with unsaturated fatty acids at position '2'	IM	Lagarde et al., 1981, 1982b
Diglyceride lipase	1,2-diacyl glycerol	IM	Lagarde et al., 1981
Diglyceride kinase	1,2-diacyl glycerol	IM, also cytosol	Lagarde et al., 1981
Fatty acyl CoA-transferase	Fatty acid CoA esters and lysophospho-lipids	IM and SM, but tenfold higher in IM	McKean et al., 1983
Cyclooxygenase	Arachidonic acid	IM	Carey et al., 1982
Thromboxane synthase	Cyclic endoperoxides	IM	Carey et al., 1982
Lipoxygenase	Arachidonic acid	MSU, also cytosol	Lagarde et al., 1984
NADH cytochrome-c-reductase (antimycin insensitive)	Cytochrome c, NADH	IM, specific	Menashi et al., 1981
Glucose-6-phosphatase	Glucose-6-P	IM, nonspecific phosphatase	Haslam, unpublished observations
Glycogen synthase	Glycogen, UDP-glucose	MSU, also cytosol	Greenberg et al., 1973

[a]Abbreviations: SM, surface membrane; IM, intracellular membrane; MSU, membrane site unknown; X, bases other than adenine.

cross-linking agents. There are potential redistribution artifacts with some of these surface probes but, with care, such problems can be substantially reduced or eliminated. Chapters 3 and 5 refer to the use of some of these membrane probes in identifying surface-oriented glycoproteins and they will not be discussed in detail here, except later in the specific context of membrane subfraction identification and domain differences.

4. DIFFERENTIAL ISOLATION OF PLATELET MEMBRANE SUBFRACTIONS

Most of this review so far has been concerned with the separation and characterization of mixed membrane fractions from platelet homogenates and most of the attempts to subfractionate these into surface membrane and intracellularly derived membrane elements have met with little, or only limited, success. Resolution of these different membranes by conventional density gradient procedures seems not to be possible probably because the physical properties of the surface and intracellular membrane vesicles are insufficiently different to express different rate sedimentation characteristics. Using zonal rotor centrifugation with sucrose gradients, two distinct membrane subpopulations were isolated from pig platelet homogenates (Taylor and Crawford, 1974a,b). These were reported to be of surface and intracellular origin on the basis of their expression of two different phosphodiesterase activities and the localization, in only one of the fractions, of a surface radioactive probe applied prior to platelet homogenization. Unfortunately, this procedure was found not to be applicable to human platelet membrane fractionation presumably because the disparity in vesicle size and/or density of the surface and intracellular membranes was less than with pig platelets. Moreover, the two phosphodiesterase activities, used to differentiate between the surface and intracellular membranes, were less well expressed in human platelet membranes than in the pig platelet fractions and could not be exploited as differential markers.

An ingenious approach to the subfractionation problem was to incubate a human platelet mixed membrane fraction with Ca^{2+} and oxalate before applying it to a sucrose density gradient (Käser-Glanzmann et al., 1978). Two membrane subfractions separated and it was presumed that the higher density calcium oxalate-loaded vesicles represented vesicles derived from the intracellular calcium-sequestering membranes. The technique looked promising, but unfortunately this fractionation was not as clear-cut nor reproducible as hoped. A careful examination of this preloading procedure in our own laboratories (unpublished observations) established that cross-contamination between surface and intracellular membranes was unacceptable for biochemical analyses. The higher density membrane fraction was considerably contaminated with inside-out–oriented surface membrane vesicles loaded with calcium oxalate. The normal platelet Ca^{2+} extrusion pump resident in the surface membrane simulates an uptake process in inside-out vesicles and cannot be differentiated kinetically or by inhibitors from the normally oriented intracellular membrane vesicles.

A technique that avoids the use of density gradients and results in the harvesting of a plasma membrane fraction reasonably free of contamination with intracellular membrane elements is that of Kinoshita et al. (1979). This method depends on the binding of whole platelets to polylysine-coated polyacrylamide beads by electrostatic interactions. The beads with the attached platelets are sonicated and large areas of surface membrane remain attached to the beads during the sonic disruption and washing procedures. Enrichment figures for marker enzymes and lactoperoxidase-catalyzed surface iodination showed reasonable purification of the surface membrane with less contamination with intracellular organelles than in most of the gradient procedures. The major shortcomings of this technique, however, are the difficulties in establishing recovery data for the construction of fractionation balance sheets and the inability to harvest intracellular membranes for comparison with the surface-derived components. Moreover, the sheets of membrane attached to the beads are of little value in investigating membrane transport functions.

5. FREE-FLOW ELECTROPHORESIS FOR THE SEPARATION OF PLATELET SURFACE AND INTRACELLULAR MEMBRANES

5.1. Initial Characterization of the System

Most of the above-described procedures have their limitations in both purity and yield, and in 1980, the present author and his colleagues developed an entirely differ-

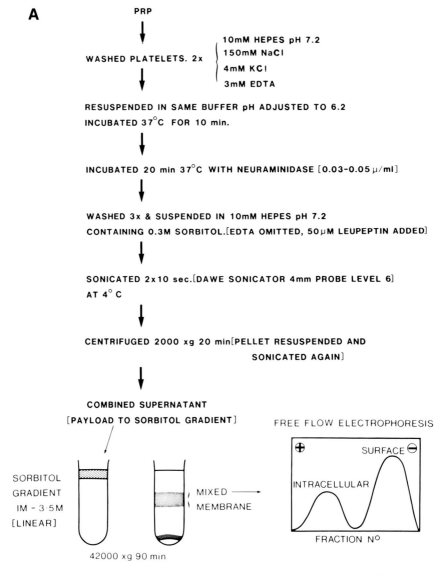

Figure 1. (A) Flow diagram of the procedure for isolating a platelet mixed membrane fraction preparatory to free-flow electrophoresis. For a two-peak profile consisting of intracellular and surface membranes, EDTA is omitted from the sorbitol density gradient. Its inclusion results in a further subfractionation of the surface membrane into two vesicle subpopulations (see Figure 3 B and C). (B) Principles of free-flow electrophoresis. The samples (whole cells, organelles, or soluble constituents) to be separated are injected continuously into the chamber holding the buffer film flowing downward between two glass plates 0.5 mm apart. An electrical field is applied at right angles to the downward flow and the components of a mixture separated with angles of deflection determined by their charge characteristics and the buffer flow rate. The chamber is cooled to dissipate heat and samples are collected in a fraction collector (90 samples across the 10-cm wide chamber).

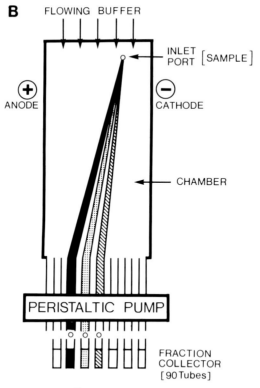

Figure 1. (Continued)

ent approach to the problem of platelet differential membrane isolation using particle electrophoresis in a free-flowing buffer film [Menashi *et al.*, 1981]. In our earlier studies, using various gradient centrifugation procedures and marker enzyme evaluation, it had become clear that if the method for platelet disruption could be carefully controlled, one could prepare a mixed membrane fraction by a density gradient procedure that could be established to be relatively free from contamination with intracellular granular organelles. In our hands, sonication proved to be the only procedure that would ensure the maximum breakage of the platelets in the shortest time (to avoid heating artifacts) and moreover could be tuned to give a cavitation energy that would not seriously affect the integrity of the intracellular organelles. A mixed membrane fraction could then be isolated free from granule contamination by density gradient sedimentation. A further consideration in our choice of sonication was that both the surface and intracellular membranes gave only vesicles without any sheet membrane structures present and this was regarded as a necessary prerequisite for both uniformity of charge expression in the separation by electrophoresis and for further studies of membrane transport properties with the isolated fractions. It was therefore decided to use this mixed membrane fraction as the starting material for further subfractionation, using the technique of high-voltage free-flow electrophoresis that had earlier been shown to separate cells (red cells, lymphocytes, etc.) on the basis of differences in surface charge characteristics. With this procedure, which is shown diagrammatically in Figure 1, cells or subcellular elements are injected into a downward-flowing buffer film through a port on the cathodal side of the chamber. When a potential difference is applied across the chamber, the sample material is deflected toward the anode at an angle that depends on a number of parameters but principally the flow rates of the sample and the buffer film and the electrophoretic mobility of the particles. The electrophoretic mobility of a whole cell or subcellular particle in a free-flowing buffer film is conditioned by the buffer flow rate, the potential applied across the film, and by the zeta potential expressed at the plane of shear between the hydrated shell around the particle (cell or organelle) and the surrounding buffer zone. Ionic strength, osmolarity, pH, temperature, and composition of the buffers used both in the chamber and the electrode compartments are, of course, additional influential factors. Preliminary experiments with glutaraldehyde-fixed red cells of different animal species with different surface charge characteristics established that the system we had chosen (a Bender Hobein Elphor VAP 5 instrument) could be operated for some hours with good temperature and electrical stability and gave uniform separations. Accordingly, in our initial experiments with the gradient-derived platelet mixed membrane fractions applied to the electrophoresis chamber, we obtained a separation that, if not ideal, was at least encouraging. A partial resolution of the mixed membranes into two subfractions was achieved (Figure 2A). A label applied at the whole platelet level to the surface membrane (usually ^{125}I-labeled whole mixed membrane antibody or ^{125}I-labeled *Lens culinaris* lectin) served to locate the fractions containing plasma membrane, and the fractions enriched in NADH cytochrome-*c*-reductase, a well-established ER marker enzyme, identified the intracellular membrane elements. Surprisingly, in terms of unit charge density, the reductase-rich intracellular membrane vesicles were significantly more electronegative than the surface membrane vesicles carrying the surface label. The electrophoresis unit used had a threshold limit to the current density that could be applied at a potential difference of 100 V/cm across the chamber because of limitations in chamber cooling.

5.2. Isolation of Different Membrane Fractions

In order to try to improve the resolution, attempts were made to attach positively charged ligands to the whole platelet surface before the isolation of the mixed membrane fraction. Unfortunately, all such chemical modifications caused some degree of platelet activation and loss of discoidicity. However, turning to enzymatic modifications, it was found that probably one of the most innocuous covalent modifications one can make to the platelet surface, which has minimal activating effects, is the removal of sialic acid moieties by exposure to neuraminidase. For this treatment, it is of course essential that the enzyme preparation used is free from any significant protease activity. Trial studies with a number of different commercial preparations of neuraminidase revealed that the Sigma type IX affinity purified preparation was the most suitable for platelet membrane modification and sufficient sialic acid could be removed after 20-min exposure to the enzyme to render the surface membrane significantly less electronegative. Figure 2B shows that following such enzyme treatment, carried out at the whole cell level before sonication and fractionation, the reductase-rich intracellular membrane vesicles, which had not been exposed to the desialyating enzyme, migrated with the same electrophoretic mobility as the most electronegative and reductase-rich

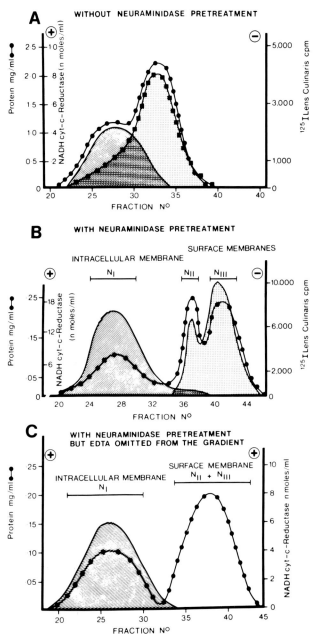

Figure 2. Free-flow electrophoresis profiles after separation of human platelet mixed membrane fractions taken from the sorbitol density gradient (see Figure 1B). (A) Represents a separation without pretreatment of the whole cells with neuraminidase and shows profiles for protein (●———●) and ^{125}I-labeled *Lens culinaris* lectin (■———■) bound to the platelet surface membrane before sonication. The hatched area between fractions 22 and 35 represents the activity of the intracellular marker enzyme NADH-cyochrome-c-reductase. Note that with respect to unit charge density, the intracellular membrane vesicles are more electronegative than the surface membrane vesicles. (B, C) Show separations of mixed membrane fractions derived from platelets pretreated at the whole cell level with neuraminidase. If EDTA is included in the sorbitol density gradient (as in B), this results in the surface membrane migrating as two separable peaks both of which contain the ^{125}I-surface probe and have similar adenylate cyclase specific activities.

vesicles from the untreated cells (compare Figure 2A with 2B). The membrane vesicles carrying the surface label are now substantially less electronegative and with suitable tuning of the electrical parameters and flow rates can be well resolved from the intracellular membrane subfraction. This has now become the routine procedure in our laboratory for preparing platelet surface and intracellular membranes and has been successfully applied to membranes of other cells too. Whether the platelet surface-derived vesicles migrate as two fractions or one (Figure 2C) depends on the presence or absence of EDTA in the sorbitol density gradient procedure. It is believed this splitting into two surface-derived fractions or domains is associated with the presence of the contractile protein myosin attached to intrinsic membrane actin and may be indicative of subpopulation heterogeneity in the whole platelet circulating pool. This possibility will be amplified and discussed later.

6. CHARACTERIZATION OF ELECTROPHORETICALLY SEPARATED SURFACE AND INTRACELLULAR MEMBRANES

Since the development of this differential isolation procedure using free-flow electrophoresis, we have accrued considerable information about the analytical, en-zymatic, and physicochemical differences between these surface and intracellular membrane fractions. Some differentiating features and particularly those which have had marker value in developing the membrane isolation procedures have already been referred to, but with certain constituents (phospholipids, fatty acids, polypeptides) a more comprehensive analytical approach has been made and most of the information available at present is summarized in Table 3. The reader is referred to the original papers (Menashi *et al.*, 1981; Lagarde *et al.*, 1982b) for the technical details of the various preparative and assay procedures used, but the major features to be emphasized in the differences between the two membrane fractions are summarized as follows since they are probably of considerable functional significance with respect to platelet behavior. First, the surface membrane has a substantially higher cholesterol content, a higher cholesterol/phospholipid ratio, a much higher sphinogomyelin content, and a lower fluidity (or higher microviscosity) than the intracellular membranes. Sphingo-myelin and cholesterol are known to be major contributors to membrane physical properties. The major ER-marker enzyme we have used, the antimycin-insensitive NADH cytochrome-*c*-reductase, is almost exclusively localized in the intracellular membrane fraction and adenylate cyclase was only detectable in the surface membrane vesicles. The major phospholipid modifying enzyme, phospholipase A_2, which liber-ates arachidonic acid from membrane phospholipids, is predominantly intracellular as are also the diglyceride lipase and diglyceride kinase enzymes (Lagarde *et al.*, 1981, 1982). Both cyclooxygenase and thromboxane synthetase are also almost exclusive to the intracellular membranes (Carey *et al.*, 1982). Thus the whole metabolic sequence from the provision of the precursor fatty acid to its conversion to endoperoxides and thromboxanes has now been established as essentially an intracellular event in human platelets. More recently the intracellular membrane vesicles have been shown to be in some ways analogous to the SR of muscle tissue in that they are capable of sequester-ing Ca^{2+} by an ATP-dependent uptake process. Some features of the uptake process

Table 3. Comparative Data for Surface and Intracellular Membranes Separated by Free-Flow Electrophoresis[a]

	Surface membranes[b]	Intracellular membranes
Vesicle size (μm)	0.3 − 0.5	0.1 − 0.2
Cholesterol (μmole/mg)	0.69 ± 0.06	0.27 ± 0.03
Phospholipids (μmole/mg)	0.93 ± 0.06	0.92 ± 0.05
Cholesterol/phospholipid ratio	0.74 ± 0.05	0.29 ± 0.01
Sphingomyelin (% total phospholipids)	24.8 ± 1.8	2.6 ± 2.6
Phosphatidylserine (% sum of glycerophospholipids)	2.8 ± 0.6	3.0 ± 0.7
Phosphatidylethanolamine (% sum of glycerophospholipids)	41.9 ± 1.5	28.5 ± 1.4
Phosphatidylcholine (% sum of glycerophospholipids)	50.7 ± 2.5	60.3 ± 2.0
Phosphatidylinositol (% sum of glycerophospholipids)	4.4 ± 1.5	8.7 ± 0.7
Microviscosity (poise) 25° (37°)	4.7 (2.5)	3.1 (2.0)
Adenylate cyclase (pmole/min per mg)	8.0	1.0
NADH cytochrome c-reductase (nmole/min per mg)	0	1600
Leucyl-amino peptidase (μmole/hr per mg)	0.2	1.4
5'Nucleotidase (nmole/hr per mg)	7	220
Ca^{2+} ATPase (high μ) (μmole/hr per mg)	12.0	0.6
Phospholipase-A_2 (nmole/hr per mg)	1.4	20.9
Diglyceride lipase (nmole/hr per mg)	2.0	23.1
Cyclooxygenase (pmol/min per mg)	< 50.0	1200
Thromboxane synthase (pmole/min per mg)	< 50.0	1600
Fatty acyl CoA transferase (nmole/hr per mg)	11.1 ± 2.9	115.6 ± 26.1

[a]All data are the mean ± SD from at least 3 values.
[b]These data are taken from Menashi et al., 1981; Lagarde et al., 1981, 1982b; Carey et al., 1982; McKean et al., 1983.

are shown in Figure 3A–D. Some years ago, White and his colleagues (Statland et al., 1969), working with a platelet mixed membrane fraction, demonstrated calcium uptake and speculated on the presence in platelets of an intracellular mechanism similar to that in other contractile tissues that could sequester calcium and thus control the cytosolic concentration of the free ion. Since there are so many Ca^{2+}-dependent metabolic and functional activities within the platelet, some of which are well established as vital to the cell's response to hemostatic stimuli, the identification of an SR-like membrane complex with a Ca^{2+} pump, as a feature of the electrophoretically separated intracellular membrane fraction, supports this earlier view of White and his colleagues. The control of cytosolic Ca^{2+} levels by this system acting alone or in concert with the plasma membrane (Ca^{2+}, Mg^{2+})ATPase extrusion pump could be of major functional importance in platelet behavior (see Chapter 6).

6.1. Protein and Glycoprotein Composition

The polypeptide profiles for the two-membrane fractions are particularly distinctive. Even in one-dimensional sodium dodecyl sulfate (SDS) polyacrylamide gel electrophoresis (PAGE) separations (Figure 4), there are few components present in both membrane fractions with coincident mobilities. It is particularly striking, too, that the intracellular membranes seem to be devoid of cytoskeletal elements, whereas the

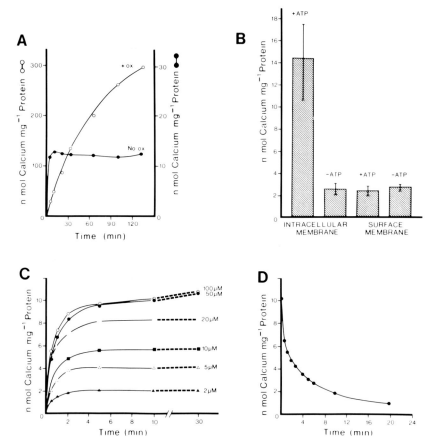

Figure 3. Some features of the Ca^{2+}-accumulating property of the isolated intracellular membrane vesicles. The vesicles were incubated in a medium containing 120 mM KCl, 5 mM $MgCl_2$, 2 mM ATP, 20 mM Tris HCl, pH 7.0 containing 50 μM $CaCl_2$ with ^{45}Ca of specific activity 70 mCi/mmole. Aliquots were removed and filtered rapidly through millipore membranes that were counted in a scintillation counter after washing three times with ice-cold buffer. (A) Shows the time course of Ca^{2+} accumulation in the presence of 2.5 mM oxalate (O———O) and the absence of oxalate (●———●). Note that in the absence of oxalate a steady-state concentration of intravesicular Ca^{2+} is reached in under 5 min, whereas with oxalate the system is not saturated in 2 hr. (B) Shows the extent of Ca^{2+} uptake after 30 min incubation by surface and intracellular membrane vesicles in the presence and absence of ATP. (C) Calcium accumulation by intracellular membrane vesicles in the absence of oxalate at different concentrations of external calcium (range 2–50 μM). (D) Shows the time course of the release of calcium from preloaded vesicles by the calcium ionophore A23187 (5.4 μM). The ionophore was added after steady-state concentration had been reached in the absence of oxalate.

surface membranes show prominent bands that have been confirmed to be actin (42 kD), myosin heavy chain (200 kD), and actin-binding protein (260 kD). Both the actin and actin-binding protein present in the surface membrane are presumed to be part of the organized network of cytoskeleton associated closely with the cytoplasmic face of the plasma membrane and resistant to detergent dissection. Myosin, however, is in

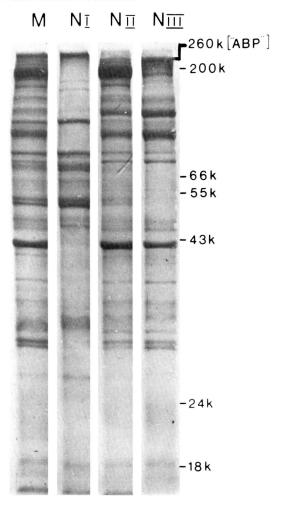

M N$_I$ N$_{II}$ N$_{III}$

260k [ABP]
- 200k
- 66k
- 55k
- 43k
- 24k
- 18k

Figure 4. One-dimensional SDS–PAGE of the polypeptides present in human platelet intracellular membranes (N$_I$) and the two-surface membrane subfractions (N$_{II}$ and N$_{III}$) separated by free-flow electrophoresis according to the profile in Figure 2B. Gels stained with Coomassie Blue. Note that the intracellular membranes (N$_I$) are devoid of contractile proteins. Actin (42 kD) is present in both surface membrane subfractions (N$_{II}$ and N$_{III}$), but only N$_{II}$ contains myosin heavy chain ~200 kD and N$_{III}$ contains most of the actin-binding protein ABP (~260 kD). The two components around 50–55 kD present in the mixed membrane and intracellular membrane fractions have been tentatively identified as the α and β subunits of tubulin.

normal circumstances predominantly a soluble phase protein occurring largely in monomeric or low oligomeric forms in the platelet cytosol (Harris and Crawford, 1973). The finding that this contractile protein is present in only one of the surface membrane fractions (N$_{II}$ in Figure 3A) when the preparative conditions are arranged to separate these two surface-derived domains suggests that different plasma membrane vesicles may arise from different platelet subpopulations in the circulating pool. We would suggest that the actomyosin-rich vesicles are derived from a slightly activated platelet subpopulation that pre-exists in the circulating pool. Alternatively, these platelets may be more sensitive to the stimulus of blood collection and platelet isolation procedures. That the 200 kD component of the surface membrane subfraction is myosin has been confirmed by comigration in PAGE with authentic muscle myosin heavy chain and by the demonstration of a high ionic strength Ca^{2+}-ATPase associated with the membrane that in the presence of 3 mM Ca^{2+} can be inhibited by low levels of Mg^{2+} (50% inhibition at 25 μM Mg^{2+}). This is a feature of myosin from other tissue sources when it is dissociated from actin by ATP under high ionic strength conditions.

From comigration studies with purified preparations of platelet microtubule proteins (two cycles of polymerization/disassembly), there are two components in the SDS–PAGE separations of intracellular membranes that have mobility characteristics closely similar to α and β tubulin. This identity requires confirmation, however, since the high-molecular-weight MAPs appear not to be present in the intracellular membranes. Tubulin as an integral membrane protein has been reported in other cells (Bhattacharyya and Wolff, 1976), and it is possible that MAPs are only a feature of the more dynamic cytosol pool of microtubule subunits and not associated with integral membrane tubulin.

Two-dimensional isoelectrofocusing/PAGE polypeptide maps for the surface and intracellular membranes are presented in Figure 5A, B. By comparing reduced and nonreduced preparations with those presented by other workers (Clemetson et al., 1979; Sixma and Schiphorst, 1980; see also Chapter 3), it is possible to identify most of the major membrane glycoproteins in the surface membrane map that are known to be present at the platelet surface. Similarly reduced and nonreduced preparations of a mixed membrane fraction prepared from platelets with and without neuraminidase treatment has provided us with an appreciation of the minor alterations in mobility characteristics of some of these major glycoproteins that result from the removal of sialic acid. The most significant features, however, of the two-dimensional polypeptide patterns are the virtual absence in the intracellular membranes of the glycoproteins GP Ib, IIb, IIIa, and IIIb, which are such prominent features of the surface membrane polypeptide separations. One-dimensional gels stained by PAS (Figure 6) have substantiated this finding and only very faint PAS-staining bands of GP Ib, GP IIb, and GP IIIa are seen in the intracellular membranes. As referred to earlier, none of the major cytoskeletal elements, actin, myosin, actin-binding protein (ABP) or α-actinin, appear to be present in the intracellular membranes. Evidence confirming the presence of glycoproteins has been provided directly by PAS and concanavalin A peroxidase-staining techniques applied to SDS–polyacrylamide gel separations of the solubilized membrane proteins.

6.2. Lipid Composition

One of the more significant differences between the two membrane fractions lies in their lipid composition. The cholesterol content of surface membranes has been frequently used as a correlative marker in fractionation studies with other cells. Coleman and Finean (1966) compared surface membrane fractions from a wide variety of different guinea pig tissues and in all cases the surface membrane fractions were rich in

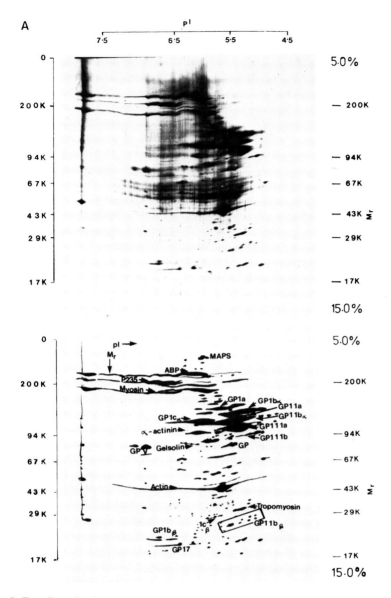

Figure 5. Two-dimensional polypeptide patterns prepared from solubilized human platelet surface membranes (A) and intracellular membranes (B). The composite tracings below each gel photograph have been made from a number of separations and some of the major constituents in the surface membrane identified. Few of the polypeptides in the intracellular membrane separation have been identified with any certainty. For the first dimension under isoelectric focusing conditions (IEF), the membrane samples (200 μg protein) were solubilized in 9 M urea containing 2% NP 40, 5% mercaptoethanol, and 2% ampholines (pH 3.5 to 10.0). The gels were prepared essentially according to O'Farrell (1975). The pH gradient was found to be linear between pH 4.0 and 7.5. For the second dimension, the IEF gels were equilibrated and run in a discontinuous SDS-Tris-glycine system with a linear acrylamide gradient (5–15% acrylamide). The polypeptides were localized using the silver staining procedure of Morrissey (1981). Note the virtual absence of the major known platelet glycoproteins and cytoskeletal proteins in the intracellular membranes which are the more prominent components of the surface membrane separation.

B

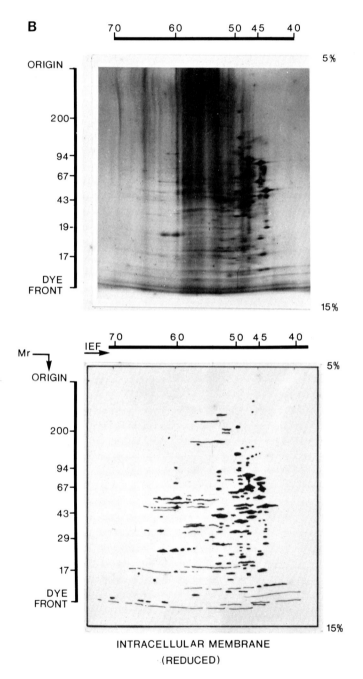

INTRACELLULAR MEMBRANE
(REDUCED)

Figure 5. (Continued)

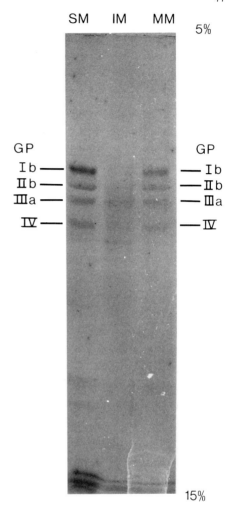

Figure 6. One-dimensional SDS polyacrylamide gel separation of solubilized mixed membrane (MM), surface membrane (SM), and intracellular membrane (IM). The major glycoproteins have been visualized with the periodic acid/Schiff procedure as described by Clemetson *et al.* (1979) and are labeled according to the usual convention.

cholesterol. This enrichment is generally reflected in the cholesterol/phospholipid molar ratio and most of plasma membrane fraction studies from many different tissues and a variety of animal species fall in the range 0.64 to 1.05. The value for this molar ratio (Table 3) in the platelet surface membrane 0.74 fits well within this range and is substantially higher than the intracellular membrane ratio for these components (0.29). This ratio difference in the platelet is entirely due to the cholesterol content since the values for phospholipids present are essentially the same for the two membrane fractions. With respect to the phospholipid profiles, the most significant difference is in their sphingomyelin content. As with other surface membranes, the platelet surface membrane is rich in sphingomyelin accounting for approximately 25% of the total phospholipids, whereas the sphingomyelin content of the intracellular membranes is extremely low, 2–3% of the total phospholipids. Since with other cell membranes it has been generally found that low cholesterol and low sphingomyelin contents are associated with decreased membrane rigidity and high fluidity, the compositional data for these components in the platelet membranes correlate well with the microviscosity measurements made using fluorescence polarization (Table 3). Although membrane proteins have not generally been regarded as significant contributors to the fluidity characteristics of membranes, it might be of significance that the presence and absence of the cytoskeletal proteins, actin, myosin, and ABP, in the membranes are associated with high- and low-microviscosity characteristics, respectively.

With respect to phospholipid and fatty acid composition of the surface and intracellular membranes, some of these data have been included in the comparative table (Table 3). A more fully comprehensive compositional study has been made of these constituents, however, and this has been presented in a recent communication (Lagarde et al., 1982b). Phosphatidylcholine (PC) and phosphatidylinositol (PI) were more enriched in the intracellular membranes than in the surface membranes and the converse was observed for phosphatidylethanolamine (PE). In labeling experiments with [^{14}C]arachidonate, the pattern of incorporation for both membranes closely followed the glycerophospholipid composition profile, although significantly more arachidonic acid was incorporated into the PE fraction of the surface membranes than into the PE of the intracellular membranes. In considering the specific radioactivities after exposure to [^{14}C]arachidonate, the analyses suggested that the turnover of arachidonyl PI was much higher than arachidonyl PE and PC, and it is known that PI contains only arachidonic acid in its "2" position, whereas other phospholipids contain, in addition to arachidonate, oleate, and linoleate.

6.3. Enzyme Activities Associated with Phospholipid and Arachidonic Acid Metabolism in Intracellular Membranes

Arachidonic acid is the key precursor fatty acid for the formation of prostanoids in cells and it is generally accepted that the enzymes phospholipase A$_2$ (PLA$_2$) and diglyceride (DG) lipase are responsible for its liberation from membrane phospholipids, the former by direct action on the phospholipid and the latter follows specific phospholipase C attack on PI. In our studies, both these cleavage enzymes (PLA$_2$ and DG lipase) are predominantly localized in the platelet intracellular membranes with little, if any, activity detectable in the surface membrane fraction or fractions. Since, as referred to earlier, the intracellular membrane is also the major site of activity of the endoperoxide-producing enzyme cyclooxygenase as well as the thromboxane synthetase that forms the proaggregatory compound thromboxane A$_2$, the complete process from arachidonate mobilization to thromboxane production appears to be an integral part of the DTS membrane system.

In 1976, Gerrard et al. provided convincing cytochemical evidence for prostaglandin production in the platelet DTS and postulated that thromboxane A$_2$ may act as a natural ionophore liberating sequestered calcium from this membrane store. With these surface and intracellular membrane subfractions isolated by free-flow electrophoresis, we have been able to provide biochemical support for the cytochemical evidence, but as yet we have been unable to confirm any ionophoric action for thromboxanes produced in situ by the membranes of Ca^{2+}-loaded intracellular vesicles. More recently we have identified the major polypeptide target for aspirin acetylation that is also present exclusively in the intracellular membrane. Earlier, Roth and Ma-

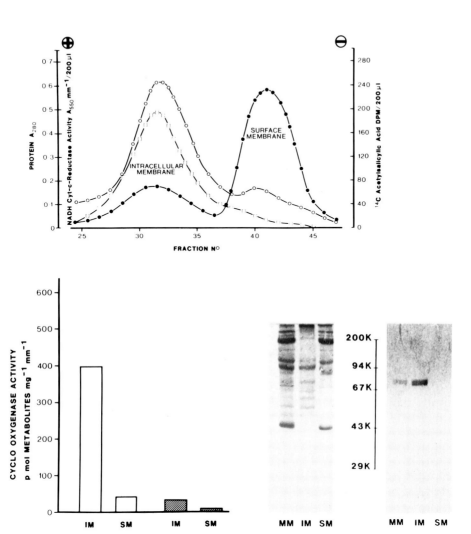

Figure 7. Identification of the platelet target for aspirin acetylation (top) shows the free-flow electrophoresis profile (protein ●————●) of separated surface and intracellular membranes from human platelets pretreated with [1-^{14}C]acetyl salicylic acid before sonication and fractionation. Note the predominant localization of the ^{14}C-label (○————○) in the fractions containing intracellular membranes identified by the presence of NADH cytochrome-*c*-reductase (□————□) (bottom right) one-dimensional electrophoresis of membrane polypeptides visualized by Coomassie Blue (left diagram) and by fluorography (right diagram). Note the presence of a highly labeled ^{14}C-component in the mixed membrane fraction (MM) and intracellular membrane fraction (IM) ~72 kDa. Inhibition of cyclooxygenase activity (bottom left) in the membrane fractions concomitant with the acetylation of the 72 kDa polypeptide (IM = intracellular, SM = surface). The hatched areas represent fractions from platelets pretreated with ^{14}C-aspirin and the open areas the fractions from the untreated control cells. Although the cyclooxygenase activity is inhibited in fractions from cells pretreated with [^{14}C]aromatic ring-labeled aspirin, there is no polypeptide-associated radioactive labeling as with [^{14}C]acetyl group-labeled aspirin.

jerus (1975) had provided evidence for the acetylation by aspirin of a polypeptide associated with a particulate fraction of platelet homogenates. In our studies, this has been further characterized as a component of $M_r \sim 72$ kD and the level of radioactivity due to acetylation with [1-^{14}C]acetyl aspirin correlates with the inhibition of cyclooxygenase [Figure 7A–C] (Hack *et al.*, 1984).

Liberated arachidonic acid is also competed for by lipoxygenase, the enzyme whose products, such as 12-hydroperoxyeicosatetraenoic acid (HPET), are inhibitors of platelet aggregation. Our preliminary studies with this enzyme suggest too that a substantial proportion of it is membrane bound adding a further dimension of complexity to our understanding of the origin and destiny of arachidonate and its metabolic products (Lagarde *et al.*, 1984).

These findings concerning the intracellular disposition of key enzymes for fatty acid mobilization and conversion do of course raise the important question of how does a surface membrane-stimulated event, such as contact with a hemostatic activator, produce a signal that is then transduced into the interior of the cell to trigger functional and metabolic processes? Certainly we need further knowledge about the dynamic aspect of the membrane protein and lipid constituents, about any important topographical changes that may occur in these following ligand binding, and also much more information will be required about the likely second messenger candidates, cAMP and Ca^{2+}, and how their production or mobilization is initiated. There are important protein kinase targets for both these messengers to operate on in a regulatory way and the number of Ca^{2+}-dependent processes that are controlled through quite minor changes in the cytosolic concentration of this ion are legion. The earlier-mentioned close proximity of the surface invaginated domains (OCS) with DTS membrane complexes, which is clearly seen in some areas in platelet electron micrographs, may also be of prime importance in our understanding of the poststimulus sequence of events and particularly if a signal to the surface involves Ca^{2+} release from internal stores. It is outside the scope of this review to consider all the many different processes that are switched on during hemostatic activation, but it is almost certain that the DTS, an SR-like membrane concerned with Ca^{2+} flux control as well as prostaglandin synthesis, is fundamental to the general metabolic health and functional competence of the platelet. That the lipoxygenase enzyme activity is also associated with the intracellular membranes (although some is present in the cytosol) raises the question of what proportion of liberated arachidonic acid may be metabolized by this pathway leading to the production of a number of known pharmacologically active fatty acid derivatives that may have a regulatory function in some of the platelet's intracellular metabolic processes. We know little, however, about the free fatty acid pools in normal or stimulated platelets, but it may also be of considerable significance that in our current collaborative studies with Drs. Mar-lee McKean and Mel Silver at the Cardeza Institute in Philadelphia, these have revealed that the platelet intracellular membranes also have significant levels of fatty acyl transferase activity capable of dealing with a range of CoA esters of long-chain unsaturated fatty acids.

7. CONCLUSIONS AND COMMENTS

As outlined in the introduction to this chapter, most of the emphasis has been placed on the ultrastructural features of the various platelet membrane systems and on technological aspects of membrane isolation and characterization with some emphasis on evaluation procedures for the final isolated material. A necessary prerequisite for the characterization of any subcellular component, and particularly true for membranes, is a suitable isolation procedure. There must be confidence in the reproducibility of such a procedure so that information accruing from detailed analytical, enzymatic, or functional studies can be unequivocally attributed to what are reasonably well-defined and morphologically identifiable subcellular elements.

This review includes a survey of most of the currently used procedures for platelet membrane isolation and, where appropriate, comments have been made about their advantages and/or limitations. Most of the procedures, however, result in a mixed membrane fraction, generally free from granular organelles, but containing membrane elements of both surface and intracellular origin. Also included in the survey is a brief introduction to a new approach to membrane isolation, the use of free-flow electrophoresis. This development has facilitated the separation of a conventionally produced mixed membrane fraction into discrete subfractions containing surface and intracellular membranes. For comparative purposes, a comprehensive list of the major distinguishing features of these two membrane classes has been incorporated into the review. Some of the better earlier procedures will continue to have their value in platelet membrane studies, however, because the preparation of mixed membranes is less time consuming than the full differential separation. Moreover, with the free-flow electrophoresis procedure, it is virtually impossible to construct full balance sheets and recovery data as is usual in conventional subcellular fractionation.

Enthusiasts for subcellular fractionations have always been targets for the criticism that artifacts can be generated by the very process of removing subcellular features from their natural environment and generally dislocating them from their many intracellular relationships with other organelles. There is of course some validity in such criticism, but it also can be strongly argued that if a careful morphological and molecular dissection is applied to the platelet in this way and if analytical biochemists and electron microscopists collaborate together in full cognizance of each other's disciplines (and technical limitations), then much exciting new information can emerge that, with cautious interpretations, will begin to consolidate our understanding of the microanatomical complexities and functional aspects of this very important and often maligned cell entity. The importance of such knowledge of the platelet for a proper appreciation of its normal behavior and changes in disease, for antithrombotic drug design, for advances in transfusion technology, for oncologists, clinical hematologists, and cell biochemists cannot be overemphasized.

ACKNOWLEDGMENTS. I am grateful to my departmental research colleagues who kindly allowed me to incorporate some of their, as yet, unpublished data into their review. I also thank my secretary, Miss Heather Watson, for preparation of the typescript. My thanks are also extended to the British Heart Foundation, I.C.I. Pharmaceuticals, Alderley Park, Cheshire, and M.S.E. of Crawley, for financial support for some of the studies incorporated in this survey.

REFERENCES

Akkerman, J. W. N., Ebberink, R. H. M., Lips, J. P. M., and Christiaens, C. M. L., 1980, Rapid separation of cytosol and particle fractions of human platelets by digitonin-induced cell damage, *Br. J. Haematol.* **44**:291–300.

Baenziger, N. L., and Majerus, P. W., 1974, Isolation of human platelets and platelet surface membranes, *Methods Enzymol.* **31**:149–156.

Barber, A. J., and Jamieson, G. A., 1970, Isolation and characterization of plasma membranes from human blood platelets, *J. Biol. Chem.* **245**:6357–6362.

Barber, A. J., and Jamieson, G. A., 1971a, Platelet collagen adhesion characterization of collagen glucosyltransferase of plasma membranes of human blood platelets, *Biochim. Biophys. Acta* **252**:533–545.

Barber, A. J., and Jamieson, G. A., 1971b, Characterization of membrane-bound collagen galactosyltransferase of human blood platelets, *Biochim. Biophys. Acta* **252**:546–552.

Barber, A. J., Pepper, D. S., and Jamieson, G. A., 1971, A comparison of methods for platelet lysis and the isolation of platelet membranes, *Thromb. Diathes. Haemorrh.* **26**:38.

Behnke, O., 1965, Further studies on microtubules: A marginal bundle in human and rat thrombocytes, *J. Ultrastruct. Res.* **13**:469–477.

Behnke, O., 1967, Electron microscopic observations on the membrane systems of the rat blood platelet, *Anat. Rec.* **158**:121–138.

Behnke, O., 1970, Microtubules in disc shaped blood cells, *Int. Res. Exp. Pathol.* **9**:1–92.

Behnke, O., and Zelander, T., 1966, Substructure in negatively stained microtubules of mammalian blood platelets, *Exp. Cell Res.* **43**:236–247.

Bennett, H. S., 1963, Morphological aspects of extracellular polysaccharides, *J. Histochem. Cytochem.* **11**:14–23.

Bhattacharyya, B., and Wolff, J., 1976, Polymerization of membrane tubulin, *Nature (London)* **264**:576–578.

Brodie, G. N., Baenziger, N. L., Chase, L. R., and Majerus, P. W., 1972, The effects of thrombin on adenyl cyclase activity and a membrane protein from human platelets, *J. Clin. Invest.* **51**:81–87.

Broekman, M. J., Westmorland, M. P., and Cohen, P., 1974, An improved method for isolating α granules and mitochondria from human platelets, *J. Cell Biol.* **60**:507–513.

Carey, F., Menashi, S., and Crawford, N., 1982, Localization of cyclooxygenase and thromboxane synthetase in human platelet intracellular membranes, *Biochem. J.* **204**:847–851.

Castle, A. G., and Crawford, N., 1976, Phosphorylation of platelet microtubule proteins by an endogenous protein kinase, *Biochem. Soc. Trans.* **4**:691–693.

Castle, A. G., and Crawford, N., 1977, The isolation and characterization of platelet microtubule proteins, *Biochim. Biophys. Acta* **494**:76–91.

Clemetson, K. J., Capitanio, A., and Lüscher, E. F., 1979, High resolution two dimensional gel electrophoresis of the proteins and glycoproteins of human blood platelets and platelet membranes, *Biochim. Biophys. Acta* **553**:11–24.

Cohen, I., and De Vries, A., 1973, Platelet contractile regulation in an isometric system, *Nature (London)* **246**:36–37.

Cohen, I., and Lüscher, E. F., 1975, The blood platelet contractile system, *Haemostasis* **4**:125–243.

Coleman, R., and Finean, J. B., 1966, Preparation and properties of isolated plasma membranes from guinea-pig tissues, *Biochim. Biophys. Acta* **125**:197–203.

Crawford, N., 1976, Platelet microfilaments and microtubules, in: *Platelets in Biology and Pathology* (J. L. Gordon, ed.) Elsevier/North Holland, Biomedical Press, Amsterdam, pp. 121–157.

Crawford, N., Amos, L. A., and Castle, A. G., 1980, Platelet microtubule subunit proteins: Assembly and disassembly factors in platelets, in: *Cellular Response Mechanisms and their Biological Significance* (A. Rotman, F. A. Meyer, C. Gitler and A. Sulberberg, eds.), John Wiley & Sons, Ltd., New York, pp. 177–178.

Cutler, L. S., Feinstein, M. B., and Christian, C. P., 1980, Cytochemical localization of oubain-sensitive K-dependent *p* nitrophenyl phosphatase (transport ATPase) in human blood platelets, *J. Histochem. Cytochem.* **28**:1183–1188.

Day, H. J., Holmsen, H., and Hovig, T., 1969, Subcellular particles of human platelets, *Scand. J. Haematol.* (suppl. 7.1):1–35.

De Duve, C., 1963–64, The separation and characterization of subcellular particles, *Harvey Lec.* **59**:49–87.

De Duve, C., 1964, Principles of tissue fractionation, *J. Theor. Biol.* **6**:33–39.

Feinberg, H., 1982, Platelet membrane proteins, in: *Membrane Abnormalities and Disease,* Volume 1 (M. Tao, ed.), C.R.C. Press, Boca Raton, Florida, pp. 91–129.

French, P. C., Holmsen, H., and Stormorken, H., 1970, Adenine nucleotide metabolism of blood platelets

VII ATPases: Subcellular localization and behavior during the thrombin platelet interaction, *Biochim. Biophys. Acta* **206**:438–448.

George, J. N., 1978, Platelet behavior and aging in the circulation, in: *Blood Platelets in Transfusion Therapy* (T. J. Greenwalt and G. A. Jamieson, eds.), Alan R. Liss Inc., New York, pp. 39–64.

George, J. N., and Lewis, P. C., 1978, Studies on platelet membranes III. Membrane glycoprotein loss from circulating platelets in rabbits: Inhibition by aspirin–dipyridamole and acceleration by thrombin, *J. Lab. Clin. Med.* **91**:301–306.

George, J. N., Lewis, P. C., and Sears, D. A., 1976, Studies in platelet plasma membranes II. Characterization of surface proteins of rabbit platelets *in vitro* and during circulation *in vivo* using diazotized ^{125}I diiodosulfanilic acid as a label, *J. Lab. Clin. Med.* **88**:247–260.

George, J. N., Thoi, L. L., and Morgan, R. K., 1981, Quantitative analysis of platelet membrane glycoproteins: Effect of platelet washing procedures and isolation of platelet density subpopulations, *Thromb. Res.* **23**:69–77.

Gerrard, J. M., White, J. G., and Rao, G. M. R., 1974, Effects of the ionophore A23187 on blood platelets 2. Influence on ultrastructure, *Am. J. Pathol.* **77**:151–166.

Gerrard, J. M., White, J. G., Rao, G. M. R., and Townsend, D., 1976, Localization of platelet prostaglandin production in the platelet dense tubular system, *Am. J. Pathol.* **83**:283–294.

Gerrard, J. M., White, J. G., and Peterson, D. A., 1978a, The platelet dense tubular system: Its relationship to prostaglandin synthesis and calcium flux, *Thromb. Haemostasis* **40**:224–231.

Gerrard, J. M., Butler, A. M., Graff, G., Stoddard, S. F., and White, J. G., 1978b, Prostaglandin endoperoxides promote calcium release from a platelet membrane preparation, *Prostaglandins Leukotrienes Med.* **1**:373–385.

Greenberg, J. H., Fletcher, A. P., and Jamieson, G. A., 1973, The presence of glycogen synthase in preparations of platelet plasma membranes, *Thromb. Diathes. Haemorrh.* **30**:307–314.

Griffith, L. M., and Pollard, T. D., 1982, The interaction of actin filaments with microtubules and microtubule associated with proteins, *J. Biol. Chem.* **257**:9143–9151.

Guccione, M. A., Packham, M. A., Kinlough-Rathbone, R. L., and Mustard, J. F., 1971, Reactions of ^{14}C ADP and ^{14}C ATP with washed platelets from rabbits, *Blood* **37**:542–551.

Hack, N., Carey, F., and Crawford, N., 1984, The inhibition of platelet cyclooxygenase by aspirin is associated with the acetylation of a 72kDa polypeptide in the intracellular membranes, *Biochem. J.* **223**:105–111.

Harker, L. A., 1971, Thrombokinetics, in: *The Platelet* (K. M. Brinkhaus, R. W. Shermer, and F. K. Mostofi, eds.), Williams and Wilkins, Baltimore, Maryland, pp. 13–25.

Harris, H., 1981, Regulation of motile activity in platelets, *Platelets in Biology and Pathology*, Vol. 2 (J. L. Gordon, ed.), Elsevier-North Holland, Amsterdam, pp. 473–500.

Harris, G. L., and Crawford, N., 1973, Isolation of pig platelet membranes and granules: Distribution and validity of marker enzymes, *Biochim. Biophys. Acta* **291**:701–709.

Haslam, R. J., 1973, Interactions of the pharmacological receptors of blood platelets with adenylate cyclase, *Ser. Hematol.* **6**:333–350.

Hovig, T., 1968, The ultrastructure of platelets in normal and abnormal states, *Ser. Hematol.* **1**:13–64.

Käser-Glanzmann, R., George, J. N., Jakabova, M., and Lüscher, E. F., 1977, Stimulation of calcium uptake into platelet membrane vesicles by cAMP and protein kinase, *Biochim. Biophys. Acta* **466**:429–440.

Käser-Glanzmann, R., Goerge, J. N., Jakabova, M., and Lüscher, E. F., 1978, Further characterization of the calcium accumulating vesicles from human blood platelets, *Biochim. Biophys. Acta* **512**:1–12.

Kaulen, H. D., and Gross, R., 1973, Metabolic properties of human platelet membranes, 1. Characterization of platelet membranes prepared by sucrose and Ficoll Density gradients, *Thromb. Diathes. Haemorrh.* **30**:307–314.

Kinoshita, T., Nachman, R. L., and Minick, R., 1979, Isolation of human platelet membranes with polylysine beads, *J. Cell Biol.* **82**:688–696.

Koteliansky, V. E., Gneushev, G. N., and Muszbek, L., 1983, Identification and isolation of vinculin from platelets, *Thromb. Haemostasis* **50**:188.

Lagarde, M., Bryon, P. A., Guichardant, M., and Dechauanne, M., 1980, A simple and efficient method for platelet isolation from their plasma, *Thromb. Res.* **17**:581–587.

Lagarde, M., Menashi, S., and Crawford, N., 1981, Localization of phospholipase A_2 and diglyceride lipase activities in human platelet intracellular membranes, *Febs Lett.* **124**:23–26.

Lagarde, M., Authi, K. S., and Crawford, N., 1982a, Fatty acid specificity of human platelet membrane phospholipase A_2 and diacylglycerol lipase, *Biochem. Soc. Trans.* **10**:241–243.

Lagarde, M., Guichardant, M., Menashi, S., and Crawford, N., 1982b, The phospholipid and fatty acid composition of human platelet surface and intracellular membranes isolated by high voltage free flow electrophoresis, *J. Biol. Chem.* **256**:3100–3104.

Lagarde, M., Croset, M., Authi, K. S., and Crawford, N., 1984, Subcellular localization and some properties of lipoxygenase activity in human blood platelets, *Biochem. J.* **222**:495–500.

Mangenstern, E., and Stark, G., 1975 Morphometric analysis of platelet ultrastructure in normal and experimental conditions, in: *Platelets: Recent Advances in Basic Research and Clinical Aspects* (O. N. Ulutin and J. Verrier Jones, eds.), Excerpta Medica, Amsterdam, pp. 37–42.

Mannuci, P. M., 1972, Methods for the preparation of wasted platelet suspension, *Adv. Exp. Med. Biol.* **34**:57–78.

Marcus, A. J., Zucker-Franklin, D., Safier, L. B., and Ullman, M. L., 1966, Studies on human platelet granules and membranes, *J. Clin. Invest.* **45**:14–21.

Martin, J. F., Trowbridge, E. A., Salmon, G. L., and Slater, D. N., 1982, The relationship between platelet and megakaryocyte volumes, *Thromb. Res.* **28**:447–459.

Martin, J. F., Slater, D. N., and Trowbridge, E. A., 1983, Abnormal intrapulmonary platelet production. A possible cause of vascular and lung disease, *Lancet* **1**:793–796.

McKean, M. L., Smith, J. B., and Silver, M., 1983, Phospholipid biosynthesis in human platelets. Formation of phosphatidyl choline from l-acyl lysophosphatidyl choline by acyl-CoA l acyl sn-glycero-3-phosphocholine transferase, *J. Biol. Chem.* **257**:11278–11283.

Menashi, S., Harwood, R., and Grant, M. E., 1976, Native collagen is not a substrate for collagen glucosyl transferase of platelets, *Nature (London)* **246**(5587):670–672.

Menashi, S., Weintroub, H., and Crawford, N., 1981, Characterization of human platelet surface and intracellular membranes isolated by free flow electrophoresis, *J. Biol. Chem.* **256**:4095–4101.

Menashi, S., Authi, K. S., Carey, F., and Crawford, N., 1984, Characterization of the calcium-sequestering process associated with human platelet intracellular membranes isolated by free-flow electrophoresis, *Biochem. J.* **222**:413–417.

Minter, B. F., and Crawford, N., 1967, The subcellular distribution of serotonin in blood platelets before and after incubation with serotonin, *Biochem. J.* **105**:22–23.

Minter, B. F., and Crawford, N., 1974, Subcellular distribution of reserpine and 5-hydroxytryptamine in blood platelets after treatment with reserpine *in vitro* and *in vivo*, *Biochem. Pharmacol.* **23**:351–367.

Moake, J. L., Ahmed, K., Bachur, N. R., and Gutfreund, D. E., 1970, Mg^{2+} dependent (Na^+,K^+)-stimulated ATPase of human platelets: Properties and inhibition by ADP, *Biochim. Biophys. Acta* **211**:337–344.

Morrissey, J. H., 1981, A silver stain for proteins in polyacrylamide gels. A modified procedure with enhanced uniform sensitivity, *Anal. Biochem.* **117**:307–310.

O'Farrell, P. H., 1975, High resolution two-dimensional electrophoresis of proteins, *J. Biol. Chem.* **250**:4007–4021.

Paulus, J. M., 1975, Platelet size in man, *Blood* **46**:321–336.

Pennington, D. G., 1981, Formation of platelets, in: *Research Monographs in Cell and Tissue Physiology*, Volume 5, *Platelets in Biology and Pathology—2* (J. L. Gordon, ed.), Elsevier/North Holland Biomedical Press, Amsterdam, pp. 19–41.

Pennington, D. G., and Streatfield, K., 1975, Heterogeneity of megakaryocytes and platelets, *Ser. Hematol.* **8**:22–48.

Pennington, D. G., Streatfield, D. G. K., and Weste, S., 1974, Megakaryocyte ploidy and ultrastructure in stimulated thrombopoiesis, in: *Platelets—Production, Function, Transfusion and Storage* (M. Baldini and S. Ebbe, eds.), Grune and Stratton, New York, pp. 115–130.

Rittenhouse-Simmons, S., and Deykin, D., 1976, Isolation of membranes from normal and thrombin-treated gel filtered platelets using a lectin marker, *Biochim. Biophys. Acta* **426**:688–674.

Roth, G. T., and Majerus, P. W., 1975, The mechanism of the effect of aspirin on human platelets. 1. Acetylation of a particulate fraction protein, *J. Clin. Invest.* **56**:624–632.

Roth, G. T., Stanford, N., and Majerus, P. W., 1975, Acetylation of prostaglandin synthase by aspirin, *Proc. Natl. Acad. Sci. U.S.A.* **72**(8):3073–3076.

Saba, S. R., Rodman, N. F., and Mason, R. G., 1969, Platelet ATPase activities II. ATPase activities of isolated platelet membrane fractions, *Am. J. Pathol.* **55**:225.

Salganicoff, L., Hebda, P. A., Yandrasitz, J., and Fukami, M. M., 1975, Subcellular fractionation of pig platelets, *Biochim. Biophys. Acta* **385:**394–411.

Schliwa, M., and van Blerkon, J., 1981, Structural interaction of cytoskeletal components, *J. Cell Biol.* **90**(1)**:**222–235.

Siegel, A., and Lüscher, E. F., 1967, Non-identity of the α-granules of human blood platelets with typical lysosomes, *Nature (London)* **215:**745–747.

Sixma, J. J., and Lips, J. P. M., 1978, Isolation of platelet membranes: A review, *Thromb. Haemostasis* **39:**328–337.

Sixma, J. J., and Schiphorst, M. E., 1980, Identification of ectoproteins of human platelets. Combination of radioactive labelling and two dimensional electrophoresis, *Biochim. Biophys. Acta* **603:**70–83.

Slater, D. N., Trowbridge, E. A., and Martin, J. F., 1983, The megakaryocyte in thrombocytopaenia: A microscopic study which supports the theory that platelets are produced in the pulmonary circulation, *Thromb. Res.* **28:**475–483.

Smith, S. K., and Limbird, L. E., 1982, Evidence that human platelet α adrenergic receptors coupled to inhibition of adenylate cyclase are not associated with the subunit of adenylate cyclase ADP ribosylated by cholera toxin, *J. Biol. Chem.* **257:**10471–10478.

Statland, B. E., Heagen, B. M., and White, J. G., 1969, Uptake of calcium by platelet relaxing factor, *Nature (London)* **223:**521–522.

Taylor, D. G., and Crawford, N., 1974a, The subfractionation of platelet membranes by zonal centrifugation: Identification of surface membranes, *Febs Lett.* **41:**317–321.

Taylor, D. G., and Crawford, N., 1974b, The subcellular fractionation of pig blood platelets by zonal centrifugation, in: *Methodological Developments in Biochemistry,* Volume 4, *Subcellular Studies* (E. Reid, ed.), Longmans Press Ltd. Burnt Hill, Harlow, Essex, England, pp. 319–326.

Trowbridge, E. A., Martin, J. F., and Slater, D. N., 1982, Evidence for a theory of physical fragmentation of megakaryocytes implying that all platelets are produced in the pulmonary circulation, *Thromb. Res.* **28:**461–475.

White, J. G., 1971, Platelet morphology, in: *The Circulating Platelet* (S. A. Johnson, ed.), Academic Press, New York, pp. 45–121.

White, J. G., 1972a, Interaction of membrane systems in blood platelets, *Am. J. Pathol.* **66:**295–372.

White, J. G., 1972b, Uptake of latex particles by human platelets. Phagocytosis or sequestration, *Am. J. Pathol.* **70:**45–56.

White, J. G., and Krivit, W., 1965, Fine structural localization of adenosine triphosphatase in human platelets and other blood cells, *Blood* **26:**554–568.

White, J. G., and Sank, J. J., 1984, Microtubule coils in spread blood platelets, *Blood* **64:**470–478.

3

Glycoproteins of the Platelet Plasma Membrane

Kenneth J. Clemetson

1. INTRODUCTION

The glycoproteins of the platelet plasma membrane, in common with those of the plasma membranes of other cells, have an important role in the interaction of platelets with their environment and in particular in the platelet's function of hemostasis.

Many of the platelet receptors for vital physiological stimulators such as thrombin, ADP, and collagen are either known to be or are thought to be glycoproteins. Similarly, the receptors for von Willebrand factor (VWF) involved in the binding of platelets to exposed subendothelium and for fibrinogen are glycoproteins. These functional aspects of various glycoproteins will be dealt with in detail in later chapters, but it is the object of this review to describe the glycoprotein composition of the platelet plasma membrane and to give the characteristics of the individual glycoproteins and what is known of their structures.

2. ANALYTICAL METHODS

2.1. Early Approaches and Nomenclature

Before a role or a function can be established for the individual platelet glycoproteins, it is necessary to have a good system of analysis so that the number and characteristics of the various membrane proteins and glycoproteins can be established and their qualitative and quantitative variations determined. Some early experiments to investigate these glycoproteins were carried out by Pepper and Jamieson (1969, 1970),

Kenneth J. Clemetson • Theodor Kocher Institute, University of Berne, CH-3000 Berne 9, Switzerland.

who looked at the effects of proteolytic enzymes on the platelet surface and separated the glycopeptides that were split off into different size categories. The first studies on undegraded platelet membrane proteins were performed by Nachman and Ferris (1972) and Phillips (1972) with the newly developed technique of gel electrophoresis. Both these studies showed the presence of a large number of polypeptides in the membranes. Carbohydrate-staining techniques indicated three bands that were termed I, II, and III. Current platelet membrane glycoprotein nomenclature is based on these initial findings.

2.2. Improved Separations

Increased resolution and improved sensitivity of detection using discontinuous buffer systems for gel electrophoresis and by comparing separations obtained under both reducing and nonreducing conditions led to an increase in the number of glycoproteins detected and to a need for an expanded nomenclature (Phillips and Poh-Agin, 1977a; Clemetson et al., 1977; George et al., 1976; Okumura and Jamieson, 1976; Jenkins et al., 1980). Since many of the new components were obtained by resolving the old glycoprotein bands on gels, some authors (George et al., 1976; Clemetson et al., 1977) introduced an added small letter after the original Roman number to describe these—thus Ia, IIb, IIIa, etc. However, other authors used a different approach with the glycoproteins resolved from the original band III and termed these III (new) and IV (Okumura and Jamieson, 1976; Phillips and Poh-Agin, 1977a). This distinction in usage still persists despite attempts to introduce a uniform system (George et al., 1981).

An additional complication was introduced by the use of nonreduced/reduced two-dimensional gel electrophoresis (Phillips and Poh-Agin, 1977a). As well as increasing the resolution of the individual glycoproteins, it was also demonstrated that several consist of subunits linked by disulfide bridges. The large subunit was termed the α and the small subunit the β; e.g., GP Ibα, Ibβ. Although the present nomenclature based on the separation obtained under reducing conditions implies decreasing molecular mass with the progression of numbers (I, II, etc.) and letters (a,b, etc.), this is not always true with nonreduced glycoproteins (e.g., Ib is larger than Ia and IIb larger than IIa under nonreducing conditions). From a practical point of view, this is not too serious since the situation is anyway more complex than this simplistic nomenclature would imply. By increasing the sensitivity of detection methods in conjunction with nonreduced/reduced two-dimensional gel electrophoresis, it has been possible to detect still more platelet membrane glycoproteins consisting of disulfide bridge-linked subunits (Sixma et al., 1982a). This method is also useful for detecting changes in glycoproteins after various treatments.

The two-dimensional electrophoresis method of O'Farrell (1975), where isoelectric focusing (IEF) is used as the first dimension giving a separation according to charge followed by sodium dodecyl sulfate–polyacrylamide gel electrophoresis (SDS–PAGE) giving a separation according to molecular weight, was also quickly applied to the analysis of platelet membrane proteins (Clemetson et al., 1979). It soon became clear that many platelet glycoproteins, in common with other glycoproteins, are markedly heterogeneous in isoelectric point and give either streaks or a series of spots on these two-dimensional gels. Here also, the tendency to use more sensitive detection techniques has led to the discovery of additional platelet glycoproteins (McGregor *et al.*, 1981; Spycher *et al.*, 1982; Sixma and Schiphorst, 1980).

2.3. Nomenclature

As mentioned above, from relatively simple beginnings, with the detection of more and more glycoproteins the nomenclature used has become more complex. Attempts to introduce a uniform nomenclature have not been markedly successful probably because competitive systems for the major glycoproteins were already well established and recognized as such and not enough was known about the minor glycoproteins for their nomenclature to be a matter of urgency. For the major glycoproteins, from the quantities found in platelets and from the length of time that they have been recognized, I shall use the nomenclature now used by most research workers in this area: GP Ib, IIb, IIIa, and IIIb. For the minor glycoproteins, where the ''minor'' refers to ease of detection and recognition and not to functional significance, the nomenclature is very heterogeneous. Some components have been recognized by several groups and there is general agreement on their names (e.g., GP V), whereas others have as yet only been reported by single groups. One solution to this problem (George *et al.*, 1981) is to name these components by their apparent isoelectric point (pI) and apparent molecular weight on two-dimensional O'Farrell gels until some sort of role or function can be established. It is not, however, completely problem free since both these parameters are dependent on the precise techniques (and even batches of reagents) that are used and it should not be thought (as is apparently often the case) that they are somehow absolute. A very useful empirical parameter is the position of a glycoprotein spot relative to defined components on a standardized O'Farrell gel.

2.4. Oligosaccharide Chains

The oligosaccharide chains of glycoproteins can be divided into two broad types, the O-linked and the N-linked. The O-linked are so named because they are linked to the protein via the hydroxyl group of a serine or threonine residue, whereas in the N-linked this occurs via the $-NH_2$ group of asparagine. These two broad groups of oligosaccharides have quite different compositions and structures. The O-linked are quite short with rarely more than six sugars and are relatively rich in galactose and sialic acid. On the other hand, they may occur often in a given glycoprotein and give rise to glycoprotein molecules with stiff, rodlike properties such as glycophorin and glycocalicin, where access to the protein backbone by enzymes is severely restricted.

The N-linked oligosaccharides are generally large, complex entities, rich in mannose and GlcNAc, often containing fucose or other sugars. They may form biantennary or triantennary structures and normally few chains are found in a given glycoprotein.

2.5. Determination of Molecular Weights
of Glycoproteins

Molecular weights of proteins are generally determined by one of three principle methods: analytical ultracentrifugation, gel permeation chromatography, and SDS–

PAGE. The degree to which these techniques can be relied on to give reliable estimates of the molecular weight of a glycoprotein depends on the extent and type of glycosylation.

In principal, these methods are based on idealized globular proteins as standard, or in the case of SDS–PAGE, on uniform binding of SDS to the protein, producing a denatured rodlike molecule with charge proportional to the SDS bound and hence to the molecular weight (Weber and Osborn, 1969).

Values obtained by analytical ultracentrifugation are the most reliable, since the shape of the molecule has less influence than with other techniques, and they provide a good basis for comparison.

Glycoproteins containing low amounts of carbohydrate do not differ much from standard proteins, but as the content of sugar increases, so does the degree of deviation. Carbohydrates differ considerably in physicochemical properties from peptides and influence the hydrodynamics of the molecules of which they form part. Thus, glycoproteins with many short oligosaccharide side chains, such as glycophorin or glycocalicin, tend to form stiff rodlike molecules that behave quite differently on gel permeation systems from globular proteins.

Oligosaccharides do not bind SDS to the same extent as proteins and may also shield the protein backbone. Thus, the glycoproteins will have a lower charge and will migrate more slowly, appearing to have a higher molecular weight than in reality. For these reasons the molecular weights of glycoproteins determined by SDS–PAGE are apparent values and should be recognized as such. More accurate values may be obtained by comparison with nonglycosylated standards at different acrylamide concentrations. As the concentration increases, the values obtained approach the real value asymptotically (Segrest *et al.,* 1971). Alternatively, to minimize the anomalies found with glycoproteins, gel electrophoresis may be performed under conditions where more time is allowed for equilibration of the system by running Weber—Osborn gels (Weber and Osborn, 1969) at low current (5 mA/gel) for about 40 hr (Banker and Cotman, 1972). This method has been applied to glycocalicin and its tryptic cleavage products (see Section 5.1) and gave results comparable with those obtained by analytical ultracentrifugation.

3. DETECTION TECHNIQUES

3.1. Staining Methods

Various protein staining techniques have been used to detect platelet membrane components in gels. The original method that is still the most popular was the use of Coomassie Blue (sodium anazolene). It is, however, not very effective for staining sialic-acid-rich glycoproteins since the dye is negatively charged itself and is prevented from binding to the protein part of the molecule because of steric hindrance by the carbohydrate and by the negative charge of the sialic acid. The only glycoproteins that stain strongly with this method are GP IIb and IIIa, although other glycoproteins may stain faintly.

Figure 1 shows a comparison of various staining and labeling methods used with

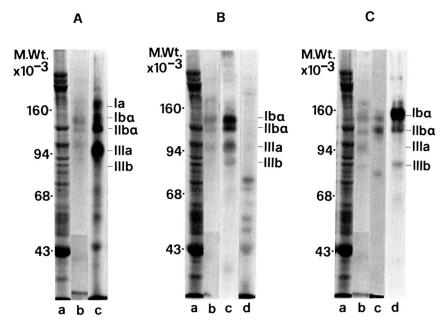

Figure 1. SDS–PAGE (7.5% acrylamide) of reduced, labeled, whole platelets. (A) Lactoperoxidase-catalyzed ^{125}I-labeling. (a) Coomassie Blue staining; (b) periodic acid-Schiff's reagent staining; (c) indirect autoradiography. (B) Sodium periodate/[³H]NaBH₄ labeling. (a) Coomassie Blue staining; (b) periodic acid-Schiff's reagent staining; (c) fluorography; (d) fluorography of pyridoxal phosphate/[³H]NaBH₄ labeling. The fluorogram (d) was exposed 75% longer than (c). (C) Neuraminidase/galactose oxidase/[³H]NaBH₄ labeling. (a) Coomassie Blue staining; (b) periodic acid-Schiff's reagent staining; (c) fluorography of galactose oxidase/[³H]NaBH₄ labeling; (d) fluorography of neuraminidase/galactose oxidase/[³H]NaBH₄ labeling. The fluorogram (c) was exposed 75% longer than (d). The material for this figure was kindly provided by Dr. J. L. McGregor. The methods used here are described in McGregor *et al.* (1979a).

one-dimensional gel electrophoresis, whereas Figure 2 shows Coomassie Blue staining of a two-dimensional (IEF/SDS–PAGE) separation of platelet proteins.

Another dye used to detect platelet proteins and glycoproteins is alcian blue (Kunicki *et al.*, 1981b), which gives a strong staining of sialoglycoproteins as well as other proteins in the presence of sodium tetradecyl sulfate.

The classic method of detecting glycoproteins that has been extensively applied in histochemistry is the use of periodic acid oxidation followed by Schiff's reagent. This is effective for staining a wide range of glycoproteins in gels run on platelets, but is a relatively insensitive technique requiring high protein loads on the gels. Recently there have been some improvements in the reliability and sensitivity of this method. We have found that soaking the stained gels in dimethyl sulfoxide, followed by water, stabilized and enhanced the color reaction and recently it was reported that treatment of the gels with 40% ethanol had a similar effect (Konat *et al.*, 1983).

3.2. Surface-Labeling Methods

Although many of these staining techniques can be used to acquire useful information about platelet membrane glycoproteins and proteins, undoubtedly the use of

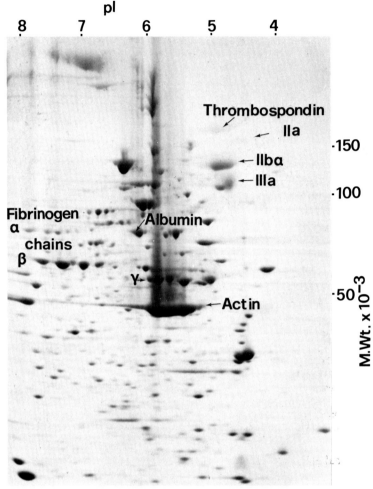

Figure 2. Two-dimensional slab gel of whole platelets, reduced. Stained with Coomassie Brilliant Blue R 250. First dimension, isoelectric focusing pH 3.5–8.5, second dimension, SDS–gradient PAGE, 5–20% acrylamide. The methods used here are described in Clemetson *et al.* (1979).

surface-labeling techniques to detect and quantitate these components has greatly added to our knowledge of their properties.

A prerequisite for all surface-labeling techniques is that the reagents should be impermeant and not label internal components. This is normally determined for newly developed techniques by comparing the results obtained by labeling intact cells with those using solubilized cells. With carbohydrate-specific methods at present levels of sensitivity of detection there is no problem, since even if some of the reagent penetrated the cell, it would not label cytoplasmic components. With protein-specific methods the situation is less clear. In most cases it is virtually impossible to differentiate between traces of reagent penetrating into the cell and labeling part of the large amount of cytoplasmic material or large amounts of reagent labeling traces of cytoplasmic material stuck to the outside of the cells and coming from lysis of a small proportion of the cells caused by the isolation and washing procedures. This is not a serious problem with the methods currently in use, but might become so as detection methods become more sensitive and as alternative labeling methods are developed.

3.2.1. Lactoperoxidase-Catalyzed Iodination

This was first applied to platelet glycoproteins by Phillips (1972) and by Nachman *et al.* (1973) and was used to demonstrate that the glycoprotein bands found in gels by periodic acid/Schiff's staining were in fact *surface* oriented. This method labels principally tyrosine and histidine residues in the protein part of the molecule and is further restricted by the accessibility of these amino acids within the three-dimensional structure of the glycoproteins. Thus, GP IIb and IIIa are strongly labeled, whereas GP Ib and IIIb, which contain large amounts of carbohydrate, label poorly presumably due to steric hindrance. Figure 3 shows a two-dimensional O'Farrell separation of platelets surface labeled by this technique.

Surface iodination of platelets has also been carried out using variations of this technique with 1,3,4,6-tetrachloro-3α,6α-diphenylglycoluril as the oxidizing agent (Tuszynski *et al.*, 1983). This latter method has the advantage of being somewhat more

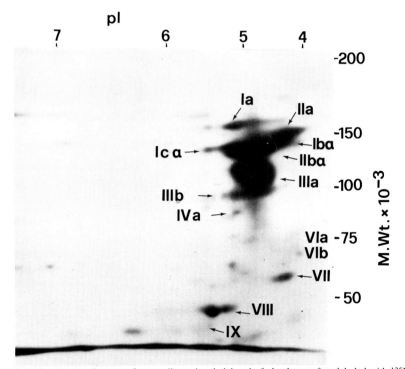

Figure 3. Indirect autoradiogram of a two-dimensional slab gel of platelets surface labeled with ^{125}I by the lactoperoxidase technique, reduced. First dimension, isoelectric focusing pH 3.5–7.5; second dimension, SDS–PAGE, 7.5% acrylamide. The material for this figure was kindly provided by Dr. J. L. McGregor. Reproduced from McGregor *et al.* (1981), with permission.

easily controlled and a higher incorporation of radioactive iodine into protein can be obtained. In addition, the reaction is easily terminated by decantation without contamination by lactoperoxidase (which can also label itself). The total number of labeled glycoproteins detected can be extended by the use of scintillation screens (indirect autoradiography) or by increasing the exposure times (Swanstrom and Shank, 1978; Laskey and Mills, 1977). This does not, however, help where the weakly labeled species is in close proximity to strongly labeled molecules due to the relatively diffuse spots obtained with this γ-emitter.

A further use of iodine-labeling techniques is that of double labeling with ^{125}I and ^{131}I in order to obtain precise information about the relative amount of a particular component in two samples. Examples of this technique applied to platelet membrane glycoproteins can be found in Jenkins et al. (1976) and Phillips et al. (1975), where pathological platelets were compared with normals in a one-dimensional technique. This method could also be applied to two-dimensional gels after using autoradiography to establish the position of the labeled glycoproteins.

3.2.2. [125I]Diazodiiodosulfanilic Acid

A surface-labeling system based on radioactive iodine that has some advantages over lactoperoxidase-catalyzed iodination is the use of diazodiiodosulfanilic acid (or

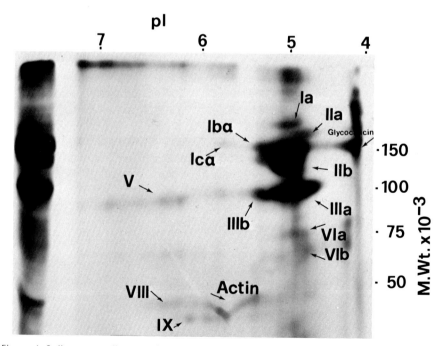

Figure 4. Indirect autoradiogram of a two-dimensional slab gel of platelets, surface labeled by the [125I]diazodiiodosulfanilic acid technique, reduced. First dimension, isoelectric focusing, pH 4–7.5; second dimension, SDS–PAGE, 7.5% acrylamide. The material for this figure was kindly provided by Prof. J. J. Sixma. This technique is described in Sixma and Schiphorst (1980).

the commercially available monoiodo compound). Again, originally used with erythrocytes (Sears *et al.*, 1977), this system was used with platelets in one-dimensional separations by George *et al.* (1976) and later in two-dimensional separations by Sixma and Schiphorst (1980). Figure 4 shows a two-dimensional separation (IEF/SDS–PAGE) of platelets labeled using this technique. The main advantage is that the labeling is not restricted to one or two amino acids, but reaction has been shown to occur with tyrosine, histidine, lysine, proline, tryptophan, phenylalanine, glycine, and cysteine residues. A problem is that this reagent is very soluble in the membrane lipid bilayer and unreacted material is difficult to remove.

3.2.3. *Periodate/[³H]NaBH₄ Labeling*

Oxidation of vicinal diol groups present in the carbohydrate (particularly the sialic acid) of surface glycoproteins followed by reduction of the aldehyde groups thus formed with tritiated sodium borohydride can be used to introduce tritium atoms into these molecules. Following the successful application of this method to surface-labeling of erythrocytes (Liao *et al.*, 1973), it was quickly applied to platelets followed by separation on one-dimensional gel electrophoresis (Phillips and Poh-Agin, 1977b) and later by two-dimensional electrophoresis (IEF/SDS–PAGE) (McGregor *et al.*, 1980). The usefulness of this method was greatly extended by the development of quantitative fluorographic methods for the detection of tritium in polyacrylamide gels (Bonner and Laskey, 1974; Laskey and Mills, 1975).

Figure 5 shows a two-dimensional (IEF/SDS) polyacrylamide gel electrophoretic fluorogram of platelets surface-labeled by this method and Figure 6 shows a fluorogram of a two-dimensional separation (nonreduced/reduced) of platelets similarly labeled. In comparison to the iodine-labeling methods, this technique gives a much stronger labeling of the glycoproteins with a high content of sialic acid such as Ib and IIIb.

Although other glycoproteins such as IIb and IIIa contain less sialic acid, they are rich in mannose which often is linked so as to leave a vicinal diol available for oxidation. Although this method also has the disadvantage that the strong labeling of some components tends to swamp nearby weaker-labeled entities, the short path length of the tritium β particles tends to produce sharper spots on the x-ray film even with fluorographic detection.

3.2.4. *Neuraminidase/Galactose Oxidase/[³H]NaBH₄ Labeling*

Galactose oxidase converts the 6-hydroxy group of terminal galactose or N-acetylgalactosamine (GalNAc) residues of glycoproteins to an aldehyde group which can then be reduced with tritiated borohydride to incorporate a tritium atom. Galactose or GalNAc residues are generally found preterminal to sialic acid residues, however, and in order to obtain an efficient incorporation of radioactivity, it is necessary to remove the terminal sialic acid by neuraminidase treatment. As a surface-labeling method for cells, the two enzymes may be added simultaneously. This method, initially developed by Gahmberg and Hakomori (1973) and Steck and Dawson (1974), has the advantage that only terminal galactose or GalNAc residues are labeled. A possible disadvantage is that the necessary removal of sialic acid residues makes

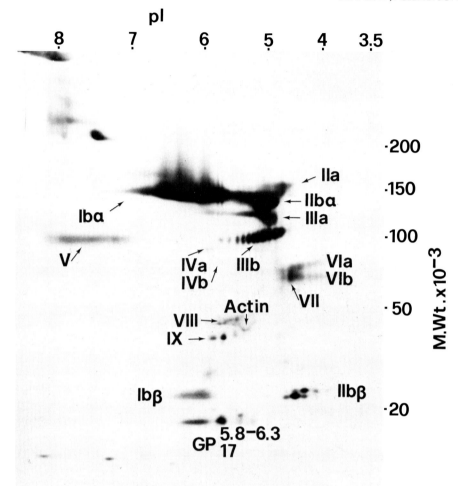

Figure 5. Fluorogram of a two-dimensional slab gel of platelets surface labeled by the sodium metaperiodate/[³H]NaBH₄ technique, reduced. First dimension, isoelectric focusing, pH 3.5–8.5; second dimension, SDS–PAGE, 5–20% acrylamide. The methods used here are described in McGregor *et al.* (1980).

glycoproteins with a high sialic acid content more basic and often also affects the apparent molecular weight on SDS–PAGE acrylamide gel electrophoresis.

Figure 7 shows a two-dimensional (IEF/SDS–PAGE) separation of platelets labeled by this method. The position of several glycoproteins and their heterogeneity is considerably altered compared with separations where sialic acid is present. In particular, GPIIb and V are shifted to a much more basic pI and are less heterogeneous. Glycoprotein Ib is less affected than might be expected—this may be due to the fact that part of this glycoprotein is so rich in sialic acid that its normal pI (without neuraminidase) is less than 3 and it is not contained within the IEF range of pH 3–8. After removal of sialic acid, this part is less acid and appears within the pH range of the gel, compensating for the loss of more basic material from the gel. Glycoprotein Ib

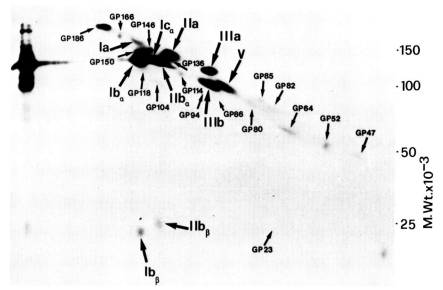

Figure 6. Fluorogram of a two-dimensional slab gel of platelets surface labeled by the sodium metaperio-date/[³H]NaBH₄ technique. First dimension, nonreduced, SDS–PAGE, 7.5% acrylamide; second dimension, reduced, SDS–PAGE, 10% acrylamide. Figure kindly provided by P. Clezardin and Dr. J. L. McGregor. These methods are described in Phillips and Poh-Agin (1977a,b).

moves also to a slightly higher apparent molecular weight after removal of sialic acid. The migration of most of the other glycoproteins is only slightly affected by neuraminidase treatment. By comparing this method with periodic acid [³H]NaBH₄ labeling, a good idea of the relative sialic acid content of the glycoproteins can be obtained. Also, indirectly, some idea of the content of O-linked oligosaccharides can be derived from this comparison.

3.2.5. Tritium-Labeling of Protein

As well as the methods mentioned above, it is also possible to label platelet membrane proteins with tritium. One technique described in the literature is the use of pyridoxal phosphate followed by reduction with NaB^3H_4 (Subbarao *et al.*, 1978), which was also used by McGregor *et al.* (1979a) for comparison with carbohydrate-labeling techniques (Figure 1).

More recently, reductive methylation, which had previously been used as a labeling method for soluble proteins, has been applied to surface labeling of platelets. The platelets are treated with formaldehyde, which forms Schiff's bases with amino groups on the protein—almost exclusively with epsilon-amino groups of lysine. No cross-linking occurs with the small amount of formaldehyde used. By reduction of these Schiff's bases with NaB^3H_4, a methyl group is formed containing a tritium atom. Alternatively, [^{14}C]formaldehyde may be used followed by $NaBH_4$, giving a methyl group containing a ^{14}C atom.

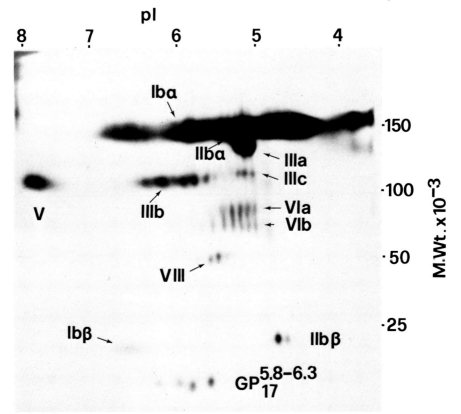

Figure 7. Fluorogram of a two-dimensional slab gel of platelets surface labeled by the neuramini-dase/galactose oxidase/[³H]NaBH₄ technique, reduced. First dimension, isoelectric focusing, pH 3.5–8; second dimension, SDS–PAGE, 5–20% acrylamide. The methods used here are described in McGregor *et al.* (1980).

Figures 8 and 9 show two-dimensional separations (IEF/SDS–PAGE) of platelets surface labeled by reductive methylation. Figure 8 shows a separation in the non-reduced state, whereas Figure 9 shows a separation in the reduced state. Although the overall patterns resemble those obtained by the methods mentioned above, the labeling of the individual glycoproteins is more uniform. This makes the general pattern easier to analyze and several minor glycoproteins are easily detected since the "swamping-out" phenomenon that occurs via GP IIb-IIIa with iodine labeling or via GP Ib with carbohydrate methods of labeling does not occur.

One obvious problem is that of the lysine content of the glycoproteins. If it is absent, then labeling will not occur, which seems to be the case with the Ibβ and IIbβ subunits. A problem that does not seem to occur with platelets, but which might be present with other cells, is the transport of formaldehyde and NaB³H₄ across the membrane so that the labeling is not restricted to surface components.

As indicated above, this method may be used as part of a double-labeling technique with ¹⁴C used in reductive methylation to label protein and ³H used in the

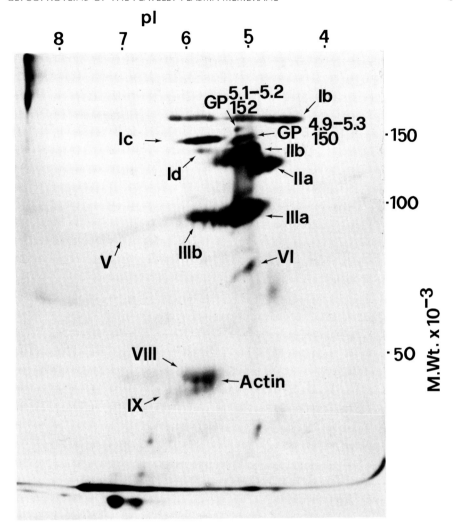

Figure 8. Fluorogram of a two-dimensional slab gel of platelets surface labeled with ³H by reductive methylation, nonreduced. First dimension, isoelectric focusing, pH 3.5–8.5; second dimension, SDS–PAGE, 5–20% acrylamide. Reproduced from Spycher *et al.* (1982), with permission. The techniques used are described in this paper.

periodate/[³H]NaBH$_4$ or neuraminidase/galactose oxidase/[³H]NaBH$_4$ techniques to label carbohydrate. Methods for qualitative and quantitative double-label fluorography and radiography with ³H and ¹⁴C have recently been described (Choo *et al.*, 1980; Goldman *et al.*, 1983).

3.2.6. Other Labeling Techniques

Several other surface-labeling techniques have been used with platelets on occasions and, in particular circumstances, may have some advantages. Thus, Rotman

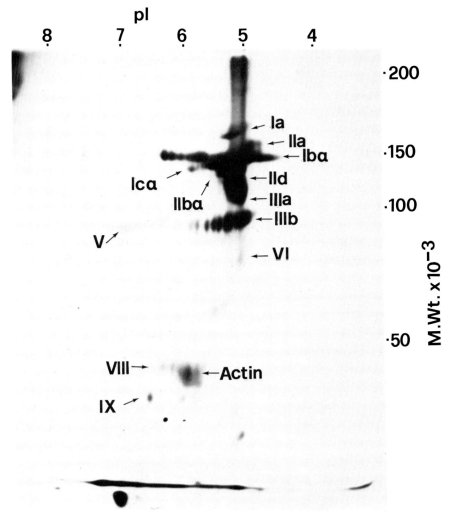

Figure 9. Fluorogram of a two-dimensional slab gel of platelets surface labeled with ³H by reductive methylation, reduced. First dimension, isoelectric focusing, pH 3.5–8.5; second dimension, SDS–PAGE, 5–20% acrylamide. Reproduced from Spycher *et al.* (1982), with permission. The techniques used are described in this paper.

(1980) has used 3,5-diiodo-L-tyrosine hydrazine in conjunction with periodate oxidation as a method of introducing an ¹²⁵I-containing carbohydrate specific label and has also used [¹²⁵I]diazodiiodoarsanilic acid as an alternative reagent to [¹²⁵I]diazodiiodosulfanilic acid and found that it gave a different labeling pattern.

A further approach is the use of nonpolar labeling reagents in order to label proteins that penetrate the lipid bilayer. Although these are not strictly surface-labeling reagents, platelet membrane glycoproteins are nevertheless among the components that are labeled, therefore giving additional information about the structure of the glycoproteins. The main compound used for these experiments has been iodonaphthylazide. This has been used to follow changes in labeling of membrane proteins during platelet activation (Rotman *et al.*, 1982). Glycoproteins Ibα, IIbα, and IIIa were all labeled with this technique indicating that they contain a hydrophobic portion that penetrates the membrane. Several other membrane proteins were also labeled, but it is not yet known whether these are surface glycoproteins or proteins either totally inbedded in the membrane or exposed only on the cytoplasmic face.

There have been several reports that platelet membrane glycoproteins can be labeled metabolically. Apitz-Castro *et al.* (1979) compared the labeling obtained using ^{14}C-labeled glucosamine, mannose, and galactose with normal, Glanzmann's thrombasthenia, and Bernard-Soulier platelets. Soslau and Rybicki (1982) showed that both [^{35}S]methionine and [^{14}C]fucose were incorporated into human platelet proteins. Although the methionine appeared to be used in *de novo* synthesis, the fucose was incorporated into pre-existing glycoproteins. Marchesi and Chasis (1979) demonstrated that some human platelet glycoproteins were phosphorylated after incubation of platelets with [^{32}P]orthophosphate.

3.3. Silver-Staining Methods

Since 1979, several staining methods based on the use of silver compounds, which are much more sensitive than Coomassie Blue, have been developed for detecting proteins and glycoproteins after their separation by gel electrophoresis. The chemistry of these techniques is not yet well understood, and for a review of some of the problems and solutions that have been used, the reader is referred to Merril *et al.* (1982).

These methods have also been applied to platelet proteins (Sammons *et al.*, 1981). A considerable problem is caused by the sensitivity of the method when techniques that detect protein are used. Thus, Figure 10 shows a two-dimensional (IEF/SDS–PAGE) separation of platelet proteins stained by the Sammons technique. Even on the original colored version it is difficult to see which spots represent glycoproteins. Proteins stain in a variety of colors from yellow through orange and brown to green, blue, and black. The color is characteristic for the protein and is reproducible. Using this Sammons technique, all the glycoproteins so far identified stain between orange and brown. Because of its particularly high sensitivity, this technique is especially useful where purified preparations are being examined, but it has the disadvantage that the staining response is not linear and it cannot be used for quantitative work. For such purposes, the method of Merril *et al.* (1982), although not showing peptide-specific colors, gives a quantitative result.

We have also used the Dubray/Oakley silver-staining technique (Dubray and Bezard, 1982; Oakley *et al.*, 1980), which involves a periodate oxidation step and in principle is specific for the carbohydrate of glycoproteins. Figure 11 shows a two-dimensional (IEF/SDS–PAGE) separation of platelets stained with this technique. Although many glycoproteins can be readily identified, some problems of characterization compared with surface labeling methods have still to be solved. One problem is the number of low-molecular-weight components that stain with this method. Either

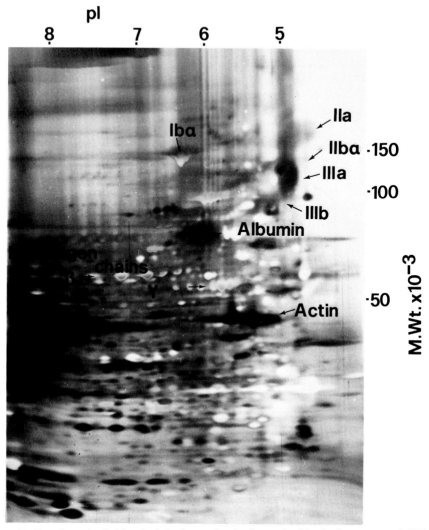

Figure 10. Two-dimensional slab gel of platelets, reduced, silver stained by the Sammons *et al.* (1981) technique. First dimensional, isoelectric focusing, pH 4–8.5; second dimension, SDS–gradient PAGE 5–20% acrylamide.

these are carbohydrate-containing components from the interior of the platelet or they are staining for other reasons, for example, due to a high content of cysteine or methionine. An additional problem is that glycoproteins with less sialic acid, e.g., GP IIb-IIIa, do not stain well.

An unfortunate complication of the use of silver-staining methods is the high quenching that this produces with ^{3}H labeling. Thus, it is difficult to examine tritium surface-labeled platelets by both silver staining and fluorography in order to correlate the two methods (Van Keuren *et al.*, 1981).

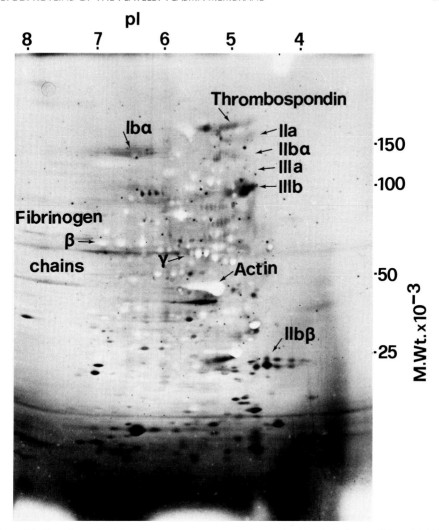

Figure 11. Two-dimensional slab gel of platelets, reduced, silver stained by the Dubray and Bezard (1982) variant of the Oakley *et al.* (1980) technique. First dimension, isoelectric focusing, pH 3.5–8; second dimension, SDS–gradient PAGE, 5–20% acrylamide.

3.4. Use of ^{125}I-Labeled Lectins to Identify Platelet Glycoproteins on Polyacrylamide Gels

This method can be used to identify platelet glycoproteins that have been separated by gel electrophoresis. The gels are fixed, washed, and then incubated with a lectin labeled with ^{125}I by the chloramine-T method. Unbound lectin is then removed by washing and the gel is dried and used for autoradiography (McGregor *et al.*, 1979b; Judson and Anstee, 1979). Figure 12 shows an autoradiograph of a two-dimensional

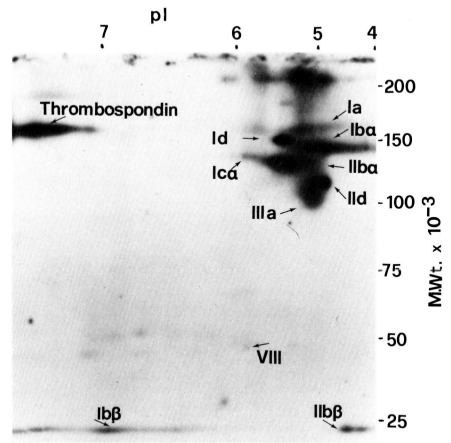

Figure 12. Indirect autoradiogram of a two-dimensional slab gel of platelets, reduced, stained with ^{125}I-labeled *Lens culinaris* lectin. First dimension, isoelectric focusing, pH 4–7.5; second dimension SDS–PAGE 7.5% acrylamide. Thrombospondin is a large molecule and migrates slowly through acrylamide gels. This is why it has a more basic pI than usual on this autoradiogram since it has not reached the equilibrium position. Material for figure kindly provided by Dr. J. L. McGregor. Reproduced from McGregor *et al.* (1981), with permission.

separation (IEF/SDS–PAGE) of platelets labeled with [125I]*Lens culinaris* lectin. As can be clearly seen by comparison with other labeling techniques, this method is relatively specific and each lectin binds to a subset of the total glycoproteins (Table 1). This provides a parameter in addition to apparent molecular weight and pI for the characterization of the various glycoproteins. It should be noted that not only surface glycoproteins will be labeled with this technique, but also granule and other glycoproteins when whole platelets are examined. Variants of this method might include the transfer of the separated glycoprotein to nitrocellulose by electrophoretic blotting techniques (Gershoni and Palade, 1983) before incubating with [125I]lectins or the use of lectins coupled to enzymes such as peroxidase together with enzyme substrates giving a color reaction as the detection method (Clegg, 1982; Moroi and Jung, 1984).

Table 1. Specificity of Lectins for Sugars, Oligosaccharides,
and Platelet Membrane Glycoproteins

Lectin	Simple sugar specificity	Oligosaccharide specificity	Platelet glycoprotein specificity[g]
Lens culinaris	Me-α-Man, Me-α-Glc	Bi-antennary complex N-linked, containing fucose[a,b]	GP Ib, IIb, IIIa (Ia, IIa, V)[d]
Concanavalin A	Me-α-Man Me-α-Glc	Trimannosidic core of N-linked, complex bi-antennary, without fucose	GP IIIa, IIIb (IIb)[d]
Wheat germ agglutinin	Sialic acid, GlcNAc	O-linked, rich in sialic acid and GlcNAc[b]	GP Ib, IIIb (IIa, IIb, V)[d]
Ricinus commu- nis I	Gal, GalNAc	O-linked, rich in Gal, GalNAc[b]	GP Ia, Ib, IIIb (asialo)[e,h]
Peanut agglutinin	Gal, lactose	D-Gal-β-[1-3]-D-GalNAc (found in O-linked chains)[c]	GP Ib (asialo)[f,h]

[a]Kornfeld *et al.* (1981).
[b]Debray *et al.* (1981).
[c]Pereira *et al.* (1976).
[d]Clezardin *et al.* (1984).
[e]Clezardin *et al.*, unpublished results.
[f]Clemetson *et al.* (1981).
[g]Main specificities are without brackets, weaker binding is indicated in brackets.
[h]After treatment with neuraminidase.

4. METHODS OF ISOLATION OF PLATELET GLYCOPROTEINS

4.1. Lectin Affinity-Chromatography

The first approaches to the isolation of platelet glycoproteins were the use of protein chemistry techniques to isolate proteolytically cleaved glycopeptides (Pepper and Jamieson, 1969). The use of detergents to solubilize intact platelet membrane glycoproteins was then followed by lectin affinity-chromatography on concanavalin A to isolate GP IIIb (Nachman *et al.*, 1973). Later studies showed that different lectins bound different glycoproteins and considerable purification could be obtained by using different lectin columns sequentially (Clemetson *et al.*, 1977). More recently, we have demonstrated that the use of glycosidases in conjunction with lectin columns allows the use of lectins that give more specific binding (Clemetson *et al.*, 1981).

In comparing lectin affinity-chromatography with results obtained with [^{125}I]lectin binding to glycoproteins separated on polyacrylamide gels, it should be borne in mind that the chromatography is "dynamic," whereas the gels are "static." In other words, the relative affinity plays a greater role in chromatography and the glycoproteins with high affinity bind preferentially. A similar phenomenon has also been observed with the glycoproteins isolated from erythrocytes (Furthmayr, 1981). Thus, although [^{125}I]lectin binding to glycoproteins separated on polyacrylamide gels provides a rough guide to the separations that may be expected using the same lectin for affinity chromatography, it may not indicate which of these glycoproteins will be preferentially bound in the chromatography system.

4.2. Immunoabsorption Systems

A natural extension of affinity chromatography for the characterization and isolation of platelet membrane glycoproteins has been the use of antibodies in either immunoabsorption techniques. The sophisticated technique of crossed immunoelectrophoresis, which has been extensively applied to the analysis of platelet membrane glycoproteins (Hagen, 1983) and which offers the advantages of relatively easy quantitation and the possibility of examining interactions and functional aspects of the glycoproteins in a nondenaturing system, will be dealt with in detail in Chapter 5. Immunoprecipitation has been generally applied to the identification and characterization of antibodies to platelet membrane glycoproteins. Earlier work was carried out with autologous antibodies found in patients with platelets deficient in membrane glycoproteins who had received multiple transfusions (Tobelem et al., 1976) or with heterologous antibodies against platelets (Nachman et al., 1977) or against purified platelet glycoprotein fractions (Jenkins et al., 1983). More recently, this method has been extensively used to characterize monoclonal antibodies to platelets (McGregor et al., 1983a; Coller et al., 1983). Figure 13 shows an example of immunoprecipitation of solubilized surface-labeled platelets by a variety of monoclonal antibodies. The glycoproteins precipitated by these antibodies can be identified by several different parameters: different migration under reducing and nonreducing conditions on polyacrylamide gels, absence or changes in their characteristics in pathological conditions, and different lectin-binding properties. In some cases the pI on two-dimensional PAGE may be a crucial factor, but studies so far have not required this. Problems may be encountered due to coprecipitation of complexed glycoproteins, for example, GPIIb and IIIa, and here crossed immunoelectrophoresis, in the presence of EDTA to separate the complex, followed by binding of radiolabeled monoclonal antibody (McEver et al., 1983) may be used to identify the glycoprotein containing the epitope. Alternatively, platelet membrane glycoproteins separated by SDS–PAGE may be transferred by electrophoretic blotting to a nitrocellulose membrane and the binding of monoclonal antibody to the separated components investigated.

Immunoabsorption chromatography using immobilized monoclonal antibodies is also being developed as an extremely practical method for characterizing the antigens to which the antibodies bind and in order to purify these platelet membrane glycoproteins in high yield by a one-step method for further biochemical study. In this way, McEver et al. (1983) have purified GP IIb and IIIa using their monoclonal antibody Tab directed to GP IIb and Coller et al. (1983) and Berndt et al. (1983) have purified GP Ib.

Nevertheless, the high immunogenicity of GP IIb-IIIa as shown by the ease of preparation of monoclonal antibodies to these glycoproteins compared with others (possibly expecting GP Ib) makes it still highly desirable to continue to develop methods for the isolation and purification of other platelet membrane glycoproteins in

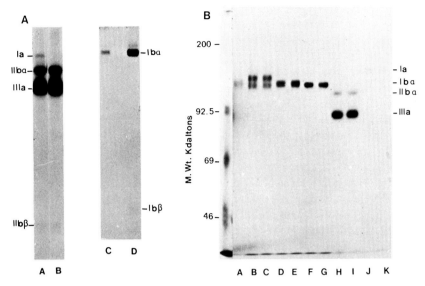

Figure 13. A. Indirect autoradiogram of ^{125}I-labeled (A,B) and fluorogram of neuraminidase/galactose oxidase/[3H]NaBH$_4$-labeled (C,D) platelet glycoproteins immunoprecipitated by monoclonal antibodies and separated on a 10% SDS–PAGE. (A) Antibody P2, epitope on GP IIb-IIIa complex; (B) antibody P4, epitope on GP IIb-IIIa complex; (C) antibody P3, epitope on GP Ia or GP Ib; (D) antibody P1, epitope on GP Ib B. Fluorogram (A–G) and indirect autoradiogram (H–K) of neuraminidase/galactose oxidase/[3H]NaBH$_4$ labeled (A–E), periodate/[3H]NaBH$_4$-labeled (F,G) and ^{125}I-labeled platelet membrane glycoproteins precipitated by monoclonal antibodies P1, epitope on GP Ib (A–G) and P2, epitope on Gy IIb-IIIa complex (H–K). Platelet extracts immunoprecipitated were normal donors: (A, D, E, H, I); patient with double band in GP Ib position (Bolin *et al.*, 1977) (B,C); Glanzmann's thrombasthenia patient (J, K). Figure kindly provided by Dr. J. L. McGregor. Reproduced from McGregor *et al.* (1982), with permission.

order to study their structure and function. Moroi *et al.* (1982) have shown that the properties of the glycoproteins may be used to devise methods for their isolation. Thus, glycocalicin could be purified using an affinity column of thrombin–Sepharose.

5. ISOLATION METHODS, STRUCTURE, AND PROPERTIES OF INDIVIDUAL PLATELET MEMBRANE GLYCOPROTEINS

5.1. Glycoprotein Ib

This is the principle sialoglycoprotein of the platelet membrane and its proteolytic fragments were among the first glycopeptides to be described from the platelet surface (Pepper and Jamieson, 1970; Nurden, 1974). The ease with which it is cleaved by endogenous platelet-calcium-activated proteases to the water-soluble glycocalicin (Solum *et al.*, 1980a,b; Clemetson *et al.*, 1981) led initially to some confusion about the properties and identity of these components (Okumura and Jamieson, 1976; Solum *et al.*, 1977), which is now largely resolved. Because of its solubility and relative ease of preparation, the glycocalicin moiety has been studied more than either the intact molecules or the residual pieces. Isolation methods have been based on classical biochemical techniques (Okumura *et al.*, 1976), use of wheat germ agglutinin– Sepharose affinity columns (Solum *et al.*, 1980a), or peanut agglutinin–Sepharose affinity columns with asialoglycocalicin (Clemetson *et al.*, 1981). The amino acid and sugar compositions have been reported and the cleavage products obtained with trypsin have been characterized (Okumura *et al.*, 1976; Jamieson *et al.*, 1980). It is known that glycocalicin has a relatively sensitive cleavage point with trypsin that splits it into two fragments. The larger fragment is a highly glycosylated peptide reported as 120 Kd or 75 Kd (Okumura *et al.*, 1976; J. W. Lawler, personal communication) depending on the technique used, which is resistent to further degradation and contains the bulk of the oligosaccharide side chains. The other piece, 45 Kd, contains a thrombin-binding site (Jamieson *et al.*, 1980) and at least one loop structure linked by disulfide bridges (Lawler *et al.*, 1980; Clemetson *et al.*, 1981). This loop structure contains a further trypsin cleavage site that can be demonstrated by the presence of two smaller peptides in the trypsin-treated fragment after reduction (J. W. Lawler, personal communication).

As mentioned above the 120 Kd fragment is rich in carbohydrate. This appears all to be in the form of O-linked mucin type oligosaccharides and the carbohydrate composition of these chains has been determined and possible structures proposed (Judson *et al.*, 1982). More recently with glycocalicin as starting material the O-linked oligosaccharides have been isolated and their structure determined by 500 mHz NMR (Korrel *et al.*, 1984) and by methylation analyses and glycosidase treatment (Tsuji *et al.*, 1983). The structures found (Figure 14) are in good agreement with those proposed earlier (Judson *et al.*, 1982). This work also found some evidence for at least one N-linked oligosaccharide chain in glycocalicin. Comparison with the 120 Kd glycopeptide sugar composition indicates that this N-linked oligosaccharide may be on the 45 Kd fragment. Cleavage of glycocalicin from the platelet surface leads to loss of the platelet aggregation response to VWF, which will be discussed in detail in Chapter 10. This would imply that the receptor site is on the glycocalicin part of the GP Ib molecule, but its precise location is not yet known.

Glycoprotein Ib has a molecular weight of 160 Kd nonreduced and 145 Kd,

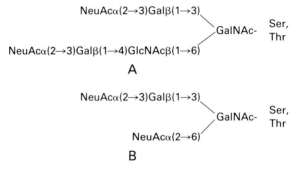

Figure 14. Structures of the main O-linked oligosaccharides of glycocalicin (Tsuji *et al.*, 1983; Korrel *et al.*, 1983).

reduced (α chain). It has a surprisingly basic pI 6.5–7.5 for a molecule so rich in sialic acid. The pI is not noticably affected by reduction. Glycocalicin has a molecular weight of 140 Kd when electrophoresed under the same conditions as GP Ib. Electrophoresis under limiting conditions (36–43 hr on 10% gels) gave a molecular weight of 108 Kd (J. W. Lawler, personal communication). Glycocalicin is much more acidic than GP Ib with a pI of 4–5.

Glycoprotein Ib consists of two chains, α and β, joined by at least one disulfide bridge (Phillips and Poh-Agin, 1977a). Both chains seem to contain hydrophobic sequences and are integral glycoproteins (Clemetson *et al.*, 1981; Solum *et al.*, 1980b). The piece of the α chain left after cleavage of glycocalicin has a molecular weight of about 20–30 Kd (K. J. Clemetson, unpublished results), and there is some evidence that it penetrates into the cytoplasm and has a phosphorylation site on the cytoplasmic side (Marchesi and Chasis, 1979). The phosphorylation of GPIb is supported by recent data (Solum and Olsen, 1984) but is now thought to occur only on the β-chain (Bienz *et al.*, 1984). It is not known definitively whether this fragment contains any carbohydrate.

The β chain has a molecular weight of 22,000 and contains carbohydrate. It seems likely that this is in the form of N-linked complex oligosaccharide because of the lectin-binding and carbohydrate-labeling properties of this chain. It binds *L. culinaris* lectin strongly but concanavalin A only weakly, and wheat germ agglutinin and peanut agglutinin (after neuraminidase treatment) binding could not be shown. This would indicate a branched structure rich in mannose but probably containing a fucose residue.

Glycoprotein Ibα and its cleavage products such as glycocalicin are classic examples of the problems encountered in determination of the molecular weight of glycoproteins and dealt with in Section 2.5. Glycocalicin and the membrane-bound fragment have apparent molecular weight of 140,000 and approximately 25,000, respectively, which are obviously greater combined than that of intact GP Ibα (145,000). This is because glycocalicin has relatively an even higher carbohydrate content than GP Ibα and therefore shows a larger anomaly in apparent molecular weight when compared with nonglycosylated standards by methods of determination such as normal SDS–PAGE. The real molecular weights of these glycoproteins are probably much closer to the values obtained by Lawler and cited above.

Figure 15 shows a diagrammatic model of the GP Ib structure with its various features indicated.

Recent evidence has shown that GP Ib may possibly be noncovalently linked to other membrane or cytoskeletal structures (Solum *et al.*, 1983; Rotman *et al.*, 1982).

5.2. The Glycoprotein IIb-IIIa Complex

In the absence of chelators (i.e., in the presence of calcium ions), these glycoproteins form a complex (Kunicki *et al.*, 1981a; Hagen *et al.*, 1982) that acts as the binding site for fibrinogen (Nachman and Leung, 1982) and may also participate in the platelet association with fibronectin (Ginsberg *et al.*, 1983) and VWF (Ruggeri *et al.*, 1982) on activated platelets (see Chapter 8). Both of these glycoproteins are absent or abnormal in platelets from patients with Glanzmann's thrombasthenia, an inherited bleeding disorder (see Chapter 15). This suggests either a common genetic origin for

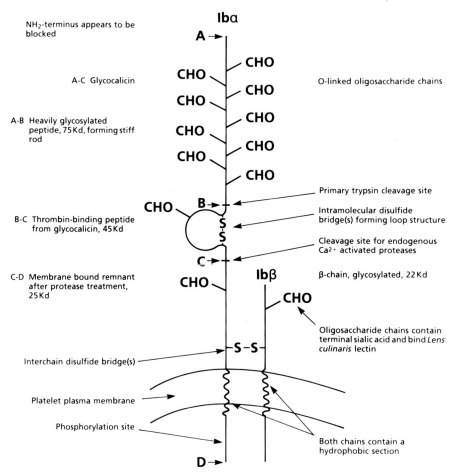

Figure 15. Structure of GP Ib. More recent work (Bienz et al., 1984) shows that the phosphorylation site is on the β-chain.

these glycoproteins or that both need to be synthesized in order to be incorporated in the plasma membrane. The latter situation has been described for other systems, e.g., HLA heavy chains and β2-microglobulin, although produced by different genes and noncovalently associated, are both necessary for incorporation of HLA molecules into cell membranes (Owen et al., 1980). The genetic relationship of GP IIb and IIIa is still unknown, although tryptic peptide maps of these glycoproteins are clearly different (McEver et al., 1982; Leung et al., 1981; McGregor et al., 1982). It has not yet been definitely established whether these glycoproteins exist as complexes in resting platelets or whether the complexes form on activation.

Treatment of intact platelets with chymotrypsin cleaves both GP IIb and IIIa to give smaller membrane-bound pieces (McGregor et al., 1983b; Kornecki et al., 1983) and at the same time exposes the fibrinogen receptor without platelet activation (Greenberg et al., 1979; Niewiarowski et al., 1981). The outer part of the GP IIb and IIIa molecular complex is therefore involved in regulation of the fibrinogen receptor during activation. The complex is resistant to leukocyte elastase, but is degraded when solubilized and dissociated by removal of Ca^{2+} (Shulman *et al.*, 1983). As after chymotrypsin treatment, there remain fragments of both glycoproteins that are capable of reassociation in the presence of calcium. Glycoproteins IIb and IIIa both contain binding sites for Ca^{2+} (Gogstad *et al.*, 1983; Brass and Shattil, 1983; Fujimura and Phillips, 1983a) that are probably involved in conformational changes controlling the association of these two molecules. Glycoproteins IIb and IIIa have been isolated by a variety of methods. Jennings and Phillips (1982) made use of the propensity of GP IIb and IIIa to associate with the triton X-100 insoluble platelet cytoskeleton after activation and to be removable from the cytoskeleton by treatment with high-salt buffer. The GP IIb-IIIa complex could then be dissociated with EDTA and separated by gel filtration. Another approach was adopted by Leung *et al.* (1981), who used lectin affinity-chromatography on *L. culinaris* lectin and concanavalin A to separate these glycoproteins from detergent solubilized membranes. *Lens culinaris* lectin binds GP IIb preferentially indicating the presence of N-linked complex oligosaccharides with fucose substitution, whereas GP IIIa binds concanavalin A preferentially indicating the presence of N-linked oligosaccharide chains with branched mannose structures (Kornfeld *et al.*, 1981). Neither binds wheat germ agglutinin indicating the absence of O-linked oligosaccharides. McEver *et al.* (1982) used affinity chromatography with a monoclonal antibody against GP IIb (Tab) coupled to Sepharose to isolate these glycoproteins. Glycoprotein IIIa could be removed from the column by washing with a buffer containing EDTA, whereas IIb required either high or low pH to be eluted. Newman and Kahn (1983) used a combination of phase partition in triton X-114 followed by high-pressure gel filtration chromatography of the detergent phase, in the presence of SDS, to purify GP IIb and IIIa.

Although it is not completely clear why GP IIb-IIIa are so immunogenic compared with the other membrane glycoproteins, they contain a considerable proportion of the protein of the platelet surface and are less glycosylated than most other membrane glycoproteins.

Figure 16 shows a diagrammatic model of the GP IIb-IIIa structure with various features indicated.

5.2.1. Glycoprotein IIb

This glycoprotein has also been described at a relatively early stage in the study of platelet membrane glycoproteins (Nachman and Ferris, 1972) and was initially called GP II (Phillips, 1972). It consists of two subunits linked by disulfide bonds. The α-subunit has a molecular weight of 130,000 and pI of 4.7–5.5 and a β-subunit has a molecular weight of 23,000 and a pI of 4.5. The β-subunit also shows a characteristic, complex, heterogeneous pattern on two-dimensional (IEF/SDS) gel electrophoresis.

The amino-acid composition of GP IIb has been determined (McEver *et al.*, 1982; Jennings and Phillips, 1982) as has also the sugar composition (McEver *et al.*, 1982). Little is known of the properties of the IIbβ subunit compared with IIbα. It seems also to have rather *N*-linked oligosaccharides than O-linked since it binds *L. culinaris* and concanavalin A rather than wheat germ agglutinin. Glycoprotein IIb can be cleaved by thrombin when it is dissociated from GP IIIa, solubilized in detergent in the absence of

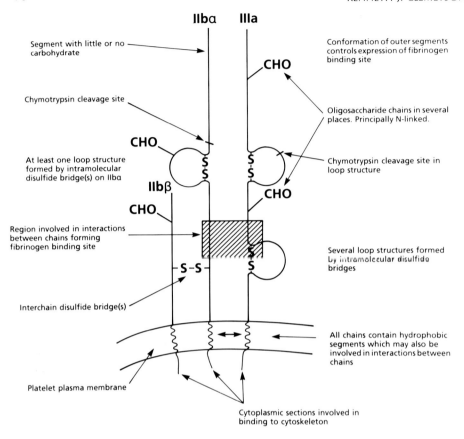

IIbα   IIIa

Segment with little or no carbohydrate

Conformation of outer segments controls expression of fibrinogen binding site

CHO

Chymotrypsin cleavage site

Oligosaccharide chains in several places. Principally N-linked.

CHO

At least one loop structure formed by intramolecular disulfide bridge(s) on IIbα

Chymotrypsin cleavage site in loop structure

IIbβ

CHO

CHO

Region involved in interactions between chains forming fibrinogen binding site

Several loop structures formed by intramolecular disulfide bridges

S–S

Interchain disulfide bridge(s)

All chains contain hydrophobic segments which may also be involved in interactions between chains

Platelet plasma membrane

Cytoplasmic sections involved in binding to cytoskeleton

Figure 16. Structure and possible interactions of GP IIb and IIIa.

calcium but is not cleavable when integrated in the membrane or in IIb/IIIa complexes (Fujimura and Phillips, 1983b).

5.2.2. Glycoprotein IIIa

The amino acid and sugar compositions have also been reported (Jennings and Phillips, 1982; McEver *et al.*, 1982). This glycoprotein consists of a single chain and shows unusual behavior on polyacrylamide gel electrophoresis since on reduction it moves to an apparent higher molecular weight (Phillips and Poh-Agin, 1977a). These authors also found evidence indicating that this glycoprotein contains several intra-chain disulfide bridges and is relatively difficult to reduce, suggesting a compact structure. The reduced form bound more *L. culinaris* lectin than the unreduced (McGregor *et al.*, 1981) suggesting that in the nonreduced form the oligosaccharide chains are tightly associated with the protein and relatively inaccessible to the lectin.

Results obtained with chymotrypsin treatment suggest that the intramolecular disulfide bridges form loop structures (McGregor *et al.*, 1983b) and that a 66,000 dalton piece left in the membrane is part of the fibrinogen receptor (Kornecki *et al.*, 1983).

5.3. Glycoprotein IIIb

Glycoprotein IIIb is the smallest of the four major platelet membrane glycoproteins with a molecular weight of 95,000 and a pI 4.5–5.7. This glycoprotein is highly glycosylated with a typical heterogeneous spot pattern on two-dimensional gel electrophoresis. Much of this heterogeneity seems to be due to sialic acid since after treatment with neuraminidase this glycoprotein becomes much more basic and the pattern of spots simplifies to four (McGregor et al., 1981). It probably does not contain intramolecular disulfide bonds since there is no change in molecular weight on reduction. It binds wheat germ agglutinin, although less strongly than GP Ib, which indicates the presence of O-linked oligosaccharides. It binds concanavalin A, but not L. culinaris lectin, indicating the presence of branched mannose structures in N-linked oligosaccharides. It is very resistant to a wide range of proteases indicating that the part of the molecule not inbedded in the membrane is highly glycosylated. No evidence of function has been found until now since antibody probably directed against this glycoprotein did not influence platelet behavior (Nachman et al., 1973).

5.4. Glycoprotein V

Although not one of the major glycoproteins, GP V has attracted attention because it is the only glycoprotein so far identified on the platelet membrane that is normally a substrate for thrombin. It has been isolated by Berndt and Phillips (1981b) using classical biochemical methods. Glycoprotein V has a molecular weight of 82,000 and a pI of 6.5–7.5. After treatment with neuraminidase, GPV moves to a more basic pI 7.5–7.9 (McGregor et al., 1981) and goes from at least nine distinct spots to one major spot. The amino acid and sugar compositions have been determined (Berndt and Phillips, 1981b). This glycoprotein has not shown a strong affinity for any lectins, but reacts weakly with several. Together with the sugar composition, this indicates the presence of both O-linked and N-linked oligosaccharides. The cleavage of GP V by thrombin gives a 69,000 piece that is water soluble and contains all the carbohydrate, (Phillips and Poh-Agin, 1977b; Mosher et al., 1979) and, presumably, an approximately 20,000 piece containing the membrane-binding segment. This latter piece has not yet been detected. There is some controversy in general about whether GP V is an integral or a peripheral glycoprotein. Berndt and Phillips (1981b) claimed that it is lost from the platelet surface on treatment with high-salt buffers, whereas Sixma et al. (1982b) reported that it was an integral glycoprotein. Recent findings indicate that GPV is lost from the membrane surface after cleavage of GP Ib by Ca^{2+}-activated proteases (Clemetson et al., 1983).

The bulk of the evidence still points to GP V being an integral glycoprotein, although there is some confusion due to this facilitated release from the membrane. However, it is not clear if direct proteolysis is involved in cleavage of GP V from the membrane or if the cleavage of GP Ib causes a membrane rearrangement releasing intact GP V. There have also been conflicting reports on the sensitivity of GP V in situ to various enzymes. Berndt and Phillips (1981b) have shown that it is susceptible to thrombocytin, α-clostripain, and trypsin. Fujimura et al. (1981) claimed that it was not cleaved by chymotrypsin, whereas Detwiler (1983) claimed that it was. It is possible that the chymotrypsin proteolysis site lies further from the membrane still allowing cleavage by thrombin to take place.

5.5. Glycoproteins Ia, Ic, Id, IIa, and Other Glycoprotein Ib Region Glycoproteins

In the region of GP Ib on two-dimensional gels, there is a large group of quantitatively "minor" glycoproteins. Glycoprotein Ia, with a molecular weight of 153,000 unreduced and 167,000 reduced and a pI of 4.5–5.5, is relatively little affected by neuraminidase treatment. As with most minor glycoproteins, little is known of its properties or functions. Recently, monoclonal antibodies have been reported that precipitate GP Ia from detergent solutions of platelet membranes along with other glycoproteins (McGregor et al., 1983a; Coller et al., 1983).

Glycoprotein Ic is a more heterogeneous glycoprotein at a molecular weight of 148,000 unreduced and pI of 5.2–5.8. Similar to Ib and IIb, this glycoprotein consists of two subunits, an α and a β. The α has a molecular weight of 145,000 and is generally concealed by GP Ibα on two-dimensional gels. The β-subunit is slightly larger than Ibβ and IIbβ and does not label well by most techniques that have been applied. However, it labels weakly with the [^{125}I]diazodiiodosulfanilic acid technique and with periodate/NaB^{3}H$_4$ (Sixma et al., 1982b) and has been tentatively assigned a pI of 5.2 on two-dimensional gels. Unlike the other glycoproteins so far described, GP Ic has properties characteristic of a peripheral membrane component and can be removed from the platelet surface by treatment with 0.1N NaOH (Sixma et al., 1982b).

Glycoprotein Id is the name assigned to a minor glycoprotein found just below GP Ic on two-dimensional PAGE of nonreduced samples of platelets surface labeled by reductive methylation (Spycher et al., 1982) and also detected by [^{125}I]-L. culinaris lectin binding (McGregor et al., 1981). Peterson et al. (1982) have detected a glycoprotein with a higher molecular weight than Ic on gels of reduced samples of platelets from Bernard-Soulier patients that they called GPId. Since GP Ic moves to a lower molecular weight on reduction, the glycoproteins labeled Id in these two studies are probably the same species.

Glycoprotein IIa (molecular weight 130,000 nonreduced, 150,000 reduced, pI 4.1–4.8), similar to GP IIIa, moves to a higher molecular weight on reduction, indicating the presence of intramolecular disulfide bonds. The fact that it also shows a diagonal streak effect on two-dimensional gels with a higher molecular weight on the acidic than the basic side probably indicates that it is highly glycosylated. Steiner et al. (1982) have proposed that GP IIa is involved in the platelet Fc receptor. McEver and Martin (1984) described a monoclonal antibody to an antigen they designated GPIIa, which reacts only with activated platelets. Subsequent immunocytochemical studies demonstrated that this glycoprotein is located in the α-granule membrane and it is currently designated GMP-140 (granule membrane protein, molecular weight 140,000) (Bainton and McEver, unpublished observations).

Several other glycoproteins have been noted in the GPI region. One has a molecular weight of approximately 150,000 and a pI of 4.9–5.3, which lies just above GP IIb and on the acid site of GP Ic (Spycher et al., 1982) (see Figure 9) and which may be that designated GP Ia by Sixma and Schiphorst (1980). A further glycoprotein has a molecular weight of approximately 152,000 and a pI of 5.1–5.2 (lying between GP Ib and GP Ic in the unreduced gel) (see Figure 8). This is probably equivalent to the spot number 9 found by Sixma and Schiphorst (1980).

5.6. Glycoproteins in the Glycoprotein IIb-IIIa Region

On reduced gels with the reductive methylation-labeling method, the spot that corresponds to GP IIIa appears to split into two (see Figure 9). This additional spot may correspond to GP IId, which can be distinguished from GP IIIa by its *L. culinaris* lectin-binding properties (Figure 12). This is clearer with platelets from patients with Glanzmann's thrombasthenia where GP IIIa is absent (McGregor *et al.*, 1981). Glycoprotein IId can also be distinguished by the fact that it moves to a lower molecular weight on reduction, unlike GP IIIa.

A further minor glycoprotein, termed IIIc, can be distinguished from IIIa after neuraminidase, galactose oxidase, $[^3H]NaBH_4$ labeling, since it then moves slightly faster and on the basic side of GP IIIa (McGregor *et al.*, 1981).

Several minor glycoproteins can be found on the basic side of GP IIb-IIIa, migrating somewhat faster (molecular weights about 75–80,000), which we termed IVa and IVb. These glycoproteins are absent in Glanzmann's thrombasthenia (McGregor *et al.*, 1981) so that the possibility exists that they are proteolytic fragments of GP IIb-IIIa.

5.7. Glycoproteins in the 43,000- to 70,000-Dalton Region

The major glycoproteins in this region are the VI group (Phillips and Poh-Agin, 1977b) consisting of two glycoproteins showing similar behavior on two-dimensional gels, but one about 5,000 daltons larger than the other. To distinguish these, we have named the larger VIa and the smaller VIb (Clemetson *et al.*, 1982) (see Figure 5). These glycoproteins have the peculiar property of becoming more heterogeneous on treatment with neuraminidase. As with GP IIIb and V, they also become much more basic (compare Figures 5 and 7). As with glycocalicin, the increase in heterogeneity may be due to very acidic material, normally not included within the pH range of the IEF gel, becoming more basic after removal of sialic acid so as to focus within the gel. The reason for the difference in molecular weight may be differential glycosylation, but this remains to be clarified.

Just beneath the VI group on two-dimensional gels (Figure 5) and slightly more basic is a minor glycoprotein that we have termed VII (McGregor *et al.*, 1981). As with the IV group (see above), this is absent in Glanzmann's thrombasthenia and may also be a proteolytic fragment of GP IIb-IIIa.

The nomenclature used for the glycoproteins of lower molecular weight is still controversial and equivalents will be given where possible. Based on one-dimensional techniques, Phillips and Poh-Agin (1977b), using neuraminidase/galactose oxidase/$[^3H]NaBH_4$ labeling, found a single glycoprotein in the actin region that they called VII. This glycoprotein is also recognizable on two-dimensional gels and has been called VIII (Figure 7). Berndt and Phillips (1981a), using periodate/$[^3H]NaBH_4$ labeling, found two bands in this region on one-dimensional gels. The upper of these is obviously VIII (our nomenclature, Figure 5), whereas the lower has been called IX (McGregor *et al.*, 1981). These were also found on two-dimensional gels by Sixma and Schiphorst (1980). who suggested that one of these (IX) might be the HLA (human leukocyte antigen) heavy chains, known to migrate just below actin.

5.8. Glycoproteins of Low Molecular Weight (Less than 43,000)

Several glycoproteins in this region on reduced gels are subunits of larger glycoproteins and have been dealt with earlier (GP Ibβ, Icβ, and IIbβ). There exists, however, a relatively major glycoprotein in this region, present also in nonreduced gels. On one-dimensional gels, this was termed GP IX (Phillips and Poh-Agin, 1977b) and on two-dimensional gels is referred to by its molecular weight and isoelectric point as $GP_{17}{}^{5.8-6.5}$(Clemetson et al., 1982). This is an interesting glycoprotein since it is absent in Bernard-Soulier syndrome and shows heterogeneity in otherwise normal individuals (Clemetson et al., 1982) and is complexed with GP Ib (Berndt et al., 1983; Coller et al., 1983).

Sixma et al. (1982a) have investigated in detail the low-molecular-weight glycoproteins and found 19 different species in this region that could be surface labeled by various techniques. Several of these represent subunits linked by disulfide bonds. It is also possible that some of these lower molecular weight species are proteolytic fragments of larger glycoproteins formed by endogeneous enzymes (McGregor et al., 1983b).

5.9. "High-Molecular-Weight" Glycoproteins

The development of more sensitive detection techniques has necessarily led to the situation that glycoproteins have been detected that have higher molecular weights than GP Ib. These have been described by Sixma and Schiphorst (1980) who found about nine glycoproteins in this region. McGregor et al. (1979a) found four glycoproteins in this higher molecular weight range. On two-dimensional gels, the majority of these glycoproteins appear basic perhaps because their migration in the first-dimension IEF is hindered by the pore size of the polyacrylamide. Almost certainly some of these components are complexes of lower molecular weight glycoproteins such as IIb-IIIa that require strong conditions for complete dissociation.

6. CONCLUSION

Progress in the study of the platelet membrane glycoproteins has been rapid and extensive. Only ten years ago the original investigations defined three glycoproteins in platelet membrane preparations and now 50 recognizable components are known. Determination of the structure of these glycoproteins and their role in platelet function has started and will constitute the exciting work of the next decade.

ACKNOWLEDGMENTS. I am very grateful to Miss Yvonne Fuhrer for typing the manuscript and to Miss Marie-Luise Zahno for help with the figures. I would like to thank Dr. J. L. McGregor and Prof. J. J. Sixma for kindly providing figures. This work was supported by the Swiss National Science Foundation, Grant No. 3.302.082.

REFERENCES

Apitz-Castro, R., Michelena, V. Sorribes, V., and Rodriguez, P., 1979, Metabolic isotopic labelling of plasma membrane glycoproteins from normal, Glanzmann's and Bernard-Soulier platelets, *Thromb. Res.* **16**:715–725.

Banker, G. A., and Cotman, C. W., 1972, Measurement of free electrophoretic mobility and retardation coefficient of protein–sodium dodecyl sulfate complexes by gel electrophoresis, *J. Biol. Chem.* **247:**5856–5861.

Berndt, M. C., and Phillips, D. R., 1981a, Platelet membrane proteins: Composition and receptor function, in: *Platelets in biology and pathology,* Volume 2 (J. Gordon, ed.), Elsevier/North Holland, Amsterdam/New York, pp. 43–75.

Berndt, M. C., and Phillips, D. R., 1981b, Purification and preliminary physiochemical characterization of human platelet membrane glycoprotein V, *J. Biol. Chem.* **256:**59–65.

Berndt, M. C., Gregory, C., Castaldi, P. A., and Zola, H., 1983, Purification of human platelet membrane glycoprotein Ib complex using a monoclonal antibody, *Thromb. Haemostasis* **50:**361.

Bienz, D. E., Clemetson, K. J., and Lüscher, E. F., 1984, Phosphorylation of platelet membrane glycoproteins, *Experientia* **40:**603 (abstr.).

Bolin, R. B., Okumura, T., and Jamieson, G. A., 1977, New polymorphism of platelet membrane glycoproteins, *Nature (London)* **269:**69–70.

Bonner, W. M., and Laskey, R. A., 1974, A film detection method for tritium-labelled proteins and nucleic acids in polyacrylamide gels, *Eur. J. Biochem.* **46:**83–88.

Brass, L. F., and Shattil, S. J., 1983, Identification of the saturable binding sites for Ca^{2+} on the surface of human platelets (abstract), *Thromb. Haemostasis* **50:**327.

Choo, K. H., Cotton, R. G. H., and Danks, D. M., 1980, Double-labeling and high precision comparison of complex protein patterns on two-dimensional polyacrylamide gels, *Anal. Biochem.* **103:**33–38.

Clegg, J. C. S., 1982, Glycoprotein detection in nitrocellulose transfers of electrophoretically separated protein mixtures using concanavalin A and peroxidase: Application to Arenavirus and Flavivirus proteins, *Anal. Biochem.* **127:**389–394.

Clemetson, K. J., Pfueller, S. L., Lüscher, E. F., and Jenkins, C. S. P., 1977, Isolation of the membrane glycoproteins of human blood platelets by lectin affinity chromatography, *Biochim. Biophys. Acta* **464:**493–508.

Clemetson, K. J., Capitanio, A., and Lüscher, E. F., 1979, High resolution two-dimensional gel electrophoresis of the proteins and glycoproteins of human blood platelets and platelet membranes, *Biochim. Biophys. Acta* **553:**11–24.

Clemetson, K. J., Naim, H. Y., and Lüscher, E. F., 1981, Relationship between glycocalicin and glycoprotein Ib of human platelets, *Proc. Natl. Acad. Sci. U.S.A.* **78:**2712–2716.

Clemetson, K. J., McGregor, J. L., James, E., Dechavanne, M., and Lüscher, E. F., 1982, Characterization of the platelet membrane glycoprotein abnormalities in Bernard-Soulier syndrome and comparison with normal by surface-labeling techniques and high-resolution two-dimensional gel electrophoresis, *J. Clin. Invest.* **70:**304–311.

Clemetson, K. J., Wicki, A., and Lüscher, E. F., 1983, Cleavage of platelet membrane glycoprotein Ib by calcium-activated protease: Effect on von Willebrand factor and thrombin-induced aggregation (abstract), *Throm. Haemostasis* **50:**187.

Clezardin, P., McGregor, J. L., James, E., Dechavanne, M., and Clemetson, K. J., 1984, Identification of cytoplasmic and membrane platelet glycoproteins using a combination of SDS-polyacrylamide gel electrophoresis and ^{125}I-labelled lectins, in: *Marker proteins in inflammation,* Volume II (P. Arnaud, J. Bienvenu and P. Laurent, eds.), Walter de Gruyter, Berlin, New York, pp. 601–604.

Coller, B. S., Peerschke, E. I., Scudder, L. E., and Sullivan, C. A., 1983, Studies with a murine monoclonal antibody that abolishes ristocetin-induced binding of von Willebrand factor to platelets: Additional evidence in support of GPIb as a platelet receptor for von Willebrand factor, *Blood* **61:**99–110.

Debray, H., Decout, D., Strecker, G., Spik, and Montreuil, J., 1981, Specificity of twelve lectins toward oligosaccharides and glycopeptides related to *N*-glycosylproteins, *Eur. J. Biochem.* **117:**41–55.

Detwiler, T. J., 1983, Interactions of thrombin with platelets (abstract), *Thromb. Haemostasis* **50:**326.

Durbray, G., and Bezard, G., 1982, A highly sensitive periodic acid-silver stain for 1,2 diol groups of glycoproteins and polysaccharides in polyacrylamide gels, *Anal. Biochem.* **119:**325–329.

Fujimura, K., and Phillips, D. R., 1983a, Binding of $^{45}Ca^{2+}$ to glycoprotein IIb from human platelet plasma membranes (abstract), *Thromb. Haemostasis* **50:**251.

Fujimura, K., and Phillips, D. R., 1983b, Calcium cation regulation of glycoprotein IIb-IIIa complex formation in platelet plasma membranes, *J. Biol. Chem.* **258:**10247–10252.

Fujimura, K., Maehama, S., and Kuramoto, A., 1981, The role of glycoprotein V in thrombin activation of human platelets (abstract), *Thromb. Haemostasis* **46**:107.

Furthmayr, H., 1981, Glycophorin A: A model membrane glycoprotein, in: *Biology of Carbohydrates,* Volume 1, (V. Ginsburg and P. Robbins, eds.), John Wiley and Sons, New York, pp. 123–198.

Gahmberg, C. G., and Hakomori, S.-I., 1973, External labeling of cell surface galactose and galactosamine in glycolipid and glycoprotein of human erythrocytes, *J. Biol. Chem.* **248**:4311–4317.

George, J. N., Potter, R. D., Lewis, P. C., and Sears, D. A., 1976, Studies on platelet plasma membranes. I. Characterization of surface proteins of human platelets labeled with diazotized (^{125}I)-diiodosulfanilic acid, *J. Lab. Clin. Med.* **88**:232–260.

George, J. N., Nurden, A. T. and Phillips, D. R. 1981, Recommendations of the International Committee on Thrombosis and Haemostasis concerning platelet membrane glycoprotein nomenclature, *Thromb. Haemostasis* **46**:764–765.

Gershoni, J. M., and Palade, G. E., 1983, Protein blotting: Principles and applications, *Anal. Biochem.* **131**:1–15.

Ginsberg, M. H., Forsyth, J., Lightsey, A., Chediak, J., and Plow, E. F., 1983, Reduced surface expression and binding of fibronectin by thrombin-stimulated thrombasthenic platelets, *J. Clin. Invest.* **71**:619–624.

Gogstad, G. O., Krutnes, M.-B., and Solum, N. O., 1983, Calcium-binding proteins from human platelets. A study using crossed immunoelectrophoresis and ^{45}Ca^{2+}, *Eur. J. Biochem.* **133**:193–199.

Goldman, R. C., Trus, B. L., and Lieve, L., 1983, Quantitative double-label radiography of two-dimensional protein gels using color negative film and computer analysis, *Eur. J. Biochem.* **131**:473–480.

Greenberg, J. P., Packham, M. A., Guccione, M. A., Harfenist, E. J., Orr, J. L., Kinlough-Rathbone, R. L., Perry, D. W., and Mustard, F. J., 1979, The effect of pretreatment of human or rabbit platelets with chymotrypsin on their responses to human fibrinogen and aggregating agents, *Blood* **54**:753–765.

Hagen, I., 1983, Electroimmunochemical characterization of platelet proteins: Their subcellular location and involvement in platelet adhesion, aggregation and secretion, in: *Electroimmunochemical Analysis of Membrane Proteins* (O. J. Bjerrum, ed.), Elsevier Science Publishers, Amsterdam, New York, pp. 209–222.

Hagen, I., Bjerrum, O. J., Gogstad, G., Korsmo, R., and Solum, N. O., 1982, Involvement of divalent cations in the complex between the platelet glycoproteins IIb and IIIa, *Biochim. Biophys. Acta* **701**:1–6.

Jamieson, G. A., Okumura, T., and Hasitz, M., 1980, Structure and function of platelet glycocalicin, *Thromb. Haemostasis* **42**:1673–1677.

Jenkins, C. S. P., Phillips, D. R., Clemetson, K. J., Meyer, D., Larrieu, M.-J., and Lüscher, E. F., 1976, Platelet membrane glycoproteins implicated in ristocetin-induced aggregation, *J. Clin. Invest.* **57**:112–124.

Jenkins, C. S. P., Ali-Briggs, E. F., Zonneveld, G. T. E., Sturk, A., and Clemetson, K. J., 1980, Human blood platelet membrane glycoproteins: Resolution in different polyacrylamide gel electrophoresis systems, *Thromb. Haemostasis* **42**:1490–1502.

Jenkins, C. S. P., Ali-Briggs, E. F., and Clemetson, K. J., 1983, Antibodies against platelet membrane glycoproteins. III. Effect on thrombin and bovine von Willebrand factor-induced aggregation, *Br. J. Haemotol.* **53**:491–501.

Jennings, L. K., and Phillips, D. R., 1982, Purification of glycoproteins IIb and III from human platelet plasma membranes and characterization of a calcium-dependent glycoprotein IIb-III complex, *J. Biol. Chem.* **257**:10488–10466.

Judson, P. A., and Anstee, D. J., 1979, Characterization of membrane glycoproteins of human platelets, in: *Proceedings of the 27th meeting, Protides of the Biological Fluids* (H. Peeters, ed.), Pergamon, Oxford, pp. 871–874.

Judson, P. A., Anstee, D. J., and Clamp, J. R., 1982, Isolation and characterization of the major oligosaccharide of human platelet membrane glycoprotein GPIb, *Biochem. J.* **205**:81–90.

Konat, G., Offner, H., and Mellah, J., 1984, Improved sensitivity for detection and quantitation of glycoproteins on polyacrylamide gels, *Experientia* **40**:303–304.

Kornecki, E., Tuszynski, G. P., and Niewiarowski, S., 1983, Inhibition of fibrinogen receptor-mediated platelet aggregation by heterologous anti-human platelet membrane antibody: Significance of an M_r = 66,000 protein derived from glycoprotein IIIa, *J. Biol. Chem.* **258**:9349–9356.

Kornfeld, K., Reitman, M. L., and Kornfeld, R., 1981, The carbohydrate-binding specificity of pea and lentil lectins. Fucose is an important determinant, *J. Biol. Chem.* **256:**6633–6640.

Korrel, S. A. M., Clemetson, K. J., van Habeek, H., Kamerling, J. P., Sixma, J. J., and Vliegenthart, J. F. G., 1984, Structural studies of the O-linked carbohydrate chains of human platelet glycocalicin, *Eur. J. Biochem.,* **140:**571–576.

Kunicki, T. J., Pidard, D., Rosa, J.-P., and Nurden, A. T., 1981a, The formation of Ca^{2+}-dependent complexes of platelet membrane GPIIb and IIIa in solution as determined by crossed immunoelectrophoresis, *Blood* **58:**268–278.

Kunicki, T. J., Christie, D. J., Winkelhake, J. L., and Aster, R. H., 1981b, Alcian blue as a protein stain for sodium alkyl sulfate-polyacrylamide gels: Application to analysis of polypeptide components of the human platelet, *Anal. Biochem.* **110:**412–419.

Laskey, R. A., and Mills, A. D., 1975, Quantitative film detection of ^3H and ^{14}C in polyacrylamide gels by fluorography, *Eur. J. Biochem.* **56:**335–341.

Laskey, R. A., and Mills, A. D., 1977, Enhanced autoradiographic detection of ^{32}P and ^{125}I using intensifying screens and hypersensitized film, *Febs Lett.* **82:**314–316.

Lawler, J. W., Margossian, S., and Slayter, H. S., 1980, Physical parameters of platelet glycocalicin (abstract), *Fed. Proc. Fed. Am. Soc. Exp. Biol.* **39:**1895.

Leung, L. L. K., Kinoshita, T., and Nachman, R. L., 1981, Isolation, purification and partial characterization of platelet membrane glycoproteins IIb and IIIa, *J. Biol. Chem.* **256:**1994–1997.

Liao, T.-H., Gallop, P. M., and Blumenfeld, O. O., 1973, Modification of sialyl residues of sialoglycoprotein(s) of the human erythrocyte surface, *J. Biol. Chem.* **248:**8247–8253.

Marchesi, S. L., and Chasis, J. A., 1979, Isolation of human platelet glycoproteins, *Biochim. Biophys. Acta* **555:**442–459.

McEver, R. P., Baenziger, J. U., and Majerus, P. W., 1982, Isolation and structural characterization of the polypeptide subunits of membrane glycoprotein IIb-IIIa from human platelets, *Blood* **59:**80–85.

McEver, R. P., Bennett, E. M., and Martin, M. N., 1983, Identification of two structurally and functionally distinct sites on human platelet membrane glycoprotein IIb-IIIa using monoclonal antibodies, *J. Biol. Chem.* **258:**5269–5275.

McEver, R. P., and Martin, M. N., 1984, A monoclonal antibody to platelet membrane GPIIa binds only to activated platelets, *J. Biol. Chem.* **259:**9799–9804.

McGregor, J. L., Clemetson, K. J., James, E., and Dechavanne, M., 1979a, A comparison of techniques used to study externally oriented proteins and glycoproteins of human blood platelets, *Thromb. Res.* **16:**437–452.

McGregor, J. L., Clemetson, K. J., James, E., Greenland, T., and Dechavanne, M., 1979b, Identification of human platelet glycoproteins in SDS-polyacrylamide gels using ^{125}I-labelled lectins, *Thromb. Res.* **16:**825–831.

McGregor, J. L., Clemetson, K. J., James, E., Lüscher, E. F., and Dechavanne, M., 1980, Characterization of human blood platelet membrane proteins and glycoproteins by their isoelectric point (pI) and apparent molecular weight using two-dimensional electrophoresis and surface-labelling techniques. *Biochim. Biophys. Acta* **599:**473–483.

McGregor, J. L., Clemetson, K. J., James, E., Capitanio, A., Greenland, T., Lüscher, E. F., and Dechavanne, M., 1981, Glycoproteins of platelet membranes from Glanzmann's thrombasthenia: A comparison with normal using carbohydrate or protein-specific labelling techniques and high-resolution, two-dimensional gel electrophoresis, *Eur. J. Biochem.* **116:**379–388.

McGregor, J. L., Clemetson, K. J., James, E., Clezardin, P., Dechavanne, M., and Lücher, E. F., 1982, Tryptic peptide map analysis of the major human blood platelet membrane glycoproteins separated by two-dimensional polyacrylamide gel electrophoresis, *Biochim. Biophys. Acta* **689:**513–522.

McGregor, J. L., Brochier, J., Wild, F., Follea, G., Trzeciak, M.-C., James, E., Dechavanne, M., McGregor, L., and Clemetson, K. J., 1983a, Monoclonal antibodies against platelet membrane glycoproteins: Characterization and effect on platelet function, *Eur. J. Biochem.* **131:**427–436.

McGregor, J. L., Clezardin, P., James, E., McGregor, L., Dechavanne, M., and Clemetson, K. J., 1983b, Identification and characterization of fragments of major membrane glycoproteins remaining on platelets after chymotrypsin treatment (abstract), *Thromb. Haemostasis* **50:**187.

Merril, C. R., Goldman, D., and Van Keuren, M. L., 1982, Simplified silver protein detection and image enhancement methods in polyacrylamide gels, *Electrophoresis* **3:**17–23.

Moroi, M., Goetze, A., Dubay, E., Wu, C., Hasitz, M., and Jamieson, G. A., 1982, Isolation of platelet glycocalicin by affinity chromatography on thrombin-Sepharose, *Thromb. Res.* **28:**103–114.

Moroi, M., and Jung, S. M., 1984, Selective staining of human platelet glycoproteins using nitrocellulose transfer of electrophoresed proteins and peroxidase-conjugated lectins. *Biochim. Biophys. Acta* **798:**295–301.

Mosher, D. F., Vaheri, A., Choate, J. J., and Gahmberg, C. G., 1979, Action of thrombin on surface glycoproteins of human platelets, *Blood* **53:**437–445.

Nachman, R. L., and Ferris, B., 1972, Studies on the proteins of human platelet membranes, *J. Biol. Chem.* **247:**4468–4475.

Nachman, R. L., and Leung, L. L. K., 1982, Complex formation of platelet membrane glycoproteins IIb and IIIa with fibrinogen, *J. Clin. Invest.* **69:**263–269.

Nachman, R. L., Hubbard, A., and Ferris, B., 1973, Iodination of the human platelet membrane: Studies of the major surface glycoprotein, *J. Biol. Chem.* **248:**2928–2936.

Nachman, R. L., Jaffe, E. A., and Weksler, B. B., 1977, Immunoinhibition of ristocetin-induced platelet aggregation, *J. Clin. Invest.* **59:**143–148.

Newman, P. J., and Kahn, R. A., 1983, Purification of human platelet membrane glycoproteins IIb and IIIa using high-performance liquid chromatography gel filtration, *Anal. Biochem.* **132:**215–218.

Niewiarowski, S., Budzynski, A. Z., Morinelli, T. A., Brudzinski, T. M., and Stewart, G. J., 1981, Exposure of fibrinogen receptor on human platelets by proteolytic enzymes, *J. Biol. Chem.* **256:**917–925.

Nurden, A. T., 1974, Platelet macroglycopeptide, *Nature (London)* **251:**151–153.

Oakley, B. R., Kirsch, D. R., and Morris, N. R., 1980, A simplified ultrasensitive silver stain for detecting proteins in polyacrylamide gels, *Anal. Biochem.* **105:**316–363.

O'Farrell, P. H., 1975, High resolution two-dimensional electrophoresis of proteins, *J. Biol. Chem.* **250:**4007–4021.

Okumura, T., and Jamieson, G. A., 1976, Platelet glycocalicin. I. Orientation of glycoproteins of the human platelet surface, *J. Biol. Chem.* **251:**5944–5949.

Okumura, T., Lombart, C., and Jamieson, G. A., 1976, Platelet glycocalicin. II. Purification and characterization, *J. Biol. Chem.* **251:**5950–5955.

Owen, M. J., Kissonerghis, A.-M., and Lodish, H. F., 1980, Biosynthesis of HLA-A and HLA-B antigens *in vivo*, *J. Biol. Chem.* **255:**9678–9684.

Pepper, D. S., and Jamieson, G. A., 1969, Studies on glycoproteins. III. Isolation of sialylglycopeptides from human platelet membranes, *Biochemistry* **8:**3362–3369.

Pepper, D. S., and Jamieson, G. A., 1970, Isolation of a macroglycopeptide from human platelets, *Biochemistry* **9:**3706–3713.

Pereira, M. E. A., Kabat, E. A., Lotan, R., and Sharon, N., 1976, Immunochemical studies on the specificity of the peanut (*Arachis hypogaea*) agglutinin, *Carbohydr. Res.* **51:**107–118.

Peterson, D. M., Hirst, A., and Wehring, B., 1982, Comparison of normal and Bernard-Soulier platelet membrane glycoproteins, *J. Lab. Clin. Med.* **100:**26–36.

Phillips, D. R., 1972, Effect of trypsin on the exposed polypeptides and glycoproteins in the human platelet membrane, *Biochemistry* **11:**4582–4588.

Phillips, D. R., and Poh-Agin, P., 1977a, Platelet plasma membrane glycoproteins: Evidence for the presence of nonequivalent disulphide bonds using nonreduced-reduced, two-dimensional gel electrophoresis, *J. Biol. Chem.* **252:**2121–2126.

Phillips, D. R., and Poh-Agin, P., 1977b, Platelet plasma membrane glycoproteins: Identification of a proteolytic substrate for thrombin, *Biochem. Biophys. Res. Commun.* **75:**940–947.

Phillips, D. R., Jenkins, C. S. P., Lüscher, E. F., and Larrieu, M.-J., 1975, Molecular differences of exposed surface proteins on thrombasthenic platelet plasma membranes, *Nature (London)* **257:**599–600.

Rotman, A., 1980, Labelling and separation fo platelet membrane proteins, in: *Platelets: Cellular Responses Mechanisms and Their Biological Significance* (A. Rotman, F. A. Meyer, C. Gitler, and A. Silberberg, eds.), John Wiley and Sons, New York, Chichester, pp. 107–118.

Rotman, A., Helman, J., and Linder, S., 1982, Association of membrane and cytoplasmic proteins with the cytoskeleton in blood platelets, *Biochemistry* **21:**1713–1719.

Ruggeri, Z. M., Bader, R., and De Marco, L., 1982, Glanzmann's thrombasthenia: Deficient binding of von Willebrand factor to thrombin-stimulated platelets, *Proc. Natl. Acad. Sci. U.S.A.* **79:**6038–6041.

Sammons, D. W., Adams, L. D., and Nishezawa, E. E., 1981, Ultrasensitive silver-based color staining of polypeptides in polyacrylamide gels, *Electrophoresis* **2:**135–141.

Sears, D. A., Friedman, J. M., and George, J. N., 1977, Topography of the external surface of the human red blood cell membrane studied with a nonpenetrating label, [^{125}I]-diazodiiodosulfanilic acid, *J. Biol. Chem.* **252:**712–720.

Segrest, J. P., Jackson, R. L., Andrews, E. P., and Marchesi, V. T., 1971, Human erythrocyte membrane glycoprotein: A re-evaluation of the molecular weight as determined by SDS-polyacrylamide gel electrophoresis, *Biocheim. Biophys. Res. Commun.* **44:**390–395.

Shulman, S., Wiesner, R., Troll, W., and Karpatkin, S., 1983, Reevaluation of the presence of the major antigen Ca^{2+} complex in Bernard-Soulier syndrome platelets, *Thromb. Res.* **30:**61–69.

Sixma, J. J., and Schiphorst, M. E., 1980, Identification of ectoproteins of human platelets: Combination of radioactive labelling and two-dimensional electrophoresis, *Biochim. Biophys. Acta* **603:**70–83.

Sixma, J. J., Schiphorst, M. E., and Verhoeckx, C., 1982a, Identification of ectoproteins of human platelets. II. Proteins of low molecular weight (10,000–43,000), *Biochim. Biophys. Acta* **687:**97–100.

Sixma, J. J., Schiphorst, M. E., Verhoeckx, C., and Jockusch, B. M., 1982b, Peripheral and integral proteins of human blood platelet membranes: α-Actinin is not identical to glycoprotein III, *Biochim. Biophys. Acta* **704:**333–344.

Solum, N. O., Hagen, I., and Peterka, M., 1977, Human platelet glycoproteins. Further evidence that the "GPI band" from whole platelets contains three different polypeptides one of which may be involved in the interaction between platelets and Factor VIII, *Thromb. Res.* **10:**71–82.

Solum, N. O., Hagen, I., Filion-Myklebust, C., and Stabaek, T., 1980a, Platelet glycocalicin: Its membrane association and solubilization in aqueous media, *Biocheim. Biophys. Acta* **597:**235–246.

Solum, N. O., Hagen, I., and Sletbakk, T., 1980b, Further evidence for glycocalicin being derived from a larger amphiphilic platelet membrane glycoprotein, *Thromb. Res.* **18:**773–785.

Solum, N. O., Olsen, T. M., Gogstad, G. O., Hagen, I., and Brosstad, F., 1983, Demonstration of a new glycoprotein Ib-related component in platelet extracts prepared in the presence of leupeptin, *Biochim. Biophys. Acta* **729:**53–61.

Solum, N. O., and Olsen, T. M., 1984, GPIb in the Triton-insoluble (cytoskeletal) fraction of blood platelets, *Biochim. Biophys. Acta* **799:**209–220.

Soslau, G., and Rybicki, A., 1982, *In vitro* incorporation of fucose and methionine into human platelets, *Biochem. Biophys. Res. Commun.* **109:**1256–1263.

Spycher, M. O., Clemetson, K. J., and Lüscher, E. F., 1982, Surface labeling of human platelets by reductive methylation, *Thromb. Haemostasis* **48:**169–172.

Steck, T. L., and Dawson, G., 1974, Topographical distribution of complex carbohydrates in the erythrocyte membrane, *J. Biol. Chem.* **249:**2135–2142.

Steiner, M., and Lüscher, E. F., 1982, Photoaffinity labeling of IgG receptor on human platelets (abstract), *Blood* **60:**192a.

Subbarao, K., Kakkar, V. V., and Ganguly, P., 1978, Binding of pyridoxal phosphate to human plateletes: Its effect on platelet function, *Thromb. Res.* **13:**1017–1029.

Swanstrom, R., and Shank, P. R., 1978, X-ray intensifying screens greatly enhance the detection by autoradiography of the radioactive isotopes ^{32}P and ^{125}I, *Anal. Biochem.* **86:**184–192.

Tobelem, G., Levy-Toledano, S., Bredoux, R., Michel, H., Nurden, A., and Caen, J. P., 1976, New approach to determination of specific functions of platelet membrane sites, *Nature (London)* **263:**427–428.

Tsuji, T., Tsunehisa, S., Watanabe, Y., Yamamoto, K., Tohyama, H., and Osawa, T., 1983, The carbohydrate moiety of human platelet glycocalicin: The structure of the major Ser/Thr-linked chain, *J. Biol. Chem.* **258:**6335–6339.

Tuszynski, G. P., Knight, L. C., Kornecki, E., and Srivastava, S., 1983, Labeling of platelet surface proteins with ^{125}I by the Iodogen method, *Anal. Biochem.* **130:**166–170.

Van Keuren, M. L., Goldman, D., and Merril, C. R., 1981, Detection of radioactively labeled proteins is quenched by silver staining methods: Quenching is minimal for ^{14}C and partially reversible for ^{3}H with a photochemical stain, *Anal. Biochem.* **116:**248–255.

Weber, K., and Osborn, M., 1969, The reliability of molecular weight determination by dodecyl sulfate-polyacrylamide gel electrophoresis, *J. Biol. Chem.* **244:**4406–4412.

4

Organization of Glycoproteins within the Platelet Plasma Membrane

Thomas J. Kunicki

I. INTRODUCTION

Many processes within the eukaryotic cell membrane require the formation of functional macromolecular complexes from individual membrane or cytoplasmic proteins. In light of accumulating evidence that specific membrane glycoproteins function as receptors in critical stages of the platelet adhesion and aggregation mechanisms, there is currently much interest in the organization of proteins within the human platelet membrane and the changes that occur in this organization during platelet function. Within the last ten years, analytical methods such as sodium dodecyl sulfate–polyacrylamide gel electrophoresis (SDS–PAGE) in one or two dimensions (Phillips and Poh Agin, 1977a) and combined isoelectric focusing–SDS/PAGE (Clemetson *et al.*, 1979) have proved to be extremely valuable for the resolution, identification, and characterization of membrane glycoproteins (see Chapter 3). Thus, Phillips and Poh Agin (1977a) were able to establish molecular criteria for the identification of seven "major" membrane glycoproteins based on their differential mobilities in the presence or absence of disulfide bond reduction. When such analytical methods were used to compare platelet proteins from normal individuals and patients with inherited abnormalities of platelet function, the link between certain membrane glycoproteins and selected platelet functions was established. For example, Glanzmann's thrombasthenia, which is characterized by defective platelet cohesion, is associated with the absence of or a qualitative defect in glycoproteins (GP) IIb and IIIa (Nurden and Caen, 1974; Phillips and Poh Agin, 1977b; Lightsey *et al.*, 1981). Another disorder, the Bernard-Soulier syndrome, which is characterized by a defective platelet-vessel wall

Thomas J. Kunicki • The Blood Center of Southeastern Wisconsin, Inc., Milwaukee, Wisconsin 53233.

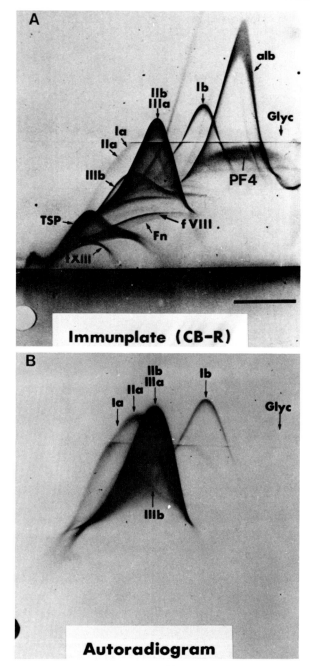

Figure 1. CIE of Triton X-100 lysates derived from ^{125}I-labeled, washed, normal human platelets. One hundred micrograms of soluble platelet protein were electrophoresed against polyspecific rabbit anti-human platelet antibody (IgG fraction). (A) CIE plate stained with Coomassie Blue-R (CB-R). (B) Autoradiograph of CIE plate in (A). Immunoprecipitin arcs containing selected platelet antigens are indicated: albumin (alb); glycocalicin (Glyc); fibrinogen (F); fibronectin (Fn); factor VIII–VWF (fVIII); factor XIII-subunit A (fXIII); platelet-factor 4 (PF4); thrombospondin (TSP); and membrane GP Ia-IIa, Ib, IIb-IIIa, and IIIb. Bar at lower right-hand corner of (A) represents 1 cm.

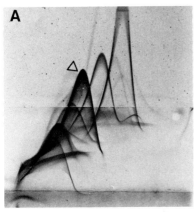
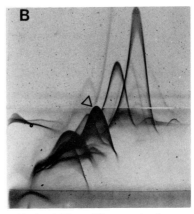

Figure 2. CIE of platelet lysates from (A) a normal individual and (B) an obligate carrier (heterozygote) of Glanzmann's thrombasthenia. Both plates are stained with Coomassie Blue-R. Open arrowhead indicates immunoprecipitin arc given by the GP IIb-IIIa complex. Triangulation of respective IIb-IIIa arcs shows that the sample from the obligate carrier of thrombasthenia (B) contains approximately 50% of the amount of IIb-IIIa detected in the sample from a control subject (A). Note that the height and area of all other immunoprecipitin arcs in both (A) and (B) are essentially identical.

adhesion and giant platelets, is associated with the absence of GP Ib, GP V, and GP IX (Nurden and Caen, 1975; Nurden *et al.,* 1981; Clemetson *et al.,* 1982). These disorders are discussed in detail in Chapter 16.

Under the denaturing conditions inherent in these analytical methods, membrane protein antigenicity is often lost, and noncovalent protein associations are irreversibly inhibited. While sacrificing, to some degree, the high resolution afforded by SDS–PAGE, the method of crossed immunoelectrophoresis (CIE) in agarose gels in the presence of nonionic detergents allows an equivalent degree of reproducibility and maintenance of both protein antigenicity and the potential of noncovalent interactions (Figure 1; see also Chapter 5). Crossed immunoelectrophoresis also offers the ability to obtain quantitative estimates of membrane antigens directly without resorting to the incorporation of radioisotopes or other extrinsic labels (Figure 2). It is thus not surprising that the first evidence for a specific interaction between platelet membrane glycoproteins was derived from CIE of Triton X-100 lysates.

2. GLYCOPROTEINS IIb–IIIa

2.1. Initial Detection of the Glycoprotein IIb–IIIa Complex in Platelet Lysates

In the initial studies of whole platelet lysates by CIE, GP IIb and GP IIIa were both eluted from a single precipitin arc, designated "protein 16" by Hagen *et al.* (1979, 1980) and "major antigen" or "protein 10" by Shulman and Karpatkin (1980). Under unspecified conditions, decreased levels of this GP IIb–IIIa precipitin arc were associated with the appearance of two new precipitin arcs (Hagen *et al.,* 1979). Kunicki *et al.* (1981a) demonstrated that protein 16 actually represented a calcium-dependent complex of GP IIb and GP IIIa that could be readily dissociated by

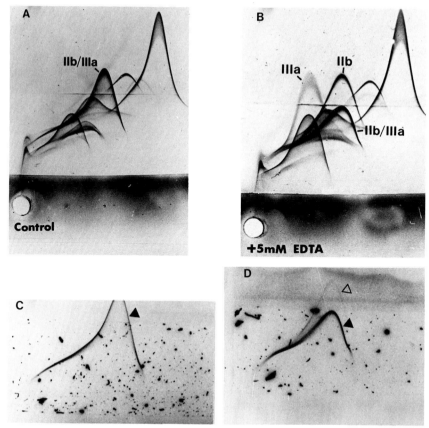

Figure 3. Effect of EDTA upon the association of GP IIb and IIIa as determined by CIE (A,B). Specificity of L . . . IgG determined by CIE (C,D). In platelet lysates containing endogenous calcium (A), IIb and IIIa exist as a complex (compare Figures 1 and 2). Lysis of platelets in the presence of excess EDTA (B) results in the appearance of distinct immunoprecipitin arcs containing free IIb, free IIIa, and residual IIb-IIIa complexes. Coomassie Blue-stained plates are shown in (A) and (B). L . . . IgG was labeled with ^{125}I and incorporated into the intermediate gels of plates depicted in (A) and (B). Autoradiographs of (A) and (B) are shown in (C) and (D), respectively. Thus, L . . . IgG binds weakly to free IIb (open arrowhead in D) but strongly to the IIb-IIIa complex (solid arrowheads in C and D).

addition of EDTA or EGTA during detergent solubilization (Figure 3). The two new precipitin arcs formed from the dissociation of the GP IIb–IIIa complex contained free GP IIb and free GP IIIa, respectively. Readdition of excess Ca^{2+}, but not Mg^{2+}, resulted in reformation of GP IIb–IIIa complexes and the reappearance of a single GP IIb–IIIa precipitin arc. This calcium-mediated interaction between GP IIb and GP IIIa was subsequently confirmed by several laboratories, using CIE as well as other analytical methods (Hagen *et al.,* 1982; Jennings and Phillips, 1982; Pidard *et al.,* 1982; Howard *et al.,* 1982). Such studies have indicated that these complexes are probably heterodimers (Jennings and Phillips, 1982; Pidard *et al.,* 1982).

The question arises whether or not such complexes pre-exist in nonactivated (and, therefore, noncohesive) platelets. The answer to this question can only be satisfactorily obtained from an analysis of the nature and extent of GP IIb–IIIa associations within the membrane of intact platelets and of the effect of platelet activation on such interrelationships. Two possibilities are immediately apparent: either (1) GP IIb and GP IIIa are not complexed in resting platelets and associate in a calcium-dependent manner as a result of platelet activation, or (2) GP IIb–IIIa complexes pre-exist in resting platelets in a nonactivated conformation that is transformed to an active conformation by physiologic agonists (see Chapter 11, Figure 5).

2.2. Glycoprotein IIb–IIIa and Fibrinogen

In view of accumulated evidence that glycoproteins IIb and/or IIIa represent the membrane receptor for fibrinogen, which is a requisite cofactor in platelet–platelet cohesion (Bennett and Vilaire, 1979; Mustard *et al.*, 1979; Nachman and Leung, 1982; Gogstad *et al.*, 1982; Bennett *et al.*, 1982; see also Chapter 9), it was logical to postulate that calcium-mediated GP IIb–IIIa complex formation represents the molecular basis of fibrinogen receptor induction. Certain lines of evidence strongly implicate the complex, and neither free GP IIb nor free GP IIIa, as the fibrinogen receptor. In CIE, labeled fibrinogen binds only to the GP IIb–IIIa complex immunoprecipitin arc (Gogstad *et al.*, 1982). In solid-phase studies, fibrinogen bound to microtiter wells binds GP IIb–IIIa complexes, but does not react with the individual GP IIb or GP IIIa components (Leung *et al.*, 1981). Finally, murine monoclonal antibodies or rabbit antibodies specific for either GP IIb or GP IIIa fail to inhibit fibrinogen binding to activated platelets and do not inhibit platelet aggregation (Leung *et al.*, 1981; McEver *et al.*, 1980), whereas GP IIb–IIIa complex-specific monoclonal antibodies, which do not bind to free GP IIb or free GP IIIa, are potent inhibitors of both fibrinogen binding and platelet aggregation (McEver *et al.*, 1983; Pidard *et al.*, 1983). Additional monoclonal antibodies that inhibit fibrinogen binding as well as fibrinogen-dependent platelet aggregation have been characterized (Bennett *et al.*, 1983; Coller *et al.*, 1983a; DiMinno *et al.*, 1983). Although it would be logical to conclude that these monoclonal antibodies also recognize GP IIb–IIIa complex-dependent epitopes, the methods used to characterize their specificity do not as yet permit this distinction. Finally, antisera produced by a patient with Glanzmann's thrombasthenia (IgG L) and previously shown to effectively inhibit fibrinogen binding and subsequent platelet aggregation (Lee *et al.*, 1981) have subsequently been shown by CIE to contain predominantly antibodies directed against GP IIb–IIIa complex-associated epitopes (Figure 3) (Rosa *et al.*, 1984).

2.3. Organization of Glycoprotein IIb and Glycoprotein IIIa in Unstimulated and Stimulated Platelets

2.3.1. Electron Microscopy Studies

Taken together, the preceding results certainly support the theory that GP IIb-IIIa complexes and not the dissociated glycoproteins are required for fibrinogen binding and platelet–platelet cohesion. Several laboratories have undertaken the study of membrane glycoprotein–glycoprotein interactions within intact human platelets, with par-

ticular emphasis on an analysis of GP IIb and GP IIIa. Using immunoelectron microscopy, Polley *et al.* (1981) analyzed the distribution of GP IIb and GP IIIa in the plasma
membrane from stimulated and unstimulated platelets. Glycoprotein IIIa was localized
by the binding of ferritin-conjugated goat anti-rabbit gamma globulin to prebound
rabbit F(ab')$_2$ specific for GP IIIa. Glycoprotein IIb was localized by the direct binding
of keyhole-limpet hemocyanin-conjugated rabbit F(ab')$_2$ specific for GP IIb. In double-label experiments, GP IIb and GP IIIa were found to be randomly distributed in
nonstimulated platelets, but to codistribute in large clusters following thrombin activation. In the same fashion, GP Ib appeared to be organized in "microclusters" in the
resting platelet and in larger clusters in thrombin-activated platelets. Double labeling of
GP Ib and GP IIb indicated no codistribution of these glycoproteins in the respective
macroclusters. This study was certainly an elegant attempt to analyze the spatial
organization of GP Ib, GP IIb, and GP IIIa within intact platelets. Nevertheless, certain
limitations of the method employed must be considered in an interpretation of the
results. First, bivalent antibody fragments were chosen as probes raising the possibility
that the probes themselves were responsible for glycoprotein crosslinking and cluster
formation. Second, the size of the probes themselves, as well as that of the macromolecular electrodense conjugates, eliminated any assurance that the binding of anti-
GP IIb and anti-GP IIIa were not mutually inhibitory. Thus, it was possible that only
one or the other probe could bind to a pre-existing GP IIb-IIIa complex at one time.
Although it was concluded from these results that thrombin activation induced the
formation of GP IIb-IIIa from dissociated GP IIb and dissociated GP IIIa, the possibility that aggregation of pre-existing GP IIb-IIIa complexes was actually observed
could not be eliminated.

2.3.2. Chemical Cross-Linking Studies

In an alternative approach, Davies and Palek (1982) elected to examine platelet
protein organization by exposing nonstimulated and thrombin-activated intact platelets
to the chemical cross-linking reagents, diamide and dithiobis (succinimidyl propionate)
(DTSP). Covalently associated complexes putatively generated by chemical bridging
were resolved by SDS–PAGE in the absence of reducing agent and their composition
determined after reductive cleavage in a second-dimension gel. This approach has the
theoretical advantage that pre-existing noncovalent interactions are artificially made
covalent and can thus be distinguished from noncovalent interactions that occur after
platelet disruption. In nonstimulated platelets, several complexes of reproducible composition were obtained and DTSP, in particular, was shown to cross-link trace amounts
of GP IIb and GP IIIa into high molecular weight ($> 10^6$ daltons). Thrombin activation
prior to DTSP treatment resulted in a net increase in the amount of GP IIb and GP IIIa
incorporated into high-molecular-weight complexes. Additional DTSP-induced complexes of intermediate size (200,000–240,000 daltons) composed of GP IIb and GP
IIIa, as well as glycoproteins Ia and IIa, were also detected. It was concluded that
thrombin activation induced the formation of heteromultimers of GP IIb and GP IIIa,
GP Ia dimers, GP Ia-IIa heteromultimers, and an association of GP IIb-IIIa with the
cytoskeleton. Since platelets were solubilized directly in SDS, nonspecific interactions
that might occur following nonionic detergent lysis would have been eliminated. These

results were interpreted to confirm the hypothesis that GP IIb and GP IIIa exist as free entities in the membrane of nonstimulated platelets and that GP IIb-IIIa complex formation is a specific consequence of activation. However, the use of chemical cross-linking reagents to determine protein associations within intact cells is restricted by the specificity, solubility, and size of the cross-linking agent itself, and the relative accessibility of the proteins in question to the probe under the different conditions tested. In order that cross-linking occur, the necessary functional groups on the proteins must be properly oriented and aligned so that the particular cross-linking reagent can link them. More importantly, in order to conclude that the degree of cross-linking between two proteins is a true reflection of their relative juxtaposition under two different sets of conditions, it is necessary to first establish that both proteins are equally reactive with the reagent under both conditions. In the preceding study (Davies and Palek, 1982), the relative binding of DTSP to GP IIb and GP IIIa in unstimulated as opposed to stimulated platelets was not determined. Therefore, these results are not conclusive evidence against the absence of GP IIb-IIIa complexes in unstimulated platelets.

One additional but not unimportant limitation of the chemical cross-linking approach is the inability to resolve high-molecular-weight complexes (e.g., $> 10^6$ daltons) in standard acrylamide electrophoresis gels. Consequently, unrelated complexes of varying size and differing protein composition may comigrate as a single dense band that enters the gel (Davies and Palek, 1982). Resolution of such complexes could be facilitated by the use of SDS–agarose or SDS–mixed agarose/acrylamide gels such as those developed for von Willebrand factor (VWF) multimer analysis (Ruggeri and Zimmerman, 1981; Montgomery and Johnson, 1982).

2.3.3. Studies Using Monoclonal Antibodies

Within the past year, the development and characterization of murine monoclonal antibodies specific for various epitopes on GP IIb and/or GP IIIa has provided an additional approach for the analysis of the juxtaposition of these glycoproteins in the membrane of intact platelets. In our laboratory, extensive analyses have been performed with one such antibody, designated AP-2, that binds to the GP IIb-IIIa complex, but not to dissociated GP IIb or dissociated GP IIIa (Pidard et al., 1983; Montgomery et al., 1983). Comparative studies have been carried out with a second monoclonal antibody specific for GP Ib, designated AP-1 (Montgomery et al., 1983). The effects of each monoclonal antibody on selected platelet functions are listed in Table 1. Both antibodies are subclass IgG1 and contain exclusively kappa chains.

The steady-state binding of purified ^{125}I-labeled monoclonal IgG to intact washed platelets was evaluated. AP-1 was shown to bind to 34,000 ± 5000 (mean ± SD; n = 7) sites with a Kd of 2.1 nM; AP-2 bound to 54,000 ± 7000 sites (n = 6) with a Kd of 0.5 nM. Identical results were obtained using platelets thrice washed in buffer or platelets isolated by albumin density gradient centrifugation. As determined by CIE, AP-1 recognizes an epitope on GP Ib and its proteolytic product, glycocalicin, whereas AP-2 binds only to the GP IIb-IIIa complex, but not to dissociated GP IIb or dissociated GP IIIa (Figures 4 and 5). The number of AP-1 and AP-2 molecules bound per platelet did not differ significantly using washed platelets derived from whole blood anticoagulated by EDTA, citrate, ACD, or heparin. Purified monoclonal IgG coupled

Table 1. Characterization of AP-1 and AP-2

	AP-1	AP-2
Binding to:		
Normal platelets	Yes	Yes
Thrombasthenic platelets	Yes	No
Bernard-Soulier platelets	No	Yes
Inhibition of:		
Ristocetin-induced agglutination	Yes	No
Fibrinogen binding	No	Yes
Ristocetin-induced VWF binding	Yes	No
Thrombin-induced VWF binding	No	Yes
Clot retraction	No	No
Shape change	Yes	No
Platelet lysis by anti-Pl^{A1}	No	No
Platelet lysis by drug-dependent antibodies	Yes (20%)	No
Direct aggregation of citrated-PRP	No	No
Inhibition of aggregation of citrated-PRP induced by:		
Arachidonic acid	No	Yes
ADP	No	Yes
Epinephrine	No	Yes
Thrombin	Yes	Yes
Collagen	Yes	Yes

to Sepharose was used to isolate respective antigens from Triton X-100 lysates of whole platelets. Thus, the GP IIb-IIIa complex was purified to homogeneity using AP-2–Sepharose.

The question arose whether AP-2 recognizes a complex-dependent epitope or a calcium-dependent epitope, neither of which can be excluded by the CIE data obtained in the presence or absence of EDTA/EGTA. To address this question, we performed

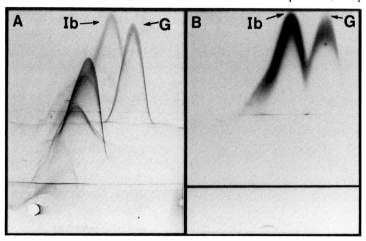

Figure 4. Specificity of AP-1 determined by CIE. ^{125}I-labeled AP-1 was incorporated into the intermediate gels of CIE plates in which lysates (100 μg protein) derived from thrice washed platelets were electrophoresed against polyspecific rabbit anti-human platelet antibody. A Coomassie Blue-R stained gel is shown in (A); the corresponding autoradiograph in (B). The positions of precipitin arcs containing GP Ib (Ib) and glycocalicin (G) are indicated.

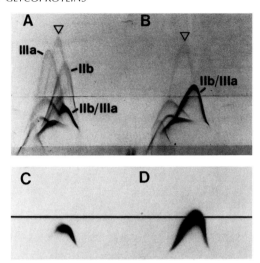

Figure 5. Specificity of AP-2 determined by CIE. ^{125}I-labeled AP-2 was incorporated into the intermediate gels of CIE plates in which soluble protein (20 μg), derived from isolated platelet membranes lysed in the presence of 1 mM EDTA (A,C) or 0.5 mM calcium chloride (B,D), were electrophoresed against poly-specific rabbit anti-human platelet antibody. (A) and (B) represent the Coomassie Blue-R stained plates; (C) and (D) represent autoradiographs of (A) and (B), respectively. Note that AP-2 binds solely to the IIb-IIIa complex immunoprecipitin arc and neither to the free IIb arc nor the free IIIa arc. Open arrowhead in (A) and (B) indicates the position of the GP Ia-IIa precipitin arc (see Figure 7) that can be used as a reference point. Reprinted from Pidard *et al.* (1984) with permission.

CIE in the presence of limiting amounts of calcium to obtain precipitates representing dissociated GP IIb, dissociated GP IIIa, and the GP IIb-IIIa complex. The unstained gels were then overlaid with a buffer solution containing [^{125}I]AP-2 plus 1 mM CaCl$_2$ and incubated for 72 hr in a humidified chamber. The binding of [^{125}I]AP-2 was then analyzed by autoradiography. Under these conditions, i.e., in the presence of dissociated GP IIb, dissociated GP IIIa, *and* calcium ions, AP-2 still bound only to the GP IIb-IIIa complex, indicating that the epitope recognized by this monoclonal antibody was dependent on GP IIb-IIIa association. In further studies using intact platelets, neither the number of AP-2 molecules bound per platelet nor the dissociation constant were shown to be affected by the presence of 3 mM CaCl$_2$ or 5 mM EDTA (Table 2) when studies were performed at 22°C. This finding is consistent with the conclusion that the epitope recognized by AP-2 on the surface of intact platelets is not calcium dependent at that temperature. On the other hand, when intact platelets are incubated at 37°C in the presence of 5 mM EDTA, one observes a time-dependent loss of AP-2 binding which parallels a decrease in the ability to bind fibrinogen and a loss of the GP IIb-IIIa complex, as determined by CIE (D. Pidard, unpublished observations). Thus, there is apparently a temperature-dependent difference in the calcium-mediated stability of the GP IIb-IIIa complex. Whether AP-2 binds to a combinatorial determinant (a site composed of segments of each of GP IIb and GP IIIa) or to a conformational determinant (a site[s] on GP IIb and/or GP IIIa that is exposed only when the one glycoprotein is bound to the other) remains to be determined and is currently under investigation.*

*D. Pidard, D. Didry, T. J. Kunicki, and A. P. Nurden, Temperature-dependent effects of EDTA on the membrane glycoprotein IIb-IIIa complex and platelet aggregability. Submitted for publication.

Table 2. Effect of Ca^{2+} on AP-2 Binding to Washed Platelets at 22°C [a]

Donor	Platelet suspension made in:		
	Buffer alone	Buffer + 1 mM CaCl$_2$	Buffer + 3 mM EDTA
1	n = 47,640	n = 49,970	ND
	Kd = 0.40	Kd = 0.41	
2	n = 49,450	n = 55,040	n = 50,780
	Kd = 0.70	Kd = 0.76	Kd = 0.70
3	ND	n = 59,470	n = 56,500
		Kd = 0.48	Kd = 0.43

[a]Washed platelets were resuspended to a final concentration of 10^8/ml in 0.02 M Tris-HCl, 0.15 M NaCl, pH 7.4, containing 5 mg/ml bovine serum albumin and 1 mg/ml glucose, and incubated for 30 min at 22°C in the presence of buffer alone or buffer containing 3 mM EDTA or 1 mM calcium chloride (final concentrations). Nonimmune mouse IgG at a final concentration of 50 μg/ml was then added. After an additional 15-min incubation, [^{125}I]-AP-2 IgG at a final concentration of 0.2 to 4.5 μg/ml was added. After a final incubation of 30 min, the platelets were centrifuged, and bound [^{125}I]-AP-2 was measured. All incubations were performed at room temperature. For platelet samples from each of three normal donors, the number of AP-2 molecules bound per platelet (n) and the calculated Kd (nM) are given. ND = not done.

McEver *et al.* (1983) recently demonstrated that the GP IIb-IIIa complex-specific monoclonal antibody, T10, bound equally well to unactivated platelets or platelets stimulated with 10 μM ADP. Recent findings in our laboratory extended this observation by showing no significant increase in the number of AP-2 molecules bound per platelet following thrombin activation. In CIE, ^{125}I-labeled T10 reacted only with the GP IIb-IIIa complex and with neither dissociated IIb nor dissociated GP IIIa. ^{125}I-labeled Tab, a murine monoclonal antibody that recognizes an epitope on GP IIb and does not inhibit fibrinogen binding or platelet aggregation (McEver *et al.*, 1982), bound to both dissociated GP IIb and the GP IIb-IIIa complex in CIE (McEver *et al.*, 1983).

In summary, the cumulative experience with GP IIb-IIIa complex-specific monoclonal antibodies from several laboratories has been that the number of molecules bound per platelet is the same using platelets in whole blood, platelet-rich plasma (PRP), gel-filtered platelets, platelets isolated by albumin density gradient centrifugation, washed platelets, formalin-fixed platelets, or platelets activated with ADP or thrombin. These results would support the hypothesis that GP IIb-IIIa complexes pre-exist in the membrane of nonstimulated platelets. Moreover, the number of molecules bound per platelet for GP IIb-IIIa complex-specific monoclonal antibodies is essentially identical to the number of fibrinogen-binding sites per platelet previously determined by several laboratories. Finally, the concentrations of complex-specific antibody (e.g., AP-2) required to abolish aggregation of citrated PRP are essentially identical to the minimum concentrations shown to maximally inhibit fibrinogen binding and to saturate all available antigenic sites.

In light of these findings, it may be necessary to rethink the mechanism of fibrinogen receptor induction as it relates to GP IIb-IIIa. A viable and testable hypothesis at this point is that platelet activation by ADP, thrombin, or other stimuli induces a conformational change in preformed GP IIb-IIIa complexes, thus generating specific sites on the complex (or on one of the glycoproteins) for fibrinogen binding. Since

AP-2 binding to intact platelets is independent of the concentration of external calcium (Pidard *et al.*, 1983; Montgomery *et al.*, 1983) and that of T10 is decreased by only 60% in the presence of EDTA (McEver *et al.*, 1983) in experiments performed at 22°C, it is likely that calcium fluxes within the plasma membrane or in the cytoplasm mediate the conformation of GP IIb-IIIa complex that is critical to fibrinogen receptor induction.

3. GLYCOPROTEIN Ib

3.1. Glycoprotein Ib Associations in Detergent Lysates

In early studies concerned with the structure of platelet membrane glycoproteins, the term "GP I complex" was rather indiscriminately applied to various PAS-positive proteins that comigrated in SDS–PAGE under either nonreduced or reduced conditions. With a gradual improvement in electrophoretic methods, these glycoproteins were eventually resolved into individual components including proteins that were eventually shown to be structurally and/or functionally unique, such as Ia, Ib, Ic, thrombospondin (Ig), and glycocalicin (Is) (Phillips and Poh Agin, 1977a; Solum *et al.*, 1977; Okumura *et al.*, 1978; Lawler *et al.*, 1978). In the context of this chapter, the preceding application of the term "GP I complex" is inappropriate since no evidence has ever been obtained that any of the above glycoproteins are physically associated. Immunochemical evidence has been obtained that GP Ib and glycocalicin are, in fact, structurally related (Okumura *et al.*, 1978; Solum *et al.*, 1980a,b; Clemetson *et al.*, 1981), and subsequent evidence has demonstrated that glycocalicin is derived from GP Ib by the action of an endogenous calcium-activated protease (Solum *et al.*, 1980b, 1983; Berndt and Phillips, 1981; Yoshida *et al.*, 1983).

In several studies, wheat germ agglutinin affinity chromatography of platelet lysates prepared in various detergents (SDS, Triton X-100, Brij 99) resulted in the copurification of GP Ib, glycocalicin, and a high-molecular-weight (210,000 daltons) polypeptide (Nachman *et al.*, 1977, 1979; Kunicki *et al.*, 1981b). Nachman *et al.* (1979) have designated this triad the GP I"complex," since a high degree of homology was found between tryptic peptide maps of GP Ib and the putative 210,000 dalton analogue. However, in that same study, no peptide homology was observed between GP Ib and glycocalicin. In view of the overwhelming evidence from other laboratories indicating a precursor–product relationship between GP Ib and glycocalicin, the proposed structural relationship between GP Ib and the above 210,000-dalton polypeptide (Nachman *et al.*, 1979) requires further study.

3.2. Glycoprotein Ib Organization in the Intact Platelet Membrane

As mentioned in Section 2, Polley *et al.* (1981) failed to show codistribution of GP Ib with membrane GP IIb (or, by inference, GP IIIa) using immunoelectron microscopy, and Davies and Palek (1982) did not observe detectable crosslinking of GP Ib to any other platelet proteins using DTSP. However, others have reported an association of GP Ib with a 17,000–22,000 dalton membrane protein (GP IX) (Clemet-

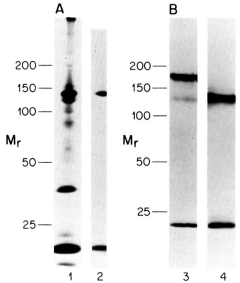

Figure 6. Fluorographic analysis of radi-olabeled platelet membrane glycoproteins that bind to a monoclonal antibody to GP Ib (termed 6D1). (A) Platelet membrane glycoproteins were labeled with sodium metaperiodate/NaB^{3}H$_4$, solubilized in Triton X-100, and immu-noprecipitated. The immunoprecipitate was sol-ubilized in SDS, reduced with 2-mercap-toethanol, and electrophoresed in a 5–15% polyacrylamide gel. After suitable preparation, the gel was dried and exposed to x-ray film for ten days at −70°C. Lane 1 contains the whole platelet preparation and lane 2 contains the im-munoprecipitate showing 2 bands of mol. wt. 135,000 and 20,000. (B) Platelet membrane glycoproteins were labeled with sodium meta-periodate/NaB^{3}H$_4$, solubilized in Triton X-100, and passed sequentially over separate columns of control mouse IgG and 6DI antibody. No ra-dioactivity bound to the control column, where-as 15% bound to the 6DI column. The latter was completely eluted by 1 M NaCl buffer. Lane 3 contains the fluorogram showing bands of mol. wt. 165,000 (GP Ib), mol. wt. 135,000 (glycocalicin), and mol. wt. 20,000. Lane 4 contains the fluorogram of a reduced sample of the 1 M NaCl eluate showing bands of mol. wt. 135,000 (GP Ibα and glycocalicin), mol. wt. 20,000, mol. wt. 22,000 (GP Ibβ) and a very light band of mol. wt. 150,000 (GP Ia). The fluorograms were exposed for eight days at −70°C. (Reproduced from Coller *et al.*, 1983b, with permission.)

son *et al.*, 1982; Coller *et al.*, 1983b; Berndt *et al.*, 1983b) that because of its staining and/or labeling properties would not have been detected in earlier studies. Clemetson *et al.* (1982) and Berndt *et al.* (1983a) found that platelets from patients with the Bernard-Soulier syndrome contained not only negligible amounts of GP Ib, but also decreased levels of a 17,000 dalton membrane protein that was surface labeled with periodate sodium borotritide in normal platelets, but not detectable by surface labeling with ^{125}I or staining with Coomassie Blue or PAS. Coller *et al.* (1983b) copurified a similar protein (20,000 daltons) using a monoclonal antibody specific for GP Ib cou-pled to Affigel 10 (Figure 6). Using a different monoclonal antibody specific for GP Ib, Berndt *et al.* (1983b) also copurified a similar protein (22,000 daltons) and showed that, once separated, this low-molecular-weight protein would reassociate with free GP Ib subsequently coupled to a solid support. These studies provide evidence for a soluble complex of GP Ib and the low-molecular-weight protein in question. Nothing is yet known about the association of these proteins within the membrane of intact platelets.

3.3. High-Molecular-Weight Forms of Glycoprotein Ib

The existence of an additional GP Ib-related platelet membrane component has also been suggested by the recent findings of Solum *et al.* (1983), comparing CIE of platelet lysates prepared in the presence or absence of leupeptin. When leupeptin, which specifically inhibits the endogenous calcium-activated protease, was present during platelet lysis, CIE revealed a slow-migrating (cathodic) immunoprecipitin arc that crossreacted with GP Ib. A time-dependent transformation of this new component to GP Ib and a subsequent conversion of GP Ib to glycocalicin was observed when intact platelets were treated with dibucaine. The authors stated that these transformations were associated with an increase followed by a decrease in the agglutinability of the platelet by VWF. No difference in the size of GP Ib was observed by SDS–PAGE when lysates prepared in the presence or absence of leupeptin were compared. From these data, Solum *et al.* (1983) concluded that the slow-migrating form of GP Ib detected by CIE represents a complex composed of GP Ib and "some other material," and that dissociation of the complex occurs as a result of proteolytic degradation of the "other material." Since considerable "transformed" GP Ib was stated to be present in platelets that had almost completely lost agglutinability in the presence of VWF, it was postulated that the precursor form of GP Ib (complexed) most likely expresses optimum VWF receptor activity.

4. OTHER GLYCOPROTEIN ASSOCIATIONS?

Kunicki *et al.* (1981c) identified the glycoprotein antigens giving rise to several immunoprecipitin arcs in CIE by direct SDS–PAGE analysis of [^{125}I]antigens contained in individually excised precipitates (Figure 7). These platelet antigens included GP Ia, GP IIa, and GP IIIb, as well as those glycoproteins that had been previously localized, namely, GP IIb-IIIa and GP Ib/glycocalicin. Of interest was the finding that GP Ia and GP IIa appeared to codistribute within a single immunoprecipitin arc, suggesting the existence of a soluble complex of these glycoproteins. This hypothesis finds support in the study of Davies and Palek (1982), in which DSTP cross-linked complexes of apparent molecular weight equivalent to 240,000 daltons and presumed to contain heteromultimers of GP Ia and GP IIa or dimers of either GP Ia or GP IIa were routinely detected.

The existence of such GP Ia-IIa complexes in detergent extracts or within the membrane of intact platelets remains to be confirmed.

5. SUMMARY

As outlined above, substantial progress has been made in characterizing complex formations between platelet membrane glycoproteins *in situ,* yet the results of such studies clearly demonstrate that we have only begun to appreciate the myriad of potential protein interactions that can occur within the intact platelet and the impact of these on the regulation of platelet function.

The calcium-mediated association of GP IIb and GP IIIa has received by far the most attention, owing to the pivotal role of fibrinogen receptor induction in platelet function. That the induction event is synonymous with complex formation *in situ* or a change in the conformation of the pre-existing complex represents an enticing model, and this mechanism could serve as a prototype for calcium-regulated stimulus–response coupling in other cell membrane systems.

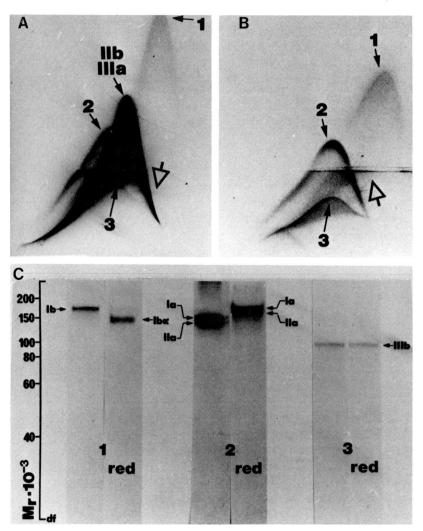

Figure 7. Identification of immunoprecipitin arcs in CIE given by platelet glycoprotein antigens. (A) Autoradiograph of lysate derived from ^{125}I-labeled normal platelets. IIb-IIIa complex is indicated; other arcs are designated Nos. 1 through 3. (B) Autoradiograph of lysate derived from ^{125}I-labeled platelets of a patient with type I Glanzmann's thrombasthenia (no detectable IIb-IIIa). Open arrowheads in (A) and (B) denote the position of an as yet unidentified trace-labeled immunoprecipitin arc. (C) Identification of ^{125}I] antigens contained in immunoprecipitates numbered 1 through 3. The regions of agarose gel compassing the peaks of precipitates numbered 1 through 3 produced by CIE of type I thrombasthenic platelet protein (see B) were individually excised, solubilized in SDS, and analyzed by SDS-PAGE in the absence of (NR) or presence of (R) 2-mercaptoethanol. An autoradiogram developed from a dried acrylamide gel is shown. Each of the ^{125}I-labeled bands present in the autoradiograph shown in this figure corresponded to bands on the original gel which were also stained by periodic acid Schiff reagent. [125]I]antigens present in each precipitate were identified by comparing the relative mobilities of ^{125}I]bands derived from each precipitate with previously established electrophoretic mobilities of the major membrane glycoproteins. In this manner, it was shown that precipitate 1 is given by GP Ib; precipitate 2, by GP Ia and IIa; and precipitate 3, by GP IIIb. Relative molecular weights are shown at left; df, dye front. Reproduced from Kunicki *et al.* (1981c) with permission.

Studies of the interactions of GP Ib with other membrane proteins will provide important insight into the mechanism of platelet–vessel wall adhesion, and the most recent findings of changes in GP Ib associations concurrent with stimulation of the calcium-activated protease introduce yet another potential mechanism for control of membrane glycoprotein receptor function.

Thus, with increasing interest in the molecular bases of transmembrane stimulus–response in various eukaryotic cell types, the information gained from the analyses of platelet glycoprotein receptor modulation represents a timely contribution to our understanding of cell membrane physiology.

REFERENCES

Bennett, J. S., and Vilaire, G., 1979, Exposure of platelet fibrinogen receptors by ADP and epinephrine, *J. Clin. Invest.* **64:**1393–1401.

Bennett, J. S., Vilaire, G., and Cines, D. B., 1982, Identification of the fibrinogen receptor on human platelets by photoaffinity labelling, *J. Biol. Chem.* **257:**8049–8054.

Bennett, J. S., Hoxie, J. A., Leitman, S. F., Vilaire, G., and Cines, D. B., 1983, Inhibition of fibrinogen binding to stimulated human platelets by a monoclonal antibody, *Proc. Natl. Acad. Sci. U.S.A.* **80:**2417–2421.

Berndt, M. C., and Phillips, D. R., 1981, The use of proteolytic probes to identify platelet membrane receptors, (abstract), *Thromb. Haemostasis* **46:**75.

Berndt, M. C., Gregory, C., Chong, B. H., Zola, H., and Castaldi, P. A., 1983a, Additional glycoprotein defects in Bernard-Soulier syndrome: Confirmation of genetic basis by parental analysis, *Blood* **62:**800–807.

Berndt, M. C., Gregory, C., Castaldi, P. A., and Zola, H., 1983b, Purification of human platelet membrane glycoprotein Ib complex using a monoclonal antibody (abstract), *Thromb. Haemostasis* **50:**361.

Clemetson, K. J., Capitanio, A., and Luscher, E. F., 1979, High resolution two-dimensional gel electrophoresis of the proteins and glycoproteins of human blood platelets and platelet membranes, *Biochim. Biophys. Acta* **553:**11–24.

Clemetson, K. J., Naim, H. Y., and Luscher, E. F., 1981, Relationship between glycocalicin and glycoprotein Ib of human platelets, *Proc. Natl. Acad. Sci. U.S.A.* **78:**2712–2716.

Clemetson, K. J., McGregory, J. L., James, E., Dechavanne, M., and Luscher, E. F., 1982, Characterization of the platelet membrane glycoprotein abnormalities in Bernard-Soulier syndrome and comparison with normal by surface-labeling techniques and high-resolution two-dimensional gel electrophoresis, *J. Clin. Invest.* **70:**304–311.

Coller, B. S., Peerschke, E. I., Scudder, L. E., and Sullivan, C. A., 1983a, A murine monoclonal antibody that completely blocks the binding of fibrinogen to platelets produces a thrombasthenic-like state in normal platelets and binds to glycoproteins IIb and/or IIIa, *J. Clin. Invest.* **72:**325–338.

Coller, B. S., Peerschke, E. I., Scudder, L. E., and Sullivan, C. A., 1983b, Studies with a murine monoclonal antibody that abolishes ristocetin-induced binding of von Willebrand factor to platelets: Additional evidence in support of GP Ib as a platelet receptor for von Willebrand factor, *Blood* **61:**99–110.

Davies, G. E., and Palek, J., 1982, Platelet protein organization: Analysis by treatment with membrane-permeable cross-linking reagents, *Blood* **59:**502–513.

DiMinno, G., Thiagarajan, P., Perussia, B., Martinez, J., Shapiro, S., Trinchieri, G., and Murphy, S., 1983, Exposure of platelet fibrinogen-binding sites by collagen, arachidonic acid, and ADP: Inhibition by a monoclonal antibody to the glycoprotein IIb-IIIa complex, *Blood* **61:**140–148.

Gogstad, G. O., Brosstad, F., Krutnes, M.-B., Hagen, I., and Solum, N. O., 1982, Fibrinogen-binding properties of the human platelet glycoproteins IIb-IIIa complex: A study using crossed-radioimmunoelectrophoresis, *Blood* **60:**663–671.

Hagen, I., Bjerrum, O. J., and Solum, N. O., 1979, Characterization of human platelet proteins solubilized with Triton X-100 and examined by crossed immunoelectrophoresis: Reference patterns of extracts from whole platelets and isolated membranes, *Eur. J. Biochem.* **99**:9–22.

Hagen, I., Nurden, A., Bjerrum, O. J., Solum, N. O., and Caen, J., 1980, Immunochemical evidence for protein abnormalities in platelets from patients with Glanzmann's thrombasthenia and Bernard-Soulier syndrome, *J. Clin. Invest.* **65**:722–731.

Hagen, I., Bjerrum, O. J., Gogstad, G., Korsmo, R., and Solum, N. O., 1982, Involvement of divalent cations in the complex between the platelet glycoproteins IIb and IIIa, *Biochim. Biophys. Acta* **701**:1–9.

Howard, L., Shulman, S., Sadanandan, S., and Karpatkin, S., 1982, Crossed immunoelectrophoresis of human platelet membranes. The major antigen consists of a complex of glycoproteins GP IIb and GP IIIa, held together by Ca^{++} and missing in Glanzmann's thrombasthenia, *J. Biol. Chem.* **257**:8331–8336.

Jennings, L. K., and Phillips, D. R., 1982, Purification of glycoproteins IIb and III from human platelet plasma membranes and characterization of a calcium-dependent glycoprotein IIb-III complex, *J. Biol. Chem.* **257**:10458–10466.

Kunicki, T. J., Pidard, D., Rosa, J.-P., and Nurden, A. T., 1981a, The formation of Ca^{++}-dependent complexes of platelet membrane glycoproteins IIb and IIIa in solution as determined by crossed immunoelectrophoresis, *Blood* **58**:268–278.

Kunicki, T. J., Russell, N., Nurden, A. T., Aster, R. H., Caen, J. P., 1981b, Further studies of the human platelet receptor for quinine- and quninidine-dependent antibodies, *J. Immunol.* **126**:398–402.

Kunicki, T. J., Nurden, A. T., Pidard, D., Russell, N. R., and Caen, J. P., 1981c, Characterization of human platelet glycoprotein antigens giving rise to individual immunoprecipitates in crossed-immunoelectrophoresis, *Blood* **58**:1190–1197.

Lawler, J. W., Slayter, H. S., and Coligan, J. E., 1978, Isolation and characterization of a high molecular weight glycoprotein from human blood platelets, *J. Biol. Chem.* **253**:8609–8616.

Lee, H., Nurden, A., Thomaidis, A., and Caen, J. P., 1981, Relationship between fibrinogen binding and the platelet glycoprotein deficiencies in Glanzmann's thrombasthenia Type I and II, *Br. J. Haematol.* **48**:47–55.

Leung, L. L. K., Kinoshita, T., and Nachman, R. L., 1981, Isolation, purification, and partial characterization of platelet membrane glycoproteins IIb and IIIa, *J. Biol. Chem.* **256**:1994–1997.

Lightsey, A. L., Thomas. W. J., Plow, E. F., McMillan, R., and Ginsberg, M., 1981, Glanzmann's thrombasthenia in the absence of glycoprotein IIb and III deficiency (abstract), *Blood* **58**:199.

McEver, R. P., Baenziger, N. L., and Majerus, P. W., 1980, Isolation and quantitation of the platelet membrane glycoprotein deficient in thrombasthenia using a monoclonal hybridoma antibody, *J. Clin. Invest.* **66**:1311–1318.

McEver, R. P., Baenziger, J. U., and Majerus, P. W., 1982, Isolation and structural characterization of the polypeptide subunits of membrane glycoprotein IIb-IIIa from human platelets, *Blood* **59**:80–85.

McEver, R. P., Bennett, E. M., and Martin, M. N., 1983, Identification of two structurally and functionally distinct sites on human platelet membrane glycoprotein IIb-IIIa using monoclonal antibodies, *J. Biol. Chem.* **258**:5269–5275.

Montgomery, R. R., and Johnson, J., 1982, Specific factor VIII-related antigen fragmentation: An *in vivo* and *in vitro* phenomenon, *Blood* **60**:930–939.

Montgomery, R. R., Kunicki, T. J., Taves, C., Pidard, D., and Corcoran, M., 1983, Diagnosis of Bernard-Soulier syndrome and Glanzmann's thrombasthenia with a monoclonal assay on whole blood, *J. Clin. Invest.* **71**:385–389.

Mustard, J. F., Kinlough-Rathbone, R. L., Packham, M. A., Perry, D. W., Harfenist, E. J., and Pai, K. R. M., 1979, Comparison of fibrinogen association with normal and thrombasthenic platelets on exposure to ADP or chymotrypsin, *Blood* **54**:987–993.

Nachman, R. L., and Leung, L. L. K., 1982, Complex formation of platelet membrane glycoproteins IIb and IIIa with fibrinogen, *J. Clin. Invest.* **69**:263–269.

Nachman, R. L., Tarasov, E., Weksler, B. B., and Ferris, B., 1977, Wheat germ agglutinin affinity chromatography of human platelet membrane glycoproteins, *Thromb. Res.* **12**:91–104.

Nachman, R. L., Kinoshita, T., and Ferris, B., 1979, Structural analysis of human platelet membrane glycoprotein I complex, *Proc. Natl. Acad. Sci. U.S.A.* **76**:2952–2954.

Nurden, A. T., and Caen, J. P., 1974, An abnormal platelet glycoprotein pattern in three cases of Glanzmann's thrombasthenia, *Br. J. Haematol.* **28:**253–260.

Nurden, A. T., and Caen, J. P., 1975, Specific roles for surface glycoproteins in platelet function, *Nature (London)* **255:**720–722.

Nurden, A. T., Dupuis, D., Kunicki, T. J., and Caen, J. P., 1981, Analysis of the glycoprotein and protein composition of Bernard-Soulier platelets by single and two-dimensional sodium dodecyl sulfate-polyacrylamide gel electrophoresis, *J. Clin. Invest.* **67:**1431–1440.

Okumura, T., Hasitz, M., and Jamieson, G. A., 1978, Platelet glycocalicin: Interaction with thrombin and role as thrombin receptor of the platelet surface, *J. Biol. Chem.* **253:**3435–3443.

Phillips, D. R., and Poh Agin, P., 1977a, Platelet membrane glycoproteins: Evidence for the presence of nonequivalent disulfide bonds using nonreduced-reduced two-dimensional gel electrophoresis, *J. Biol. Chem.* **252:**2121–2126.

Phillips, D. R., and Poh Agin, P., 1977b, Platelet membrane defects in Glanzmann's thrombasthenia: Evidence for decreased amounts of two major glycoproteins, *J. Clin. Invest.* **60:**535–545.

Pidard, D., Rosa, J.-P., Kunicki, T. J., and Nurden, A. T., 1982, Further studies on the interaction between human platelet membrane glycoproteins IIb and IIIa in Triton X-100, *Blood* **60:**894–904.

Pidard, D., Montgomery, R. R., Bennett, J. S., and Kunicki, T. J., 1983, Interaction of AP-2, a monoclonal antibody specific for the human platelet glycoprotein IIb-IIIa complex, with intact platelets, *J. Biol. Chem.* **258:**12582–12586.

Polley, M. J., Leung, L. L. K., Clark, F. Y., and Nachman, R. L., 1981, Thrombin-induced platelet membrane glycoprotein IIb and IIIa complex formation: An electron microscope study, *J. Exp. Med.* **154:**1058–1068.

Rosa, J.-P., Kieffer, N., Didry, D., Pidard, D., Kunicki, T. J., and Nurden, A. T., 1984, The human platelet membrane glycoprotein complex GP IIb-IIIa expresses antigenic sites not exposed on the dissociated glycoproteins, *Blood* **64:**1246–1253.

Ruggeri, Z. M., and Zimmerman, T. S., 1981, The complex multimeric composition of Factor VIII/von Willebrand factor, *Blood* **57:**1140–1143.

Shulman, S., and Karpatkin, S., 1980, Crossed immunoelectrophoresis of human platelet membranes. Diminished major antigen in Glanzmann's thrombasthenia and Bernard-Soulier syndrome, *J. Biol. Chem.* **255:**4320–4327.

Solum, N. O., Hagen, I., and Peterka, M., 1977, Human platelet glycoproteins: Further evidence that the ''GP I band'' from whole platelets contains three different polypeptides, one of which may be involved in the interaction between platelets and Factor VIII, *Thromb. Res.* **10:**71–82.

Solum, N. O., Hagen, I., Filion-Myklebust, C., and Stabaek, T., 1980a, Platelet glycocalicin: Its membrane association and solubilisation in aqueous media, *Biochim. Biophys. Acta* **597:**235–246.

Solum, N. O., Hagen, I., and Sletbakk, T., 1980b, Further evidence for glycocalicin being derived from a larger amphiphilic platelet membrane glycoprotein, *Thromb. Res.* **18:**773–785.

Solum, N. O., Olsen, T. M., Gogstad, G. O., Hagen, I., and Brosstad, F., 1983, Demonstration of a new glycoprotein Ib-related component in platelet extracts prepared in the presence of leupeptin, *Biochim. Biophys. Acta* **729:**53–61.

Yoshida, N., Weksler, B., and Nachman, R., 1983, Purification of human platelet calcium-activated protease: Effect on platelet and endothelial function, *J. Biol. Chem.* **258:**7168–7174.

5

Structure and Function of Platelet Membrane Glycoproteins as Studied by Crossed Immunoelectrophoresis

Inger Hagen and Nils Olav Solum

1. INTRODUCTION

1.1. General

This chapter presents studies of platelet membrane glycoproteins based on crossed immunoelectrophoresis (CIE). Crossed immunoelectrophoresis has been a critical method in assessing the structure of membrane glycoproteins, their organization within the membrane, their interactions with other intrinsic proteins and external ligands, and their abnormalities in disease states. Such studies with human platelets have provided new insights into membrane glycoprotein structure, function, and organization that were not previously appreciated.

1.2. Crossed Immunoelectrophoresis

Crossed, or two-dimensional, immunoelectrophoresis is a procedure in which proteins are separated in the first dimension by agarose electrophoresis and in the second dimension by electrophoresis into a gel containing precipitating antibodies. Figure 1 illustrates the basic technique. Bjerrum and Lundahl (1973) were the first to modify this technique for examination of membrane proteins by solubilizing and elec-

---

Inger Hagen and Nils Olav Solum • Research Institute for Internal Medicine, Section on Hemostasis and Thrombosis, Rikshospitalet, University of Oslo, Oslo 1, Norway.

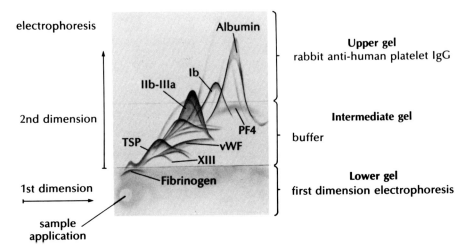

Figure 1. The method of CIE. Platelets solubilized in a nonionic detergent are applied to the agarose gel as indicated, and the proteins are separated by the first-dimension electrophoresis in the lower gel section. The second-dimension electrophoresis moves the separated platelet proteins into the antibody-containing gel sections. The upper gel section contains rabbit anti-human platelet IgG in a concentration sufficient to produce immunoprecipitation arcs. The intermediate gel section contains only buffer in this example, but can be used to include specific antibodies or other materials as described in this chapter. Abbreviations used for the arcs identified in this figure are for platelet factor 4 (PF4), von Willebrand factor (VWF), thrombospondin (TSP), and coagulation factor XIII, subunit a (XIII). Reproduced from George *et al.* (1984) with permission.

trophoresing proteins in the presence of a nonionic detergent. Nonionic detergents are suitable for this procedure since they solubilize the majority of membrane proteins without altering their antigenicity. This antigenic stability probably occurs because the nonionic detergents selectively bind to the hydrophobic part of membrane proteins, leaving antigenic determinants on the exposed hydrophilic regions unaffected and free to react with antibodies (Helenius and Simons, 1975; Robinson and Tanford, 1975; Bjerrum, 1983). The great advantage of this method is that it permits functional and structural characterization of individual protein antigens in complex protein mixtures without prior purification. Another advantage is that contamination with other types of cells is not critical as long as cell-specific antibodies are used. The applicability of CIE techniques in the analysis of membrane proteins in general has been reviewed by Bjerrum and Bøg-Hansen (1976), Bjerrum (1983), and Bjerrum and Hagen (1983).

2. IDENTIFICATION OF ANTIGENS

2.1. General

A reference pattern obtained by CIE of Triton X-100-solublized platelet proteins against whole platelet antibodies from our laboratory is shown in Figure 2A. Described

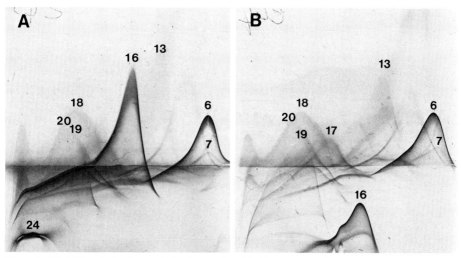

Figure 2. (A) CIE analysis of normal platelet proteins. Platelets were solubilized in 38 mM Tris, 100 mM glycine buffer (pH 8.6) containing 1% (v/v) Triton X-100. The agarose 1% (w/v) was dissolved in the same buffer, but in some instances 0.5% Triton was used. The first-dimension electrophoresis of 100 μg of platelet proteins was performed at 10 V/cm for 45 min, and the second dimension was performed overnight at 1–2 V/cm. Antibodies raised against whole platelets were incorporated into the top agarose section. No antibodies were incorporated in the intermediate gel. The electrode buffer contained 38 mM Tris and 100 mM glycine. Immunoplates were washed and pressed three times in 154 mM NaCl, dried, and stained with Coomassie Brilliant Blue. The identity of the numbered immunoprecipitates is given in Table I. (B) This Coomassie Brilliant Blue-stained immunoplate was performed together with the plate shown in Figure 1A, and by identical methods except that 400 μl of antiserum (IgG L) from a polytransfused patient with Glanzmann's thrombasthenia was incorporated in the intermediate gel. Note the specific precipitation of the GP IIb-IIIa complex (arc 16) in the intermediate gel compared with the control. Reproduced from Hagen *et al.* (1980) with permission.

below are the variety of techniques that we and others have used to identify the antigens responsible for these individual arcs. The proteins identified in our studies are: platelet factor 4 (PF4), α_2-antiplasmin, von Willebrand factor (VWF), GP Ib, GP IIb-IIIa, GP IIIb, thrombospondin (TSP), factor XIII (subunit a), β_2-microglobulin containing the histocompatibility antigen (HLA-ABC), and fibrinogen (Hagen *et al.*, 1979, 1980, 1981, 1982b; Gogstad *et al.*, 1983b). Table 1 lists the identity of the antigen arcs using the numerical system of our laboratory. Not all arcs are seen in Figure 2A because different preparations of antibodies yield different patterns of immunoprecipitation arcs. This is due to the different specificities and titers of the precipitating rabbit antibodies and may complicate comparisons of reference patterns obtained in different laboratories. In particular, it may be difficult to identify immunoprecipitates near the origin since there are so many arcs in this region. However, several antigens are consistently and easily recognized due to their position as well as the characteristic shape and appearance of their immunoprecipitates. For example, albumin and GP IIb-IIIa complex both form dominant immunoprecipitates with strong protein staining with all antibody preparations tested. Platelet factor 4 is easily identified since the immunoprecipitate forms a ''line'' rather than the usual arc (Figure 1)

Table 1. Identification of Immunoprecipitates
Obtained by CIE of Whole Platelets with
Anti-whole Platelet Antibodies[a]

Precipitate numbers	Platelet protein
6	Albumin
7	α_2-antiplasmin
13	GP Ib
16	GP IIb-IIIa
14	Thrombospondin
19	Factor XIII (subunit a)
20	β_2-microglobulin
24	Fibrinogen

[a]The immunoprecipitates labeled in Figure 2 that have been
identified as specific platelet proteins are listed here.

(Gogstad *et al.*, 1981; Hagen *et al.*, 1982a) (see also Chapter 4, Figure 1). Factor XIII
and fibrinogen form characteristic immunoprecipitates close to the application point.
Glycoprotein Ib and TSP form diffuse immunoprecipitates and some antibody preparations that we have tested lack precipitating antibodies against these two proteins.

Although the whole platelet reference pattern appears complicated, there are
several ways to characterize further the individual proteins. These include identification of antigens with monospecific antibodies, characterization of amphiphilic properties, content of the carbohydrate moiety of glycoproteins, enzyme and receptor function, susceptibility to degradation, and information about supramolecular organization.
These aspects as applied to human blood platelets will be treated in the following
sections.

2.2. Monospecific Antibodies

One method often used for the identification of platelet proteins in the reference
pattern is to incorporate the monospecific antibody in an intermediate gel, inserted
between the first-dimension electrophoresis gel and the gel containing the precipitating
anti-whole-platelet antibodies. This causes the antigen to be precipitated in the intermediate gel and lowers the corresponding arc. This is illustrated in Figure 2B using an
alloantibody from a polytransfused thrombasthenic patient (IgG L) (Hagen *et al.*,
1980). The immunoprecipitate containing the GP IIb-IIIa complex was formed in the
intermediate gel, whereas all the other arcs were unaffected. By this technique we were
able to demonstrate that this antibody contained specificities against GP IIb-IIIa and no
other solubilized platelet proteins. Further studies regarding the specificity of this
antibody are reported in Chapter 4 (Figure 3). A similar approach has been employed
for identification of immunoprecipitates containing PF4, albumin, α_2-antiplasmin,
VWF, TSP, factor XIII (subunit a), and fibrinogen (Hagen *et al.*, 1979, 1980, 1981;
Gogstad *et al.*, 1983b).

2.3. Purified Antigens

Various immunoelectrophoretic techniques can be used to identify directly an antigen in a complex pattern of immunoprecipitates provided the purified protein is available. These include crossed line immunoelectrophoresis, tandem CIE, and addition experiments whereby a mixture of the solubilized proteins and the purified antigen is examined and the resulting pattern of immunoprecipitates is compared with the patterns obtained with the separate samples. The principles of these methods have been described in more detail elsewhere (Axelson *et al.*, 1973).

Addition experiments have been employed to verify the identity of certain immunoprecipitates in the platelet reference pattern, i.e., glycocalicin, TSP, and PF4 (Solum *et al.*, 1980b; Hagen *et al.*, 1982a). This is based on the fact that the area of the arc is proportional to the amount of the respective antigen that is present. Thus, when the purified platelet protein is electrophoresed together with the Triton X-100 extract of whole platelets, the area of the immunoprecipitate corresponding to the purified antigen is increased compared with the area obtained on CIE of an equal amount of the purified antigen or the platelet extract alone. This is illustrated in Figure 3 for the identification of the glycocalicin immunoprecipitate after CIE of a Triton X-100 extract of whole platelets.

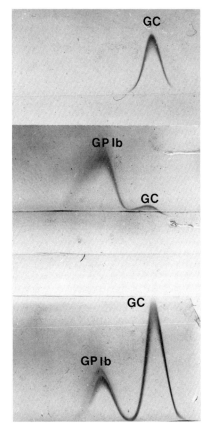

Figure 3. CIE of purified glycocalicin (upper panel), solubilized platelet proteins (middle panel), and a mixture of purified glycocalicin and solubilized platelet proteins (lower panel) using precipitating antibodies raised against purified glycocalicin in the top gel section of each panel (GC, glycocalicin). Note the continuous double-peak immunoprecipitate in the middle panel and the elevation of the fast-migrating precipitate when purified glycocalicin is added to the platelet extract in the bottom panel. The experimental conditions were the same as given in Figure 2 except that the agarose gels always contained 1% Triton X-100 and the antiglycocalicin antiserum concentration was 100–150 μl per ml agarose. Reproduced from Solum *et al.* (1980b) with permission.

2.4. Identification of Antigens by Sodium Dodecyl Sulfate–Polyacrylamide Gel Electrophoresis (SDS–PAGE) of Excised Immunoprecipitates

Another method to identify the antigen in an immunoprecipitate is to excise the arc and characterize the proteins present by SDS–PAGE (Norrild *et al.*, 1977). A prerequisite for this method is that the immunoprecipitate can be cut free of contaminating proteins from other arcs, and this is often difficult when polyspecific antibodies are used. The problem can be partially circumvented by employing different techniques to eliminate the undesired immunoprecipitates. For instance, when platelet membrane proteins in an extract of whole platelets are to be specifically examined, the prior absorption of the antibody preparation with cytosol proteins removes antibodies against these proteins from the original antibody preparation. Alternatively, adding an excess amount of cytosol proteins together with the extracted proteins from whole platelets causes antigen excess and migration of the precipitates containing cytosol proteins off the plate. Kunicki *et al.* (1981a) used another approach. To avoid contamination from the GP IIb-IIIa precipitate, they used solubilized proteins from thrombasthenic platelets which lack these glycoproteins (see Chapter 16). Immunoprecipitates in this region representing other membrane proteins could then be excised and examined by SDS–PAGE. In this way, they were able to identify GP Ia, GP IIa, GP Ib, and GP IIIb in their reference pattern (Kunicki *et al.*, 1981a). The polypeptide bands obtained after SDS–PAGE represent polypeptides from the immunoglobulin as well as the platelet antigen. By radiolabeling the platelet proteins prior to the immunoelectrophoresis, the platelet origin of the bands seen on the SDS gels can be confirmed by autoradiography (Kunicki *et al.*, 1981a,b). Using these techniques we have identified GP IIb and GP IIIa in immunoprecipitate No. 16 (Hagen *et al.*, 1980) and have shown that a major, slow migrating peak that cross-reacts with glycocalicin represents GP Ib (Solum *et al.*, 1980b; Kunicki *et al.*, 1981a).

2.5. Identification of Antigens by the Use of Radioactively Labeled Antibodies

Antigens can be identified by the use of radiolabeled antibodies with identification of the corresponding antigen by autoradiography of the immunoplate. Kunicki and co-workers (1981b) used ^{125}I-labeled IgG L, an alloantibody from a polytransfused thrombasthenic patient (see Chapter 4) in concentrations below that capable of precipitating the GP IIb-IIIa complex to show that this antibody bound specifically to the GP IIb-IIIa complex subsequently precipitated by the polyspecific rabbit antibodies. We have used ^{125}I-labeled antibody to β_2-microglobulin-bearing histocompatibility antigen (HLA-ABC). The immunoplate obtained after CIE of Triton X-100-solubilized platelet proteins against anti-platelet antibodies was incubated with [^{125}I]antibodies to β_2-microglobulin followed by washing of the plate and autoradiography. This revealed heavy labeling of an immunoprecipitate close to the application well termed No. 20 (Hagen *et al.*, 1982b). The experiment illustrates the use of radioiodinated antibodies to increase the sensitivity for identification of proteins that stain only faintly with Coomassie Blue.

3. CHARACTERIZATION OF ANTIGENS BY CIE

3.1. Identification of Amphiphilic Proteins

The function of many membrane proteins depends on their capacity to engage in direct contact with substances in the aqueous fluid at the outer or inner side of the membrane while their body is firmly anchored within the lipid membrane. This means that their functions will be directly dependent on their amphiphilic character. The amphiphilic properties of a protein can be demonstrated by various applications of the CIE technique (Bjerrum, 1978; Bjerrum *et al.*, 1979, 1982).

One technique that we have used is hydrophobic adsorption to phenyl-Sepharose. The inclusion of phenyl-Sepharose in the intermediate gel section retards the migration of hydrophobic proteins during the second-dimension electrophoresis. This has been particularly useful to study the relationship between GP Ib and glycocalicin. These were first believed to represent two different but related membrane proteins. Whereas GP Ib required a detergent for its extraction, glycocalicin was found in the soluble fraction after lysis of platelets in nondetergent solutions. However, using antiserum to purified glycocalicin, we observed that CIE of platelet extracts gave one continuous immunoprecipitate occurring as two peaks, one representing glycocalicin and the other GP Ib, demonstrating that both proteins contained identical antigenic sites (Figure 3) (Solum *et al.*, 1980a,b). Using phenyl-Sepharose in an intermediate gel, we could see that gylcocalicin behaved like a perfect hydrophilic protein, not being affected by the hydrophobic matrix, whereas GP Ib was totally retained by the phenyl-Sepharose (Figure 4). Similar results have also been obtained with purified GP Ib and glycocalicin (Clemetson *et al.*, 1981). Subsequent experiments have shown that glycocalicin is a product of the GP Ib α-chain as a consequence of the activation of a calcium-dependent protease during platelet lysis (see Section 3.4).

A different method of characterizing amphiphilic proteins, called the charge-shift technique, is based on the idea that the addition of a small amount of a negatively or positively charged ionic detergent to a nonionic detergent will lead to the formation of "mixed micelles" and that this will affect the overall electric charge of an amphiphilic protein carrying detergent molecules (Bhakdi *et al.*, 1977). As it has been shown that hydrophilic proteins bind almost no Triton X-100 under the conditions used in these experiments (Helenius and Simons, 1975), they will be unaffected, whereas amphiphilic proteins will exhibit shifts in charge and thus in electrophoretic mobility. As the amount of the ionic detergent used is low, the ability of the amphiphilic proteins to form immunoprecipitates is not lost. This is demonstrated in Figure 5 where the amphiphilic behavior of GP Ib and the hydrophilic nature of glycocalicin are illustrated using a combination of Triton X-100 plus deoxycholate as a negatively charged detergent. The electrophoretic mobility of glycocalicin was unaffected by the addition of deoxycholate, whereas that of GP Ib increased. Apparently the hydrophobic groups on GP Ib bound the deoxycholate-Triton X-100 micelles, causing it to have an increase in negative charge, whereas such hydrophobic, detergent-binding regions are lacking in glycocalicin.

One possible pitfall of the charge-shift technique is that even hydrophilic proteins may show reduced electrophoretic mobility when positively charged detergents are

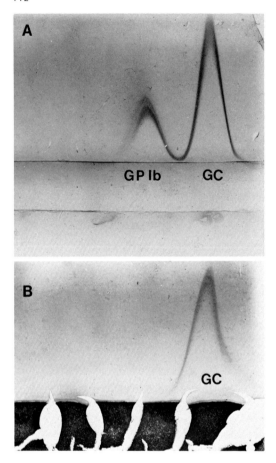

Figure 4. Demonstration of the hydrophobic character of GP Ib and hydrophilic character of glycocalicin (GC) by crossed hydrophobic interaction immunoelectrophoresis with phenyl-Sepharose in the intermediate gel. A mixture of a Triton X-100 extract of whole platelets and purified glycocalicin was subjected to CIE under standard conditions. (A) Control CIE with the intermediate gel containing only agarose and aqueous buffer in the second-dimension electrophoresis. (B) Hydrophobic interaction with phenyl-Sepharose (20% v/v) incorporated into the intermediate gel. Note the retardation of GP Ib but not of glycocalicin during passage through the hydrophobic matrix. An arc for GP Ib is not visible in the dense intermediate gel. The phenyl-Sepharose-containing agarose gel sections are often disrupted during drying and staining. Reproduced from Solum *et al.* (1980b) with permission.

used, due to a direct charge neutralization. This may be the case for glycoproteins carrying considerable amounts of sialic acid and has been seen with glycocalicin using cetyltrimethylammoniumbromide in combination with Triton X-100. This causes a reduced electrophoretic mobility also of glycocalicin, albeit to a lesser extent than of GP Ib. However, as a corresponding effect on glycocalicin could be seen by the addition of high concentrations of positively charged nondetergent molecules such as glucosamine and galactosamine, this effect is considered a nonspecific one.

The third principle used to study the amphiphilic properties of platelet membrane proteins is that of phase separation of extracts prepared in the presence of Triton X-114 instead of Triton X-100. This is based on the fact that the nonionic detergents have a "cloud point" temperature above which the microscopic micelles will start to aggregate and form a detergent-rich phase spearated by a distinct boundary from a water phase on top (Bordier, 1981). Whereas the "cloud point" of Triton X-100 is above 60 °C, that of Triton X-114 is near 20 °C. If an extract that has been prepared by solubilization in Triton X-114 at 4 °C is left for a short period of time at 37 °C, the two phases are formed, and the most hydrophobic proteins will show a tendency to partition

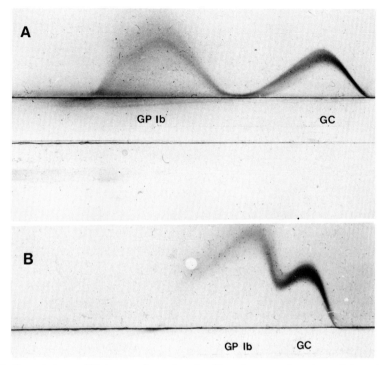

Figure 5. Crossed charge shift immunoelectrophoresis with antiserum against glycocalicin performed on Triton X-100 extracts of whole platelets with added glycocalicin. Washed platelets (2×10^{10} cells/ml) were extracted with (A) 2% Triton X-100 in Tris-glycine buffer pH 8.7, or (B) a mixture of 2% Triton X-100 and 0.8% sodium deoxycholate in the same buffer. The platelet extracts were incubated with purified glycocalicin (GC) (37 μg/ml) for 1 hr at 4°C prior to the first-dimension electrophoreses. The first-dimension electrophoreses were performed simultaneously in two different chambers for (A) and (B) until the hemoglobin marker had migrated 20 mm in both systems. Antiserum-free intermediate gels were used. Both first-dimension gels were cut 18 mm from the application well. Note the reduced distance between the GP Ib peak and the glycocalicin peak in (B) compared with (A), demonstrating the increased electrophoretic mobility of GP Ib and the unchanged mobility of glycocalicin. Reproduced from Solum *et al.* (1980b) with permission.

into the detergent phase. As seen from Figure 6, the GP IIb-IIIa complex of platelet extracts is absent from the water phase, whereas GP Ib is recovered almost quantitatively in the water phase. This occurs in spite of the fact that GP Ib is characterized as a clearly amphiphilic protein by the other principles described above. This has also been observed by others (Newman *et al.*, 1982). However, as it can be shown that the GP Ib of the water phase still has hydrophobic domains available for interactions with phenyl-Sepharose, this "anomalous" behavior or GP Ib probably means that the phase-separation technique is less suited for detection of all of the amphiphilic proteins. On the other hand, this may also reveal interesting structural features of GP Ib. It is probable that with all membrane proteins, the balance between strongly hydrophobic and strongly hydrophilic areas give the molecule a predefined shape. It may be expected that the high content of carbohydrate of the glycocalicin end of the GP Ib α-

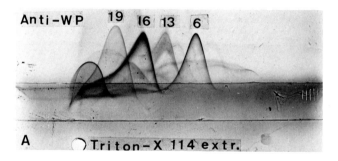

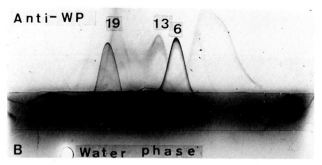

Figure 6. Partition of platelet proteins after phase separation of a Triton X-114 extract of platelets. CIE using antibodies to whole platelet proteins of (A) a Triton X-114 extract prepared at 4 °C, and (B) the water phase from the same extract after phase separation at 37 °C. The immunoprecipitates are numbered as follows: 6, albumin; 13, GP Ib; 16, GP IIb-IIIa; and 19, factor XIII (subunit a). Electrophoretic conditions are essentially as in Figure 3. Note the disappearance of the GP IIb-IIIa immunoprecipitate, but not those of GP Ib, albumin, and factor XIII, from the water phase after phase separation.

chain (around 60% by weight) will contribute significantly to the hydrophilic part of this balance and keep the molecule stretched out from the surface of the platelet.

A fourth application of CIE to identify amphiphilic proteins, as well as glycolipids, takes advantage of the ability of the hydrophobic dye, Sudan black, to bind to detergent micelles. Thus the immunoprecipitates of amphiphilic proteins will be stained by Sudan black and can be visualized on the immunoplates immediately after electrophoresis. Employing this technique, Bjerrum et al. (1982) were able to demonstrate that three immunoprecipitates were stained with Sudan black following CIE of solubilized platelet proteins. These were GP IIb-IIIa, β_2-microglobulin, and an immunoprecipitate termed No. 15. All of these antigens have also been shown to represent amphiphilic proteins by charge-shift immunoelectrophoresis (Hagen et al., 1979).

3.2. Carbohydrate-Related Reactions of Glycoproteins

Sialic acid residues are present in the majority of membrane glycoproteins and represent a major contribution to the overall negative charge of platelets. Glycoproteins that contain sialic acid residues can often be identified with CIE by using neuramini-

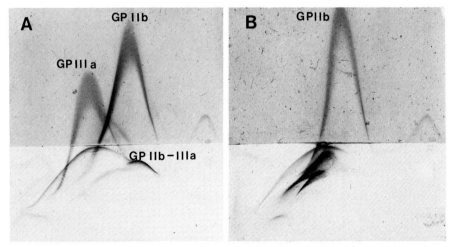

Figure 7. Crossed affinity-immunoelectrophoresis of platelet membranes washed in the presence of 1 mM EDTA and solubilized in Triton X-100. The GP IIb-IIIa complex is dissociated in the presence of EDTA and individual immunoprecipitin arcs are seen for GP IIb and GP IIIa. (A) Control, no lectin incorporated in the lower, first-dimension gel. (B) Concanavalin A (100 μg/cm²) was incorporated in the first dimension gel and added to the sample before application. The immunoprecipitates representing GP IIb-IIIa and GP IIIa were retarded by concanavalin A, whereas GP IIb was unaffected. Reproduced from Hagen *et al.* (1982b) with permission.

dase to reduce their electrophoretic mobility. The immunoprecipitates corresponding to the desialylated glycoproteins will thus appear closer to the cathode compared with the unmodified molecules. This has been demonstrated for glycocalicin, GP IIb-IIIa, and β_2-microglobulin as well as several arcs representing glycoproteins not yet identified (Hagen *et al.*, 1979; Solum *et al.*, 1980a).

The combination of immunoelectrophoresis and affinity electrophoresis with lectins has been employed for quantitation and characterization of glycoproteins, and for the prediction of preparative separations (Bøg-Hansen, 1981). In our studies on platelet glycoproteins, we have either incorporated soluble lectins in the gel used in the first-dimension electrophoresis or employed immobilized lectins either in the first-dimension gel or in an intermediate gel during the second-dimension electrophoresis. These techniques have been employed to identify glycoproteins in the reference pattern of whole platelets and α-granules as well as secreted proteins (Hagen *et al.*, 1979; Gogstad *et al.*, 1981, 1982b). An illustration of the application of crossed affinity immunoelectrophoresis and the interaction of concanavalin A with solubilized membrane proteins is shown in Figure 7. Concanavalin A was added to the sample just prior to electrophoresis and was also incorporated into the first-dimension gel. Glycoproteins interacting with the lectin will exhibit reduced electrophoretic mobility and/or reduced area of their immunoprecipitates as demonstrated for the GP IIb-IIIa complex and GP IIIa, respectively. Glycoprotein IIb was relatively unaffected by the presence of concanavalin A. Kunicki and Aster (1979) had previously shown that GP IIIa reacts with concanavalin A, but not with lentil lectin, whereas the opposite was the case for GP IIb.

3.3.   Macromolecular Interactions

In our first analysis of Triton X-100-solubilized platelet proteins by CIE, we observed an antigen forming a dominant immunoprecipitate termed No. 16 (Figure 2A). This was shown to represent a heterogeneous, surface-oriented sialoglycoprotein based on the nonsymmetrical shape of the immunoprecipitate, its labeling following lactoperoxidase-catalyzed iodination of the platelets, its reduced electrophoretic mobility after neuraminidase treatment, and its interaction with different lectins. This immunoprecipitate was also present in extracts of isolated membranes. However, some variability of this immunoprecipitate was seen on examination of membrane extracts in that increased levels of two antigens were associated with decreased levels of protein 16 (Hagen et al., 1979). Sodium dodecyl sulfate–PAGE of the excised immunoprecipitate revealed that immunoprecipitate 16 contained both GP IIb and GP IIIa (Hagen et al., 1980), suggesting that these glycoproteins exist together in the membrane as a complex. Subsequent work by Kunicki and co-workers (1981b) showed that addition of EDTA or EGTA to the Triton X-100 extract caused the loss of arc 16 with the appearance of two new arcs. The identity of the proteins present in the newly formed arcs was established by SDS–PAGE of the excised immunoprecipitates. Glycoprotein IIb was identified in the arc with a similar migration as the complex, whereas GP IIIa was present in the arc closer to the application point (Kunicki et al., 1981b) (see Figure 7 and Chapter 4). Similar results were obtained by Hagen et al. (1982b). It thus appeared that GP IIb and GP IIIa existed in a complex and that this complex was stabilized by Ca^{2+}. This is illustrated in Figure 7, which demonstrates the splitting of the GP IIb-IIIa complex and the interaction of the separated glycoproteins with concanavalin A. The effect of cations and pH on GP IIb-IIIa complex formation has also been investigated by CIE. Kunicki et al. (1981b) and Howard et al. (1982) observed that Mg^{2+} was unable to support the formation of the GP IIb-IIIa complex. In contrast, Gogstad et al. (1982c) concluded that Mg^{2+} as well as Ca^{2+} were able to maintain the complex. The reason for this discrepency is not known. Both the separate antigens, GP IIb and GP IIIa, as well as the GP IIb-IIIa complex were able to bind $^{45}Ca^{2+}$ (Gogstad et al., 1983a).

In conclusion, the application of CIE in combination with SDS–PAGE has proved to be a useful approach to identify and characterize the GP IIb-IIIa complex in platelets and to study under which conditions the complex is formed and dissociated. Glycoprotein complexes are discussed in detail in Chapter 4.

3.4.   Proteolytic Precursor–Product Relationships

Much attention has been paid in the past to the possibility that platelet extracts contain active proteolytic enzymes that may degrade endogenous substrates (for example, see Nachman and Ferris, 1968; Beese et al., 1966). Despite these observations, most studies using the CIE technique have been performed with Triton X-100-solubilized platelet extracts prepared without proteolytic inhibitors. No effect could be observed on the immunoprecipitate pattern of whole platelet extracts, at least not for the major immunoprecipitates, when serine esterase inhibitors were included in the Triton X-100-containing buffer used for extraction (I. Hagen and N. O. Solum, unpublished observations). These were phenylmethylsulphonylfluoride (PMSF) (4 mM), soybean trypsin inhibitor (SBTI) (1 mg/ml), and trasylol (500 IU/ml). The cathepsin inhibitor, *N*-carbobenzoxy-L-glutamyl-L-tyrosine (5 mM), also had no effect. However, the recently described group of proteolytic enzymes, the neutral calcium-activated proteases, is of particular interest (Phillips and Jakabova, 1977; Truglia and Stracher, 1981). This protease is known to degrade proteins involved in the formation and maintenance of the cytoskeleton, particularly actin-binding protein and P235 (White, 1980; Rosenberg *et al.*, 1981). This degradation may secondarily affect membrane proteins associated with the cytoskeleton. In addition, this protease also cleaves GP Ib (Solum *et al.*, 1980a,b; Ali-Briggs *et al.*, 1981; Berndt and Phillips, 1981; Clemetson *et al.*, 1981). Glycoprotein Ib consists of two disulfide-linked polypeptide chains, termed GP Ibα and GP Ibβ (Phillips and Poh Agin, 1977). Hydrolysis of GP Ibα by the calcium-activated protease produces at least two fragments: a short, hydrophobic piece that is disulfide linked to GP Ibβ, and glycocalicin, a water-soluble one-chain glyco-protein with as much as 60% of its weight accounted for by carbohydrate residues (Okumura *et al.*, 1976; see also Chapter 3). Estimations of molecular weight by SDS–PAGE have given a value of about 145,000 daltons for both GP Ibα and glycocalicin (Solum *et al.*, 1977; Phillips and Poh Agin, 1977). As discussed in Section 3.1, CIE of platelet extracts using an antibody to purified glycocalicin gave one continuous immu-noprecipitate consisting of two peaks. By a combination of CIE and SDS–PAGE, it could be shown that the faster-moving antigen represented glycocalicin and the more slowly moving antigen represented GP Ib. The relative concentrations of glycocalicin and GP Ib can be modified in two ways. One is by the concentration of Triton X-100 present during platelet solubilization. At low Triton X-100 concentrations, practically all of the solubilized GP Ib-related material is in the form of glycocalicin, whereas at higher concentrations (e.g., 1%) the bulk of the extracted material is present as GP Ib (Figure 8). Thus, by the use of Triton X-100 at a concentration of 1%, one avoids most of the action of the calcium-dependent proteases on GP Ib. The reason for this is unknown, but it may simply be that the proteases require an ordered phospholipid structure for maximal activity and that these structures are rapidly taken apart by the detergent at these concentrations. The appearance of glycocalicin can also be prevented by the addition of calcium-dependent protease inhibitors, including leupeptin, N-ethylmaleimide, EDTA, and EGTA. When such inhibitors were added to the buffer used for extraction, glycocalicin was present in only trace amounts. Crossed immu-noelectrophoresis of these extracts using an antiserum to glycocalicin showed a total of four peaks (or defined positions) along a continuous immunoprecipitate (Figure 9). The peak of the most fast-moving component, glycocalicin, was now located in the inter-mediate gel, indicating its trace quantity. This was followed by the part of the immu-noprecipitate corresponding to the previously defined GP Ib. This may either be seen as a low peak or just as a line connecting the glycocalicin precipitate to a third, pronounced peak representing what we have called "peak III-GP Ib." Finally, the fourth peak is seen as a rocket-shaped immunoprecipitate emerging from the applica-tion well, representing material that is immobile in the first-dimension electrophoresis (Figure 9) (see also Solum *et al.*, 1983b). When antibodies to the whole platelet proteins are used in the second-dimension gel instead of antiglycocalicin antiserum, the regular GP Ib immunoprecipitate is not seen under these conditions (Solum *et al.*,

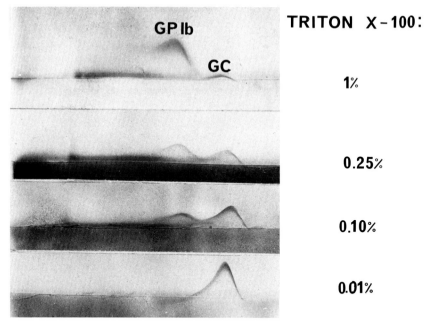

Figure 8. Effect of variation in the concentration of Triton X-100 used for extraction of platelets on the relative amounts of GP Ib and glycocalicin in the extracts. 200 μl Triton X-100 containing standard Tris-glycine buffer (pH 8.7) were added to each of four aliquots of sedimented, washed platelets (1.75 × 10⁹ cells per aliquot). The concentrations of Triton X-100 were 0.01, 0.10, 0.25, and 1.0%. The platelets were suspended and shaken at 20 °C for 30 min followed by centrifugation at 8000 × g for 4 min. The supernatants were subjected to CIE using antiserum to glycocalicin. Migration of glycocalicin (GC) in the first dimension was 3.8 cm in all electrophoreses. Note that only a small amount of glycocalicin was formed at a Triton X-100 concentration of 1%. Reproduced from Solum *et al.* (1983b) with permission.

1983). Peak III-GP Ib is difficult to observe with the anti-whole-platelet antibodies because the corresponding area shows a very complex immunoprecipitate pattern being dominated by GP IIb-IIIa. ''Peak III-GP Ib'' may be observed, however, either by dissociating the GP IIb-IIIa complex by chelation of divalent cations, as described in Section 3.3, or by inclusion of antiglycocalicin antibodies in the intermediate gel according to the principles described in Section 2.2.

Three additional observations are considered important in an attempt to identify these additional electrophoretic forms of GP Ib. No difference in molecular weight of GP Ib could be observed by SDS–PAGE of our platelet extracts whether these were prepared in the presence or absence of leupeptin in addition to Triton X-100 (Solum *et al.*, 1983b). Sodium dodecyl sulfate–PAGE of such extracts showed the presence of the actin-binding protein in such extracts only if these were prepared in the presence of leupeptin, not in its absence, and this cytoskeletal protein is known to be a sensitive substrate for the calcium-activated proteases (White, 1980; Rosenberg *et al.*, 1981). Incubation of intact platelets with the local anesthetic dibucaine, which is known to ''switch on'' a calcium-dependent proteolysis in ''intact'' platelets (Coller, 1982; Solum *et al.*, 1983b), induced a continuous transformation of the GP Ib-related mate-

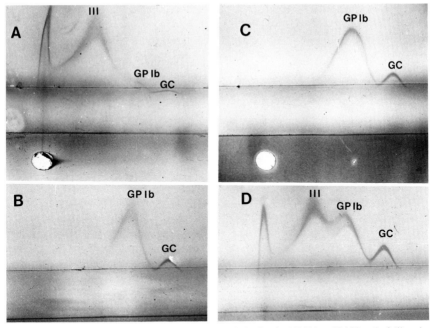

Figure 9. The effect of leupeptin during solubilization of platelets in 1% Triton X-100 and of dibucaine on the GP Ib-related components of platelet extracts. Antiserum to glycocalicin was used in all plates. (A) Platelets incubated only with buffer (control platelets), leupeptin present during solubilization. (B) Control platelets, leupeptin absent during solubilization. (C) Dibucaine-incubated platelets, leupeptin absent during solubilization. (D) Dibucaine-incubated platelets, leupeptin present during solubilization. Washed platelets, suspended in Tris-buffered saline (pH 7.4), were divided into two portions. One portion was diluted with buffer containing dibucaine (final concentration, 1 mM), and the other was diluted with buffer only. Both portions were then incubated at 37 °C. After 20 min, two aliquots were removed from each portion and centrifuged at 8000 × g for 2 min. The sedimented platelets of one of the aliquots were then solubilized in Tris-glycine buffer (pH 8.7) that contained both 1% Triton X-100 and 4.2 mM leupeptin, whereas the platelets of the other aliquot were solubilized in the same buffer containing Triton X-100 only. Solubilization was performed by stirring at 20 °C followed by centrifugation at 8000 × g for 2 min. CIE of the extracts was then performed on the supernatant material. For technical reasons, the extracts used in (B) and (C) were diluted one third as compared with the other extracts prior to electrophoresis. Note that four components can be recognized when leupeptin has been present during solubilization versus the two observed in its absence. Further, that incubation of "intact" platelets with dibucaine leads to a conversion from the peak III form into the "regular" form of GP Ib (GC, glycocalicin). Reproduced from Solum *et al.* (1983b) with permission.

rial from the peak III material to the previously defined GP Ib and further to extracellular glycocalicin (Figure 9).

These observations indicate to us that GP Ib in the intact platelet may be associated with structures that are particularly sensitive to proteolysis by calcium-dependent proteases and that complexes between GP Ib and proteins of such structures may exist even in solution. They also suggest that cytoskeletal proteins may be involved in this phenomenon. Preliminary studies suggest that in the presence of leupeptin some GP Ib may be associated with the cytoskeletal fraction insoluble in Triton X-100 (Solum *et al.*, 1983a).

4. SUPRAMOLECULAR ORGANIZATION

Various approaches have been employed to study membrane topography, i.e., the exposure of the membrane proteins on the inner and outer membrane surfaces.

4.1. Absorption of Antibodies

The immunoelectrophoretic technique itself may provide information about the topographical arrangement of membrane proteins. Antibodies are first absorbed with whole, intact cells and subsequently used for CIE analysis of platelet proteins. Those precipitates present in CIE with control antiserum but not with adsorbed antiserum represent proteins located on the outer membrane surface. Those proteins penetrating the membrane will exhibit an increased area of their immunoprecipitates if they contain antigenic sites both on the outer and the inner surface of the membrane. Proteins exposed only to the inner membrane surface will remain unchanged (Bjerrum and Bøg-Hansen, 1976). This approach was first applied to the study of erythrocyte membrane proteins, but has also been employed on platelets (Hagen et al., 1979). The results obtained using antibodies absorbed with whole platelets confirm the conclusions reached by platelet-surface-radiolabeling experiments and treatment of platelets with enzymes.

4.2. Subcellular Localization

Crossed immunoelectrophoresis has also been employed to examine proteins present in the cytoplasm or α-granules. This has mainly been done by examination of the antigens in isolated fractions obtained after disruption of the platelets. However, since it is difficult to prepare pure subcellular fractions, this technique is often only suggestive. Possible artifacts may also arise from nonspecific adsorption of antigens to membrane or organelle structures. Using an improved method for isolation of α-granules, the solubilized proteins from this organelle fraction have been characterized by CIE (Gogstad et al., 1981, 1982b). More than 20 immunoprecipitates from this fraction were seen when antibodies to whole platelets were used. Several of these were identified by the use of monospecific antibodies or purified antigens as described in Section 2, i.e., TSP, PF4, VWF, albumin, and fibrinogen. In addition, six glycoproteins were observed by crossed-affinity immunoelectrophoresis where different lectins were included in the gel during the first-dimension electrophoresis. The complex consisting of GP IIb-IIIa was clearly present in the pattern obtained after CIE of Triton X-100-solubilized α-granules. These glycoproteins have previously been demonstrated in the plasma membrane. Their presence in the α-granules could either represent a true α-granule pool of these glycoproteins or they could originate from the surface membrane by adsorption of membrane fragments to the α-granules during homogenization of the platelets. To clarify this issue, whole platelets were labeled by lactoperoxidase-catalyzed iodination. From previous work, it was well known that GP IIb-IIIa in the plasma membrane would become labeled by this treatment. This is also demonstrated by autoradiography of immunoplates obtained by CIE of lactoperoxidase-labeled platelets. However, a possible α-granule pool of glycoprotein IIb-IIIa would not be available to ^{125}I by iodination of intact platelets. Autoradiography of the immunoplates obtained after CIE of solubilized membranes and α-granules isolated from ^{125}I-labeled whole platelets revealed labeling of GP IIb-IIIa in the membrane fraction, but not in the α-granule fraction. It was thus concluded that the α-granules contain a separate pool of GP IIb-IIIa. Further separation of the isolated α-granules into a particulate and a soluble fraction revealed that GP IIb-IIIa was present in the particulate fraction and thus probably as part of the α-granule membrane (Gogstad *et al.*, 1982b). However immunocytochemical studies in conjunction with electron microscopy have indicated that all GP IIb-IIIa is restricted to the surface plasma (Steinberg *et al.*, unpublished observations). Most of the immunoprecipitates present in the α-granule pattern could be recognized in the pattern obtained with the released proteins, and an elevation of the arcs representing common antigens is seen when a mixture of the two samples is analyzed (Gogstad *et al.*, 1982b). The immunoprecipitate containing GP IIb-IIIa was only seen in the α-granule pattern and not among the released proteins. Thus, these experiments provide strong evidence that the release reaction occurs by exocytosis and that the α-granule membrane is left with the platelet after secretion has taken place.

4.3. Functional Integrity of Antigens

As mentioned in the introduction, the use of Triton X-100 in CIE permits normal antigen–antibody interactions to occur. Use of this nondenaturing detergent also allows for determination of functional activities of resolved antigens. As will be discussed in Section 4.4, we have used immunoprecipitates of assumed receptor mole-

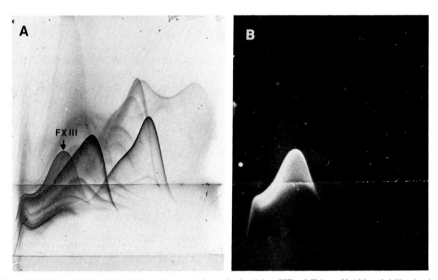

Figure 10. Factor XIII activity in an immunoplate obtained by CIE of Triton X-100 solubilized whole platelets with anti-whole-platelet antibodies. (A) Immunoplate stained with Coomassie Brilliant Blue. (B) Immunoplate that had been incubated with casein and monodansylcadaverine and exposed to ultraviolet light. The immunoplate was washed three times after incubation and dried prior to overnight incubation in a medium containing casein, monodansylcadaverine, dithiothreitol, and calcium at pH 7.4. After incubation, the immunoplate was washed for 15 min in 0.15 M NaCl and then for 15 min in ethanol-ether (1:1). Note the fluorescence along the immunoprecipitin arc of factor XIII. Reproduced from Gogstad and Brosstad (1983) with permission.

cules to study the binding of ligands to their receptors (Gogstad *et al.*, 1982a). Enzymatic activities may also persist in immunoprecipitates. Gogstad and Brosstad (1983) demonstrated that immunoprecipitates of factor XIII were able to catalyze the incorporation of fluorescent monodansylcadaverine into casein by a transglutaminase reaction (Figure 10). This shows that functional activities may be expressed even after immunoprecipitation has occurred. However, this should not be considered an absolute rule. Thus, the presence of the antibody molecule may inhibit enzyme function by sterically hindering substrate accessibility. Additional factors that may inactivate enzyme function by CIE are the presence of endogenous inhibitors, proteolysis, or disruption of macromolecular organizations.

4.4. Binding of Ligands and Identification of Receptors

One of the most fascinating aspects of platelet membrane research has been that of receptor identifications. As this problem will be discussed in separate chapters of this book, we will restrict our presentation to a brief description of how CIE has been applied to the direct study of ligand bindings, particularly the binding of thrombin, heparin, and fibrinogen. Principally, two approaches have been used. The first of these uses the ligand and the other uses the ''receptor'' in an immobilized state. In the first approach, immobilized ligand coupled to beads of Sepharose 4-B is incorporated into the intermediate gel in the CIE. Binding of an antigen to a ligand is observed as the reduction of the peak height of the corresponding immunoprecipitate (or reduction in the area encompassed by the immunoprecipitate). Using this approach, four thrombin-binding proteins were identified in platelet extracts. These represent GP Ib (as well as glycocalicin), PF4, factor XIII, and an antigen that could not be identified (No. 2a) (Hagen *et al.*, 1981). Binding of GP Ib to Sepharose-coupled thrombin is illustrated in Figure 11. Thrombin–GP Ib interactions may be of physiological significance since

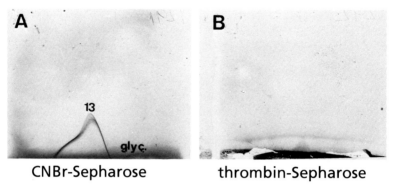

CNBr-Sepharose   thrombin-Sepharose

Figure 11. CIE of platelet proteins solubilized in 1% Triton X-100 with antiglycocalicin antiserum. The intermediate gel contained (A) CNBr-activated Sepharose 4B that had been inactivated by 1 M ethanolamine, and (B) thrombin coupled to Sepharose 4-B. The minor, fast-moving component of the double-peak immunoprecipitate seen in (A) represents the hydrophilic glycocalicin (glyc), whereas the major, slow-migrating component is GP Ib. Note the absence of the immunoprecipitates in (B) as compared with (A) due to binding of the antigens to the thrombin-Sepharose present in the intermediate gel. Reproduced from Hagen *et al.* (1981) with permission.

GP Ib is the only one of these four proteins present on the surface of unstimulated platelets. These observations are consistent with those of Okumura *et al.* (1978) who showed that glycocalicin inhibited thrombin-induced platelet activation.

Using immobilized heparin in the intermediate gel, six antigens of platelet extracts demonstrated binding to heparin. These were PF4, GP Ib, TSP, and three antigens termed G4, 17, and 25 (Gogstad *et al.*, 1983c). The binding of TSP to heparin has previously been demonstrated by Lawler *et al.* (1978).

Fibrinogen-binding proteins have also been identified by CIE. Following CIE of whole platelets with anti-whole-platelet antibodies, fibrinogen-binding proteins were identified by incubating ^{125}I-labeled fibrinogen with the plates containing immunoprecipitated proteins. Using this procedure, five immunoprecipitates were identified that bound fibrinogen: albumin, fibrinogen, factor XIII (subunit a), a granule-located antigen termed G4, and GP IIb-IIIa (Gogstad *et al.*, 1982a) (Figure 12). Since the labeling of albumin was prevented by the addition of unrelated proteins to the incubation mixture, this binding was considered unspecific. The labeling of the fibrinogen precipitate was probably due to an exchange of labeled for unlabeled fibrinogen in the immunoprecipitate. The binding of fibrinogen to the factor XIII immunoprecipitate probably occurs because the activated form of this enzyme can use fibrinogen as a substrate (Ly *et al.*, 1974; Kanaide and Shainoff, 1975). The reason for the labeling of G4 is unexplained. Glycoprotein IIb-IIIa is the only antigen labeled by the radioactive fibrinogen in our studies that is derived from the membrane surface (Gogstad *et al.*, 1982b), demonstrating the important role of GP|IIb-IIIa as the fibrinogen receptor (see Chapter 9).

Binding of ^{125}I-labeled degradation products of fibrinogen and fibrin to the GP IIb-IIIa immunoprecipitate may provide information on the structural features of the fibrinogen molecule responsible for binding. Such studies have been performed with fibrinogen degradation products derived either by plasmin digestion or CNBr-treatment (Brosstad *et al.*, 1983). We have found at least two sites in the E-domain and one site in each of the two D-domains capable of binding to GP IIb-IIIa (Brosstad *et al.*, 1983). These results were supported by concomitant studies on the binding of the same degradation products to ADP-activated whole platelets, their ability to function as ADP cofactor, or to inhibit ADP-fibrinogen-induced platelet aggregation (Thorsen *et al.*, 1983). Thus fibrinogen may interact with its receptor by more than one domain.

4.5. Metal-Binding Proteins

The binding of metal ions to platelet proteins can also be studied by applications of the CIE procedures described in Section 4.4. Calcium-binding platelet proteins were thus identified by incubation of CIE immunoplates with ^{45}Ca^{2+} after a prior removal of intrinsic calcium by EDTA followed by autoradiography to detect the ^{45}Ca^{2+} distribution. Three immunoprecipitates were markedly labeled with ^{45}Ca^{2+}. These corresponded to GP IIb-IIIa, GP Ia, and a presently unidentified antigen termed No. 5. As judged from studies on lactoperoxidase-catalyzed iodination of intact platelets and subcellular fractionation, all of these antigens were surface-oriented membrane proteins (Gogstad *et al.*, 1983a). When the EDTA-dissociated glycoproteins were subjected to CIE and ^{45}Ca^{2+} binding, the immunoprecipitates of both GP IIb and GP IIIa

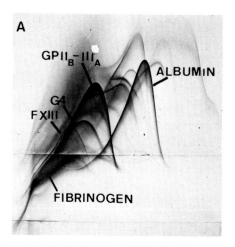

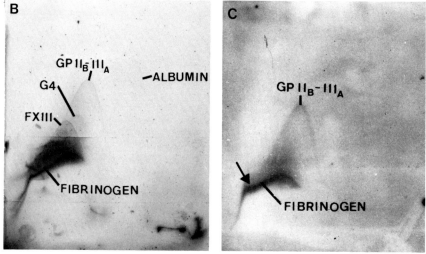

Figure 12. CIE of solubilized platelets with antibodies to whole platelets followed by incubation of the immunoplates with [125I]fibrinogen. (A) Stained immunoplate. (B) Autoradiography of immunoplate incubated with 0.1 mg [125I]fibrinogen/ml. (C) Autoradiography of immunoplate incubated with 0.01 mg [125I]fibrinogen/ml. The platelets were solubilized in Tris-glycine buffer containing 1% Triton X-100 and aliquots of about 60 μg protein were applied to each agarose gel. After electrophoresis, the gels were pressed and swelled in 0.154 M NaCl three times and finally dried. The immunoprecipitates were either stained with Coomassie Brilliant Blue or incubated with [125I]fibrinogen in 10 ml Tris-HCl buffer (pH 7.4)-0.154 M NaCl-1 mM CaCl₂-1 mM MgCl₂ for 18 hr at 20 °C. After incubation, excess of [125I]fibrinogen was removed by washing in 0.154 M NaCl. The immunoplates were finally dried and exposed to x-ray films for (B) 3 days, or (C) 5 days. Note the binding of radioactive fibrinogen to the immunoprecipitates marked in (B) and (C). Reproduced from Gogstad *et al.* (1982a) with permission.

bound $^{45}Ca^{2+}$ (Gogstad *et al.*, 1983a). Although divalent cations are required in the formation and maintenance of the GP IIb-IIIa complex, it is not known whether binding to one or both of those subunits is responsible for complex formation.

An observation that we believe to be important is the finding that incubation of citrated platelet-rich plasma with $^{45}Ca^{2+}$ followed by isolation of the platelets, solubilization in Triton X-100, CIE, and autoradiography revealed radioactivity in a previously unidentified immunoprecipitate that was not stained by Coomassie Blue. No increase in radioactivity of this immunoprecipitate was observed in connection with ADP-induced aggregation of the platelets in the $^{45}Ca^{2+}$-containing citrated plasma. However, if the platelets were first incubated with EDTA in the citrated plasma and then isolated and incubated with $^{45}Ca^{2+}$ after resuspension in Tris-buffered saline at pH 7.4, the radioactivity associated to this immunoprecipitate was markedly increased (Gogstad *et al.*, 1983a). It is generally accepted that washed platelets require added calcium ions for aggregation by ADP and fibrinogen (Born and Cross, 1964). One possible explanation of this may be that the added calcium ions are needed to replenish a pool of protein-bound calcium lost during the platelet washing, and that the calcium-binding antigen described here is involved in this phenomenon.

REFERENCES

Ali-Briggs, E. F., Clemetson, K. J., and Jenkins, C. S. P., 1981, Antibodies against platelet membrane glycoproteins. I. Crossed immunoelectrophoresis studies with antibodies that inhibit ristocetin-induced platelet aggregation, *Br. J. Haematol.* **48**:319–324.

Axelson, N. H., Kroll, J., and Weeke, B., 1973, A manual of quantitative immunoelectrophoresis, *Scand. J. Immunol.* **2**(suppl. 1):15–87.

Beese, J., Farr, W., Gruner, E., and Haschen, R. J., 1966, Proteolytische Enzyme in normalen menschlichen Blutplätchen, *Klin. Wschr.* **44**:1049–1053.

Berndt, M. C., and Phillips, D. R., 1981, The use of proteolytic probes to identify platelet membrane receptors (abstract), *Thromb. Haemostasis* **46**:75.

Bhakdi, S., Bhakdi-Lehnen, B., and Bjerrum, O. J., 1977, Detection of amphiphilic proteins and peptides in complex mixtures. Charge shift crossed immunoelectrophoresis and two-dimensional charge-shift electrophoresis, *Biochim. Biophys. Acta* **470**:35–44.

Bjerrum, O. J., 1978, Crossed hydrophobic interaction immunoelectrophoresis: An analytical method for detection of amphiphilic proteins in crude mixtures and for prediction of the result of hydrophobic interaction chromatography, *Anal. Biochem.* **90**:331–348.

Bjerrum, O. J., 1983, Detergent-immunoelectrophoresis. General principles and methodology, in: *Electroimmunochemical Analysis of Membrane Proteins* (O. J. Bjerrum, ed.), Elsevier Science Publisher, Amsterdam, pp. 3–44.

Bjerrum, O. J., and Bøg-Hansen, T. C., 1976, The immunochemical approach to the characterization of membrane proteins. Human erythrocyte membrane proteins analysed as a model system, *Biochim. Biophys. Acta* **455**:66–89.

Bjerrum, O. J., and Hagen, I., 1983, Biomolecular characterization of membrane antigens, in: *Electroimmunochemical Analysis of Membrane Proteins* (O. J. Bjerrum, ed.), Elsevier Science Publishers, Amsterdam, pp. 77–116.

Bjerrum, O. J., and Lundahl, P., 1973, Detergent-containing gels for immunological studies of solubilized erythrocyte membrane components, *Scand. J. Immunol.* **2**(suppl. 1):139–143.

Bjerrum, O. J., Helle, K. B., and Bock, E., 1979, Immunochemically identical hydrophilic and amphiphilic forms of the bovine adrenomedullary dopamine-β-hydroxylase, *Biochem. J.* **181**:231–237.

Bjerrum, O. J., Gerlach, J. H., Bøg-Hansen, T. C., and Hertz, J. B., 1982, Electroimmunochemical analysis of amphiphilic proteins and glycolipids stained with Sudan Black-containing detergent micelles, *Electrophoresis* **3:**89–98.

Bøg-Hansen, T. C., 1983, Affinity electrophoresis of glycoproteins, in: *Solid Phase Biochemistry Analytical and Synthetic Aspects* (W. H. Scowten, ed.), J. Wiley & Sons, Inc., New York, pp. 223–252.

Bordier, C., 1981, Phase separation of integral membrane proteins in Triton X-114 solution, *J. Biol. Chem.* **256:**1604–1607.

Born, G. V. R., and Cross, M. J., 1964, Effects of inorganic ions and of plasma proteins on the aggregation of blood platelets by adenosine diphosphate, *J. Physiol. (London)* **170:**397–414.

Brosstad, F., Thorsen, L., Gogstad, G., Sletten, K., and Solum, N. O., 1983, Crossed immunoelectrophoretic studies on the binding of plasmin- and CNBr-mediated fibrin(ogen) fragments to the fibrinogen-platelet receptor (The GP IIb-IIIa complex) (abstract), *Thromb. Haemostasis* 50:85.

Clemetson, K. J., Naim, H. J., and Lüscher, E. F., 1981, Relationship between glycocalicin and glycoprotein Ib of human platelets, *Proc. Natl. Acad. Sci. U.S.A.* **78:**2712–2716.

Coller, B. S., 1982, Effects of tertiary amine local anesthetics on von Willebrand factor-dependent platelet function: Alteration of membrane reactivity and degradation of GP Ib by a calcium-dependent protease(s), *Blood* **60:**731–743.

George, J. N., Nurden, A. T., and Phillips, D. R., 1984, Molecular defects in interactions of platelets with the vessel wall, *N. Engl. J. Med.* **311:**1084–1096.

Gogstad, G. O., and Brosstad, F., 1983, Platelet factor XIII is an active enzyme after solubilization and crossed immunoelectrophoresis, *Thromb. Res.* **29:**237–241.

Gogstad, G. O., Hagen, I., Korsmo, R., and Solum, N. O., 1981, Characterization of the proteins of isolated human platelet α-granules. Evidence for a separate α-granule-pool of the glycoprotein IIb and IIIa, *Biochim. Biophys. Acta* **670:**150–162.

Gogstad, G. O., Brosstad, F., Krutnes, M. B., Hagen, I., and Solum, N. O., 1982a, Fibrinogen-binding properties of the human platelet glycoproetin IIb-IIIa complex: A study using crossed radio-immunoelectrophoresis, *Blood* **60:**663–671.

Gogstad, G. O., Hagen, I., Korsmo, R., and Solum, N. O., 1982b, Evidence for release of soluble, but not of membrane-integrated proteins from human platelet α-granules, *Biochim. Biophys. Acta* **702:**81–89.

Gogstad, G. O., Hagen, I., Krutnes, M.-B., and Solum, N. O., 1982c, Dissociation of the glycoprotein IIb-IIIa complex in isolated platelet membranes. Dependence of pH and divalent cations, *Biochim. Biophys. Acta* **689:**21–30.

Gogstad, G. O., Krutnes, M. B., and Solum, N. O., 1983a, Calcium-binding proteins from human platelets. A study using crossed immunoelectrophoresis and $^{45}Ca^{2+}$, *Eur. J. Biochem.* **133:**193–199.

Gogstad, G. O., Stormorken, H., and Solum, N. O., 1983b, Platelet α_2-antiplasmin is located in the platelet-α-granules. *Thromb. Res.* **31:**387–390.

Gogstad, G. O., Solum, N. O., and Krutnes, M. -B., 1983c, Heparin-binding platelet proteins demonstrated by crossed affinity immunoelectrophoresis, *Brit. J. Haematol.* **53:**563–573.

Hagen, I., Bjerrum, O. J., and Solum, N. O., 1979, Characterization of human platelet proteins solubilized with Triton X-100 and examined by crossed immunoelectrophoresis. Reference patterns of extracts from whole platelets and isolated membranes, *Eur. J. Biochem.* **99:**9–22.

Hagen, I., Nurden, A., Bjerrum, O. J., Solum, N. O., and Caen, J. P., 1980, Immunochemical evidence for protein abnormalities in platelets from patients with Glanzmann's thrombasthenia and the Bernard-Soulier syndrome, *J. Clin. Invest.* **65:**722–731.

Hagen, I., Brosstad, F., Solum, N. O., and Korsmo, R., 1981, Crossed immunoelectrophoresis using immobilized thrombin in intermediate gel. A method for demonstration of thrombin-binding platelet proteins, *J. Lab. Clin. Med.* **97:**213–220.

Hagen, I., Brosstad, F., Gogstad, G., Solum, N. O., and Korsmo, R., 1982a, Demonstration of variable forms of the platelet factor 4 immunoprecipitate using crossed immunoelectrophoresis, *Thromb. Res.* **27:**77–82.

Hagen, I., Bjerrum, O. J., Gogstad, G., Korsmo, R., and Solum, N. O., 1982b, Involvement of divalent cations in the complex between the platelet glycoproteins IIb and IIIa, *Biochim. Biophys. Acta* **701:**1–6.

Helenius, A., and Simons, K., 1975, Solubilization of membranes by detergents, *Biochim. Biophys. Acta* **415:**29–79.

Howard, L., Shulman, S., Sandanandan, S., and Karpatkin, S., 1982, Crossed immunoelectrophoresis of human platelet membranes, *J. Biol. Chem.* **257:**8331–8336.

Kanaide, H., and Shainoff, J. R., 1975, Crosslinking of fibrinogen and fibrin by fibrin stabilizing factor (factor XIIIa), *J. Lab. Clin. Med.* **85**:574–597.

Kunicki, T. J., and Aster, R. H., 1979, Isolation and immunologic characterization of the human platelet alloantigen, Pl^{A1}, *Mol. Immunol.* **16**:353–360.

Kunicki, T. J., Nurden, A. T., Pidard, D., Russell, N. R., and Caen, J. P., 1981a, Characterization of human platelet glycoprotein antigens giving rise to individual immunoprecipitates in crossed immunoelectrophoresis, *Blood* **58**:1190–1197.

Kunicki, T., Pidard, D., Rosa, J.-P., and Nurden, A. T., 1981b, The formation of Ca^{++}-dependent complexes of platelet membrane glycoproteins IIb and IIIa in solution as determined by crossed immunoelectrophoresis, *Blood* **58**: 268–278.

Lawler, J. W., Slayter, H. S., and Coligan, J. E., 1978, Isolation and characterization of a high molecular weight glycoprotein from human blood platelets, *J. Biol. Chem.* **253**:8609–8616.

Ly, B., Kierulf, P., and Jakobsen, E., 1974, Stabilization of soluble fibrin/fibrinogen complexes by fibrin stabilizing factor (FSF), *Thromb. Res.* **4**:509–522.

Nachman, R. L., and Ferris, B., 1968, Studies on human platelet protease activity, *J. Clin. Invest.* **47**:2530–2540.

Newman, P. J., Knipp, M. A., and Kahn, R. A., 1982, Extraction and identification of human platelet integral membrane proteins using Triton X-114, *Thromb. Res.* **27**:221–224.

Norrild, B., Bjerrum, O. J., and Vestergaard, B. F., 1977, Polypeptide analysis of individual immunoprecipitates from crossed immunoelectrophoresis, *Anal. Biochem.* **81**:432–441.

Okumura, T., Lombart, C., and Jamieson, G. A., 1976, Platelet glycocalicin. II. Isolation and purification, *J. Biol. Chem.* **251**:5950–5955.

Okumura, T., Hasitz, M., and Jamieson, G. A., 1978, Platelet glycocalicin. Interaction with thrombin and role as thrombin receptor on the platelet surface, *J. Biol. Chem.* **253**:3435–3443.

Phillips, D. R., and Jakabova, M., 1977, Ca^{2+}-dependent protease in human platelets. Specific cleavage of platelet polypeptides in the presence of added Ca^{2+}, *J. Biol. Chem.* **252**:5602–5605.

Phillips, D. R., and Poh Agin, P., 1977, Platelet plasma membrane glycoproteins. Evidence for the presence of nonequivalent disulfide bonds using nonreduced-reduced two-dimensional gel electrophoresis, *J. Biol. Chem.* **252**:2121–2126.

Robinson, N. C., and Tanford, C., 1975, The binding of deoxycholate, Triton X-100, sodium dodecyl sulfate and phosphatidylcholine vesicles to cytochrome b_5, *Biochemistry* **14**:369–378.

Rosenberg, A., Stracher, A., and Lucas, R. C., 1981, Isolation and characterization of actin and actin-binding protein from human platelets, *J. Cell. Biol.* **91**:201–211.

Solum, N. O., Hagen, I., and Peterka, M., 1977, Human platelet glycoproteins. Further evidence that the "GP I band" from whole platelets contains three different polypeptides, *Thromb. Res.* **10**:71–82.

Solum, N. O., Hagen, I., Filion-Myklebust, C., and Stabaek, T., 1980a, Platelet glycocalicin. Its membrane association and solubilization in aqueous media, *Biochim. Biophys. Acta* **597**:235–246.

Solum, N. O., Hagen, I., and Sletbakk, T., 1980b, Further evidence for glycocalicin being derived from a larger amphiphilic platelet membrane glycoprotein, *Thromb Res.* **18**:773–785.

Solum, N. O., Olsen, T., and Gogstad, G., 1983a, GP Ib in the Triton-insoluble (cytoskeletal) fraction of platelets (abstract), *Thromb. Haemostasis* **50**:372.

Solum, N. O., Olsen, T. M., Gogstad, G. O., Hagen, I., and Brosstad, F. 1983b, Demonstration of a new glycoprotein Ib-related component in platelet extracts prepared in the presence of leupeptin, *Biochim. Biophys. Acta* **729**:53–61.

Thorsen, L. I., Brosstad, F., Gogstad, G., Sletten, K., and Solum, N. O., 1983, Fibrin(ogen) degradation products: Interference with binding of fibrinogen to ADP-stimulated platelets and their aggregation (abstract), *Thromb. Haemostasis* **50**:133 (Abstr.).

Truglia, J. A., and Stracher, A., 1981, Purification and characterization of a calcium dependent sulfhydryl protease from human platelets, *Biochem. Biophys. Res. Commun.* **100**:814–822.

White, G. C., 1980, Calcium-dependent proteins in platelets. Response of calcium-activated protease in normal and thrombasthenic platelets to aggregating agents, *Biochim. Biophys. Acta* **631**:130–138.

6

Platelet Membrane Electrical Potential
Its Regulation and Relationship to Platelet Activation

Avner Rotman

1. INTRODUCTION

The electrical potential across the platelet plasma membrane changes in response to specific stimuli (e.g., thrombin, ADP) and, conversely, alteration of the trans-membrane potential affects the platelet sensitivity to these activating agents. These electrical changes are mediated by a redistribution of cations across the plasma membrane, and therefore it may be assumed that conformational changes of membrane proteins are involved in this membrane transport process. Therefore, even though a discussion of platelet membrane potential cannot yet be related to the membrane glycoproteins, an understanding of this phenomenon is essential for a complete description of platelet membrane structure and function.

2. PLATELET MEMBRANE POTENTIAL

2.1. Resting Potential of the Platelet Membrane

The transmembrane electrical potential is the difference in the electric charge across the plasma membrane bilayer resulting from a more negative charge within the cell (Figure 1). This charge gradient is primarily controlled by the plasma membrane

Avner Rotman • Department of Membrane Research, Weizmann Institute of Science, Rehovot, Israel.

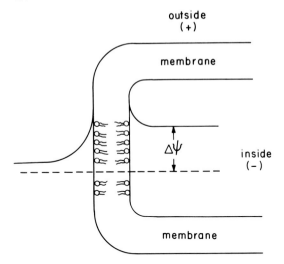

Figure 1. A schematic representation of transmembrane electrical potential. The plasma membrane is diagrammed as a lipid bilayer and the electrical potential (ψ) is presented as a measurement in millivolts, with the change in potential across the plasma membrane approximating 60 mV, inside negative.

enzyme, (Na^+, K^+)ATPase, that maintains the relatively low intracellular concentration of sodium and high intracellular concentration of potassium, relative to plasma. Human platelets have intracellular concentrations of sodium and potassium of about 40 and 120 μeq, respectively, maintaining a transmembrane resting potential 60 to 48 mV, inside negative (MacIntyre and Rink, 1982; Friedhoff and Sonenberg, 1983; Greenberg-Sepersky and Simons, 1984).

Due to the size limitation of platelets, it was (and still is) impossible to use microelectrodes and therefore the first studies on platelet membrane potential (Horne and Simons, 1976, 1978a) used a lipophilic cation dye (for example, see Figure 2) that is distributed between the cell and the medium in response to the potential difference (Sims et al., 1974; Hoffman and Laris, 1974; Bramhall et al., 1976). Friedhoff and Sonenberg (1983) used the probe [^{3}H]triphenylmethylphosphonium in addition to the cyanine dye, DiS, used by Horne and Simons (1976, 1978a) and measured the membrane potential of resting platelets as 52–60 mV, inside negative. The same value, 60 mV, inside negative, had been obtained by MacIntyre and Rink (1982) using DiS. The possibility of "side-effects" of the cyanine dye was also evaluated by MacIntyre and Rink (1982). Such a side-effect could be the inhibition at site I of the mitochondrial respiratory chain as shown by Montecucco et al. (1979). However, inhibitors of mitochondrial respiration such as cyanide and rotenone had no effect on the depolarization studies in platelets. This is in fact not surprising since platelets can maintain their normal ATP levels by either glycolytic or oxidative metabolism (Murer et al., 1976).

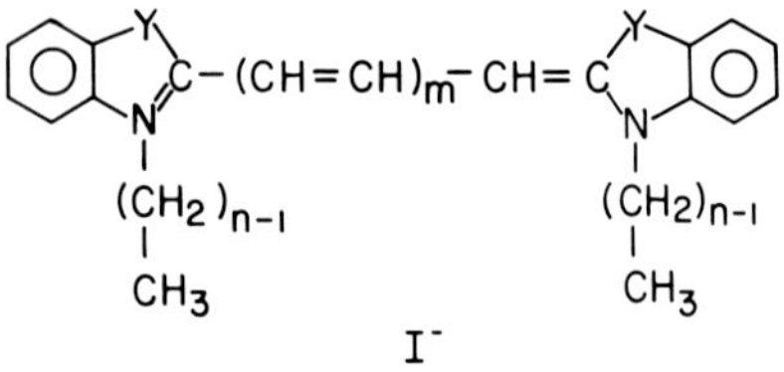

Figure 2. A general structural formula for the cyanine dyes commonly used to study changes in transmembrane potential in cells, where Y = O, S, or $(CH_3)C(CH_3)$; n varies from 2–18; m is 1–3; and the counter-ion for the positively charged N is iodide. The relationship between structure and function of these dyes is discussed in Sims et al. (1974).

Manipulation of the platelet suspension buffer demonstrated that the resting potential of the platelet plasma membrane is most sensitive to the extracellular potassium and hydrogen ion concentrations. Alterations of sodium ion, chloride ion, or calcium ion concentrations have no measurable effect on the platelet membrane resting potential (Friedhoff and Sonenberg, 1983; Greenberg-Sepersky and Simons, 1984).

In contrast to human platelets, Varečka et al. (1978) showed that in pig platelets, alteration of Ca^{2+} ion concentrations caused changes of the resting membrane potential. This difference between pig and human platelets may reflect the observation that porcine platelets are more dependent on extracellular Ca^{2+} than human platelets for response to agonists (Kinlough-Rathbone et al., 1974; MacIntyre and Gordon, 1975).

2.2. Effect of Platelet-Activating Agents on Platelet Membrane Potential

In their pioneering study, Horne and Simons (1976, 1978a) found that fluorescence of the cyanine dye, DiS, changed in response to stimulation of human platelets by ADP or thrombin, but not with collagen. This change in fluorescence reflected a decrease in the potential difference across the membrane, according to the method of Hoffman and Laris (1974), and was independent of the extracellular Ca^{2+} concentration or of aggregation. Similar results were reported by Bramhall et al. (1976), who studied suspensions of thymic lymphocytes, splenic lymphocytes, and platelets on treatment with valinomycin, a potassium ionophore, and A-23187, a calcium ionophore. Larsen et al. (1979) also reported that thrombin caused a dose-dependent depolarization of the membrane potential of washed platelets and that thrombin inactivated by tosyllysylchloromethylketone (TLCK) was unable either to induce aggregation or to change the platelet membrane potential. In contrast to these observations, MacIntyre and Rink (1982) could not observe any effect on platelet membrane potential with thrombin, ADP, or collagen. One possible explanation for this discrepancy is the different conditions used by the two groups: Horne and Simons (1978a) used much higher concentrations of platelets and cyanine dye than did MacIntyre and Rink (1982).

In a recent study, Greenberg-Sepersky and Simons (1984) demonstrated that thrombin, in the low concentrations present in clotting blood, alters the platelet membrane sodium gradient by stimulating uptake of sodium ions. The time course of this uptake and its inhibition by amiloride, a sodium channel blocker, was the same as the decrease in the platelet membrane electrical potential. Therefore, Greenberg-Sepersky and Simons (1984) concluded that platelet membrane depolarization caused by thrombin is directly dependent on the transmembrane sodium gradient and is primarily due to a dose-dependent sodium uptake by the platelets.

2.3. Effect of Platelet Membrane Potential on the Sensitivity of
Platelets to Activating Agents

The converse situation, alteration of platelet sensitivity to activating agents by prior manipulation of ion concentrations in the suspending buffer, seems dependent on potassium rather than sodium, similar to the maintenance of the platelet membrane resting potential. Greil et al. (1972) first recognized the potentiation of ADP-induced platelet aggregation by elevated potassium concentrations. Friedhoff and Sonnenberg

(1981, 1983) demonstrated that both elevating the extracellular postassium concentration and inhibiting the plasma membrane (Na^+, K^+)-dependent ATPase with ouabain caused parallel changes of platelet membrane depolarization and enhanced sensitivity to ADP-induced aggregation. Membrane depolarization caused by high extracellular potassium is not alone sufficient to initiate platelet aggregation (MacIntyre and Rink, 1982; Friedhoff and Sonenberg, 1983). Rather in platelets, as well as in quite different tissues such as brain (Wagner and Davis, 1979) and parotid (Strittmatter *et al.,* 1977), elevated extracellular potassium appears to potentiate the responsiveness to cell-surface ligands.

Thrombin-induced platelet secretion is also potassium-ion dependent, but different from ADP-induced aggregation, is not affected by the extracellular potassium concentration. Instead, this secretion requires the maintenance of potassium gradients across intracellular organelle membranes and/or maintenance of lysosomal potassium stores. Both of these parameters are decreased by the potassium ionophore, valinomycin, that also inhibits thrombin-induced secretion (Greenberg-Sepersky and Simons, 1984).

3. INTRACELLULAR pH IN PLATELETS

Several methods have been described in the last few years that enable the measurement of the internal pH of cells. These methods include microelectrodes for large cells, spectral shifts (McDonald and Jobsis, 1976), fluorescent determination of fluorescein coupled to cytochrome *c* (Thomas and Johnson, 1975) or to dextran (Ohkuma and Poole, 1978), pH-sensitive dyes (Johnson *et al.,* 1980; Bamburger *et al.,* 1973; Schuldiner *et al.,* 1972), determination of ^{14}C-labeled methylamine (Pick and Avron, 1976), or the application of 5,5-dimethyl-2,4-oxazolidinadione (Addanski *et al.,* 1968). However, due to limitations of these methods and other difficulties exhibited by the platelets themselves (e.g., small cell size), these methods are either unsuccessful in the platelet system or have not been utilized.

Only a few studies on the intracellular pH of blood platelets and its possible change during or after activation have been reported. Horne *et al.* (1981) measured changes in intraplatelet pH upon thrombin activation using two different fluorescent probes. One probe, 9-aminoacridine, is a weak base that is distributed across the membrane in a ratio that is a function of the difference in pH of the aqueous solutions on both sides of the membrane (Deamer *et al.,* 1972). The other method used by Horne *et al.* (1981) is based on the use of 6-carboxyfluorescein delivered into the cell as the diacetate ester, which is then cleaved by cytoplasmic esterases to yield the highly polar and therefore impermeable 6-carboxyfluorescein (Thomas *et al.,* 1979). Using these two methods, Horne *et al.* (1981) found that stimulation of platelets by α-thrombin leads to a dose-dependent rise in cytoplasmic pH of approximately 0.3 units when a saturating dose of 0.025 U/ml thrombin is used.

Recently, Rotman and Heldman (1982) measured the intracellular pH of blood platelets using a photolabel fluorescent probe. This probe, azidofluorescein diacetate, being hydrophobic, will penetrate into the cell very easily (Rotman and Heldman, 1980), and be enzymatically hydrolyzed there to yield the water-soluble azidofluorescein. This compound, upon photochemical activation, will insert into intracellular proteins via the active nitrene derivative. Using this method, Rotman and co-workers found that blood platelets rapidly adjusted their internal pH to that of the external medium, but they observed no change in internal pH as a result of thrombin activation. It is difficult to explain these discrepancies since only a few studies were done. However, possible explanations might be the fact that different probes were used by the two groups. Thus, a large portion of the 9-aminoacridine is not released by lysing the plasma membrane and may therefore be associated with organelles rather than the cytoplasm. 6-Carboxyfluorescein has the disadvantage of exhibiting a relatively small change in fluorescence between pH 7.0 and 7.4, a fact that decreases the sensitivity. Certainly, all three methods can be extended and further developed in order to expand the studies on intracellular pH and its possible changes as a result of platelet activation.

4. PLATELETS AND CATION FLUX

Transmembrane movement of ions down their electrochemical gradient is the basis of excitability for many cell types. Movement of ions across the various platelet membranes has been studied extensively for Ca^{2+} and much less in the case of Na^+ or K^+.

All agonists inducing platelet aggregation seem to act via a common pathway: the mobilization of intracellular calcium (Detwiler *et al.*, 1978). A comprehensive discussion of the role of Ca^{2+} in platelet activation is beyond the scope of this chapter and therefore only transport processes of calcium and other ions related to activation will be discussed.

4.1. Ca^{2+} Influx via the Plasma Membrane

Extracellular Ca^{2+} plays an important role in the process of platelet activation (Miletich *et al.*, 1978; Bennett and Vilaire, 1979; Lages and Weiss, 1981; Massini *et al.*, 1978). Free calcium concentration in the cytoplasm of resting platelets does not exceed 0.1 μM, whereas the calcium concentration in the plasma is 1–3 mM, 10,000-fold greater. There is substantial evidence that the intracellular calcium concentration rises very rapidly after platelet stimulation (Massini and Lüscher, 1976; Robblee and Shepro, 1976). The mechanism of this intracellular Ca^{2+} mobilization is still obscure. However, there is evidence that Ca^{2+} ions are mobilized from different subcellular sites, such as the stimulated plasma membrane, the dense tubular system of intracellular membranes, and perhaps also the mitochondria (Lüscher *et al.*, 1980). However, it seems that at least part of this increase in intracellular calcium concentration is due to an influx from the external medium through the plasma membrane during activation (Massini and Lüscher, 1976). It is not clear if this calcium influx is balanced by Ca^{2+} secretion and/or exchange processes. Owen and Le Breton (1980) reported that epinephrine can induce a flux of Ca^{2+} into the platelets. However, a report by Brass and Shattil (1982) provides some evidence that platelet activation by epinephrine and ADP results in exchange of Ca^{2+} between intracellular pools rather than an increase in the net amount of intracellular calcium.

4.2. Ca^{2+} Mobilization from Inner Face of Plasma Membrane

There is accumulating evidence that one of the storage pools of calcium ions in platelets is the inner surface of the plasma membrane. Evidence for membrane-bound calcium in blood platelets has come from many laboratories using different techniques. One of the first was Skaer et al. (1974), who provided electron microscopic evidence for the existence of membrane-bound Ca^{2+}. Le Breton et al. (1976) used the calcium-specific fluorophore, chlortetracycline, to identify the membrane-bound calcium pool. They also showed that shape change occurred simultaneously with release of Ca^{2+} from the membrane. There is some evidence that the calcium ion binding sites are located on membrane proteins or glycoproteins rather than on phospholipids (Kretsinger, 1979). Membrane glycoprotein IIb-IIIa is the major calcium binding site on the platelet's external surface, an observation that may be related to the calcium regulation of the GP IIb-IIIa complex formation (Peerschke et al., 1980; Fujimura and Phillips, 1983; Brass and Shattil, 1982, 1984; Gogstad et al., 1983). Additional surface binding site(s) are exposed during platelet activation that are not on GP IIb-IIIa (Brass and Shattil, 1982, 1984). Platelet membrane proteins other than GP IIb-IIIa, whose function is also calcium-dependent, such as phospholipase A_2 (Rittenhouse-Simmons et al., 1977), factor Xa receptor (Miletich et al., 1978), and calmodulin (Grinstein and Furuya, 1982), may also be potential calcium binding sites and serve as a possible membrane reservoir for calcium. However, no binding sites on the internal cytoplasmic surface of the platelet plasma membrane have yet been identified.

4.3. Mobilization of Ca^{2+} from the Dense Tubular System

The dense tubular system may be the platelet analogue to the sarcoplasmic reticulum in muscle (White, 1972a,b; see Chapter 2). The dense tubular system is the site of the (Ca^{2+}, Mg^{2+})-dependent ATPase activity (Cutler et al., 1978), of prostaglandin synthesis (Käser-Glanzmann et al., 1978; Carey et al., 1982), and of phospholipase A_2 and diglyceride lipase activities (Lagarde et al., 1982). There is also accumulating evidence that the dense tubular system is the major site of calcium storage. Thus, Käser-Glanzmann et al. (1978) reported that membrane vesicles, presumably isolated from the dense tubular system, had the highest ability to take up calcium. The same group showed that fractions enriched with plasma membrane showed very low calcium uptake. Using the technique of free-flow electrophoresis, Crawford and co-workers were able to isolate different platelet membrane fractions and study their functions and composition (Lagarde et al., 1982; Menashi et al., 1981). Recently, this group reported that an ATP-dependent, Ca^{2+}-accumulating process is almost exclusively associated with the intracellular membrane fraction (Menashi et al., 1982). They also confirmed an earlier report by Käser-Glanzmann et al. (1978) that the surface membrane vesicles have relatively low Ca^{2+} uptake capability.

4.4. Removal of Calcium from the Cytoplasm and Calcium Efflux

Due to its properties, the dense tubular system is also the likely candidate to serve as the calcium-sequestering system. Käser-Glanzmann et al. (1977) showed that

cAMP can enhance the sequestration of calcium into the intracellular membrane fraction assumed to be enriched in the dense-tubular system. The dense tubular system
contains (Ca^{2+}, Mg^{2+})ATPase activity and it was suggested that this Ca^{2+}-ac-
cumulating property might be linked to ability to produce prostaglandin endoperoxides
and thromboxanes.

There is also some evidence that the platelets contain a calcium pump in the
plasma membrane that might also remove calcium from the cytoplasm (Cutler *et al.*,
1978). This is supported by results reported by Brass and Shattil (1982), who found
that the rate of Ca^{2+} efflux from both unstimulated and stimulated platelets was
independent of the extracellular free Ca^{2+} concentration over a range of 10^{-9}–
10^{-3} M.

4.5. Na^+ and K^+ Movement across the Platelet Membrane

In addition to movement of Ca^{2+} across the platelet membranes, we will discuss
fluxes of two other ions: Na^+ and K^+. There are few reports dealing with the influx of
Na^+ into platelets. Feinberg *et al.* (1977) reported that the intracellular concentration
of Na^+ in platelets is in the range of 42 mM and that this Na^+ was exchangeable with
external Na^+ at 37 °C (measured by the use of $^{22}Na^+$). They also observed a rapid
influx of external Na^+ after ADP aggregation. This increase reached its peak at 60 sec
and diminished toward control levels at 360 sec. Under the same experimental conditions, no change in the level of $^{36}Cl^-$ was observed. Similar observations were made
by Born (1970) and by Feinberg *et al.* (1974), who reported also that ADP activation
was not followed by an increase in platelet volume. Ouabain, which inhibits influx of
K^+ into platelets (Gorstein *et al.*, 1967; Cooley and Cohen, 1967), caused a rapid
increase in intracellular Na^+ concentration, probably at the expense of the decrease in
K^+ concentration. Adenosine diphosphate caused a similar effect, namely an increase
in intracellular Na^+ concentration and a decrease in K^+ concentration. Recently,
Sandler *et al.* (1980) extended these studies by using other agonists and by blocking the
aggregation during ADP stimulation. Thus, when platelets were activated by ADP
under conditions where aggregation was inhibited (e.g., lack of Ca^{2+} in the external
medium), the induction of shape change was followed by a movement of Na^+ into the
cell. When the shape change was inhibited (e.g., by pretreatment of platelets with
prostaglandin E_1), the Na^+ uptake was also blocked. It is noteworthy to mention that
ADP activation does not induce uptake of Ca^{2+} (Le Breton and Feinberg, 1974;
Massini and Luscher, 1976). No Na^+ uptake was observed when platelets were
aggregated by epinephrine or vasopressin, which is a further indication that Na^+ influx
is associated with ADP activation and might even be an integral part of the ADP
activation process, whereas epinephrine activation is somehow related to the Ca^{2+}
influx. Inhibition by verapamil, which inhibits transmembrane Ca^{2+} flux, blocks
epinephrine-induced aggregation (Sandler *et al.*, 1980). On the other hand, ADP-
induced aggregation was not influenced by Ca^{2+} flux. Thus, amiloride, which blocks
Na^+ transport across membranes (Cuthberg and Shum, 1974; Bentley, 1968), inhibits
both ADP-induced aggregation and $^{22}Na^+$ uptake by platelets (Sandler *et al.*, 1980).
As discussed above in Section 2, Simons *et al.* (Horne and Simons, 1978b; Greenberg-
Sepersky and Simons, 1984) have shown that inhibition of Na^+ influx by amiloride

was accompanied by an inhibition of the thrombin-induced membrane depolarization, but not of secretion. Therefore, the thrombin-induced increase in sodium permeability of platelet membrane may be an initial step in the mechanism of stimulus–response coupling in platelets. However, an increase of intracellular sodium concentration per se is not sufficient to cause aggregation. This is clear from the fact that various reagents causing an increase in intracellular Na^+ do not alone cause platelet aggregation, e.g., ouabain, nigericin, and monensin (Feinstein et al., 1977).

The high intracellular K^+ concentration in platelets is maintained by (Na^+, K^+)-stimulated ATPase. Inhibition of this enzyme by ouabain or removal of K^+ from the medium results in a significant fall in platelet K^+ concentration (Cooley and Cohen, 1957; Gorstein et al., 1967; Buckingham and Maynert, 1964). Thrombin induces a marked loss of K^+ from platelets (Buckingham and Maynert, 1964; Zieve et al., 1964). The mechanism or the source of this K^+ loss is unclear. In a study performed by Wiley et al. (1976), it was observed that ADP or epinephrine aggregation was not accompanied by a significant potassium loss (only 5–10% of all K^+). On the other hand, thrombin stimulation resulted in release of about 30% of platelet K^+ independently of aggregation, which is thought to be localized in the lysosomal granules with the acid hydrolases (Wiley et al., 1976).

5. SEROTONIN TRANSPORT

5.1. Serotonin Transport across the Plasma Membrane

Blood platelets possess a unique transport system for serotonin (Rotman, 1983). This transport system has many characteristics of the presynaptic nerve ending uptake system for biogenic amines (Rotman, 1980; Sneddon, 1971). Due to this similarity, platelets have been used as a model for nerve endings for the purpose of serotonin transport studies (Modai et al., 1979; Rotman et al., 1980, 1982; Tuomisto and Tukiainen, 1976). Most of these studies were pharmacological comparisons of serotonin uptake by platelets of healthy and mentally ill people and were usually carried out in platelet-rich plasma. Only a few studies have been done where the kinetics of serotonin uptake by isolated platelets were investigated (Rotman et al., 1979; Tuomisto et al., 1979). These studies indicated that the K_m of serotonin uptake by human platelets is in the range of $0.1–1.0$ μM. Most of this serotonin is being taken up by and stored in the dense granules (Pletscher, 1968), possibly as a complex with adenine nucleotides (Pletscher and Laubscher, 1980; Costa and Murphy, 1980). The transport of serotonin across the plasma membrane of platelets (as well as nerve terminals and most other cells) is Na^+-dependent (Sneddon, 1971; Gripenberg, 1976). Recently, many studies were reported by Rudnick and co-workers on the transport of serotonin into isolated platelet membrane vesicles. Many characteristics of this transport into the isolated vesicles are similar to intact platelets and therefore these sealed membrane vesicles serve as a very suitable model for the platelet themselves. Using the plasma membrane vesicles, Rudnick and co-workers observed that K^+ stimulates the serotonin transport, possibly by accelerating the rate-limiting step of the transport cycle

(Nelson and Rudnick, 1979). Internal protons can substitute for the K^+ and a pH difference (acid inside) across the plasma membrane will enhance serotonin accumulation by the plasma membrane vesicles (Keyes and Rudnick, 1982). However, the high internal K^+ concentration blocks the stimulation by internal protons. Indeed, high proton concentrations inhibit stimulation by internal K^+, but given that the intracellular K^+ concentration is high, the physiological role of pH-driven serotonin transport may be negligible.

Rudnick and co-workers suggest that serotonin is cotransported with Na^+ (Nelson and Rudnick, 1979) and that one K^+ (or H^+) is countertransported with serotonin (Rudnick, 1977). The serotonin transport is only partially inhibited by ouabain. As the overall transport process is electroneutral and as serotonin is transported in its cationic form, it seems likely that Cl^- is the cotransported anion (Rudnick and Nelson, 1978; Keyes and Rudnick, 1982). This is in agreement with reports by Sneddon (1971) and Lingjarde (1960) that external Cl^- is required to maintain serotonin transport by intact platelets.

5.2. Serotonin Transport across the Dense Granule Membrane

The storage of serotonin in the dense granules was first observed by Pletscher and co-workers (Tranzer *et al.*, 1966; Pletscher, 1968). Recently, Johnson *et al.* (1978) demonstrated that the dense granules had an acidic pH. Carty *et al.* (1981) purified platelet dense granules and showed that they accumulated serotonin in response to both a pH gradient and electrochemical gradient. Wilkins and Salganicoff (1981) showed that both serotonin accumulation and granule acidification are stimulated by ATP. Recently, Fishkes and Rudnick (1982) reconfirmed these observations and showed that the serotonin transport into dense granules is derived from both the pH difference (interior acid) and the membrane potential (interior positive) generated by ATP hydrolysis.

6. CONCLUSION

This review of the platelet membrane regulation of electrical charge, pH, and cation concentrations demonstrates a different aspect of membrane complexity from the other chapters in this volume. It is clear that the rapid changes of membrane potential and ion fluxes discussed here are an integral part of platelet involvement in hemostasis and thrombosis. The interaction of the platelet surface with external ligands both causes, and is modulated by, electrochemical changes across the plasma membrane and the internal organelle membranes. Future research will further define the structural alterations of the platelet membrane glycoproteins that are involved in these reactions.

ACKNOWLEDGMENTS.   Most of the work from the author's laboratory was generously supported by the Gatsby Foundation (London, U.K.). Avner Rotman is the incumbent of the Samuel and Isabelle Friedman Career Development Chair.

REFERENCES

Addanski, S., Cahill, R. O., and Sotos, J. F., 1968, Determination of intramitochondrial pH and intra-mitochondrial-extramitochondrial pH gradient of isolated heart mitochondria by use of 5,5-di-methyl-2,4-oxazolidine dione, *J. Biol. Chem.* **243**:2337–2348.

Bamburger, E., Rottenberg, H., and Avron, M., 1973, Internal pH, pH and the kinetics of electron transport in chloroplasts, *Eur. J. Biochem.* **34**:557–563.

Bennett, J. S., and Vilaire, G., 1979, Exposure of platelet fibrinogen receptors by ADP and epinephrine, *J. Clin. Invest.* **64**:1393–1401.

Bentley, P. J., 1968, Amiloride: A potent inhibitor of sodium transport across the toad bladder, *J. Physiol.* **195**:317–330.

Born, G. V. R., 1970, Observations on the change in shape of blood platelets brought about by adenosine diphosphate, *J. Physiol.* **209**:487–511.

Bramhall, J. S., Morgan, J. I., Perris, A. D., and Britten, A. Z., 1976, The use of a fluorescent probe to monitor alterations in transmembrane potentials in a single cell suspensions, *Biochem. Biophys. Res. Commun.* **72**:654–662.

Brass, L. F., and Shattil, S. J., 1982, Changes in surface-bound and exchangeable calcium during platelet activation, *J. Biol. Chem.* **257**:14000–14005.

Brass, L. F., and Shatill, S. J., 1984, Identification and function of the high-affinity binding sites for Ca^{2+} on the surface of platelets, *J. Clin. Invest.* **73**:626–632.

Buckingham, S., and Maynert, E. W., 1964, The release of 5-hydroxytryptamine, potassium and amino-acids from platelets, *J. Pharmacol. Exp. Ther.* **143**:332–339.

Carey, F., Menashi, S., and Crawford, N., 1982, Localization of cyclooxygenase and thromboxane syn-thetase in human platelet intracellular membrane, *Biochem. J.* **204**:847–851.

Carty, S. E., Johnson, R. G., and Scarpa, A., 1981, Serotonin transport in isolated platelet granules, *J. Biol. Chem.* **256**:11244–11250.

Cooley, M. H., and Cohen, P., 1967, Potassium transport in human blood platelets, *J. Lab. Clin. Med.* **70**:69–79.

Costa, J. L., and Murphy, D. L., 1980, Unique specializations for the subcellular compartmentalization of amines in pig and human platelets, in: *Platelets: Cellular Response Mechanisms and their Biological Significance* (A. Rotman, F. A. Meyer, C. Gitler, and A. Silberg, eds.), Wiley, Chichester, pp. 233–247.

Cuthberg, A. W., and Shum, W. K., 1974, Amiloride and the sodium channel, *Arch. Pharmacol.* **281**:261–269.

Cutler, L., Rodan, G., and Feinstein, M. B., 1978, Cytochemical localization of adenylate cyclase and of calcium ion, magnesium ion-activated ATPase in the dense tubular system of human blood platelets, *Biochim. Biophys. Acta* **542**:357–371.

Deamer, D. W., Prince, R., and Crofts, A., 1972, The response of fluorescent amines to gradients across liposome membranes, *Biochim. Biophys. Acta* **274**:323–335.

Detwiler, T. L., Caro, I. F., and Feinman, R. D., 1978, Evidence that calcium regulates platelet function, *Thromb. Haemostasis* **40**:207–211.

Feinberg, H., Michel, H., and Born, G. V. R., 1974, Determination of the fluid volume of platelets by their separation through silicon oil, *J. Lab. Clin. Med.* **84**:926–934.

Feinberg, H., Sandler, W. C., Scorer, M., Le Breton, G. C., Grossman, B., and Born, G. V. R., 1977, Movement of sodium in human platelets induced by ADP, *Biochim. Biophys. Acta* **470**:317–324.

Feinstein, M. B., Henderson, E. G., and Sha'afi, R. I., 1977, The effects of alterations of transmembrane Na^+ and K^+ gradients by ionophores (nigericin, monensin) on serotonin transport in human blood platelets, *Biochim. Biophys. Acta* **468**:284–295.

Fishkes, H., and Rudnick, G., 1982, Bioenergetics of serotonin transport by membrane vesicles derived from platelet dense granules, *J. Biol. Chem.* **257**:5671–5677.

Friedhoff, L. T., and Sonenberg, M., 1981, The effect of altered transmembrane ion gradients on membrane potential and aggregation of human platelets in blood plasma, *Biochem. Biophys. Res. Commun.* **102**:832–837.

Friedhoff, L. T., and Sonenberg, M., 1983, The membrane potential of human platelets, *Blood* **61**:180–185.

Fujimura, K., and Phillips, D. R., 1983, Calcium cation regulation of glycoprotein IIb-IIIa complex formation in platelet plasma membranes, *J. Biol. Chem.* **258**:10247–10252.

Gogstad, G. O., Krutnes, M. B., and Solum, N. O., 1983, Calcium-binding proteins from human platelets, *Eur. J. Biochem.* **133**:193–199.

Gorstein, F., Carrol, H. J., and Puszkin, E., 1967, Electrolyte concentrations, potassium flux kinetic, and the metabolic dependence of potassium transport in human platelets, *J. Lab. Clin. Med.* **70**:92–103.

Greenberg-Seperksy, S. M., and Simons, E. R., 1984, Cation gradient dependence of the steps in thrombin stimulation of human platelets, *J. Biol. Chem.* **295**:1502–1508.

Greil, H., Patschek, H., and Brossmer, R., 1972, Effect of lithium and other monovalent cations on the ADP-induced platelet aggregation in human platelet-rich plasma, *Febs Lett.* **26**:271–273.

Grinstein, S., and Furuya, W., 1982, Calmodulin binding to platelet plasma membranes. *Biochim. Biophys. Acta* **686**:55–64.

Gripenberg, J., 1976, Inhibition by reserpine, guanethidine and imipramine of the uptake of 5-hydroxytryptamine by rat peritoneal mast cells *in vitro, Acta. Physiol. Scand.* **96**:407–416.

Hoffman, J. F., and Laris, P. C., 1974, Determination of membrane potentials in human and amphiuma red blood cells by means of a fluorescent probe, *J. Physiol.* **239**:519–552.

Horne, W. C., and Simons, E. R., 1976, The effect of aggregating agents and drugs on the membrane potential of washed human platelets (abstract), *Fed. Proc.* **35**:1451.

Horne, W. C., and Simons, E. R., 1978a, Probes of transmembrane potentials in platelets: changes in cyanine dye fluorescence in response to aggregation stimuli, *Blood* **51**:741–749.

Horne, W. C., and Simons, E. R., 1978b, Effects of amiloride on the response of human platelets to bovine α-thrombin, *Thromb. Res.* **13**:599–607.

Horne, W. C., Norman, N. E., Schwartz, D. B., and Simons, E. R., 1981, Changes in cytoplasmic pH and in membrane potential in thrombin-stimulated human platelets, *Eur. J. Biochem.* **120**:295–302.

Johnson, R. G., Scarpa, A., and Salganicoff, L., 1978, The internal pH of isolated serotonin-containing granules of pig platelets, *J. Biol. Chem.* **253**:7061–7068.

Johnson, R. G., Carty, S. E., Fingerhood, B. J., and Scarpa, A., 1980, The internal pH of mast cell granules, *Febs Lett.* **120**:75–79.

Käser-Glanzmann, R., George, J. N., Jakabova, M., and Lüscher, E. F., 1977, Stimulation of calcium uptake into platelet membrane vesicles by adenosine 3'5'-cyclic monophosphate and protein kinase, *Biochim. Biophys. Acta* **512**:1–12.

Käser-Glanzmann, R., Jakabova, M., George, J. N., and Lüscher, E. F., 1978, Further characterization of calcium accumulating vesicles from human blood platelets, *Biochim. Biophys. Acta* **512**:1–12.

Keyes, S. R., and Rudnick, G., 1982, Coupling of transmembrane proton gradients to platelet serotonin transport, *J. Biol. Chem.* **257**:1172–1176.

Kinlough-Rathbone, R. L., Chahil, A., and Mustard, J. F., 1974, Divalent cations and the release reaction of pig platelets, *Am. J. Physiol.* **226**:235–239.

Kretsinger, R. H., 1979, The informational role of calcium in the cytosol, in: *Advances in Cyclic Nucleotides Research* (P. Greengard, and G. A. Robinson, eds.), Raven Press, New York, pp. 1–26.

Lagarde, M., Quichardant, M., Menashi, S., and Crawford, N., 1982, The phospholipid and fatty acid composition of human platelet surface and intracellular membranes isolated by high voltage free from electrophoresis, *J. Biol. Chem.* **257**:3100–3104.

Lages, B., and Weiss, H. J., 1981, Dependence of human platelet functional response on divalent cations: Aggregation and secretion in heparin—and hirudin—anticoagulated platelet-rich plasma and the effects of chelating agents, *Thromb. Haemostasis* **45**:173–179.

Larsen, N. E., Horne, W. C., and Simons, E. R., 1979, Platelet interaction with active and TLCK-inactivated α-thrombin, *Biochem. Biophys. Res. Commun.* **87**:403–409.

Le Breton, G. C., and Feinberg, H., 1974, ADP-induced changes in intracellular Ca^{++} ion concentration, *Pharmacologist* **16**:699.

Le Breton, G. C., Dinerstein, R. J., Roth, L. J., and Feinberg, H., 1976, Direct evidence for intracellular divalent cation redistribution associated with platelet shape change, *Biochim. Biophys, Res. Commun.* **71**:362–370.

Lingjarde, O., 1960, Uptake of serotonin in blood platelets: Dependence on sodium and chloride, and inhibition by choline, *Febs Lett.* **3**:103–106.

Lüscher, E. F., Massini, P., and Käser-Glanzmann, R., 1980, The role of calcium ions in the induction of platelet activities, in: *Platelets: Cellular Response Mechanisms and their Biological Significance* (A. Rotman, F. A. Meyer, C. Gitler, and A. Silberberg, eds.), Wiley, Chichester, pp. 66–77.

MacIntyre, D. E., and Gordon, J. L., 1975, Calcium-dependent stimulation of platelet aggregation by PGE$_2$, *Nature (London)* **258**:337–339.

MacIntyre, D. E., and Rink, T. J., 1982, The role of platelet membrane potential in the inhibition of platelet aggregation, *Thromb. Haemostasis* **47**:22–26.

Massini, P., and Lüscher, E. F., 1976, On the significance of the influx of calcium into stimulated human blood platelets, *Biochim. Biophys. Acta* **436**:652–663.

Massini, P., Kaser-Glanzmann, R., and Luscher, E. F., 1978, Movement of calcium ions and their role in the activation of platelets, *Thromb. Haemostasis* **40**:212–218.

McDonald, V. W., and Jobsis, F. F., 1976, Spectrophotometric studies on the pH of frog skeletal muscle, *J. Gen. Physiol.* **68**:179–195.

Menashi, S., Weintroub, H., and Crawford, N., 1981, Characterization of human platelet surface and intracellular membranes isolated by free flow electrophoresis, *J. Biol. Chem.* **256**:4095–4101.

Menashi, S., Davis, C., and Crawford, N., 1982, Calcium uptake associated with intracellular membrane fraction prepared from human blood platelets by high-voltage, free flow electrophoresis, *Febs Lett.* **140**:298–302.

Miletich, J. P., Jackson, C. M., and Magerus, P. W., 1978, Properties of the factor Xa binding site on human platelets. *J. Biol. Chem.* **253**:6908–6916.

Modai, I., Rotman, A., Munitz, H., Tijano, S., and Wijsenbeek, H., 1979, Active uptake of serotonin by blood platelets of acute schizophrenic patients, *Psychopharmacology* **64**:193–196.

Montecucco, D., Pozzan, T., and Rink, T. J., 1979, Dicarbocyanine fluorescent probes of membrane potential block lymphocyte capping, deplete cellular ATP and inhibit respiration of isolated mitochondria, *Biochim. Biophys. Acta* **552**:552–557.

Murer, E. H., Hellem, A. J., and Rosenberg, M. C., 1976, Energy metabolism and platelet function, *Scand. Lab. Invest.* **19**:280–282.

Nelson, P. J., and Rudnick, G., 1979, Coupling between platelet 5-hydroxytryptamine and potassium transport, *J. Biol. Chem.* **254**:10084–10089.

Ohkuma, S., and Poole, B., 1978, Fluorescence probes measurements of the intralysosomal pH in living cells and the perturbation of pH by various agents, *Proc. Natl. Acad. Sci. U.S.A.* **75**:3327–3331.

Owen, N. E., and Le Breton, G. C., 1980, The involvement of calcium in epinephrine or ADP potentiation of human platelet aggregation, *Thromb. Res.* **17**:855–863.

Peerschke, E. I., Grant, R. A., and Zucker, M. B., 1980, Decreased association of [45]Calcium with platelets unable to aggregate due to thrombasthenia or prolonged calcium deprivation, *Br. J. Haematol.* **46**:247–256.

Pick, U., and Avron, M., 1976, A method for measuring the internal pH in illuminated chloroplasts based on the stimulation of protein uptake by amines, *Eur. J. Biochem.* **70**:569–576.

Pletscher, A., 1968, Metabolism, transfer and storage of 5-hydroxytryptamine in blood platelets, *Br. J. Pharmacol. Chemother.* **32**:1–16.

Pletscher, A., and Laubscher, A., 1980, Use and limitation of platelets as a model for neuron: Amine release and shape change reactions, in: *Platelets: Cellular Response Mechanisms and their Biological Significance* (A. Rotman, F. A. Meyer, C. Gitler, and A. Silberberg, eds.), Wiley, Chichester, pp. 267–276.

Rittenhouse-Simmons, S., Russell, F. A., and Deykin, D., 1977, Mobilization of arachidonic acid in human platelets kinetics and Ca^{++} dependency, *Biochim. Biophys. Acta* **488**:370–380.

Robblee, L. S., and Shepro, D., 1976, The effect of external calcium and lanthanum on platelet calcium content and on the release reaction, *Biochim. Biophys. Acta* **436**:448–459.

Rotman, A., 1980, The use of blood platelets serotonin uptake as a model in the study of mental illness, in: Enzymes and Neurotransmitters in Mental Disease (E. Esdin, T. S. Sourkes, and M. B. H. Youdim, eds.), Wiley, Chichester, pp. 65–76.

Rotman, A., 1983, Blood platelets in psychopharmacological research, *Prog. Neuropsychopharmacol.* **7**:135–151.

Rotman, A., and Heldman, J., 1980, Azidofluorescein diacetate: A novel intracellular labelling reagent, *Febs Lett.* **122**:215–218.

Rotman, A., and Heldman, J., 1982, Measurement of the intracellular pH of blood platelets using a photolabel fluorescent probe, *Biochim. Biophys. Acta* **720**:75–80.

Rotman, A., Modai, I., Munitz, H., and Wijsenbeek, H., 1979, Active uptake of serotonin by blood platelets of schizophrenic patients, *Febs Lett.* **101**:134–136.

Rotman, A., Caplan, R., and Szekely, G. A., 1980, Platelet uptake of serotonin in autistic and other psychotic children, *Psychopharmacology* **65**:245–248.

Rotman, A., Zemishlany, Z., Munitz, H., and Wijsenbeek, H., 1982, The active uptake of serotonin by platelets of schizophrenic patients and their families: Possibility of a genetic marker, *Psychopharmacology* **77**:171–174.

Rudnick, G., 1977, Active transport of 5-hydroxytryptamine by plasma membrane vesicles isolated from human blood platelets, *J. Biol. Chem.* **252**:2170–2174.

Rudnick, G., and Nelson, P. J., 1978, Platelet 5-hydroxytryptamine transport, an electroneutral mechanism coupled to potassium, *Biochemistry* **17**:4739–4742.

Sandler, W. C., Le Breton, G. C., and Feinberg, H., 1980, Movement of sodium into human platelets, *Biochim. Biophys. Acta* **600**:448–455.

Schuldiner, S., Rottenberg, H., and Avron, M., 1972, Determination of pH in chloroplasts. 2. Fluorescent amines as a probe for the determination of pH in chloroplasts, *Eur. J. Biochem.* **25**:64–70.

Sims, P. J., Waggoner, A. S., Wang, C. H., and Hoffman, J. F., 1974, Studies on the mechanism by which cyanine dyes measure membrane potential in red blood cells and phosphatidylcholine vesicles, *Biochemistry* **13**:3315–3330.

Skaer, R. J., Peters, P. D., and Emmines, J. P., 1974, The localization of calcium and phosphorus in human platelets, *J. Cell Sci.* **15**:679–692.

Sneddon, J. M., 1971, Relationship between internal Na^+/K^+ and the accumulation of ^{14}C-5-hydroxytryptamine by rat platelets, *Br. J. Pharmacol.* **43**:834–844.

Strittmatter, W. J., Davis, J. N., and Lefkowitz, R., 1977, Alpha-adrenegic receptors in rat parotid cells. Desensitization of receptor binding sites and potassium release, *J. Biol. Chem.* **252**:5478–5482.

Thomas, J. A., and Johnson, D. L., 1975, Fluorescein conjugates of cytochrome *c* as internal pH probes in submitochondrial particles, *Biochem. Biophys. Res. Commun.* **65**:931–939.

Thomas, J. A., Buchsbaum, R. N., Zimnial, A., and Racker, E., 1979, Intracellular pH measurements in Ehrlich ascites tumor cells utilizing spectroscopic probes generated in situ, *Biochemistry* **18**:2210–2218.

Tranzer, J. P., DaPrada, M., and Pletscher, A., 1966, Ultrastructural localization of 5-hydroxytryptamine in blood platelets, *Nature (London)* **212**:1574–1575.

Tuomisto, J., and Tukiainen, E., 1976, Decreased uptake of 5-hydroxytryptamine in blood platelets from depressed patients, *Nature (London)* **262**:596–598.

Tuomisto, J., Tukiainen, E., and Ahlfors, U. G., 1979, Decreased uptake of 5-hydroxytryptamine in blood platelets from patients with endogenous depression, *Psychopharmacology* **65**:141–147.

Varecka, L., Kovac, L., and Pogady, J., 1978, Trypsin and calcium ions elicit changes in the membrane potential in pig blood platelets, *Biochem. Biophys. Res. Commun.* **85**:1233–1238.

Wagner, H. R., and Davis, J. N., 1979, Beta-adrenergetic receptor regulation by agonist and membrane depolarization in rat slices. *Proc. Natl. Acad. Sci. U.S.A.* **76**:2057–2066.

White, J. G., 1972a, Interaction of membrane systems in blood platelets, *Am. J. Pathol.* **66**:295–372.

White, J. G., 1972b, The sarcoplasmic reticulum of platelets, *Fed Proc.* **31**:654.

Wiley, J. S., Kuchibhotla, J., Shaller, C. C., and Colman, R. W., 1976, Potassium uptake and release by human blood platelets, *Blood* **48**:185–197.

Wilkins, J. A., and Salganicoff, L., 1981, Participation of a transmembrane proton gradient in 5-hydroxytryptamine transport by platelet dense granules and dense granules ghosts, *Biochem. J.* **198**:113–123.

Zieve, P. D., Gamble, J. L., and Jackson, D. P., 1964, Effects of thrombin on the potassium and ATP content of platelets, *J. Clin. Invest.* **43**:2063–2069.

III

Interaction of Platelet Membrane Glycoproteins with the Extracellular Environment

Receptors for Platelet Agonists

David R. Phillips

1. INTRODUCTION

Platelet function is initiated by agonist interactions with specific receptors on the membrane surface. A broad spectrum of physiological agonists is capable of activating human platelets. These include a protease (thrombin), an adenine nucleotide (ADP), a structural protein (fibrillar collagen), and α-adrenergic agonists (e.g., epinephrine). Many agents not normally found in blood will also elicit an activation response. These include calcium ionophores (A23187), proteases (trypsin and thermolysin), phorbol esters (e.g., phorbol 12-myristate 13–acetate), lectins (wheat germ agglutinin), and antiplatelet antiserum. A variety of responses can occur, depending on the strength and concentration of the agonist. With a weak agonist such as ADP, platelets will change shape and aggregate. With stronger agonists such as collagen and thrombin, platelets will secrete the contents of their storage organelles in addition to changing shape and aggregating.

The objective of the present chapter is to summarize current data on the identification of the receptor(s) for four of the physiological platelet agonists: thrombin, ADP, collagen, and epinephrine. The structural diversity of the agonists indicates that these receptors are distinct, a speculation supported by observations showing that platelets rendered resistant to thrombin activation remain sensitive to collagen and that platelets respond synergistically to agonists. The structural requirements of the receptors predict that they are proteins, or glycoproteins in the case of platelets, since the majority of surface proteins are glycosylated (Phillips, 1980). A membranous location for agonist receptors has also been demonstrated: platelet-bound thrombin has been localized to the plasma membrane surface; ADP, like most polyphosphates, is relatively impermeable to membranes; collagen fibrils have been visualized on the plasma membrane surface; and α-adrenergic agonists have been shown to bind to membranes.

David R. Phillips • Gladstone Foundation Laboratories for Cardiovascular Disease, Cardiovascular Research Institute, and Department of Pathology, University of California, San Francisco, San Francisco, California 94140.

The reader should be forewarned that in no case are definitive data available on the identification of the receptors for any of the agonists listed above. Significant progress, however, has been achieved that permits speculation on the nature of some of these receptors. The stimulus–response coupling mechanisms for the initiation of these responses of platelets was the subject of a recent review (Detwiler and Huang, 1985).

2. THROMBIN

Thrombin exists in blood as prothrombin, an inactive zymogen with a molecular weight of 76,000. The catalytically active form, α-thrombin (M_r = 36,000), is a proteolytic product and is generated by a prothrombinase complex found in platelets (Miletich et al., 1978; Nesheim et al., 1979), monocytes (Tracy et al., 1983), lymphocytes (Tracy et al., 1983), and endothelial cells (Rodgers and Shuman, 1983). Purified α-thrombin undergoes autolysis to two additional forms, β-thrombin and γ-thrombin. The β- and γ-thrombins do not appear to be generated under physiological conditions, however. α-Thrombin generated in blood preferentially reacts with protease inhibitors preventing the formation of lytic products (Fenton et al., 1977). α-Thrombin also participates in many reactions in coagulation, e.g., clot formation and the activation of factor VIII, factor XIII and protein C. It also serves as a stimulus for many cells. In addition to activating platelets, thrombin also induces prostacyclin production (Weksler et al., 1978) and factor VIII release (Levine et al., 1982) from endothelial cells. It elicits a chemotactic response from monocytes (Bar-Shavit et al., 1983) and it acts as a mitotic agent for a variety of cells (see Section 2.5). Although the interactions of thrombin with proteins in coagulation have been well characterized, a precise description of the interaction of thrombin with receptors on cell surfaces has not yet been achieved. Discussed below are studies aimed at characterizing the reaction of thrombin with a receptor on the platelet membrane surface.

2.1. Thrombin–Platelet Interactions

α-Thrombin is the most potent stimulus of human platelets on a molar basis and is capable of fully activating human platelets at thrombin concentrations of 0.1 to 1 nM. Because micromolar prothrombin exists in human plasma (Fenton et al., 1977), it has been assumed that sufficient thrombin can be generated to activate platelets and initiate thrombus formation in vivo. It is known that thrombin acts on the platelet membrane surface, since it has been shown that ^{125}I-labeled thrombin is not internalized, but localizes at the plasma membrane surface (Tollefsen et al., 1974). The activation response of platelets to thrombin is rapid (Gear and Burke, 1983) and includes shape change, aggregation, secretion of the contents of α-granules, dense bodies, and lysosomes, and expression of the endogenous platelet lectin (Holmsen and Weiss, 1979; see Chapter 12).

It appears that α-thrombin hydrolyzes its receptor to induce platelet activation. Early observations by Davey and Lüscher (1967) showed that treatment of thrombin with diisopropyl fluorophosphate inhibited the proteolytic activity and platelet-activating capability of this enzyme. These authors also found that other proteases, including trypsin, papain, Pronase, and several snake venom proteases, also activated platelets. In contrast, chymotrypsin did not induce activation, even at concentrations of 1 mg/ml. These observations indicate that the proteolytic activity of thrombin is required to induce platelet activation. Subsequent studies by others support this thesis. For example, proteases with specificity similar to thrombin also activate platelets, even though these proteases are obtained from divergent evolutionary origins, i.e., papain (plant source) (Martin *et al.*, 1975), thrombocetein (*Bothrops atrox* venom), and α-clostripain (Cl. *histolyticum*) (M.C. Berndt and D.R. Phillips, unpublished results). Likewise, other inhibitors of the catalytic activity of thrombin, including phenylmethylsulfonyl fluoride, tosylchloromethylketone, and 1-chloro-3-tosylamido-7-amino-L-2-heptanone, also inhibit the platelet-activating capability of thrombin (Phillips, 1974; Larsen *et al.*, 1979; Workman *et al.*, 1977).

A proteolytic mechanism is also indicated by experiments showing that thrombin modifies its receptor. In one experiment, exposure of platelets to thrombin at 4 °C caused platelets to spontaneously activate upon warming, even though thrombin was no longer present (Mürer, 1972; Wallace and Bensusan, 1980). In another experiment, sequential additions of thrombin and hirudin permitted the continuation of metabolic responses to the stimulus, even though active thrombin was no longer present (Holmsen *et al.*, 1981). In a third experiment, Shuman *et al.* (1979) found that platelets treated with thrombin in the presence of prostacyclin were not activated by subsequent additions of thrombin after prostacyclin was removed. However, these platelets retained reactivity to other platelet agonists. This suggests that the thrombin receptor was selectively hydrolyzed by the first addition of thrombin.

2.2. Glycoprotein V Hydrolysis

To identify thrombin substrates on platelet membranes, surface membrane glycoproteins are radiolabeled to see which ones are cleaved by thrombin. In one study (Phillips and Agin, 1977), platelets were labeled by sequentially treating them with neuraminidase, galactose oxidase, and $[^3H]NaBH_4$. More than 30 labeled glycoproteins were identified on sodium dodecyl sulfate (SDS) gels. One of these, termed glycoprotein V (GP V, $M_r = 82,000$), was lost from the platelet plasma membrane following thrombin hydrolysis, resulting in the appearance of a hydrolytic fragment in the supernatant, which has been designated GP V_{f1} ($M_r = 67,500$). Subsequent work by Mosher and co-workers (1979) confirmed this observation and showed that other agonists, including ADP, collagen, or the calcium ionophore A23187, had no effect on the content of GP V in platelet membranes. Hydrolysis of GP V required that thrombin was catalytically active, since inhibition with 1-chloro-3-tosylamido-7-amino-L-2-heptanone inhibited this reaction (Knupp and White, 1981).

Glycoprotein V has been purified to > 98% homogeneity and has been shown to be a single–chain polypeptide (Berndt and Phillips, 1981a,b). It is classified as a peripheral protein, since it is eluted from the platelet plasma membranes by variations in the ionic strength of the suspending medium. This glycoprotein is heavily glycosylated and contains ~48% carbohydrate by weight, consisting of neutral hexose, hexosamine, and sialic acid in a molar ratio of ~8:2:1. In intact platelets, it exists in at least eight distinct isoelectric forms, with isoelectric focusing (pI) values from 5.85–

6.55. It appears that this microheterogeneity is due to variations in carbohydrate structure. Purified GP V has been found to be a thrombin substrate and on hydrolysis yields a major fragment, GP V_{f1}, which is identical in molecular weight to the glycoprotein observed previously in the supernatant of thrombin-treated, periodate-labeled platelets. Thus, a precursor product relationship between GP V and GP V_{f1} has been established, and it appears that thrombin acts on GP V directly and does not require the activation of another protease to affect hydrolysis.

Recently, McGowan and co-workers (1983) used an interesting series of experiments to test the hypothesis that hydrolysis of GP V by thrombin leads to platelet activation. In one experiment, the release of GP V_{f1} induced by various concentrations of thrombin was compared with the secretion of ATP. They found no consistent relationship between the amount of GP V_{f1} released and ATP secretion. In other experiments, platelets were pretreated either with the calcium-dependent protease or with chymotrypsin, both of which affected the concentration of platelet-associated GP V. These platelets showed a limited response to thrombin, but there was little GP V_{f1} in the supernatant. In a final experiment, these authors found that platelets pretreated with thrombin in the presence of PGI_2 still released GP V_{f1} when subsequently treated with thrombin in the absence of PGI_2, even though secretion was inhibited. From these experiments, McGowan *et al.* (1983) suggested that hydrolysis of GP V is neither necessary nor sufficient to induce platelet activation. However, although these experiments showed that no simple relationship exists between the thrombin-induced release of GP V_{f1} and platelet activation, they do not necessarily establish the relationship between thrombin-catalyzed *hydrolysis* of GP V and the activation response. It is entirely possible that the rate-determining step is the release of GP V_{f1} from the platelet membrane surface and not GP V proteolysis. Experiments examining the effects of the proteolytic removal of GP V by the calcium-dependent protease or chymotrypsin are difficult to interpret since these proteases may not remove the part of the glycoprotein that is cleaved by thrombin. It would thus appear that correlations between thrombin-induced responses and *proteolysis* of GP V must be made before a cause and effect relationship can be proved or disproved.

2.3. Equilibrium Binding of Thrombin to Platelets

Equilibrium binding of thrombin to platelets has been demonstrated in several laboratories using ^{125}I-labeled thrombin as the ligand (Tollefsen and Majerus, 1976; Tam and Detwiler, 1978; Workman *et al.*, 1977; White *et al.*, 1977; Ganguly and Sonnichsen, 1976). These studies have shown that platelets contain approximately 500 high-affinity thrombin-binding sites with an equilibrium dissociation constant (K_d) ≈ 1 nM and approximately 50,000 low-affinity sites with a K_d ≈ 100 nM (see Berndt and Phillips, 1981c, for review). The use of pyridoxyl 5'-phosphate-modified thrombins has shown that these two binding affinities occur on two different regions of thrombin (White *et al.*, 1981). In support of this finding, it has been shown that heparin affects the high-affinity binding of thrombin to platelets, whereas thrombomodulin inhibits both the high- and low-affinity binding of thrombin to platelets and apparently binds to thrombin in such a way that both domains on thrombin are affected (Esmon *et al.*, 1983).

Two binding sites are also suggested from the platelet-activating properties of thrombin Quick I (Henriksen and Brotherton, 1983). Thrombin Quick I has esterolytic and amidolytic activities similar to normal thrombin but has $\sim 1\%$ of the fibrinogen-clotting activity. The platelet-aggregating activity and thromboxane B_2-releasing activity of the abnormal thrombin are also markedly diminished. Henriksen and Brotherton (1983) suggest that this lack of aggregating activity results from the inability of thrombin Quick I to activate platelets and induce fibrinogen binding.

A basic question is whether equilibrium binding of thrombin is a measure of the functional receptor. This is obviously relevant in finding strategies to identify the thrombin receptor, since isolation of thrombin-binding proteins or identification of membrane proteins reactive with photoactivatable derivatives would presumably single out these thrombin-binding sites. Although several lines of evidence are in agreement with the view that the physiological response correlates with thrombin binding, some experiments are not. Detwiler and Feinman (1973a, b) showed that the extent of dense body secretion was a function of the thrombin concentration. In addition, Jamieson and Okumura (1978) found that Bernard-Soulier platelets have a decreased number of thrombin receptors and a decreased responsiveness to thrombin. Support for the role of thrombin binding has come from studies of the thrombin responsiveness of platelets enriched in cholesterol. Kramer *et al.* (1982) incubated platelets with cholesterol-rich phosphatidylcholine vesicles and enriched the cholesterol:phospholipid molar ratio in platelets 1.3-fold. This caused an increase in the thrombin-induced lipid metabolism, measured by hydrolysis of phosphatidylinositol and mobilization of arachidonic acid. Tandon *et al.* (1983) found that such enrichment in cholesterol content also increased the number of thrombin receptors and the affinity of thrombin for these receptors. They also found that these effects were reversible, in that the depletion of platelet cholesterol by incubating platelets with lecithin liposomes decreased the platelet's responsiveness to thrombin and decreased the thrombin receptor number and affinity.

There are classic experiments, however, that argue against the view that receptor occupancy induces platelet activation. These experiments compare the binding properties of proteolytically active α-thrombin with that of proteolytically inactive α-thrombin. The inactive derivatives were prepared by treating thrombin with reagents that react with the active site of the enzyme. It was found that platelets have the same number of high- and low-affinity binding sites for the inhibited thrombins as for active thrombin and, within experimental error, identical dissociation constants (Tollefsen *et al.*, 1974; Workman *et al.*, 1977). It would appear that the inhibited α-thrombins bind to the same site on platelets as active α-thrombin, because excess inhibited thrombin prevents binding of active thrombin. This competition, however, is not reflected in the functional response of platelets. Platelets equilibrated with a 1000-fold excess of inhibited thrombin show an identical or slightly enhanced time course and extent of platelet activation (Tollefsen *et al.*, 1974; Phillips, 1974). This demonstrates that the prevention of measurable α-thrombin binding has no effect on the ability of α-thrombin to activate platelets, suggesting that the measured binding does not take place at the site that activates platelets.

Inhibited α-thrombin also fails to compete for hydrolysis of GP V by active α-thrombin (Knupp and White, 1981). This discordance between thrombin binding and GP V proteolysis is also illustrated by the experiments of Fujimura *et al.* (1980), who showed that thrombin treated with 2-hydroxy-5-nitrobenzyl bromide demonstrated altered binding to platelets, yet retained platelet-activating activity and still hydrolyzed > 98% of the GP V. In another approach, Alexander *et al.* (1983) compared the binding and platelet-activating activities of α-thrombin, γ-thrombin, nitro α-thrombin, and trypsin and concluded that there was no correlation between the amount of protease saturably bound and the extent of platelet activation.

These observations are incompatible with the idea that equilibrium binding initiates platelet activation and further suggest that a nonequilibrium interaction of thrombin with platelets, such as proteolysis, initiates the physiological response.

2.4. Covalent Linkage of Thrombin to Platelets

Two studies have appeared in which investigators used reagents to covalently cross-link thrombin to its receptor on the platelet membrane surface. Tollefsen and Majerus (1976) used formaldehyde to covalently link ^{125}I-labeled, inhibited thrombin to platelets and identified a $M_r = 200,000$ adduct. This was the only product observed over a range of thrombin concentrations, indicating a single class of receptors. Larsen and Simons (1981) used thrombin modified with N,N'-bis(2-nitro-4-azidophenyl)cystamine S,S-dioxide (DNCO), a photoactivatable arylazide linked to a cleavable disulfide bond. This derivative retained proteolytic, esterolytic, and platelet-activating properties. At low concentrations of a proteolytically inactive form of ^{125}I-labeled DNCO-α-thrombin (0.01 U/ml), a single iodinated product of $M_r = 200,000$ was detected. With the proteolytically active form of ^{125}I-labeled DNCO-α-thrombin, a $M_r = 120,000$ adduct formed. It is not known whether this was a proteolytic product of the $M_r = 200,000$ adduct or represented the labeling of a new protein. At high concentrations of thrombin (0.1 U/ml and above), the $M_r = 200,000$ complex predominated, but two additional complexes were also present, one of $M_r = 400,000$ and one of $M_r = 46,000$. Neither of these additional proteins appeared to be thrombin substrates. These authors concluded, therefore, that the high-affinity receptor has an apparent molecular weight of ~160,000 (200,000 minus thrombin), with a residual molecular weight after thrombin cleavage of ~80,000.

On the basis of these two cross-linking studies, it would appear that thrombin binds to a surface protein on platelets that has a molecular weight of ~160,000. Glycoprotein Ib, a platelet membrane glycoprotein with approximately this molecular weight, has several other properties that suggest it may bind thrombin on the platelet surface: (1) it is localized on the platelet membrane surface (Phillips, 1972); (2) it binds to thrombin and is a competitive inhibitor of thrombin-induced platelet activation (Okumura and Jamieson, 1976; Okumura *et al.*, 1978; Ganguly and Gould, 1979); (3) thrombin retains GP Ib when this protease is incorporated in the intermediate gel of a crossed immunoelectrophoretic separation of platelet proteins (Hagen *et al.*, 1981) and (4) platelets from patients with Bernard-Soulier syndrome, which have deficient or abnormal GP Ib, demonstrate diminished thrombin binding and diminished thrombin-induced aggregation (Jamieson and Okumura, 1978).

Although GP Ib appears to satisfy the requirements for a thrombin-binding site on the membrane surface, there is no evidence that the interaction of thrombin with GP Ib causes platelet activation. Glycoprotein Ib is not a thrombin substrate, it inhibits the proteolytic activity of thrombin, and it reacts with proteolytically inactive thrombin. In addition, although treatment of intact platelets with chymotrypsin causes a parallel loss of GP Ib from the membrane surface and thrombin-binding activity, the time dependency of these decreases contrasts sharply with the almost instantaneous loss of the physiological response of platelets to thrombin (Okumura et al., 1978).

It is only possible, at present, to speculate on the function of the interaction of thrombin with GP Ib. One possibility is that GP Ib (or glycocalicin, a soluble derivative of GP Ib; see Chapter 3) is a thrombin inhibitor and serves to decrease the activity of this enzyme at thrombus foci. Another possibility, suggested by Fujimura et al. (1980), is that thrombin binds to GP Ib and is then positioned for catalytic hydrolysis of another substrate. This model, however, does not account for observations showing that GP Ib is a thrombin inhibitor; inhibited thrombin binds to platelets and can be covalently linked to a $M_r = 160,000$ protein, yet this binding does not inhibit the activation of platelets by active α-thrombin. Further studies are clearly required to determine the function of the interaction of thrombin with GP Ib. A $M_r = 74,000$ glycoprotein has been identified by Ganguly and Fossett (1981) that blocks thrombin-induced serotonin secretion, but which has no effect on platelet activation induced by ADP, trypsin, or ristocetin. It is not known, at present, whether this protein is identical to the one identified by the cross-linking studies described above, or whether this represents yet another thrombin-binding protein.

2.5. Thrombin Receptors on Other Cells

Many cells are responsive to thrombin, and it is instructive to compare their thrombin reactivities with the thrombin reactivity of platelets. One example is peripheral blood monocytes, which have been shown to bind 31,100 thrombin molecules with high affinity ($K_d = 3.4 \times 10^{-9}$ M) and an undetermined number of thrombin molecules with low affinity ($K_d = 1.3 \times 10^{-7}$ M) (Goodnough and Saito, 1982). Monocytes respond to thrombin by eliciting a chemotactic response, but unlike in platelets, this response still occurs with thrombin that is proteolytically inactive (Bar-Shavit et al., 1983). Thrombin has also been shown to be a potent mitogen for fibroblast cells in culture (Chen and Buchanan, 1975), a response that requires proteolytically active thrombin (Glenn et al., 1980). The thrombin receptor on fibroblasts has not yet been identified, but it has been shown that thrombin and a protein released into growth medium form a complex. This protein, nexin, is joined to the active site of thrombin by a covalent acyl linkage (Baker et al., 1980). It has been proposed that nexin regulates the mitogenic activity of thrombin by inhibiting its proteolytic activity (Low et al., 1982). Chinese hamster lung cells also undergo a mitogenic response to thrombin. A $M_r = 150,000$ thrombin receptor has been identified on these cells in studies using photoreactive cross-linking conjugates of ^{125}I-labeled α-thrombin (Moss et al., 1983), but the functional role of this receptor has not been established. Thrombin has also been shown to stimulate guanosine $3',5'$-monophosphate formation in murine neuroblastoma cells (Snider and Richelson, 1983) and it is taken up into mouse embryo cells, apparently by noncoated pit mechanisms (Carney and Bergmann, 1982).

Thrombin receptors have also been demonstrated on endothelial cells (Awbrey et al., 1979; Lollar et al., 1980; Isaacs et al., 1981a,b; Bauer et al., 1983). One thrombin receptor that has been identified is thrombomodulin, a protein that enhances the rate of activation of protein C (Owen and Esmon, 1981). A $M_r = 30,000–40,000$ protein has also been identified as a thrombin receptor (a covalent adduct to active α-thrombin) (Isaacs *et al.*, 1981a; Lollar *et al.*, 1980), but the relationship of this receptor to endothelial cell activation is not known. The activation responses of endothelial cells, such as division (Gospodarowicz *et al.*, 1978), production of prostacyclin (Weksler *et al.*, 1978), and secretion of proteins (Mosher and Vaheri, 1978; Levine *et al.*, 1982) may be elicited by additional thrombin receptors. These activation-dependent responses all appear dependent on the proteolytic activity of thrombin.

Megakaryocytes have been shown to respond to thrombin by increasing their rate of spreading (Leven *et al.*, 1983). Studies of thrombin binding to these cells have not been performed, but it is reasonable to assume that the functional thrombin receptor is similar to that which exists on platelets.

2.6. Summary

Thrombin appears to activate platelets by a proteolytic mechanism and with properties that do not correlate with equilibrium-binding parameters. Several thrombin-binding proteins have been identified on platelets, and one of these, GP V, has been identified as a thrombin substrate. Thrombin hydrolysis of GP V causes the release of a soluble fragment known as GP V_{f1}. Treatment of purified GP V with thrombin produces the same hydrolytic fragment as is produced from intact cells, indicating that thrombin acts directly on GP V.

3. ADENOSINE DIPHOSPHATE

In 1961, Gaarder *et al.* showed that ADP induces platelet shape change and subsequent aggregation. The importance of this finding was immediately recognized, since ADP is released from damaged tissues and erythrocytes following vascular injury. It is also secreted from the platelets themselves in response to collagen and thrombin stimulation (Grette, 1962) in concentrations sufficient (~μM) to induce platelet activation (Born, 1962). Because ADP is not taken into platelets (Salzman and Chambers, 1965), it must operate through a receptor on the platelet membrane surface. It is noteworthy that secreted ADP is not necessary for the aggregation response of an isolated population of platelets. Removal of secreted ADP by an enzyme-coupled system or antagonism of secreted ADP by added ATP has little effect on secretion and aggregation induced by other agonists (Macfarlane and Mills, 1975; Huang and Detwiler, 1980; Parise *et al.*, 1984). It is likely, therefore, that secreted ADP takes part in the perpetuation of the platelet response; that is, secreted ADP activates additional platelets and recruits them into the formation of a thrombus. Comprehensive reviews of platelet response to ADP have been published (Mills and Macfarlane, 1976; Haslam and Cusack, 1981).

3.1. Functional ADP Receptors

It is now known that ADP reacts with at least two functionally distinct receptors on intact platelets. One of the receptors induces the activation response of platelets.

This receptor is specific for ADP, except for 2-substituted ADP analogues that, in general, have activities equal to or greater than ADP itself (e.g., 2-azido ADP and 2-methylthio ADP). Adenosine triphosphate is an antagonist of ADP-induced responses. Phosphates other than ATP have little or no activity (Macfarlane and Mills, 1975). Although adenosine and many of its analogues inhibit platelet responses (Haslam and Cusack, 1981), this inhibition does not result from competition with ADP for the receptor that elicits platelet activation. Rather, adenosine appears to inhibit because it activates adenylate cyclase: the resulting increase in cytoplasmic cAMP inhibits activation by most stimuli.

The other membrane receptor for ADP is coupled to adenylate cyclase. Binding of ADP to this receptor causes an inhibition of the activity of this enzyme (Mills and Macfarlane, 1976). This ADP receptor can be distinguished from the one that induces platelet activation because the two receptors are affected differently by thiol reagents and substituted ADP derivatives. The thiol reagent paramercuribenzene sulfonate, which does not permeate membranes, renders platelets insensitive to ADP-induced inhibition of adenylate cyclase, but not to ADP-induced activation of platelets (Mills and Macfarlane, 1977). 2-Substituted derivatives of ADP also distinguish the two receptors. 2-Methylthio ADP is fourfold more potent than ADP in inducing platelet activation, but 300-fold more potent in inhibiting adenylate cyclase (Macfarlane *et al.*, 1979).

Platelet membranes also contain adenine nucleotide-binding proteins that are not expressed on the membrane surface and appear to be distinct from the two surface-oriented receptors. Actin and myosin, proteins that exist in platelet plasma membranes (Taylor *et al.*, 1975), are known to bind ADP (Mannherz and Goody, 1976). The activity of these proteins appears to be unrelated to the binding of ADP to surface-oriented receptors, since they are only associated with the inner aspect of the membranes of intact platelets (see Chapter 13). The presence of these and possibly other adenine nucleotide-binding proteins has complicated the interpretation of studies in which ADP binding to isolated membranes is measured. Furthermore, the existence of these proteins has frustrated attempts to identify an ADP receptor by isolation of an ADP-binding protein from isolated membranes. Since ADP-binding proteins exist on the cytoplasmic face of the membrane, it is essential that any ADP-binding protein in membranes originate from the membrane exterior surface before it can be considered the receptor for ADP-induced activation of platelets.

3.2.   Equilibrium Binding of ADP to Platelets

Attempts have been made to measure receptors on isolated membranes or intact platelets from the binding of radiolabeled ADP or ADP analogues. The binding parameters observed for ^{14}C-labeled ADP binding to isolated membranes vary widely, ranging from 200 to 1000 pmole ADP/mg of membrane protein, with dissociation constants from 0.2 to 2.5 μM (Nachman and Ferris, 1974; C. Legrand *et al.*, 1980; Lips *et al.*, 1980a). As discussed above, it seems likely that much of this binding is due to proteins such as actin and myosin. The variability observed in ADP binding to membranes is most likely caused from different procedures for membrane isolation, which yield membrane preparations containing variable amounts of these adenine nucleotide-binding proteins.

The binding of ADP to intact platelets has proved difficult to quantitate and associate with physiological responses. In initial studies using ^{14}C-labeled ADP, rapid binding was observed, which correlated with the aggregation response (Boullin *et al.*, 1972; Born and Feinberg, 1975). A subsequent, slower accumulation of label was also observed; this was due to the uptake of adenosine, a degradation product of ADP. Lips *et al.* (1980b) quantitated ADP binding and observed 30,000 binding sites with a dissociation constant of 5.5 μM. Since this dissociation constant correlated with the Michaelis constant for ADP-induced shape change, these authors concluded that this receptor is responsible for ADP-induced activation. At odds with this conclusion are the data of Macfarlane *et al.* (1982, 1983), who measured the binding of 2-substituted derivatives of ADP, 2-azido ADP and 2-methylthio ADP, compounds more active than ADP in inducing platelet activation. These authors found that platelets contain ~500 binding sites for these compounds, with a dissociation constant of ~150 nM. It was shown that these compounds bound to an adenine nucleotide receptor; ADP competed for binding with half-maximal displacement at ~5 μM (ATP at ~10 μM). To determine whether these derivatives were binding to the platelet-activating site or to the adenylate cyclase site, Macfarlane and co-workers studied the effects of *p*-chloromercuribenzene sulfonate (a reagent that does not permeate membranes), which discriminates between these two receptors. It was found that this reagent blocked the binding of 2-methylthio ADP and 2-azido ADP and inhibited the action of ADP on the cyclase. *p*-Chloromercuribenzene had no effect, however, on the ability of ADP to induce platelet shape change. Thus, these experiments show that the receptor for 2-substituted derivatives of ADP is coupled to adenylate cyclase and is not the receptor that mediates platelet activation.

3.3. Identification of ADP-Binding Proteins

Proteins in platelet membranes that bind ADP have been identified either by derivatization with chemically reactive derivatives of ADP or from observations of direct binding to purified proteins. One derivative studied was 2-azido [β-^{32}P]ADP, a photoactivatable compound. Macfarlane *et al.* (1982) photolysed this derivative (when bound to platelets) in an attempt to covalently link it to the adenylate cyclase-linked ADP receptor. Although radiolabeled membrane glycoproteins were observed, similar labeling occured in the presence of excess adenine, ADP, adenosine, or ATP, agonists of ADP binding and/or ADP-mediated responses. It thus appears that covalent labeling occurred at nonspecific sites on the platelet membrane surface.

Colman and co-workers (Bennett *et al.*, 1978; Colman *et al.*, 1980; Figures *et al.*, 1981) have used another analogue of ADP, 5'-*p*-fluorosulfonylbenzoyl adenosine (FSBA), which covalently reacts with amino and other functional groups on proteins. It was found that FSBA irreversibly inhibits ADP-induced shape change and aggregation, as well as the ADP-induced binding of fibrinogen. In contrast to ADP and the 2-substituted derivatives of ADP, FSBA does not inhibit PGE[1]-induced cyclic AMP accumulation. It also does not compete for binding of 2-methylthio ADP to platelets. Consequently, it has been suggested that FSBA binds to the receptor that mediates ADP-induced platelet activation without binding to the receptor inhibiting adenylate cyclase. When added to isolated membranes, FSBA was shown to react covalently with proteins having apparent molecular weights of 200,000, 135,000, 100,000, and 43,000 (Bennett *et al.*, 1978). Both ADP and ATP inhibited the labeling of each of these proteins. Comparison of these molecular weights to those of known ADP-binding proteins suggested that the M_r = 200,000 protein was myosin and the M_r = 43,000 protein was actin. However, when added to intact cells, FSBA only labeled the M_r = 100,000 protein. Approximately 2,500 sites were labeled per platelet, but a comparison of the kinetics of labeling to that of platelet activation indicated that as few as 200 molecules had to react before platelet activation occurred.

A different approach that has been taken to identify the ADP receptor has been that of identifying ADP-binding proteins in platelet membranes. Adler and Handin (1979) solubilized isolated platelet membranes and determined ADP-binding activity in this preparation. This binding activity had an apparent molecular weight of 61,000, as calculated from hydrodynamic measurements, and was clearly separable from nucleoside diphosphate kinase by sucrose density gradient centrifugation. It was shown that ADP bound to this protein with a K_d = 3.8 × 10^{-7} M. Binding was inhibited by ATP.

3.4. Summary

Two potential candidates for the ADP receptor are the M_r = 100,000 protein identified by FSBA labeling and the M_r = 61,000 protein identified in detergent-solubilized preparations. It is not possible, at present, to state whether either of these proteins is (1) the ADP-activation receptor, or (2) the receptor responsible for inhibition of adenylate cyclase, or (3) an unrelated ADP-binding protein. 5′-*p*-Fluorosulfonylbenzoyl adenosine contains a reactive sulfonyl fluoride derivative, and it is not known whether derivatization of membrane proteins is kinetically controlled (so that the most abundant protein is labeled), or whether reactivity is receptor controlled (so that only ADP-binding proteins are labeled). The protein isolated by Adler and Handin is of potential interest, but future studies are required to document whether the protein is located on the platelet membrane surface and whether it interacts with ADP during ADP-induced activation.

4. COLLAGEN

Platelets adhering to subendothelial tissue become activated; this has been demonstrated by observations showing that platelets on subendothelium change shape, spread, secrete the contents of their storage organelles, and aggregate. The adhesion of circulating platelets to subendothelial fibers *in vivo* is mediated by von Willebrand factor (VWF) and platelet membrane GP Ib (George *et al.*, 1984). The role of VWF in mediating platelet adhesion to collagen is unclear. Von Willebrand factor may only enhance the rate of adhesion, allowing the rapid contact interaction that is necessary in swiftly flowing blood (Aihara *et al.*, 1984). This interaction is discussed in Chapter 10. Von Willebrand factor does not affect the number of platelets adhering to collagen in more static conditions, since washed platelets can bind directly to collagen without added plasma proteins (Lyman *et al.*, 1971). Platelet–collagen interactions result directly in platelet activation; indeed, collagen is the most thrombogenic protein yet identified in the blood vessel wall (Baumgartner, 1977). Although considerable information is available on the structure of collagen and on the platelet response to interaction with this agonist, characterization of platelet-collagen interactions remains a challenging area of research (for reviews see Beachey *et al.*, 1979; Santoro and Cunningham, 1981).

4.1. Platelet-Activating Collagens

Collagen has been isolated and characterized from many tissues and species and it has been found that the monomeric forms consist of three polypeptide chains in a triple helix. Each chain has a molecular weight of $\sim$100,000. The amino acid composition is characterized by high proline ($\sim$12%), hydroxyproline ($\sim$9%), and glycine (35%) contents. Many collagens are glycosylated and contain glucosylgalactose linked to hydroxylysine residues. At least five genetically distinct collagens have been identified, each of which have unique amino acid sequences and differ in their tissue distributions (Eyre, 1980). The vessel wall contains three of these collagens, all of which readily form fibrils: type I, which consists of two $\alpha 1(I)$ chains and one $\alpha 2(I)$ chain; type II, which consists of three similar $\alpha 1(II)$ chains; and type III, which consists of three identical $\alpha 1(III)$ chains. The two additional collagens are found in basal laminae and do not form fibrils: type IV, which consists of three $\alpha 1(IV)$ chains and type V, which consists of two $\alpha 1(V)$ chains and one $\alpha 2(V)$ chain.

It was originally thought that the different types of collagens vary considerably in their reactivities with platelets. Subsequent work has shown that the monomeric forms, with a few interesting exceptions, have low activity, and that the forms of collagen that activate platelets are fibrillar. Much of the variation in reactivity is due to differences in the extent of fibril formation. For example, the differences between types I, II, and III collagens can be accounted for by differences in amounts of fibrils present (Barnes *et al.*, 1976; Santoro and Cunningham, 1977). The repeating units displayed by collagen fibrils presumably allow for multimeric interactions with collagen receptors on the platelet surface, interactions required for optimal platelet activation by most collagens (Santoro and Cunningham, 1977, 1979).

Two notable exceptions have been demonstrated. First, Kang and co-workers (Katzman *et al.*, 1973; Kang *et al.*, 1974) showed that the denatured $\alpha 1(I)$ chain of chick skin collagen induced platelet secretion and activation. The platelet-aggregating activity was confined to a distinct region of the molecule, termed $\alpha 1(I)$-CB5, containing 36 amino acids and one residue of glucosylgalactose. Although the relationship between chick skin and mammalian collagens is not clear, this finding permitted quantitative measurement of the binding of this form of collagen as well as the isolation of a receptor (see Section 4.3). A second exception is a peptide containing 149 amino acids, designated $\alpha 1(III)$-CB4, which results from the cleavage of type III collagen. This peptide retained platelet-binding activity (Fauvel *et al.*, 1978a, b). A nine-amino acid segment in this fragment, which inhibits collagen-induced platelet activation, has been identified (Y. J. Legrand *et al.*, 1980). The platelet-binding activity of this peptide, however, has not been demonstrated. It is therefore not clear whether inhibition results from binding to a platelet receptor or from disruption of collagen structure.

4.2. Measurement of Platelet–Collagen Interactions

Platelet-collagen interactions have been quantitated primarily by measuring the adhesion of platelets to insoluble matrices of collagen. In this approach, platelets are absorbed to collagen in suspension; the amount of platelets adhering to the collagen is determined either by counting adherent platelets or by using radiolabeled platelets and measuring the amount of adherent radioactivity. The collagen supports that have been used include glass coverslips (Meyer and Weisman, 1978), glass rods (Cazenave et al., 1976), glass test tubes (Cazenave et al., 1973), and Sepharose gel filtration media (Brass et al., 1976). There are two problems associated with these methods. One is that the surface area exposed to the platelet suspension is small. The second is that artifacts may be induced by an alteration of collagen structure that occurs on binding to the solid support. More recent modifications have circumvented these problems. In these experiments, a suspension of fibrillar collagen is mixed with platelets and adherent platelets are separated from nonadherent ones by filtration. One filtration procedure employs Sepharose 4B (Fauvel et al., 1978b). Another procedure allows for the processing of a large number of samples using polycarbonate membranes (Santoro and Cunningham, 1979).

The assays described above only permit qualitative evaluation of platelet–collagen interactions and do not yield precise information on the number of collagen receptors or on the affinity of these receptors for collagen. Soluble collagens that retain platelet-activating activity have been identified, however, and have been used to characterize the interactions of platelets with these forms of collagen. The binding of the soluble $\alpha 1$ chain of chick skin collagen was found to be saturable, reversible, and specific (Chiang et al., 1977). Analysis of binding data showed that $\sim 600,000$ $\alpha 1$ chains could be accommodated on the platelet surface with $K_d = 1.7$ μM, suggesting an extremely high concentration of collagen receptors. Assuming the surface area of a platelet is ~ 50 μm^2, these values only allow for ~ 100 nm^2 per collagen receptor, and predict either that single membrane glycoproteins have multiple receptors or that collagen receptors are expressed on more than one glycoprotein. An alternative hypothesis, however, is that collagen–collagen interactions also occurred in this study because of their high concentrations on the membrane surface and that this accounts for the large number of molecules bound.

4.3. The Search for the Collagen Receptor

Despite extensive investigations in many laboratories, the identity of the collagen receptor remains unclear. Three proteins found in α-granules of platelets have been shown to bind collagen and, accordingly, have been implicated in collagen receptor activity. One of these proteins is fibronectin, which has a collagen-binding domain near the amino terminus of the molecule (Ruoslahti et al., 1979; Furie and Rifkin, 1980). The experiments of Bensusan et al. (1978) were the first to suggest that fibronectin is the collagen receptor on platelets. These authors found that fibronectin was the primary platelet protein to remain adherent to collagen fibrils when a platelet–collagen suspension was sonicated. From these experiments, it was concluded that fibronectin is the cell-surface receptor for collagen and that it remained with the insoluble fibrils after sonication because of its direct association with collagen. In support of this conclusion, Lahav *et al.* (1982) found that a photoactivatable derivative of collagen was cross-linked to fibronectin when platelets were allowed to interact with substrata coated with the derivatized protein. Furthermore, Santoro and Cunningham (1979) observed that antifibronectin antibodies inhibited platelet adhesion to collagen by 25%. A second α-granule protein with collagen-binding activity is VWF. Santoro and Cowan (1982) found that fibrillar collagen adsorbed VWF from plasma. This appeared to result directly from VWF–collagen interactions and occurred only with the fibrillar form of collagen. Thrombospondin is yet another α-granule protein that binds collagen. This was demonstrated by Lahav *et al.* (1982), who observed direct cross-linking between collagen and thrombospondin.

These findings, which show direct interactions between α-granule proteins and collagen, suggest that α-granule proteins may mediate platelet–collagen interactions. It is known that collagen is capable of activating washed platelets, however, and it would therefore appear that the collagen receptor must pre-exist on the membrane surface of unstimulated platelets. The three proteins described above have an intracellular location, however. Platelet fibronectin is localized in the α-granules of platelets (Plow *et al.*, 1979; Zucker *et al.*, 1979a) and is bound on the platelet surface only following platelet activation (Plow and Ginsberg, 1981). The trace amount of fibronectin that has been detected on the surface of unstimulated platelets (Holderbaum *et al.*, 1982) appears insufficient to account for the interaction of collagen with these platelets. Von Willebrand factor is also in α-granules (Zucker *et al.*, 1979b) and binds to the platelet surface only after platelet stimulation (George and Onofre, 1982) or in the presence of a ristocetinlike activity (Ruggeri *et al.*, 1983). Thrombospondin is secreted from α-granules in response to platelet activation (Phillips, 1972) and binds to the surface of activated platelets (George *et al.*, 1980; Phillips *et al.*, 1980). It would thus appear that these proteins are not involved in the initial collagen receptor activity, but may mediate the adhesion of activated platelets to collagen on the blood vessel wall.

Chiang and Kang (1982) have identified a platelet membrane protein that binds to the soluble α1(I) chain of chick skin collagen. The purified protein ($M_r = 65,000$) was shown to bind three molecules of soluble α1(I) collagen and to inhibit the adhesion of platelets to type I fibrillar collagen. A membrane protein with a similar molecular weight was isolated by affinity chromatography, using fibrillar collagen coupled to the supporting matrix. Thus, it was shown that the $M_r = 65,000$ protein binds fibrillar collagen as well as soluble collagen. Antisera against this receptor does not cross-react with fibronectin. Recent studies have shown that the purified receptor inhibits α1(I)-chain-induced platelet aggregation (Chiang and Kang, 1983). It would thus appear that the membrane protein that has been identified binds the α1(I) chain of chick skin collagen on the platelet surface. Further studies of this potentially significant protein are required to determine whether it represents the physiologically important receptor for fibrillar collagen from mammalian species.

4.4. Summary

Platelets can be activated by various collagens. With few exceptions, the collagens capable of activating platelets are fibrillar, allowing for multimeric interactions with platelets. A M_r = 65,000 protein has been identified that binds soluble collagen from chick skin. Further studies are required to determine whether this is the functional receptor. Three α-granule proteins—fibronectin, VWF, and thrombospondin—have been shown to bind collagen. Since these proteins are expressed only on activated platelets, they may be involved in secondary platelet–collagen interactions.

5. EPINEPHRINE

Epinephrine induces platelet activation, causing the secretion of granule contents and subsequent aggregation. Although the initial responses of platelets to other stimuli involve a change in shape, epinephrine is unique in that it induces aggregation without altering the discoid shape of platelets. This interaction indicates that shape change is not required for the expression of aggregation sites on activated platelets. The concentrations of epinephrine required to initiate a platelet response (~ 1 μM) are higher than those that exist under physiological conditions. Lower concentrations (~ 10 nM) markedly potentiate responses involving other agonists, however (Grant and Scrutton, 1979), suggesting that if epinephrine has a physiological function in eliciting a platelet response, it most likely acts in concert with other agonists.

5.1. Classification of Epinephrine Receptors

Typically, the catecholamine-mediated responses of cells are mediated by one of two receptor systems on membranes, α- or β-adrenergic receptors. α-Adrenergic receptors respond to catecholamines in the order of epinephrine > norepinephrine > isoproterenol, and are antagonized by phentolamine. This is in contrast to β-adrenergic receptors, which respond in the order of isoproterenol > epinephrine > norepinephrine, and are antagonized by propranolol but not phentolamine. Since epinephrine induces platelet aggregation—a response that is antagonized by phentalomine—the platelet response to epinephrine appears to be mediated through α-adrenergic receptors (O'Brien, 1963; Mills and Roberts, 1967).

Studies on the effects of Na^+ also support the idea that epinephrine-induced platelet responses are mediated by α-adrenergic receptors. It has been shown that Na^+ affects the affinity of agonists for α-receptors (Tsai and Lefkowitz, 1978; Michel *et al.*, 1980). Since the response of platelets to epinephrine is reduced when platelets are placed in Na^+-free buffers, recent studies have examined whether this results from the influence of Na^{2+} on the platelet α-adrenergic receptors (Connolly and Limbird, 1983; Motulsky and Insel, 1983). In these studies, the Na^+ content of platelets was modified from that in saline (38 nmole/10^8 platelets) either by incubation of platelets with the Na^+-selective ionophore, monensin (to 138 nmole/10^8 platelets), or by incubation in a Na^+-free buffer (to 13 nmole/10^8 platelets) (Motulsky and Insel, 1983). Decreases in platelet Na^+ caused a decrease in the affinity of epinephrine for its receptor, as measured by the competition for [^{3}H]yohimbine binding, and also a decrease in the responsiveness of platelets to epinephrine. Increases in platelet Na^+ caused an increase in these parameters. These changes in reactivity do not appear to result from platelet damage, since platelet morphology was maintained after Na^+ fluctuations and the platelets also retained responsiveness to thrombin (Connolly and Limbird, 1983). These data suggest that intracellular Na^+ modulates the reactivity of the α-adrenergic receptor on platelets, and the effects of such modulations indicate that α-adrenergic receptors mediate the action of epinephrine.

5.2. Ligand Binding to α-Adrenergic Receptors

Several radiolabeled ligands have been used to measure α-adrenergic receptors on platelets. One is [^{3}H]dihydroergocryptine, an antagonist of α-adrenergic receptors. It was found that [^{3}H]dihydroergocryptine binds to α-adrenergic receptors on platelets, that the binding is reversible, and that receptors persist on isolated membranes (Williams et al., 1976; Newman et al., 1978; Alexander et al., 1978). Platelets contain approximately 200 receptors for this ligand, with a $K_d \sim$ 1–3 nM. Catecholamine agonists compete for binding in the order of epinephrine > norepinephrine > phenylephrine > isoproterenol, an order similar to that which induces platelet aggregation and the order expected for an α-adrenergic receptor. Phentolamine, the classical α-adrenergic agonist, also competes for binding, as does phenoxybenzamine and yohimbine. Computer modeling of binding data has indicated the existence of interconvertible high- and low-affinity states for agonist binding (Tsai and Lefkowitz, 1979; Hoffman et al., 1980).

Smith and Limbird (1981) also characterized α-adrenergic receptors on isolated platelet membranes in experiments quantitating the binding of [^{3}H]epinephrine and a radiolabeled antagonist, [^{3}H]yohimbine. Dissociation constants calculated either from equilibrium associations or kinetic analysis showed that both ligands bound with similar affinity, K_d = 2–4 nM.

5.3. Solubilization of α-Adrenergic Receptors

Soluble α-adrenergic receptors have been obtained by extraction of isolated platelet plasma membranes with digitonin. The soluble receptors were found to have similar properties to the receptors in membranes (Smith and Limbird, 1981). The affinity of the receptor in solution for [^{3}H]yohimbine (K_d = 7.3 nM) was similar to that in membranes (K_d = 5.7 nM). The concentrations of agonists and antagonists that inhibited [^{3}H]yohimbine binding to the soluble receptor were similar to those that inhibited binding to the receptor in isolated membranes or in intact cells. Although guanine nucleotides decreased the affinity of epinephrine for membranes, they did not modulate the affinity of epinephrine for the solubilized receptor. This suggests that solubilization disrupts the interactions of the receptor with other membrane components. Examinations of the physical properties of the soluble receptor have suggested that such interactions of membrane proteins could be stabilized if solubilization were performed in the presence of an agonist. Sedimentation of the digitonin-solubilized receptors through sucrose gradients showed that agonist occupancy altered the sedimentation coefficient of the receptor. Solubilization in the presence of epinephrine (agonist-prelabeled receptors) gave a sedimentation coefficient of 13.4 S, whereas solubilization in the presence of yohimbine (antagonist-prelabeled receptors) gave a value of 11.4 S. A similar difference in sedimentation properties was observed by Michel et al. (1981). The lack of responsiveness to guanine nucleotides and the differences in sedimentation properties suggest that the slower sedimentary form has

lost the guanine regulatory protein. Such an agonist-induced change in intramembranous interactions have been used to explain the variable affinities of β-adrenergic receptors (DeLean *et al.*, 1980). As has been previously discussed, however,
the data on the digitonin-solubilized α-adrenergic receptor do not exclude the possibility that agonists affect the sedimentation coefficient either by inducing a conformational change in the receptor or by increasing the lipid or detergent binding to the
receptor (Michel *et al.*, 1981).

5.4. α-Adrenergic-Mediated Responses

α-Adrenergic receptors in cells are linked to adenylate cyclase via a guanine
nucleotide-binding protein and, in many cells, the responses of α-adrenergic agents are
transmitted by alterations in cyclic AMP (Rodbell, 1975). The receptor in platelets is
linked biochemically in a similar way; the addition of epinephrine to isolated platelet
plasma membranes produces an inhibition of adenylate cyclase, an effect potentiated
by GTP (Steer and Wood, 1979; Aktories *et al.*, 1981; Jakobs 1983). Comparison of
the effects of agonist and antagonist concentrations required to achieve adenylate
cyclase inhibition indicates that the α-adrenergic receptor does not form a permanent
intramembranous complex with adenylate cyclase; such interactions are most likely
transitory (Macfarlane and Stump, 1982). It does not appear, however, that this is the
event that activates platelets. Although cyclic AMP levels do fall in activated platelets
(Salzman *et al.*, 1972), such decreases do not necessarily induce platelet activation
(Haslam and Rossen, 1975). Moreover, it has been possible to dissociate α-adrenergic-
induced aggregation responses from changes in cyclic AMP concentration (Connolly
and Limbird, 1983). It would appear, therefore, that some mechanism in addition to
the inhibition of adenylate cyclase transmits the α-adrenergic response in platelets.

One mechanism for the stimulus–response coupling of α-adrenergic agents has
been suggested by the research of Owen *et al.* (1980), who showed that epinephrine
selectively increases the Ca^{2+} permeability of the plasma membrane. A dose-dependent uptake of Ca^{2+}, measured by the uptake of $^{45}Ca^{2+}$, was observed in the range of
10^{-7} to 10^{-5} M epinephrine. Antagonism of the platelet α-receptor by phentolamine
resulted in the inhibition of platelet aggregation and of Ca^{2+} uptake. Verapamil also
inhibited both responses. Although Owen and co-workers attributed the verapamil
effect to a direct inhibition of Ca^{2+} uptake, subsequent work by Barnathan *et al.*
(1982) showed that the effect of verapamil was due to the inhibition of the binding of
epinephrine to α-adrenergic receptors. These data suggest that α-adrenergic platelet
responses may be initiated by an uptake of extracellular Ca^{2+}. This idea is supported
by observations showing that epinephrine-mediated responses are dependent on extracellular Ca^{2+}, a requirement not observed for other platelet agonists. It is not supported, however, by the observations of Clare and Scrutton (1984), who found that
Ca^{2+} uptake only occurred when platelets secreted. It would appear that further work
is required to clarify the role of Ca^{2+} in epinephrine-induced platelet responses.

5.5. Identification of β-Adrenergic Receptors

Although the binding of epinephrine to α-adrenergic receptors appears to be
responsible for the ability of epinephrine to induce platelet activation, this agent has
also been shown to bind β-adrenergic receptors. β-Adrenergic receptors have been

measured on platelets by the binding of the radiolabeled ligands [^{125}I]cyanopindolol
and [^{125}I]hydroxybenzylpindolol (Steer and Atlas, 1982). Binding was reversible and
saturable, with a binding capacity of ~25 molecules per platelet. The binding was
inhibited by adrenergic ligands in the order of propanolol $>>$ isoproterenol $>$ epinephrine $>$ practolol $>$ norepinephrine $>$ phenylephrine, the order expected of β-
adrenergic agonists. β-Adrenergic agents have been shown to inhibit platelet function
by elevating cyclic AMP, an effect mediated by activation of adenylate cyclase (Jakobs
et al., 1978).

5.6. Summary

Platelets contain receptors for epinephrine and other α-adrenergic agonists.
α-Adrenergic receptors in platelets appear to be linked to adenylate cyclase and a Ca^{2+}
pump in a way that may mediate the physiological responses of α-adrenergic agents.
Although the α-adrenergic receptor in platelets has not been identified, a soluble
preparation has been obtained that has similar binding properties to those of the
receptor in membranes.

6. CONCLUSIONS

Platelets are activated by a wide spectrum of physiological agonists, which appear
to interact with discrete receptors on the platelet membrane surface. Although morphological and biochemical responses of platelets to agonists have been well characterized, identification and characterization of receptors for agonists remains a challenging area of research. The most potent of the physiological agonists, thrombin, appears
to activate platelets by a proteolytic mechanism. Although several thrombin-binding
proteins have been identified, only one of these, termed GP V, has been shown to be a
thrombin substrate. Hydrolysis of this glycoprotein by thrombin demonstrates many of
the properties attributed to thrombin activation, raising the possibility that GP V may
be the thrombin receptor.

Adenosine diphosphate is a weaker platelet agonist and interacts with two sites on
the platelet surface, a receptor to induce platelet activation and a receptor to inhibit
adenylate cyclase. Two membrane proteins that bind ADP have been identifed.

Collagen is a potent activator of platelets and interacts at multiple sites on the
platelet membrane surface. Collagen-binding glycoproteins have been isolated from
membranes and from α-granules. The membrane glycoprotein that has been identified
binds soluble chick skin collagen; however, it is not known whether this is the receptor
for fibrillar collagen. The collagen-binding proteins in α-granules also bind to platelet
membranes, indicating that they may play a secondary role as a bridge for platelet–
collagen interactions in activated platelets.

Epinephrine induces platelet aggregation by binding to α-adrenergic receptors. A
soluble preparation of these receptors from platelet membranes retains epinephrine-
binding activity. It can be anticipated that this preparation will be useful for isolating
the α-adrenergic receptor from platelets.

ACKNOWLEDGMENTS.   I thank Drs. Joan Fox, Leslie Parise, and Larry Fitzgerald for their helpful comments. I would also like to thank Russell Levine for editorial assistance and Michele Prator for preparation of the manuscript. This research was supported in part by research grants (HL 28947 and HL) from the National Heart, Lung and Blood Institute.

REFERENCES

Adler, J. R., and Handin, R. I., 1979, Solubilization and characterization of a platelet membrane ADP-binding protein, *J. Biol. Chem.* **254:**3866–3872.

Aihara, M., Cooper, H. A., and Wagner, R. H., 1984, Platelet–collagen interactions: Increase in rate of adhesion 37 of fixed washed platelets by factor VIII-related antigen, *Blood* **63:**495–501.

Aktories, K., 1981, Epinephrine inhibits adenylate cyclase and stimulates a GTPase in human platelet membrane via alpha-adrenoceptors, *FEBS Lett.* **130:**235–238.

Alexander, R. J., Fenton, II, J. W., and Detwiler, T. C., 1983, Thrombin–platelet interactions: An assessment of the roles of saturable and nonsaturable binding in platelet activation, *Arch. Biochem. Biophys.* **222:**266–275.

Alexander, R. W., Cooper, B., and Handin, R. I., 1978, Characterization of the human platelet α-adrenergic receptor, *J. Clin. Invest.* **78:**1136–1144.

Awbry, B. J., Hoak, J. C., and Owen, W. G., 1979, Binding of thrombin to cultured human endothelial cells, *J. Biol. Chem.* **254:**4092–4095.

Baker, J. B., Low, D. A., Simmer, R. L., and Cunningham, D. D., 1980, Protease-Nexin: A cellular component that links thrombin and plasminogen activator and mediates their binding to cells, *Cell* **21:**37–45.

Barnathan, E. S., Addonizio, V. P., and Shattil, S. J., 1982, Interaction of verapamil with human platelet receptors, *Am. J. Physiol.* **242:**H19–H23.

Barnes, M. J., Gordon, J. L., and MacIntyre, D. E., 1976, Platelet aggregating activity of type I and type III collagens from human aorta and chicken skin, *Biochem. J.* **160:**647–651.

Bar-Shavit, R., Kahn, A., Fenton, II, J. W., and Wilner, G. D., 1983, Chemotactic response of monocytes to thrombin, *J. Cell Biol.* **96:**282–285.

Bauer, P. I., Machovich, R., Arányi, P., Kálmán, G. B., Eva, C., and Horráth, I., 1983, Mechanism of thrombin binding to endothelial cells, *Blood* **62:**368–372.

Baumgartner, H. R., 1977, Platelet interaction with collagen fibrils in flowing blood. I. Reaction of human platelets with alpha chymotrypsin-digested subendothelium, *Thromb. Haemost.* **37:**1–16.

Beachey, E. H., Chiang, T. M., and Kaug, A. H., 1979, Collagen platelet interaction, in: *International Review of Connective Tissue Research,* Volume 8, (D. Hall and D. S. Jackson, eds), Academic Press, Inc., New York, pp. 1–21.

Bennett, J. S., Colman, R. F., and Colman, R. W., 1978, Identification of adenine nucleotide binding proteins in human platelet membranes by affinity labeling with 5′-P-fluorosulfonylbenzoyl adenosine, *J. Biol. Chem.* **253:**7346–7354.

Bensusan, H. B., Koh, T. L., Henry, K. G., Murray, B. A., and Culp, L. A., 1978, Evidence that fibronectin is the collagen receptor on platelet membranes, *Proc. Natl. Acad. Sci. U.S.A.* **75:**5864–5868.

Berndt, M. C., and Phillips, D. R., 1981a, Interaction of thrombin with platelets: Purification of the thrombin substrate, *Ann. N.Y. Acad. Sci.* **370:**87–95.

Berndt, M. C., and Phillips, D. R., 1981b, Purification and preliminary physicochemical characterization of human platelet membrane glycoprotein V, *J. Biol. Chem.* **256:**59–65.

Berndt, M. C., and Phillips, D. R., 1981c, Platelet membrane proteins: Composition and receptor function, in: *Platelets in Biology and Pathology,* Volume 2 (J. L. Gordon, ed.), Elsevier/North-Holland Biomedical Press, Amsterdam, pp. 43–75.

Born, G. V. R., 1962, Aggregation of blood platelets by adenosine diphosphate and its reversal, *Nature (London)* **192:**927–929.

Born, G. V. R., and Feinberg, H., 1975, Binding of adenosine diphosphate to intact human platelets, *J. Physiol. (London)* **251**:803–816.

Boullin, D. J., Green, A. R., and Price, K. S., 1972, The mechanism of adenosine diphosphate induced platelet aggregation: Binding to platelet receptors and inhibition of binding and aggregation by prostaglandin G$_1$, *J. Physiol.* **221**:415–426.

Brass, L. F., Faile, D., and Bensusan, H. B., 1976, Direct measurement of the platelet:collagen interaction by affinity chromatography on collagen/Sepharose, *J. Lab. Clin. Med.* **87**:525–534.

Carney, D. H., and Bergmann, J. S., 1982, ^{125}I-thrombin binds to clustered receptors on noncoated regions of mouse embryo cell surfaces. *J. Cell Biol.* **95**:697–703.

Cazenave, J. P., Packham, M. A., and Mustard, J. F., 1973, Adherence of platelets to a collagen-coated surface: Development of a quantative method, *J. Lab. Clin. Med.* **82**:978–990.

Cazenave, J. P., Reimers, H. J., Kinlough-Rathbone, R. L., Packham, M. A., and Mustard, J. F., 1976, Effects of sodium periodate on platelet function, *Lab. Invest.* **34**:471–481.

Chiang, T. M., and Kang, A. H., 1982, Isolation and purification of collagen α_1(I) receptor from human platelet membrane, *J. Biol. Chem.* **257**:7581–7586.

Chiang, T. M., and Kang, A. H., 1983, Immunochemical studies of the purified chick skin collagen α1(I) chain receptor of human platelets (abstract), *Fed. Proc.* **42**:1368.

Chiang, T. M., Beachy, E. H., and Kang, A. H., 1977, Binding of collagen α1 chains to human platelets, *J. Clin. Invest.* **59**:405–411.

Chen, L. B., and Buchanan, J. M., 1975, Mitogenic activity of blood components. I. Thrombin and prothrombin, *Proc. Natl. Acad. Sci. U.S.A.* **72**:131–135.

Clare, K. A., and Scrutton, M. C., 1984, The role of Ca^{2+} uptake in the response of human platelets to adrenaline and to 1-0-alkyl-2-acetyl-*sn*-glycero-3-phosphocholine (platelet-activating factor), *Eur. J. Biochem.* **140**:129–136.

Colman, R. W., Figures, W. R., Colman, R. F., Morinelli, T. A., Niewiarowski, S., and Mills, D. C. B., 1980, Identification of two distinct adenosine diphosphate receptors in human platelets, *Trans. Assoc. Am. Physicians* **XCIII**:305–316.

Connolly, T. M., and Limbird, L. E., 1983, The influence of Na$^+$ on the α_2–adrenergic receptor system of human platelets. A method for removal of extraplatelet Na$^+$. Effect of Na$^+$ removed on aggregation, secretion, and cAMP accumulation, *J. Biol. Chem.* **258**:3907–3912.

Davey, M. G., and Lüscher, E. F., 1967, Actions of thrombin and other coagulant and proteolytic enzymes on blood platelets, *Nature (London)* **216**:857–858.

DeLean, A., Stadel, J. M., and Lefkowitz, R. J., 1980, A ternary complex model explains the agonist-specific binding properties of the adenylate cyclase-coupled β-adrenergic receptor, *J. Biol. Chem.* **255**:7108–7117.

Detwiler, T. C., and Feinman, R. D., 1973a, Kinetics of the thrombin-induced release of calcium (II) by platelets, *Biochemistry* **12**:282–289.

Detwiler, T. C., and Feinman, R. D., 1973b, Kinetics of the thrombin-induced release of adenosine triphosphate by platelets. Comparison with release of calcium, *Biochemistry* **12**:2462–2468.

Detwiler, T. C., and Huang, E. M., 1985, Stimulus–response coupling mechanisms, in: *Biochemistry of Platelets* (D. R. Phillips and M. Shuman, eds.), Academic Press, New York (in press).

Esmon, N. L., Carroll, R. C., and Esmon, C. T., 1983, Thrombomodulin blocks the ability of thrombin to activate platelets, *J. Biol. Chem.* **258**:12238–12242.

Eyre, D. R., 1980, Collagen: Molecular diversity in the body's protein scaffold, *Science* **207**:1315–1322.

Fauvel, F., Legrand, Y. J., and Caen, J. P., 1978a, Platelet adhesion to type I collagen and alpha 1 (I)3 trimers: Involvement of the C-terminal alpha 1(I) CB6A peptide, *Thrombin. Res.* **12**:273–285.

Fauvel, F., Legrand, Y. J., Bentz, H., Fietzek, P. P., Kuhn, K., and Caen, J. P., 1978b, Platelet collagen interaction. Adhesion of human blood platelets to purified CB-4 peptide from type III collagen, *Thromb. Res.* **12**:841–850.

Fenton, II, J. W., Walz, D. A., and Finlayson, J. S., 1977, Human thrombins, in: *Chemistry and Biology of Thrombin* (R. L. Lundblad, J. W. Fenton, II, and K. G. Mann, eds.), Ann Arbor Science Publishers, Inc., Ann Arbor, MI, pp. 43–70.

Figures, W. R., Niewiarowski, S., Morinelli, T. A., Colman, R. F., and Colman, R. W., 1981, Affinity labeling of a human platelet membrane protein with 5'-P-fluorosulfonylbenzoyl adenosine. Concomitant inhibition of ADP–induced platelet aggregation and fibrinogen receptor exposure, *J. Biol. Chem.* **256**:7789–7795.

Fujimura, K., Maehama, S., and Kuramoto, A., 1980, The effect of protease on platelet plasma membrane proteins. The mode of thrombin action on glycoprotein I and V and on platelet functions, *Acta Haematol. Jpn.* **43:**198–207.

Furie, M., and Rifkin, D. B., 1980, Proteolytically derived fragments of human plasma fibronectin and their localization within the intact molecule, *J. Biol. Chem.* **255:**3134–3140.

Gaarder, A., Jonsen, J., Laland, S., Hellem, A., and Owren, P. A., 1961, Adenosine diphosphate in red cells as a factor in the adhesiveness of human blood platelets, *Nature (London)* **192:**531–532.

Ganguly, P., and Fossett, N. G., 1981, Inhibition of thrombin-induced platelet aggregation by a derivative of wheat germ agglutinin. Evidence for a physiologic receptor of thrombin in human platelets, *Blood* **57:**343–352.

Ganguly, P., and Gould, N. L., 1979, Thrombin receptors of human platelets. Thrombin binding and antithrombin properties, *Br. J. Haematol.* **42:**137–145.

Ganguly, P., and Sonnichsen, W. J. 1976, Binding of thrombin to human platelets and its possible significance, *Br. J. Haematol.* **34:**291–301.

Gear, A. R. L., and Burke, D., 1983, Thrombin-induced secretion of serotonin from platelets can occur in seconds, *Blood* **60:**1231–1234.

George, J. N., Lyons, R. M., and Morgan, R. K., 1980, Membrane changes associated with platelet activation. Exposure of actin on the platelet surface after thrombin-induced secretion, *J. Clin. Invest.* **66:**1–9.

George, J. N., Nurden, A. T., and Phillips, D. R., 1984, Molecular defects in interactions of platelets with the vessel wall, *N. Engl. J. Med.* **311:**1084–1098.

Glenn, K. C., Carney, D. H., Fenton, II, J. W., and Cunningham, D. D., 1980, Thrombin active site regions required for fibroblast receptor binding and initiation of cell division, *J. Biol. Chem.* **255:**6609–6616.

Goodnough, L. T., and Saito, H., 1982, Specific binding of thrombin by human peripheral blood monocytes: Possible role in the clearance of activated factors from the circulation, *J. Lab. Clin. Med.* **99:**873–884.

Gospodarowicz, D., Brown, K. D., Birdwell, C. R., Zetter, B. R., 1978, Control of proliferation of human vascular endothelial cells. Characterization of the response of human umbilical vein endothelial cells to fibroblast growth factor, epidermal growth factor, and thrombin, *J. Cell Biol.* **77:**774–788.

Grant, J. A., and Scrutton, M. C., 1979, Novel α_2-adrenoreceptors primarily responsible for inducing human platelet aggregation, *Nature (London)* **277:**659–661.

Grette, K., 1962, Studies on the mechanism of thrombin-catalyzed hemostatic reactions in blood platelets, *Acta Physiol. Scand.* 56 (Suppl.) **195:**1–93.

Hagen, I., Brosstad, F., Solum, N. O., and Korsmo, R., 1981, Crossed immunoelectrophoresis using immobilized thrombin in intermediate gel. A method for demonstration of thrombin-binding platelet proteins, *J. Lab. Clin. Med.* **97:**213–220.

Haslam, R. J., and Cusack, N. J., 1981, Blood platelet receptors for ADP and for adenosine, in: *Purinergic Receptors (Receptors and Recognition,* Series B. Volume 12) (G. Burnstock, ed.), Chapman and Hall, London, pp. 223–285.

Haslam, R. J., and Rossen, G. M., 1975, Effects of adenosine on levels of adenosine 3':5'-cyclic monophosphate in human blood platelets in relation of adenosine incorporation and platelet aggregation. *Mol. Pharmacol.* **11:**528–544.

Henriksen, R. A., and Brotherton, A. F. A., 1983, Evidence that activation of platelets and endothelium by thrombin involves distinct sites of interaction. Studies with the dysthrombin, thrombin Quick I, *J. Biol. Chem.* **258:**13717–13721.

Hoffman, B. B., Michel, T., Kilpatrick, D. M., Lefkowitz, R. J., Tolbert, M. E. M., Gilman, H., and Fain, J. N., 1980, Agonist versus antagonist binding to α-adrenergic receptors, *Proc. Natl. Acad. Sci. U.S.A.* **77:**4569–4578.

Holderbaum, D., Culp, L. A., Bensusan, H. B., and Gershman, H., 1982, Platelet stimulation by antifibronectin antibodies requires the Fc region of antibody, *Proc. Natl. Acad. Sci. U.S.A.* **79:**6537–6540.

Holmsen, H., and Weiss, H. J., 1979, Secretable storage pools in platelets, *Annu. Rev. Med.* **30:**119–134.

Holmsen, H., Dangelmaier, C. A., and Holmsen, H.-K., 1981, Thrombin-induced platelet responses differ in requirement for receptor occupancy, *J. Biol. Chem.* **256:**9393–9396.

Huang, E. M., and Detwiler, T. C., 1980, Reassessment of the evidence for the role of secreted ADP in biphasic platelet aggregation. *J. Lab. Clin. Med.* **95:**59–68.

Isaacs, J. D., Savion, N., Gospodarowicz, D., Fenton, II, J. W., and Shuman, M. A., 1981a, Covalent binding of thrombin to specific sites on corneal endothelial cells, *Biochemistry* **20**:398–403.

Isaacs, J. D., Savion, N., Gospodarowicz, D., and Shuman, M. A., 1981b, Effect of cell density on thrombin binding to a specific site on bovine vascular endothelial cells, *J. Cell Biol.* **90**:670–674.

Jakobs, K. H., 1983, Determination of the turn-off reaction for the epinephrine-inhibited human platelet adenylate cyclase, *Eur. J. Biochem.* **132**:125–130.

Jakobs, K. H., Saur, W., and Shultz, G., 1978, Characterization of alpha- and beta-adrenergic receptors linked to human platelet adenylate cyclase, *Arch. Pharmacol.* **302**:285–291.

Jamieson, G. A., and Okumura, T., 1978, Reduced thrombin binding and aggregation in Bernard-Soulier platelets, *J. Clin. Invest.* **61**:861–864.

Kang, A. H., Beachey, E. H., and Katzman, R. L., 1974, Interaction of an active glycopeptide from chick skin collagen (α-1-CB5) with human platelets, *J. Biol. Chem.* **249**:1054–1059.

Katzman, R. L., Kang, A. H., and Beachey, E. H., 1973, Collagen-induced platelet aggregation: Involvement of an active glycopeptide fragment (α1-CB5), *Science* **181**:670–672.

Knupp, C. L., and White II, G. C., 1981, Effect of active site-modified thrombin on the hydrolysis of platelet-associated glycoprotein V by native thrombin, *Blood* **58**:198a.

Kramer, R. M., Jakubowski, J. A., Vaillancourt, R., and Deykin, D., 1982, Effect of membrane cholesterol on phospholipid metabolism in thrombin-stimulated platelets. Enhanced activation of platelet phospholipase(s) for liberation of arachidonic acid, *J. Biol. Chem.* **257**:6844–6849.

Lahav, J., Schwartz, M. A., and Hynes, R. O., 1982, Analysis of platelet adhesion using a radioactive chemical crosslinking reagent: Interaction of thrombospondin with fibronectin and collagen, *Cell* **31**:253–262.

Larsen, N. E., and Simons, E. R., 1981, Preparation and application of a photoreactive thrombin analog: Binding to human platelets, *Biochemistry* **20**:4144–4147.

Larsen, N. E., Horne, W. C., and Simons, E. R., 1979, Platelet interaction with active and TLCK-inactivated α-thrombin, *Biochem. Biophys. Res. Commun.* **87**:403–409.

Legrand, C., Dubernard, V., and Caen, J., 1980, Further characterization of human platelet ADP binding sites using 5′ AMP. Demonstration of a highly reactive population of sites, *Biochem. Biophys. Res. Commun.* **96**:1–9.

Legrand, Y. J., Karniguian, A., LeFrancier, P., Fauvel, F., and Caen, J. P., 1980, Evidence that a collagen-derived nonapeptide is a specific inhibitor of platelet-collagen interaction, *Biochem. Biophys. Res. Commun.* **96**:1579–1585.

Leven, R. M., Mullikin, W. H., and Nachmias, V. T., 1983, Role of sodium in ADP- and thrombin-induced megakaryocyte spreading, *J. Cell Biol.* **96**:1234–1240.

Levine, J. D., Harlan, J. M., Harker, L. A., Joseph, M. L., and Counts, R. B., 1982, Thrombin-mediated release of Factor VIII antigen from human umbilical vein endothelial cells in culture, *Blood* **60**:531–534.

Lips, J. P. M., Sixma, J. J., and Schiphorst, M. E., 1980a, Binding of adenosine diphosphate to human blood platelets and to isolated blood platelet membranes, *Biochim. Biophys. Acta* **628**:451–467.

Lips, J. P. M., Sixma, J. J., and Schiphorst, M. E., 1980b, The effect of ticlopidine administration to humans on the binding of adenosine diphosphate to blood platelets, *Thromb. Res.* **17**:19–27.

Lollar, P., Hoak, J. C., and Owen, W. G., 1980, Binding of thrombin to cultured human endothelial cells. Nonequilibrium aspects, *J. Biol. Chem.* **255**:10279–10283.

Low, D. A., Scott, R. W., Baker, J. B., and Cunningham, D. D., 1982, Cells regulate their mitogenic response to thrombin through release of protease nexin, *Nature (London)* **298**:476–478.

Lyman, B., Rosenberg, L., and Karpatkin, S., 1971, Biochemical and biophysical aspects of human platelet adhesion to collagen fibers, *J. Clin. Invest.* **50**:1854–1863.

Macfarlane, D. E., and Mills, D. C. B., 1975, The effects of ATP on platelets: Evidence against the central role of released ADP-primary aggregation, *Blood* **46**:309–320.

Macfarlane, D. E., and Stump, D. C., 1982, Parallel observation of the occupancy of the alpha-2 adrenergic receptor in intact platelets and its ability to inhibit adenylate cyclase, *Mol. Pharmacol.* **22**:574–579.

Macfarlane, D. E., Srivastova, P., and Mills, D. C. B., 1979, 2-Methylthioadenosine-5′-diphosphate (2MeSADP), a high affinity probe for ADP receptors on the human platelet (abstract), *Thromb. Haemost.* **42**:185.

Macfarlane, D. E., Mills, D. C. B., and Srivastova, P. C., 1982, Binding of 2-azidoadenosine [beta-^{32}P] diphosphate to the receptor on intact human blood platelets which inhibits adenylate cyclase, *Biochemistry* **21**:544–549.

Macfarlane, D. E., Srivastava, P. C., and Mills, D. C. B., 1983, 2-Methylthioadenosine [β-^{32}P] diphosphate. An agonist and radioligand for the receptor that inhibits the accumulation of cyclic AMP in intact blood platelets, *J. Clin. Invest.* **71**:420–428.

Mannherz, T. G., and Goody, R. S., 1976, Proteins of contractile systems, *Annu. Rev. Biochem.* **45**:427–465.

Martin, B. M., Feinman, R. D., and Detwiler, T. C., 1975, Platelet stimulation by thrombin and other proteases, *Biochemistry* **14**:1308–1314.

McGowan, E. B., Duig, A., and Detwiler, T. C., 1983, Correlation thrombin-induced glycoprotein V hyrolysis and platelet activation, *J. Biol. Chem.* **258**:11243–11248.

Meyer, F. A., and Weisman, Z., 1978, Adhesion of platelets to collagen: The nature of the binding site from competitive inhibition studies, *Thromb. Res.* **12**:431–436.

Michel, T., Hoffman, B. B., and Lefkowitz, R. J., 1980, Differential regulation of the α_2-adrenergic receptor by Na$^+$ and guanine nucleotides, *Nature (London)* **288**:709–711.

Michel, T., Hoffman, B. B., Lefkowitz, R. J., and Caron, M. G., 1981, Different sedimentation properties of agonist- and antagonist-labelled platelet alpha 2 adrenergic receptors, *Biochem. Biophys. Res. Commun.* **100**:1131–1136.

Miletich, J. P., Jackson, C. M., and Majerus, P. W., 1978, Properties of the Factor Xa binding site on human platelets, *J. Biol. Chem.* **253**:6908–6916.

Mills, D. C. B., and Macfarlane, D. E., 1976, Platelet receptors, in: *Platelets in Biology and Pathology* (J. L. Gordon, ed.), Elsevier/North Holland Biomedical Press, Amsterdam, pp. 159–202.

Mills, D. C. B., and Macfarlane, D. E., 1977, Attempts to define a platelet ADP receptor with ^{203}Hg-p-mercuribenzene sulfonate (MES) (abstract), *Thromb. Haemost.* **38**:82.

Mills, D. C. B., and Roberts, G. C. K., 1967, Effects of adrenaline on human blood platelets, *J. Physiol. (London)* **193**:443–453.

Mosher, D. F., and Vaheri, A., 1978, Thrombin stimulates the production and release of a major surface-associated glycoprotein (fibronectin) in cultures of human fibroblasts, *Exp. Cell Res.* **112**:323–334.

Mosher, D. F., Vaheri, A., Choate, J. J., and Gahmberg, C. G., 1979, Action of thrombin on surface glycoproteins of human platelets, *Blood* **53**:437–445.

Moss, M., Wiley, H. S., Fenton, II, J. W., and Cunningham, D. D., 1983, Photoaffinity labeling of specific α-thrombin binding sites on Chinese hamster lung cells, *J. Biol. Chem.* **258**:3996–4002.

Motulsky, H. J., and Insel, P. A., 1983, Influence of sodium on the α_2-adrenergic receptor system of human platelets. Role for intraplatelet sodium in receptor binding, *J. Biol. Chem.* **258**:3913–3919.

Mürer, E. H., 1972, Factors influencing the initiation and the extrusion phase of the platelet release reaction, *Biochim. Biophys. Acta* **261**:435–443.

Nachman, R. L., and Ferris, B., 1974, Binding of adenosine diphosphate by isolated membranes from human platelets, *J. Biol. Chem.* **249**:704–710.

Nesheim, M. E., Taswell, J. B., and Mann, K. G., 1979, The contribution of bovine Factor V and Factor Va to the activity of prothrombinase, *J. Biol. Chem.* **254**:10952–10962.

Newman, K. D., Williams, L. T., Bishopric, N. H., and Lefkowitz, R. J., 1978, Identification of α-adrenergic receptors in human platelets by [^{3}H]dihydroergocryptine binding, *J. Clin. Invest.* **61**:395–402.

O'Brien, J. R., 1963, Some effects of adrenalin and anti-adrenaline compounds on platelets *in vitro* and *in vivo*, *Nature (London)* **200**:763–764.

Okumura, T., and Jamieson, G. A., 1976, Platelet glycocalicin. A single receptor for platelet aggregation induced by thrombin or ristocetin, *Thromb. Res.* **8**:701–706.

Okumura, T., Hasitz, M., and Jamieson, G. A., 1978, Platelet glycocalicin. Interaction with thrombin and role as thrombin receptor of the platelet surface, *J. Biol. Chem.* **253**:3435–3443.

Owen, N. E., Feinberg, H., and LeBreton, G. C., 1980, Epinephrine induces Ca^{2+} uptake in human blood platelets, *Am. J. Physiol.* **239**:H483–H488.

Owen, W. G., and Esmond, C. T., 1981, Functional properties of an endothelial cell cofactor from thrombin-catalyzed activitation of protein C, *J. Biol. Chem.* **256**:5532–5535.

Parise, L. V., Venton, D. L., and LeBreton, G. C., 1984, Arachidonic acid-induced platelet aggregation is mediated by a thromboxane A_2/prostaglandin H_2 receptor interaction, *J. Pharmacol. Exp. Ther.* **228**:240–244.

Phillips, D. R., 1972, Effects of trypsin on the exposed polypeptides and glycoproteins in the human platelet membrane, *Biochemistry* **11**:4582–4588.

Phillips, D. R., 1974, Thrombin interaction with human platelets. Potentiation of thrombin-induced aggregation and release by inactivated thrombin, *Thromb. Diath. Haemorrh.* **32**:207–215.

Phillips, D. R., 1980, An evaluation of membrane glycoproteins in platelet adhesion and aggregation, in: *Progress in Hemostasis and Thrombosis,* Volume 5 (T. H. Spaet, ed.), Grune and Stratton, New York, pp. 81–109.

Phillips, D. R., and Agin, P. P., 1977, Platelet plasma membrane glycoproteins. Identification of a proteolytic substrate for thrombin, *Biochem. Biophys. Res. Commun.* **75**:940–947.

Phillips, D. R., Jennings, L, K., and Prasanna, H. R., 1980, Ca^{2+}-mediated association of Glycoprotein G (thrombin-sensitive protein, thrombospondin) with human platelets, *J. Biol. Chem.* **255**:11629–11632.

Plow, E. F., and Ginsberg, M. H., 1981, Specific and saturable binding of plasma fibronectin to thrombin-stimulated platelets, *J. Biol. Chem.* **256**:9477–9482.

Plow, E. F., Birdwell, C., and Ginsberg, M. H., 1979, Identification and quantitation of platelet-associated fibronectin antigen, *J. Clin. Invest.* **63**:540–543.

Rodbell, M., Lin, M. C., Solomon, Y., Londos, C., Harwood, J. P., Martin, B. R., Nendell, M., and Berman, M., 1975, Role of adenine and guanine nucleotides in the activity and response of adenylate cyclase systems to hormones: Evidence for multisite transition states, *Adv. Cyclic Nucleotide Res.* **5**:3–29.

Rogers, G. M., and Shuman, M. A., 1983, Prothrombin is activated on vascular endothelial cells by Factor X and calcium, *Proc. Natl. Acad. Sci. U.S.A.* **80**:7001–7005.

Ruggeri, Z. M., DeMarco, L., Gatti, L., Bader, R., and Montgomery, R. R., 1983, Platelets have more than one binding site for von Willebrand factor, *J. Clin. Invest.* **72**:1–12.

Ruoslahti, E., Hayman, E. G., Kuusela, P., Shively, J. E., and Engvall, E., 1979, Isolation of a tryptic fragment containing the collagen-binding site of plasma fibronectin, *J. Biol. Chem.* **254**:6054–6059.

Salzman, E. W., and Chambers, D. A., 1965, Incorporation by blood platelets of adenosine diphosphate labelled with carbon-14, *Nature (London)* **206**:727–728.

Salzman, E. W., Kensler, P. C., and Levine, L., 1972, Cyclic 3′,5′-adenosine monophosphate in human blood platelets. IV. Regulatory role of cyclic AMP in platelet function, *Ann. N.Y. Acad. Sci.* **201**:61–71.

Santoro, S. A., and Cowan, J. F., 1982, Adsorption of von Willebrand factor by fibrillar collagen. Implications concerning the adhesion of platelets to collagen, *Collagen Relat. Res.* **2**:31–44.

Santoro, S. A., and Cunningham, L. W., 1977, Collagen-mediated platelet aggregation. Evidence for multivalent interactions of intermediate specificity between collagen and platelets, *J. Clin. Invest.* **60**:1054–1060.

Santoro, S. A., and Cunningham, L. W., 1979, Fibronectin and the multiple interaction model for platelet-collagen adhesion, *Proc. Natl. Acad. Sci. U.S.A.* **76**:2644–2648.

Santoro, S. A., and Cunningham, L. W., 1981, The interaction of platelets with collagen, in: *Platelets in Biology and Pathology,* Volume 2 (J. L. Gordon, ed.), Elsevier/North-Holland Biomedical Press, Amsterdam, pp. 249–264.

Shuman, M. A., Botney, M., and Fenton, II, J. W., 1979, Thrombin-induced platelet secretion. Further evidence for a specific pathway, *J. Clin. Invest.* **63**:1212–1218.

Smith, S. K., and Limbird, L. E., 1981, Solubilization of human platelet α-adrenogenic receptors: Evidence that agonist occupancy of the receptor stabilizes receptor-effector interactions, *Proc. Natl. Acad. Sci. U.S.A.* **78**:4026–4030.

Snider, R. M., and Richelson, E., 1983, Thrombin stimulation of guanosine 3′,5′-monophosphate formation in murine neuroblastoma cells (Clone NIE-115), *Science* **221**:566–568.

Steer, M. L., and Atlas, D., 1982, Demonstration of human platelet β-adrenergic receptors using ^{125}I-labeled cyanopindolol and ^{125}I-labeled hydroxybenzylpindolol, *Biochim. Biophys. Acta* **686**:240–244.

Steer, M. L., and Wood, A., 1979, Regulation of human platelet adenylate cyclase by epinephrine, prostaglandin E_1, and gaunine nucleotides. Evidence for separate guanine nucleotide sites mediating stimulation and inhibition, *J. Biol. Chem.* **254**:10791–10797.

Tam, S. W., and Detwiler, T. C., 1978, Binding of thrombin to human platelet plasma membranes, *Biochim. Biophys. Acta* **543**:194–201.

Tandon, N., Harmon, J. T., Rodbard, D., and Jamieson, G. A., 1983, Thrombin receptors define responsiveness of cholesterol-modified platelets, *J. Biol. Chem.* **258**:11840–11845.

Taylor, D. G., Mapp, R. J., and Crawford, N., 1975, The identification of actin associated with pig platelet membranes and granules, *Biochem. Soc. Trans.* **3**:161–164.

Tollefsen, D. M., and Majerus, P. W., 1976, Evidence for a single class of thrombin-binding sites on human platelets, *Biochemistry* **15**:2144–2149.

Tollefsen, D. M., Feagler, J. R., and Majerus, P. W., 1974, The binding of thrombin to the surface of human platelets, *J. Biol. Chem.* **249**:2646–2651.

Tracy, P. B., Rohrbach, M. S., and Mann, K. G., 1983, Functional prothrombinase complex assembly on isolated monocytes and lymphocytes, *J. Biol. Chem.* **258**:7264–7267.

Tsai, B.-S., and Lefkowitz, R. J., 1978, Agonist-specific effects of monovalent and divalent cations on adenylate cyclase-coupled alpha adrenergic receptors in rabbit platelets, *Mol. Pharmacol.* **14**:540–548.

Tsai, B.-S., and Lefkowitz, R. J., 1979, Agonist-specific effects of guanine nucleotides on alpha-adrenergic receptors in human platelets, *Mol. Pharmacol.* **16**:61–68.

Wallace, W. C., and Bensusan, H. B., 1980, Protein phosphorylation in platelets stimulated by immobilized thrombin at 37° and 4 °C, *J. Biol. Chem.* **255**:1932–1937.

Weksler, B. B., Ley, C. W., Jaffe, E. A., 1978, Stimulation of endothelial cell prostacyclin production by thrombin, trypsin and the ionophore A23187, *J. Clin. Invest.* **62**:923–930.

White, II, G. C., Workman, E. F., Jr., and Lundblad, R. L., 1977, Platelet–thrombin interactions. The platelet as a substrate for thrombin, in: *Chemistry and Biology of Thrombin* (R. L. Lundblad, J. W. Fenton, II, and K. G. Mann, eds.), Ann Arbor Science Publishers, Inc., Ann Arbor, MI, pp. 479–498.

White, G. C., Lundblad, R. L., and Griffith, M. J., 1981, Structure-function relations in platelet-thrombin reactions. Inhibition of platelet–thrombin interactions by lysine modification, *J. Biol. Chem.* **256**:1763–1766.

Williams, L. T., Mullikin, D., and Lefkowitz, R. J., 1976, Identification of alpha-adrenergic receptors in uterine smooth muscle membranes by [^{3}H]dihydroergocryptine binding, *J. Biol. Chem.* **251**:6915–6923.

Workman, Jr., E. F., White, II, G. C., and Lundblad, R. L., 1977, Structure–function relationships in the interaction of α-thrombin with blood platelets, *J. Biol. Chem.* **252**:7118–7123.

Zucker, M. B., Broekman, M. J., and Kaplan, K. L., 1979a, Factor VIII-related antigen in human blood platelets. Localization and release by thrombin and collagen, *J. Lab. Clin. Med.* **94**:675–682.

Zucker, M. B., Mosesson, M. W., Broekman, M. J., and Kaplan, K. L., 1979b, Release of platelet fibronection (cold-insoluble globulin) from alpha granules induced by thrombin or collagen; lack of requirement for plasma fibronectin in ADP-induced platelet aggregation, *Blood* **54**:8–12.

8

Secreted Alpha Granule Proteins
The Race for Receptors

Deane F. Mosher, Donna M. Pesciotta, Joseph C. Loftus, and Ralph M. Albrecht

1. INTRODUCTION

Alpha granules are the specific, protein-containing granules of platelets. Contents of alpha granules are released during platelet activation and blood coagulation. Of the released proteins, some bind to the surface of activated platelets, some participate in fibrin formation, and some prepare and activate traumatized or inflamed tissue for subsequent repair. In the present chapter, we describe these processes. We pay particular attention to the concentration of certain proteins within alpha granules, which may be up to a millionfold greater than concentrations in plasma and a thousandfold greater than concentrations in serum. We emphasize our recent studies of surface-activated platelets using correlative scanning and high-voltage electron microscopy. Finally, we consider the high surface concentration of glycoprotein (GP) IIb-IIIa complexes on unactivated platelets and hypothesize that loosening of the tight packing of the complexes during platelet activation may explain why proteins such as fibrinogen bind to activated but not unactivated platelets.

2. OVERVIEW OF ALPHA GRANULES

2.1. Morphologic Description

Secretion in platelets can be subdivided into two main categories: (1) the release of material from secretory or storage granules and (2) the release of substances, such as

Deane F. Mosher and Donna M. Pesciotta • Department of Medicine, University of Wisconsin, Madison, Wisconsin 53706.   *Joseph C. Loftus and Ralph M. Albrecht* • School of Pharmacy, University of Wisconsin, Madison, Wisconsin 53706.

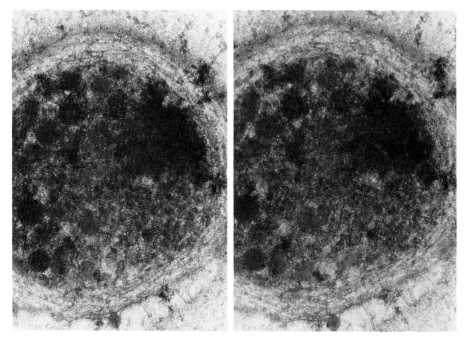

Figure 1. High-voltage electron microscopic stereo pair of an unactivated platelet. Platelets were fixed with glutaraldehyde in suspension and then adhered to polylysine-coated grids. Whole mounts were examined after postfixation with osmium tetroxide and staining with uranyl acetate. The circumferential microtubular band is present at the margin of the platelet and numerous organelles, including alpha granules, are seen in the cytoplasm (magnification 29,800×).

metabolites of arachadonic acid, which are not stored in granules, but are synthesized in response to activating stimuli. Materials are released from three types of secretory granules: (1) alpha granules, (2) dense bodies, and (3) lysosomes. The division of granules into these three groups is based on the ultrastructural appearance, density, and content of each granule. Alpha granules are membrane-enclosed vesicles 300 to 500 nm in diameter; there are 20 to 30 alpha granules per platelet (Figure 1). The granules are seen as blue dots against a grey background when blood smears are examined by Wright's stain. Thus, patients with deficiency of platelet alpha granules are said to have the gray platelet syndrome (Raccuglia, 1971). During subcellular fractionation, alpha granules can be distinguished from other granules by their ultrastructural appearance, content of platelet-specific proteins, and lack of mitochondrial (e.g., cytochrome oxidase), lysosomal (e.g., hexosaminidase), endoplasmic reticulum (e.g., glucose-6-phosphatase), and dense granule (e.g., serotonin) markers (Van der Meulen *et al.,* 1983, and references therein).

2.2. Alpha Granule Membranes

The surface area of the membranes of 25 alpha granules, each 400 nm in diameter, would be approximately 12.5×10^{-12} m². Membranes of alpha granules have been isolated, and their protein and lipid composition have been determined (Gogstad *et al.*, 1983; Van der Meulen *et al.*, 1983). Normalized to phospholipid content, the membranes contain one third as much cholesterol and twice as much protein as platelet plasma membranes isolated by the glycerol lysis method (Gogstad *et al.*, 1983). Crossed immunoelectrophoresis and sodium dodecyl sulfate–polyacrylamide gel electrophoresis indicate that the membranes contain both unique proteins and proteins that are found on the plasma membrane, such as the GP IIb-IIIa complex (Gogstad *et al.*, 1983), although recent immunoelectronmicroscopic studies have indicated the GP IIb-IIIa is restricted to the surface plasma membrane (Steinberg *et al.*, unpublished observations). As ascertained by differential surface labeling and proteolytic digestion of intact alpha granules and disrupted granular membranes, the membrane proteins are thought to be, for the most part, on the inner (noncytoplasmic) face of the membrane (Van der Meulen, 1983).

2.3.   Alpha Granule Contents

Fibrinogen, thrombospondin, beta-thromboglobulin precursor, and platelet factor 4, together with albumin, are major components of alpha granules and dominate electrophoregrams of proteins released from washed stimulated normal, but not alpha granule-deficient, platelets (Gerrard *et al.*, 1980; Nurden *et al.*, 1982). Thrombospondin, beta-thromboglobulin precursor, and platelet factor 4 are serum and not plasma proteins in the sense that almost all of the blood content of those proteins is in alpha granules (Kaplan, 1980; Files *et al.*, 1981; Saglio and Slayter, 1982; Dawes *et al.*, 1983). Fibrinogen, in contrast, is a major plasma protein, and both plasma and released platelet fibrinogen are nearly completely consumed during blood coagulation and the formation of serum. Immunochemical localization of alpha granule proteins at the fluorescent microscopic and ultrastructural levels (Giddings *et al.*, 1982; Sander *et al.*, 1983; Pham *et al.*, 1983; Stenberg *et al.*, 1984; Wenzel-Drake *et al.*, 1984) and studies of patients partially deficient in alpha granules (Weiss *et al.*, 1979) indicate that contents of individual granules are similar, i.e., there are not some granules filled with one protein and other granules filled with a second protein. The nucleoid of condensed material seen in some granules, however, stains particularly well for platelet factor 4 when platelets are examined by immunoelectron microscopy (Sander *et al.*, 1983).

The concentrations of fibrinogen, thrombospondin, beta-thromboglobulin precursor, and platelet factor 4 in alpha granules, calculated on the basis of the concentrations of these proteins in serum or platelet lysates and the estimated volume of alpha granules in intact platelets, are large (Table 1). Thus, thrombospondin in alpha granules is estimated to be at a millionfold greater concentration than thrombospondin in plasma. In the case of thrombospondin, fibrinogen, and platelet factor 4, the calculated concentrations of the proteins in alpha granules are considerably greater than the solubilities of the isolated proteins in physiologic saline. It seems likely that there are intermolecular interactions among alpha granule components that allow the components to be concentrated in the granule. For instance, platelet factor 4 in alpha granules is apparently complexed to a chondroitin-4-sulfate-containing proteoglycan (Huang *et al.*, 1982). The complex presumably has greater solubility than platelet factor 4 alone.

There are a number of plasma proteins in addition to fibrinogen and albumin that are found in alpha granules: von Willebrand factor (VWF) (Slot *et al.*, 1978), fibronectin (Plow *et al.*, 1979; Zucker *et al.*, 1979), factor V (Chesney *et al.*, 1981), alpha-2-

Table 1. Concentrations of the More Numerous Alpha Granule Proteins[a]

	Concentration (μg/ml)			
Protein	Plasma	Serum	Platelets	Alpha granules
Beta-thromboglobulin	0.006[b]	13.7[b]	4,300	51,400
Platelet factor 4	0.002[b]	5.3[b]	1,650	19,800
Thrombospondin	0.16[c]	15.7[c]	4,900	58,900
Fibrinogen	3,000	<10	7,300	87,500
Sum				217,600

[a]Platelet concentrations of the first three proteins were calculated from serum concentrations assuming a platelet count of 250,000 per μl whole blood and a mean platelet volume of 9.6 fl (Files et al., 1981) and also a packed red cell volume of 40% and 80% release of the proteins from the clot during production of serum. The platelet concentration of fibrinogen was calculated assuming that 2.5×10^8 platelets contain 17.5 μg of fibrinogen (Gerrard et al., 1980) and that the volume of a platelet is 9.6 fl. Granule concentrations were calculated assuming 25 alpha granules per platelet, each with a diameter of 400 nm and a volume of 0.03 fl, i.e., alpha granules account for approximately 8% of the volume of a platelet.
[b]Files et al. (1981).
[c]Saglio and Slayter (1982).

plasmin inhibitor (Plow and Collen, 1981), histidine-rich glycoprotein (Leung et al., 1983), alpha-2-macroglobulin (Nachman and Harpel, 1976), alpha-1-antitrypsin (Nachman and Harpel, 1976), vitronectin or spreading factor (Barnes et al., 1983), and others. Von Willebrand factor in alpha granules accounts for approximately 10% of VWF in blood, the other 90% being in plasma (Nachman and Jaffe, 1975). For the other proteins, the amounts in alpha granules are only small fractions of the amounts in plasma.

In studies of histidine-rich glycoprotein (Leung et al., 1983), four criteria were put forth as evidence that the small amount of the protein associated with isolated platelets does not represent a trapped plasma contaminant in the platelet pellet: (1) platelets isolated from platelet-rich plasma spiked with a radiolabeled derivative of histidine-rich glycoprotein did not contain significant amounts of the radiolabeled protein; (2) studies with anti-histidine-rich glycoprotein antibody revealed only traces of histidine-rich glycoprotein on the platelet surface; (3) supernates of isolated platelets contained histidine-rich glycoprotein only after stimulation of platelet granule release; and (4) antibodies to histidine-rich glycoprotein localized the antigen in cytoplasm of megakaryocytes.

Among the most interesting components of alpha granules are several growth factors. Platelet-derived growth factor has been the best studied. It supports the pro-liferation of a variety of cell types in the presence of plasma proteins (D. L. Kaplan et al., 1979; K. L. Kaplan et al., 1979). Platelet-derived growth factor initiates cells to enter the cell cycle, whereas the plasma proteins allow the cells to progress through the cell cycle (Pledger et al., 1978). Beta-thromboglobulin precursor also has growth-promoting activities (Castor et al., 1983). Platelets store a third factor called trans-forming growth factor-beta. This protein allows nonneoplastic cells to undergo an-chorage-independent growth in soft agar (Assoian et al., 1983), but inhibits growth of cells in monolayer culture (Shipley et al., 1984). The localization of this factor is not known, but it presumably is found in alpha granules.

2.4.   Functions of Alpha Granule Proteins

The functions of alpha granule proteins, once released, must be diverse. The growth factors, of course, would induce growth of sensitive cells. Platelet-derived growth factor (Seppä et al., 1982), platelet factor 4 (Senior et al., 1983), and beta-thromboglobulin (Senior et al., 1983) are chemotactic and would attract cells. Several of the proteins, including fibrinogen, VWF, fibronectin (Plow and Ginsberg, 1981), thrombospondin (Gartner and Dockter, 1983), platelet factor 4 (George and Onofre, 1982), and factor V (Tracy and Mann, 1983) become associated with the platelet surface and are thought to modulate platelet aggregation, platelet adhesion, and platelet prothrombinase activity. Fibronectin (McKeown-Longo and Mosher, 1983), thrombospondin (McKeown-Longo et al., 1984), and VWF (Wagner et al., 1982; Hormia et al., 1983) may enter the extracellular matrix.

Alpha granule proteins that are also present in plasma would be present around platelets in extra high concentrations following secretion. For instance, the concentration of initially released fibrinogen could be as high as 87.5 mg/ml as opposed to 3 mg/ml in plasma (Table 1). Leung et al. (1983) have calculated that even a trace platelet protein such as histidine-rich glycoprotein would be significantly concentrated in a platelet plug; the concentration in the microenvironment of a plug would be approximately 160 μg/ml, as contrasted to 100 μg/ml in plasma. Thus, the trace alpha granule proteins would reach high enough concentrations to be biologically important, e.g., to regulate fibrinolysis in the case of histidine-rich glycoprotein.

2.5.   Genesis of Alpha Granules and Alpha Granule Contents

Because platelets have a poor capacity to synthesize proteins, it is likely that the contents of alpha granules are synthesized and processed into granules in the megakaryocyte. Enriched cultures of megakaryocytes have been demonstrated to synthesize VWF (Nachman et al., 1977) and platelet factor 4 (Ryo et al., 1983). Isolated megakaryocytes have been found to contain VWF (Nachman et al., 1977), fibrinogen (Rabellino et al., 1979), and platelet-derived growth factor activity (Rabellino et al., 1979; Chernoff et al., 1980). Unfractionated magakaryocytes can be stained by immunofluorescence for a variety of alpha granule proteins, including histidine-rich glycoprotein (Leung et al., 1983), platelet factor 4 (Ryo et al., 1980; Rabellino et al., 1981), thrombospondin (McLaren, 1983), fibronectin (Rabellino et al., 1981), and von Willebrand factor (Rabellino et al., 1981). Platelet fibrinogen lacks the gamma-chain heterogeneity characteristic of plasma fibrinogen (Mosesson et al., 1984); such heterogeneity arises by differences in processing of mRNA for gamma chains (Crabtree and Kant, 1982). These diverse observations indicate that alpha granule proteins are indeed synthesized by megakaryocytes. The possibility exists, however, that one or more alpha granule proteins are synthesized elsewhere and then taken up by the magakaryocytes and/or platelet.

Insight into the development of alpha granules has been gained by comparative ultrastructural and cytochemical studies of normal megakaryocytes and megakaryocytes from patients with deficiencies of platelet alpha granules, the gray platelet syndrome (Breton-Gorius et al., 1981). In both normal and gray platelet syndrome megakaryocytes, 50–100 nm granules with an electron-dense core, thought to be precursors of alpha granules, are present in the Golgi region. In gray platelet syndrome megakaryocytes, these granules seem unable to mature and to be lost during maturation of the megakaryocytes. Instead, distended vacuolar structures are found that may contain alpha granule contents. The Golgi zone of the gray platelet syndrome mega-karyocytes is otherwise normally developed. Thus, gray platelet syndrome can be thought of as a defect in the megakaryocytes in which precusors of alpha granules are produced but then lost. In support of this concept, somewhat increased plasma levels of beta-thromboglobulin and platelet factor 4 have been found in gray platelet syndrome patients (Gerrard *et al.*, 1980; Levy-Toledano *et al.*, 1981), and patients have bone marrow fibrosis, a predicted result of growth factor release from megakaryocytes.

It is remarkable that a protein such as fibrinogen can be efficiently and quickly secreted by one cell type, i.e., hepatocytes, whereas another cell type, i.e., mega-karyocytes, synthesizes and packages the same protein into storage granules. Recently synthesized proteins are thought to be processed and targeted for cellular compartments in the Golgi apparatus; the processing events often involve modification of carbohy-drate side chains (Rothman, 1981). As culturing of megakaryocytes becomes more routine, there should be considerable interest in comparisons of protein processing by megakaryocytes and other cell types.

3. ALPHA GRANULE SECRETION

3.1. Overview

Secretion as we know it today is a composite of two mechanisms: (1) the mem-brane event known as stimulus–secretion coupling and (2) the transportation of intra-cellular components destined for release, the so-called secretory pathway. The secretory pathway, originally described from electron microscopic images of pan-creatic acinar cells, involves the sequestering of the secretory product into membrane-limited granules or secretory vesicles. These secretory vesicles then migrate toward the cell surface where they fuse with the plasma membrane and discharge their contents outside the cell (Jamieson and Palade, 1977).

It seems likely, however, that alpha granule secretion is only somewhat analagous to secretion in other systems. Two features distinguish platelets and alpha granule secretion. The first is the numerous invaginations of the platelet membrane to form the channels of the surface-connected canalicular system. The second is the remarkable change in platelet shape and redistribution of cytoskeletal components, so-called platelet contraction, which is necessary for release of alpha granule contents.

3.2. Secretion of Alpha Granules by Platelets in Suspension

Platelet secretion in suspension has been extensively studied at the ultrastructural level by White (1983) and others. The redistribution of alpha granules and their contents in thrombin-stimulated platelets was recently investigated in detail by Sten-berg *et al.* (1984) using both ultrastructural and immunocytochemical methods.

Thrombin stimulation causes alpha granules to cluster toward the center of platelets, to fuse with one another, and to fuse with the surface-connected canalicular system to form large surface-connected structures. Immunocytochemical localization of beta-thromboglobulin, platelet factor 4, and fibrinogen shows the predicted redistribution of these antigens on thrombin stimulation (Stenberg *et al.*, 1984). The antigens first move from alpha granules to large intracytoplasmic vacuoles. The vacuoles then fuse with the surface-connected canalicular system. These fused structures appear to communicate with the extracellular space by wide necks. The three antigens are found on the surface of stimulated but not unstimulated platelets. Because fusion of alpha granules or the large vacuoles with the plasma membrane is found only rarely, it has been concluded that most of the alpha granule contents gain access to the external milieu through the wide necks. Thus, Stenberg *et al.* (1984) described the following pattern of secretion from platelet alpha granules:

> The normally discoid platelet first becomes round and filopodia develop after stimulation with thrombin. Concomitantly, the alpha granules and other organelles move to the center of the platelet and are surrounded by a ring of microtubules. Granules fuse with other granules and with elements of the SCCS [surface-connected canalicular system] to form large vacuoles, situated near the center and periphery of the platelet, and containing alpha-granule proteins. As these vacuoles form in the cytoplasm the necks connecting the SCCS to the extracellular space become wider. The enlargement of these openings in the plasma membrane permits the rapid secretion of alpha granule proteins to the exterior of the platelet and binding of some of this secreted protein to the plasma membrane, where they carry out their various physiological functions.

This summary is consistent with a number of studies. The conclusion that platelet products are released through the surface-connected canalicular system was proposed by White on the basis of his extensive ultrastructural studies of platelet secretion in response to a variety of agents (White, 1983). Droller (1973a,b) and Holme *et al.* (1973) identified fibrillar material that they thought was fibrin in the surface-connected canalicular system of thrombin-stimulated platelets. In immunofluorescence studies, Ginsberg *et al.* (1980) noted that platelet factor 4 moved from a granular distribution in resting platelets to large masses near the surface of stimulated platelets. These large masses, for the most part, were not accessible to antibodies unless platelet membranes were permeabilized.

3.3.   Secretion of Alpha Granules in Response to Contact Activation

The tranformations that take place during platelet adhesion to siliconized glass has been examined by time-lapse differential interference contrast microscopy (Allen *et al.*, 1979). Tranformations included a disk-to-sphere change of shape, formation of sessile protuberances, extension and retraction of pseudopodia, and spreading, ruffling, and occasional regression of the hyalamere. Exocytosis of dense bodies could be observed easily by this technique; by accelerated time-lapse cinematography, platelets resembled "miniature volcanoes spewing forth particles." The conclusion that the particles represented contents of dense bodies was confirmed by parallel fluorescent microscopic studies of mepacrine-loaded platelets. After release of dense bodies, craters of variable size were left behind for variable amounts of time. These craters extended deep into the platelet and were thought to communicate with the open ca-

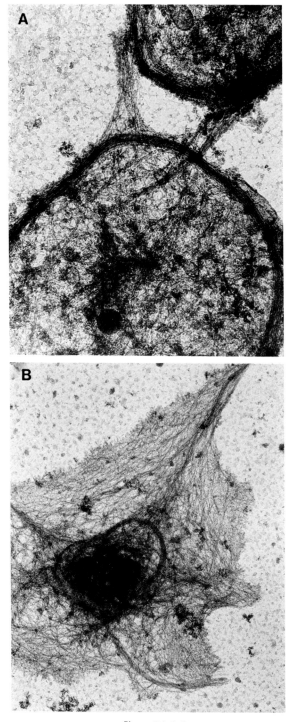

Figure 2A & B

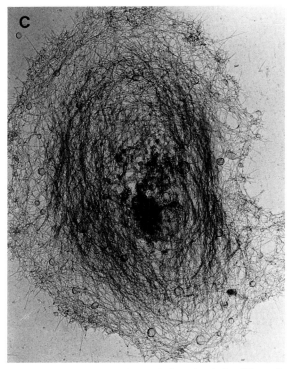

Figure 2. High-voltage electron microscopy of platelets before (A), during (B), or after (C) surface activation. Pictures are of whole mounts of platelets adhering directly to grids and then extracted with Triton X-100 to highlight the insoluble cytoskeletal networks. A prominent band of microtubules can be seen around the circumference of unactivated platelets (A, compare with Figure 1). There is also a network of short, disorganized microfilaments in the cytoplasm of the unactivated platelets. Upon activation (B), the microfilaments organize into long bundles that extend into the pseudopods. The remnant of the circular band of microtubules can be seen in the center of the platelet. At the completion of surface activation (C), the microfilaments lose their radial orientation and are organized around the central granulomere in a tightly packed central zone and a more loosely packed peripheral zone (also see Figure 3). Parallel studies of unextracted platelets (not shown) demonstrated that the granules cluster toward the center of the platelets during activation [magnifications (A) 20,000×; (B and C) 10,000×].

nalicular system. Smaller particles, thought to be alpha granules, also "disappeared" when single platelets were studied by cinematography, but it was not possible to describe their release. The degranulation events seemed to occur randomly in time once the hyalamere began to spread. Degranulated materials often bound to the surface as particles, as ascertained by both differential interference contrast and scanning election microscopy.

The cytoskeletal changes that drive the shape changes can be appreciated by high-voltage electron microscopy of whole mounts of adhering platelets (Loftus *et al.,* 1984), with or without extraction with Triton to highlight the Triton-insoluble cytoskeletal network (Figures 1–3). Resting platelets contain a circumferential band of microtubules and a small number of microfilaments randomly arranged through the dense cytoplasmic matrix (Figures 1 and 2A). Upon surface activation, prominent

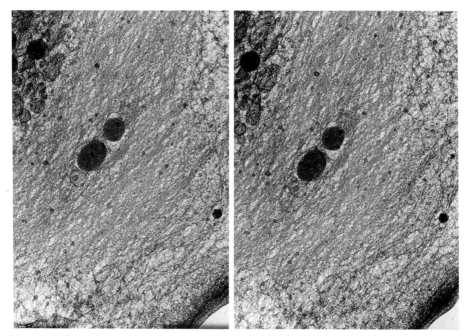

Figure 3. High-voltage electron microscopic pair of a fully spread activated platelet. Platelets were studied as described in Figure 2, except that the sample was not extracted with Triton X-100. The platelet is at the same stage of activation as the extracted platelet in Figure 2C. From lower right to upper left, note the following four zones: dense peripheral web, loosely packed outer filamentous zone, more tightly packed inner filamentous zone, and the granulomere (magnification 30,000×).

microfilament bundles develop that radiate from the center out into pseudopodia and also form a netlike arrangement in the cell body (Figure 2B). The central region is occupied by tightly clustered granules encompassed by the microtubular band, now irregular and fragmented. The cytoskeleton of a well-spread platelet has the overall round-to-oval shape of the intact spread platelet (Figures 2C and 3). It is composed entirely of a branched filamentous network. The structure of the cytoskeleton (Figure 2C) can be divided into four major domains that correlate well with the four structurally distinct zones within unextracted whole-mount preparations (Figure 3). The peripheral web is composed of a densely packed meshwork of fine filaments. This arrangement is seen more clearly in the cytoskeleton preparation. Microtubules are absent from this zone, and microfilaments are the predominant filament type. Interior to the terminal web is the outer filamentous zone with its more loosely interwoven array of filaments. Two major classes of filaments can be differentiated in this zone: microtubules 19 to 21 nm in diameter and microfilaments 6 to 8 nm in diameter. All filaments are discrete and uniform in diameter throughout their length. Numerous associations between filaments are visible at higher magnification. The adjacent inner filamentous zone has a "trabecularlike" appearance in unextracted whole mounts. Filaments appear anastomotic and variable in diameter. The course and interactions of individual filaments are difficult to discern. From the cytoskeleton preparations, it is evident that the underlying structure of the inner filamentous zone is a densely packed overlapping network of discrete filaments. The filaments are oriented predominately parallel to the long axis of the platelet, but bend to completely encircle the granulomere region. There are numerous associations between filaments within the zone and with other filaments that extend into the inner zone from the outer filamentous zone. The fourth zone is the granulomere region, composed of remaining dense granules, alpha granules, and mitochondria suspended within a microfilament net.

3.4. Comparison of Alpha Granule Secretion and Dense Granule Secretion

Dense bodies were named because of their electron dense appearance in transmission electron microscopy (Tanzer *et al.*, 1966). These granules vary in diameter from a few nanometers to 400 nm and can easily be separated out from other platelet granules on the basis of their density. Much of this electron density is due to the large quantities of calcium and phosphate. Holmsen and Weiss (1979) have calculated the following concentrations of constituents within human dense granules: ATP, 436 mM; ADP, 653 mM; pyrophosphate, 326 mM; calcium ion, 2181 mM; and serotonin 65 mM.

Upon stimulation to secrete with ADP or the ionophore A23187, several changes in the morphology of the dense bodies occur. The dense bodies enlarge and develop a clear center with an electron-dense periphery (Skaer, 1981). This enlargement and development of clear centers signifies which dense bodies are to be released, because dense bodies not being secreted retain their electron-dense appearance (Costa *et al.*, 1977) and are found deeper in the interior of the cytoplasm (Skaer, 1981).

The enlarged dense bodies are found at the surface of the plasma membrane, almost protruding through the plasma membrane and giving the appearance of a pear-shaped body (Skaer, 1981). The pear-shaped bodies have been called spheroids. The formation of spheroids occurs at the platelet surface within 30 sec after stimulation with ADP and more quickly if other secretogogues such as thrombin or the calcium ionophore, A23187, are used (Skaer, 1981). Spheroids, therefore, are believed to represent the discharging stage of dense bodies during secretion. Thus, dense body secretion in response to stimuli such as ADP and A23187 involves the migration of dense bodies to the plasma membrane, followed by fusion between dense body membranes and plasma membranes (Skaer, 1981). The granule then enlarges, presumably by hydration (Pollard *et al.*, 1977), and releases its contents into the surrounding medium, either by exocytosis or lipocytosis (Skaer, 1981).

In studies in which secretion is blocked with formaldehyde at various times after stimulation of suspended platelets with thrombin, release of beta-thromboglobulin and/or platelet factor 4 from alpha granules is fivefold to tenfold slower than release of serotonin from dense granules (Ginsberg *et al.*, 1980, Akkerman *et al.*, 1982). As an example, when washed platelets were incubated with thrombin, 1 μg/ml, at 22 °C, serotonin release reached a maximum after 2 min, whereas platelet factor 4 release was still increasing after 12 min (Ginsberg *et al.*, 1980). Discharge of dense granule constituents and extrusion of alpha granule proteins can be dissociated by varying the stimulator; ADP, epinephrine, and low doses of thrombin induce mainly discharge of dense granule constituents (White, 1983). Such kinetic and dose-response differences indicate that the pathways for degranulation from the two types of granules are different. On the basis of the microcopic studies described above, it seems likely that individual dense granules fuse with the plasma membranes and release their contents in discrete, quantal events, whereas alpha granules fuse with one another and then with the surface-connected canalicular system so that alpha granule release is both more leisurely and more orderly.

Further evidence for the existence of distinct secretory pathways has come from studies in which platelets were treated with phorbol myristate acetate. Alpha granules coalesce and fuse with the surface-connected canalicular system, but do not discharge their contents (White and Estensen, 1974; Estensen and White, 1974). This suggests that membrane fusion alone is not sufficient for relase and that alpha granule secretion must be dependent on some other factor besides fusion. Platelet contraction to expel granule contents into the canalicular system might be this factor (White *et al.*, 1974). Indeed, dense bodies can release their contents without platelet contraction (Kirkpatrick *et al.*, 1980).

4. INTERACTIONS OF ALPHA GRANULE PROTEINS WITH THE SURFACE OF ACTIVATED PLATELETS

4.1. Interactions of Alpha Granule Proteins with Suspended Platelets

As described in Sections 2.4 and 3.2 and in other chapters of this volume, a number of alpha granule proteins, including fibrinogen, VWF, thrombospondin, fibronectin, and platelet factor 4, bind to surfaces of activated platelets but not to surfaces of resting platelets. Such studies are usually done with radiolabeled proteins. The labeled probe is added to the suspension of appropriately stimulated platelets, and bound and free fractions are separated by centrifugation after a suitable incubation. Unstimulated platelets serve as an excellent control for nonspecific binding and trapping of ligand. Binding of fibrinogen, VWF, and fibronection to thrombasthenic platelets is decreased (Bennett and Vilaire, 1979; Peerschke *et al.*, 1980; Niewiarowski *et al.*, 1981; Ruggeri *et al.*, 1982; Ginsberg *et al.*, 1983a), whereas binding of fibronectin to alpha-granule-deficient platelets is normal (Ginsberg *et al.*, 1983b). Binding of fibrinogen, VWF, fibronectin, and thrombospondin to normal platelets is inhibited by monoclonal antibodies to GP IIb-IIIa complex (Bennett *et al.*, 1983; Coller *et al.*, 1983; DiMinno *et al.*, 1983; McEver *et al.*, 1983; Pidard *et al.*, 1983; Ruggeri *et al.*, 1983; Ginsberg *et al.*, 1984). Thus, it seems likely that the four proteins bind to the same protein complex on the surface-activated platelets. Further, although endogenous thrombospondin, platelet factor 4, and fibrinogen can be detected on the surfaces of activated platelets (Phillips *et al.*, 1980; George *et al.*, 1980; George and Onofre, 1982; Stenberg *et al.*, 1984), it is unlikely that externalized alpha granule membrane proteins or constituents enhance binding of exogenous proteins because binding to alpha granule-deficient platelets is normal.

4.2. Interactions of Alpha Granule Proteins with Spread Platelets

Upon surface activation of platelets, fibrinogen, VWF, fibronectin, and thrombospondin can be demonstrated on the surface of adhered and spread platelets in an

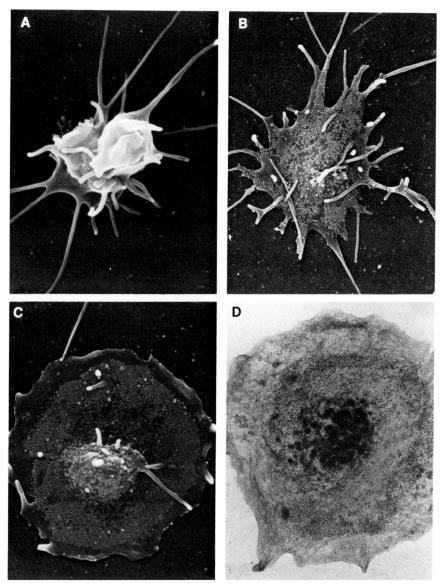

Figure 4. Scanning (A–C) and high-voltage (D) electron microscopy of fibrinogen–gold bead complexes bound to surface-activated platelets. Washed platelets were mixed with fibrinogen bound to 19-nm gold beads and then incubated with grids. Adherent platelets at various stages of activation were examined. At an early stage of activation (A), few fribrinogen–gold complexes are bound to the platelets. At an intermediate stage of activation (B), the surface of the platelet is almost completely covered by complexes. At the completion of activation (C), the center of the platelet is covered by complexes, whereas few complexes are seen on the periphery. The sharp demarcation between covered and uncovered surface corresponds to the boundary between the tightly packed central zone and loosely packed peripheral zone of microfilaments (D) (magnifications 10,000×).

enzyme-linked immunosorbent assay (Lahav and Hynes, 1981; M. J. Doyle and D. F. Mosher, unpublished data). When exogenous fibronectin or VWF is added, 50 to 100% more antigen can be detected on platelet surfaces (Lahav and Hynes, 1981). Thus, as with activation of platelets in suspension, surface activation results in surface binding of both endogenous and exogenous pools of alpha granule proteins.

The localization of fibrinogen-labeled colloidal gold in whole mounts of surface-activated platelets has been studied by scanning and high-voltage electron microscopy (Loftus and Albrecht, 1984) (Figure 4). Platelets in the early stages of activation possess a dendritic shape characterized by the presence of long, thin pseudopods and minimal spreading of the hyalomere (Figure 4A). Fibrinogen-gold markers are present in very low numbers. Gradual flattening of the hyalomere as activation proceeds is accompanied by marked increase in the binding of fibrinogen-gold (Figure 4B). The markers are visible over the entire surface of the platelet, including the pseudopodia. The transition between activated platelets with few fibrinogen-gold markers and activated platelets with near maximal numbers of markers must be sudden, because few examples of platelets with intermediate numbers of markers are seen. Further extension of the hyalomere results in flattened oval or circular platelets with no pseudopodia (Figure 4C). These platelets bind large numbers of fibrinogen-gold complexes. Binding, however, is predominantly in central regions, and markers are nearly absent from peripheral regions.

By high-voltage electron microscopic examination of well-spread platelets (Figure 4D), fibrinogen-gold complexes are localized on the platelet membrane over the inner filamentous and granulomere zones, whereas the outer filamentous zone is nearly free of labels. Some labels are also seen by stereo-pair imaging (not shown) to be on underside of peripheral web at the cell margin. These complexes may be trapped beneath the free edge of the platelet. The sharp delineation between the inner and outer filamentous zones and concomitant fibrinogen binding can be seen very clearly at high magnification (not shown, but see Figure 3). Clearing of fibrinogen-gold markers from the periphery is only seen in platelets in which the filamentous region has differentiated in inner and outer zones.

Fibrinogen, VWF, fibronectin, and thrombospondin all enhance adherence and activation of platelets when preadsorbed onto polymeric arteriovenous shunts prior to implantation (Ihlenfeld *et al.*, 1978; Young *et al.*, 1982a,b). This finding suggests that receptors for alpha granule proteins are also expressed on the ventral surface of spreading platelets, i.e., the surface in immediate contact with the protein-coated polymer. Further studies are needed, however, to critically evaluate just how molecular events on the ventral surface of activated platelets are related to events on the dorsal surface or on surfaces of activated platelets in suspension. Particular attention should be paid to the possibility that receptors migrate to and become fixed on the ventral surface, as happens with macrophage Fc receptors when macrophages are plated on surfaces precoated with immune complexes (Michl *et al.*, 1983a,b).

4.3. Mechanism of Binding of Alpha Granule Proteins to Activated but Not Unactivated Platelets: The Rainforest Hypothesis

One of the central questions of platelet biology is why unactivated platelets do not bind large protein ligands, whereas activated platelets do. As described in Section 4.1

Table 2. Surface Areas of Unstimulated and Stimulated Platelets and Calculated
Maximal Numbers of GP IIb-IIIa Complexes

Platelet shape and dimensions	Surface area (m² × 10¹²)	Maximum number of GP IIb-IIIa complexes[a] (× 10⁻³)
Unstimulated		
Sphere, volume of 9.6 fl	22	146
Right circular cylinder, diameter of 2.5 μm, height of 1.95 μm	25	166
Stimulated		
Disk, diameter of 4.5 μm, negligible height, many pseudopods (as in Figure 4B)	> 32	> 213
Disk, diameter of 7.6 μm, negligible height (as in Figure 4C)	91	606

[a]Calculated assuming that the complex occupies a square on the surface with an area of 1.5×10^{-16} m². This area was calculated based on molecular dimensions described by Phillips *et al.* (D. R. Phillips, personal communication).

and elsewhere in this volume, there is ample evidence that fibrinogen, fibronectin, VWF, and thrombospondin bind to the GP IIb-IIIa complex. Monoclonal antibodies to the complex bind to 50 to 60×10^3 sites on both unactivated and activated platelets (McEver *et al.*, 1983; Pidard *et al.*, 1983). These monoclonal antibodies block fibrinogen binding to activated platelets and do not recognize unassociated GP IIb and IIIa subunits (McEver *et al.*, 1983; Pidard *et al.*, 1983). Thus, some other event in addition to association of GP IIb and IIIa is required for fibrinogen binding to proceed.

Because we were impressed by the tight packing of fibrinogen-gold complexes on the surface of activated platelets (Figure 4), we calculated the maximal densities of GP IIb-IIIa complexes on unactivated and activated platelets (Table 2). The calculated numbers for unactivated platelets are a little more than twice the number of complexes estimated by antibody binding (McEver *et al.*, 1983; Pidard *et al.*, 1983). In other words, GP IIb-IIIa complexes would be predicted to occupy almost 50% of the surface area of the unactivated platelet. The surface area of an activated platelet is considerably larger. Thus, the percentage of the area occupied by GP IIa-IIIa complexes would be considerably smaller after activation.

These calculations suggest a hypothesis to explain binding of large protein ligands to activated platelets. We would like to call it the "rainforest hypothesis." As shown in Figure 5, GP IIb-IIIa complexes may be packed so tightly on the surfaces of unactivated platelets that large ligands cannot find their way to binding sites underneath the "canopy." It is only when the GP IIb-IIIa complexes diffuse into a larger area (Figure 5) that binding can proceed. This hypothesis has the potential of explaining several phenomena. According to the hypothesis, fibrinogen would bind to chymotrypsin-treated platelets (compare Kornecki *et al.*, 1983) because the complexes are "chopped down" to expose the binding site, perhaps in the chymotrypsin-resistant 66 kd fragment of GP IIIa (Bennett *et al.*, 1982; Kornecki *et al.*, 1983). Antigenic sites, such as GP IIa, which are hidden on unactivated platelets but exposed on activated platelets (McEver and Martin, 1984) are part of the "understory" in the

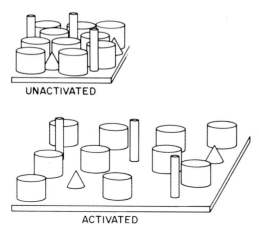

UNACTIVATED

ACTIVATED

Figure 5. Possible redistribution of platelet surface glycoproteins during activation. Calculations based on surface area and the size and number of GP IIb-IIIa complexes (cylinders) indicate that the complexes must be tightly packed on the surface of the unactivated platelets. Of the other surface glycoproteins, some (e.g., GP Ib), shown as rods, extend above the surface of the GP IIb-IIIa complexes, whereas others (including, perhaps, GP IIa), shown as cones, are shielded by the GP IIb-IIIa complexes. During activation, the surface area of the platelet increases several-fold, allowing the surface glycoproteins to diffuse apart. As a consequence, shielded glycoproteins and shielded parts of GP IIb-IIIa complexes become accessible and available to bind macromolecular ligands.

unactivated platelet and therefore not accessible to antibody. Activation of platelets would have to proceed further for efficient binding of one ligand [e.g., fibronectin, which binds to thrombin-stimulated but not ADP-stimulated platelets (Plow and Ginsberg, 1981)] than for a second ligand (e.g., fibrinogen, which binds well to both thrombin-stimulated and ADP-stiumlated platelets) if the first ligand required greater accessibility to its receptor than the second ligand. In addition, several predictions can be made based on the hypothesis. For instance, the synthetic pentadecapeptide that binds to activated platelets and blocks subsequent fibrinogen binding (Kloczewiak *et al.*, 1984; Plow *et al.*, 1984) may be small enough to diffuse and bind to the cryptic fibrinogen binding site on unactivated platelets. As another instance, there may be congenital disorders in which platelets cannot mobilize surface membrane upon activation; such a disorder would be very similar to Glanzmann's thromboasthenia in its phenotype.

The calculations in Table 2 suggest that up to $65 \times 10^{-12} \ m^2$ of new membrane is present on the surface of activated platelets. This is several-fold greater than the total surface area of membranes in alpha granules calculated in Section 2.2. Thus, it is likely that a large proportion of the new membrane originates from the surface-connected canalicular system. In any case, the contractile events that drive alpha granule release are probably also instrumental in the redistribution of membrane to the platelet surface. As a consequence, secretion of alpha granule proteins and expression of receptors for the proteins may be tightly coupled in the stimulated platelet.

ACKNOWLEDGMENTS. Experimental work described in this paper was supported by the National Institutes of Health (HL 29586) and the Wisconsin Affiliate of the American Heart Association. We thank our many colleagues, especially Dr. Mark Ginsberg, for stimulating discussion.

REFERENCES

Akkerman, J. W. N., Gorter, G., and Kloprogge, E., 1982, Kinetic analysis of alpha-granule secretion by platelets. A methodological report, *Thrombos. Res.* **27:**59–64.

Allen, R. D., Zacharski, L. R., Widirstky, S. T., Rosenstein, R., Zaitlin, L. M., and Burgess, D. R., 1979, Transformation and motility of human platelets. Details of the shape change and release reaction observed by optical and electron microscopy, *J. Cell Biol.* **83:**126–142.

Assoian, R. K., Komoriya, A., Meyers, C. A., Miller, D. M., and Sporn, M. B., 1983, Transforming growth factor-β in human platelets. Identification of a major storage site, purification, and characterization, *J. Biol. Chem.* **258:**7155–7160.

Barnes, D. W., Silnutzer, J., See, C., and Shaffer, M., 1983, Characterization of human serum spreading factor with monoclonal antibody, *Proc. Natl. Acad. Sci. U.S.A,* **80:**1362–1366.

Bennett, J. S., and Vilaire, G., 1979, Exposure of platelet fibrinogen receptors by ADP and epinephrine, *J. Clin. Invest.* **64:**1393–1398.

Bennett, J. S., Vilaire, G., and Cines, D. B., 1982, Identification of the fibrinogen receptor on human platelets by photoaffinity labeling, *J. Biol. Chem.* **257:**8049–8054.

Bennett, J. S., Hoxie, J. A., Leitman, S. F., Vilaire, G., and Cines, D.B., 1983, Inhibition of fibrinogen binding to stimulated human platelets by a monoclonal antibody, *Proc. Nat. Acad. Sci. U.S.A.* **80:**2417–2421.

Breton-Gorius, J., Vainchenker, W., Nurden, A., Levy-Toledano, S., and Caen, J., 1981, Defective α-granule production in megakaryocytes from gray platelet syndrome: Ultrastructural studies of bone marrow cells and megakarocytes growing in culture from blood precursors, *Am. J. Pathol.* **102:**10–19.

Castor, C. W., Miller, J. W., and Walz, D. A., 1983, Structural and biological characteristics of connective tissue activating peptide (CTAP-III), a major human platelet-derived growth factor, *Proc. Natl. Acad. Sci. U.S.A* **80:**765–769.

Chernoff, A., Levin, R. F., and Goodman, D. S., 1980, Origin of platelet-derived growth factor in megakaryocytes in guinea pigs, *J. Clin. Invest.* **65:**926–930.

Chesney, C. M., Pifer, D., and Colman, R. W., 1981, Subcellular localization and secretion of factor V from human platelets, *Proc. Natl. Acad. Sci. U.S.A.* **78:**5180–5184.

Coller, B. S., Peerschke, E. I., Scudder, L. E., and Sullivan, C. A., 1983, A murine monoclonal antibody that completely blocks the binding of fibrinogen to platelets produces a thrombasthenic-like state in normal platelets and binds to glycoproteins IIb and/or IIIa, *J. Clin. Invest.* **72:**325–338.

Costa, J. L., Detwiler, T. C., Feinman, R. D., Murphy, D. L., Patlak, C. S., and Pettigrew, K. D., 1977, Quantitative evaluation of the loss of human platelet dense bodies following stimulation by thrombin or A23187, *J. Physiol.* **264:**297–306.

Crabtree, G. R., and Kant, J. A., 1982, Organization of the rat γ-fibrinogen gene: Alternative mRNA splice patterns produce the γA and γB(γ') chains of fibrinogen, *Cell* **31:**159–166.

Dawes, J., Clemetson, K. J., Gogstad, G. O., McGregor, J., Clezardin, P., Prowse, C. V., and Pepper, D. S., 1983, A radioimmunoassay for thrombospondin, used in a comparative study of thrombospondin, β-thromboglobulin and platelet factor 4 in healthy volunteers, *Thromb. Res.* **29:**569–581.

DiMinno, G., Thiagarajan, G. P., Perussia, B., Martinez, J., Shapiro, S., Trinchieri, G., and Murphy, S., 1983, Exposure of platelet fibrinogen binding sites by collagen, arachidonic acid, and ADP: Inhibition by a monoclonal antibody to the glycoprotein IIb-IIIa complex, *Blood* **61:** 140–148.

Droller, M. J., 1973a, Ultrastructural visualization of the thrombin-induced platelet release reaction, *Scand. J. Haematol.* **11:**35–49.

Droller, M. J., 1973b, Ultrastructure of the platelet release reaction in response to various aggregating agets and their inhibitors, *Lab. Invest.* **29:**595–606.

Estensen, R. D., and White, J. G., 1974, Ultrastructural features of the platelet response to phorbol myristate acetate, *Am. J. Pathol.* **74:**441–452.

Files, J. C., Malpass, T. W., Yee, E. K., Ritchie, J. L., and Harker, L. A., 1981, Studies of human platelet alpha granule release in vivo, *Blood* **58:**607–618.

Gartner, T. K., and Dockter, M. E., 1983, Secreted platelet thrombospondin binds monovalently to platelets and erythrocytes in the absence of free Ca^{2+}, *Thromb. Res.* **33:**19–30.

George, J. N., and Onofre, A. R., 1982, Human platelet surface binding of endogenous secreted factor VIII-von Willebrand factor and platelet factor 4, *Blood* **58:**194–197.

George, J. N., Lyons, R. M., and Morgan, R. K., 1980, Membrane changes associated with platelet activation. Exposure of actin on the platelet surface after thrombin-induced secretion, *J. Clin. Invest.* **66:**1–9.

Gerrard, J. M., Phillips, D. R., Rao, G. H. R., Plow, E. F., Walz, D. A., Ross, R., Harker, L. A., and

White, J. G., 1980, Biochemical studies of two patients with the Gray Platelet Syndrome. Selective deficiency of platelet alpha graunules, *J. Clin. Invest.* **66:**102–109.

Giddings, J. C., Brookes, L. R., Piovella, F., and Bloom, A. L., 1982, Immunohistological comparison of platelet factor 4 (PF4), fibronectin (Fn) and factor VIII related antigen (VIII R:Ag) in human platelet granules, *Br. J. Haematol.* **52:**79–88.

Ginsberg, M. H., Taylor, L., and Painter, R. G., 1980, The mechanism of thrombin-induced platelet factor 4 secretion, *Blood* **55:**661–668.

Ginsberg, M. H., Forsyth, J., Lightsey, A., Chediak, J., and Plow, E. F., 1983a, Reduced surface expression and binding of fibronectin to thrombin-stimulated thromboasthenic platelets, *J. Clin. Invest.* **71:**619–624.

Ginsberg, M. H., Wencel, J. D., White, J. G., and Plow, E. F., 1983b, Binding of fibronectin to α-granule-deficient platelets, *J. Cell Biol.* **97:**571–573.

Ginsberg, M. H., Wolff, R., Marguerie, G., Coller, B., McEver, R., and Plow, E. F., 1984, Thrombospondin binding to thrombin-stimulated platelets: Evidence for a common adhesive protein binding mechanism, *Clin. Res.* **32:**308A.

Gogstad, G. O., Krutnes, M. B., Hetland, O., and Solum, N. O., 1983, Comparison of protein and lipid composition of the human platelet α-granule membranes and glycerol lysis membranes, *Biochim. Biophys. Acta* **732:**519–530.

Holme, R., Sixma, J. J., Murer, E. H., and Hovig, T., 1973, Demonstration of platelet fibrinogen secretion via the surface connecting system, *Thromb. Res.* **3:**347–356.

Holmsen, H., and Weiss, H. J., 1979, Secretable storage pools in platelets, *Annu. Rev. Med.* **30:**119–134.

Hormia, M., Lehto, V.-P., and Virtanen, I., 1983, Factor VIII-related antigen. A pericellular matrix component of cultured human endothelial cells, *Exp. Cell Res.* **149:**483–497.

Huang, S. S., Huang, J. S., and Deuel, T. F., 1982, Proteoglycan carrier of human platelet factor 4. Isolation and characterization, *J. Biol. Chem.* **257:**11546–11550.

Ihlenfeld, J. V., Mathis, T. R., Barber, T. A., Mosher, D. F., Riddle, L. M., Hart, A. P., Updike, S. J., and Cooper, S. L., 1978, Transient *in vivo* thrombus deposition onto polymeric biomaterials. Role pf plasma fibronectin, *Trans. Am. Soc. Artif. Intern. Organs* **24:**727–735.

Jamieson, J. D., and Palade, G. E., 1977, Production of secretory proteins in animal cells, in: *International Cell Biology 1976–1977* (B. R. Brinkley and K. R. Porter, eds.), The Rockefeller University Press, New York, pp. 308–317.

Kaplan, D. L., Chao, F. C., Stiles, C. D., Antoniades, M. N., and Scher, C. D., 1979, Platelet α-granules contain a growth factor for fibroblasts, *Blood* **53:**1043–1052.

Kaplan, K. L., 1980, β-thromboglobulin, *Prog. Hemostas. Thrombos.* **5:**153–178.

Kaplan, K. L., Broekman, M. J., Chernoff, A., Lesznik, G. R., and Drillings, M., 1979, Platelet α-granule proteins: Studies on release and subcellular localization, *Blood* **53:**604–618.

Kirkpatrick, J. P., McIntire, L. V., Moake, J. L., and Cimo, P. L., 1980, Differential effects of cytochalasin B on platelet release, aggregation and contractility: Evidence against a contractile mechanism for the release of platelet granuler contents, *Thromb. Haemost.* **42:**1483–1489.

Kloczewiak, M., Timmons, S., Lukas, T. J., and Hawiger, J., 1984, Platelet receptor recognition site on human fibrinogen. Synthesis and structure–function relationships of peptides corresponding to the carboxy-terminal segment of the γ chain, *Biochemistry* **23:**1767–1774.

Kornecki, E., Tuszynski, G. P., and Niewiarowski, S., 1983, Inhibition of fibrinogen receptor-mediated platelet aggregation by heterologous anti-human platelet membrane antibody, *J. Biol. Chem.* **258:**9349–9356.

Lahav, J., and Hynes, R. O., 1981, Involvement of fibronectin, von Willebrand factor, and fibrinogen in platelet interaction with solid substrata, *J. Supramolec. Struct. Cell. Biochem.* **17:**299–311.

Leung, L. L. K., Harpel, P. C., Nachman, R. L., and Rabellino, E. M., 1983, Histidine-rich glycoprotein is present in human platelets and is released following thrombin stimulation, *Blood* **62:**1016–1021.

Levy-Toledano, S., Caen, J. P., Breton-Gorius, J., Rendu, F., Cywiner-Golenzer, C., Dupuy, E., Legrand, Y., and Maclouf, J., 1981, Gray platelet syndrome: α-Granule deficiency. Its influence on platelet function, *J. Lab. Clin. Med.* **98:**831–848.

Loftus, J. C., and Albrecht, R. M., 1984, Redistribution of the fibrinogen receptor of human platelets after surface activation, *J. Cell Biol.* **99:**822–829.

Loftus, J. C., Choate, J., and Albrecht, R. M., 1984, Platelet activation and cytoskeletal reorganization: HVEM examination of intact and Triton extracted whole mounts, *J. Cell Biol.* **98:**2019–2025.

McEver, R. P., and Martin, M. N., 1984, A monoclonal antibody to a membrane glycoprotein binds only to activated platelets. *J. Biol. Chem.* **289:**9799–9804.

McEver, R. P., Bennett, E. M., and Martin, M. N., 1983, Identification of two structurally and functionally distinct sites on human platelet membrane glycoprotein IIb-IIIa using monoclonal antibodies, *J. Biol. Chem.* **258:**5269–5275.

McKeown-Longo, P. J., and Mosher, D. F., 1983, Binding of plasma fibronectin to cell layers of human skin fibroblasts, *J. Cell Biol.* **97:**466–472.

McKeown-Longo, P. J., Hanning, R., and Mosher, D. F., 1984, Binding and degradation of platelet thrombospondin by cultured fibroblasts, *J. Cell Biol.* **98:**22–28.

McLaren, K. M., 1983, Immunohistochemical localization of thrombospondin in human megakaryocytes and platelets, *J. Clin. Pathol.* **36:**197–199.

Michl, J., Pieczonka, M. M., Unkeless, J. C., Bell, G. I., and Silverstein, S. C., 1983a, Fc receptor modulation in mononuclear phagocytes maintained on immobilized immune complexes occurs by diffusion of the receptor molecule, *J. Exp. Med.* **157:**2121–2139.

Michl, J., Unkeless, J. C., Pieczonka, M. M., and Silverstein, S. C., 1983b, Modulation of Fc receptors of mononuclear phagocytes by immobilized antigen-antibody complexes. Quantitative analysis of the relationship between ligand number and Fc receptor response, *J. Exp. Med.* **157:**1746–1757.

Mosesson, M. W., Homandberg, G. A., and Amrani, D. L., 1984, Human platelet fibrinogen gamma chain structure, *Blood* **63:**990–995.

Nachman, R. L., and Harpel, P. C., 1976, Platelet α_2-macroglobulin and α_1-antitrypsin, *J. Biol. Chem.* **251:**4514–4521.

Nachman, R. L., and Jaffe, E. A., 1975, Subcellular platelet Factor VIII antigen and von Willebrand factor, *J. Exp. Med.* **141:**1101–1112.

Nachman, R. L., Levine, R., and Jaffe, E. A., 1977, Synthesis of factor VIII antigen by cultured guinea pig megakaryocytes, *J. Clin. Invest.* **60:**914–921.

Niewiarowski, S., Budzynski, A., Morinelli, T., Budzynski, T., and Stewart, G. J., 1981, Exposure of fibrinogen receptor on human platelets by proteolytic enzymes, *J. Biol. Chem.* **256:**917–925.

Nurden, A. T., Kunicki, T. J., Dupuis, D., Soria, C., and Caen, J. P., 1982, Specific protein and glycoprotein deficiencies in platelets isolated from two patients with the gray platelet syndrome, *Blood* **59:**709–718.

Peerschke, E. I., Zucker, M. B., Grant, R. A., Egan, J. J., and Johnson, M., 1980, Correlation between fibrinogen binding to human platelets and platelet aggregability, *Blood* **55:**841–847.

Pham, T. D., Kaplan, K. L., and Butler, V. P., 1983, Immunoelectron microscopic localization of platelet factor 4 and fibrinogen in the granules of human platelets, *J. Histochem. Cytochem.* **31:**905–910.

Phillips, D. R., Jennings, L. K., and Prasanna, H. R., 1980, Ca^{2+}-mediated association of glycoprotein G (thrombin-sensitive protein, thrombospondin) with human platelets, *J. Biol. Chem.* **255:**11629–11632.

Pidard, D., Montgomery, R. R., Bennett, J. S., and Kunicki, T. J., 1983, Interaction of AP-2, a monoclonal antibody specific for the human platelet glycoprotein IIb-IIIa complex, with intact platelets, *J. Biol. Chem.* **258:**12582–12586.

Pledger, W. J., Stiles, C. D., Antoniades, H. N., and Scher, C. D., 1978, An ordered sequence of events is required before BALB/C-3T3 cells becomes committed to DNA synthesis, *Proc. Natl. Acad. Sci. U.S.A.* **75:**2839–2943.

Plow, E. F., and Collen, D., 1981, The presence and release of α_2-antiplasmin from human platelets, *Blood* **58:**1069–1074.

Plow, E. F., and Ginsberg, M. H., 1981, Specific and saturable binding of plasma fibronectin to thrombin-stimulated human platelets, *J. Biol. Chem.* **256:**9477–9482.

Plow, E. F., Birdwell, C., and Ginsberg, M. H., 1979, Identification and quantitation of platelet-associated fibronectin antigen, *J. Clin. Invest.* **63:**540–543.

Plow, E. F., Srouji, A. H., Meyer, D., Marguerie, G., and Ginsberg, M. H., 1984, Evidence that three adhesive proteins interact with a common recognition site on activated platelets, *J. Biol. Chem.* **259:**5388–5391.

Pollard, H. B., Tack-Goldman, K., Pazoles, C. J., Creutz, C. E., and Shulman, N. C., 1977, Evidence for control of serotonin secretion from human platelets by hydroxyl ion transport and osmotic lysis, *Proc. Natl. Acad. Sci. U.S.A.* **74:**5295–5299.

Rabellino, E. M., Nachman, R. L., William, N. Winchester, R. J., and Ross, G. D., 1979, Human megakaryocytes. I. Characterization of the membrane and cytoplasmic components of isolated marrow megakaryocytes, *J. Exp. Med.* **149:**1273–1287.

Rabellino, E. M., Levene, R. B., Leung, L. L. K., and Nachman, R. L., 1981, Human megakaryocytes. II. Expression of platelet proteins in early marrow megakaryocytes, *J. Exp. Med.* **154:**88–100.

Raccuglia, G., 1971, Gray platelet syndrome: A variety of qualitative platelet disorder, *Am. J. Med.* **51:**818–828.

Rothman, J. E., 1981, The Golgi apparatus. Two organelles in tandem, *Science* **213:**1212–1219.

Ruggeri, Z. M., Bader, R., and DeMarco, L., 1982, Glanzmann thrombasthenia: Deficient binding of von Willebrand factor to thrombin-stimulated platelets, *Proc. Natl. Acad. Sci. U.S.A.* **79:**6038–6041.

Ruggeri, Z. M., DeMarco, L., Gatti, L., Bader, R., Montgomery, R. R., 1983, Platelets have more than one binding site for von Willebrand factor, *J. Clin. Invest.* **72:**1–12.

Ryo, R., Proffitt, R. T., Poger, M. E., O'Bear, R., and Deuel, T. F., 1980, Platelet factor 4 antigen in megakaryocytes, *Thromb. Res.* **17:**645–652.

Ryo, R., Nakoff, A., Huang, S. S., Ginsberg, M., and Deuel, T. F., 1983, New synthesis of a platelet-specific protein: Platelet factor 4 synthesis in a megakaryocyte-enriched rabbit bone marrow culture system, *J. Cell Biol.* **96:**515–520.

Saglio, S. D., and Slayter, H. S., 1982, Use of a radioimmunoassay to quantify thrombospondin, *Blood* **59:**162–166.

Sander, H. J., Slot, J. W., Bouma, B. N., Bolhuis, P. A., Pepper, D. S., and Sixma, J. J., 1983, Immunocytochemical localization of fibrinogen, platelet factor 4, and beta thromboglobulin in thin frozen sections of human blood platelets, *J. Clin. Invest.* **72:**1277–1287.

Senior, R. M., Griffin, G. L., Huang, J. S., Walz, D. A., and Deuel, T. F., 1983, Chemotactic activity of platelet alpha granule proteins for fibroblasts, *J. Cell Biol.* **96:**382–385.

Seppä, H., Grotendorst, G., Seppä, S., Schiffmann, E., and Martin, G. R., 1982, Platelet derived growth factor in chemotactic for fibroblasts, *J. Cell Biol.* **92:**584–588.

Shipley, G. D., Tucker, R. F., and Moses, H. L., 1984, Platelet transforming growth factor activity on AKR-2B cells in serum-free medium, *Fed. Proc.* **43:**373.

Skaer, R. J., 1981, Platelet degranulation, in: *Platelets in Biology and Pathology,* Volume 2 (J. L. Gordon, ed.), Elsevier/North Holland, Amsterdam, pp. 321–348.

Slot, J. W., Bouma, B. N., Montgomery, R., and Zimmerman, T. S., 1978, Platelet factor VIII-related antigen: Immunofluorescent localization, *Thromb. Res.* **13:**871–881.

Stenberg, P. E., Shuman, M. A., Levine, S. P., and Bainton, D. F., 1984, Redistribution of alpha-granules and their contents in thrombin-stimulated platelets, *J. Cell Biol.* **98:**748–760.

Tanzer, J. P., DaPrada, M., and Pletscher, A., 1966, *Nature (London)* Ultrastructural localization of 5-hydroxy-tryptamine in blood platelets, **212:**1574–1575.

Tracy, P. B., and Mann, K. G., 1983, Prothrombinase complex assembly on the platelet surface is mediated through the 74,00 dalton component of factor V_a, *Proc. Natl. Acad. Sci. U.S.A.* **80:**2380–2384.

Van der Meulen, J., Furuyo, W., and Grinstein, S., 1983, Isolation and partial characterization of platelet α-granule membranes, *J. Membr. Biol.* **71:**47–59.

Wagner, D. D., Olmsted, J. B., and Marder, V. J., 1982, Immunolocalization of von Willebrand protein in Weibel-Palade bodies of human endothelial cells, *J. Cell Biol.* **95:**355–360.

Weiss, H. J., Witte, L. D., Kaplan, K. L., Lages, B. A., Chernoff, A., Nossel, H. L., Goodman, D. S., and Baumgartner, H. R., 1979, Heterogeneity in storage pool deficiency: Studies on granule-bound substances in 18 patients including variants deficient in α-granules, platelet factor 4, β-thromboglobulin and platelet-derived growth factor, *Blood* **54:**1296–1319.

Wenzel-Drake, J. D., Plow, E. F., Zimmerman, T. S., Painter, R. G., and Ginsberg, M. H., 1984, Immunofluorescent localization of adhesive glycoproteins in resting and thrombin-stimulated platelets. *Am. J. Pathol.* **115:**156–164.

White, J. G., 1983, The ultrastructure and regulatory mechanisms of blood platelets, in: *Blood Platelet Function and Medicinal Chemistry* (A. Lasslo, ed.), Elsevier Biomedical, New York, pp. 15–59.

White, J. G., and Estensen, R., 1974, Cytochemical electron microscopic studies of the action of phorbol myristate acetate on platelets, *Am. J. Pathol.* **74:**453–466.

White, J. G., Rao, G. H. R., and Estensen, R. D., 1974, Investigation of the release reaction in platelets exposed to phorbol myristate acetate, *Am. J. Pathol.* **75:**301–314.

Young, B. R., Lambrecht, L. K., Mosher, D. F., and Cooper, S. L., 1982a, Plasma proteins: Their role in initiating platelet and fibrin deposition on biomaterials, *Adv. Chem. Ser.* **199:**317–350.

Young, B. R., Doyle, M. J., Collins, W. E., Lambrecht, L. K., Jordan, C. A., Albrecht, R. M., Mosher, D. F., and Cooper, S. L., 1982b, Effect of thrombospondin and other platelet α-granule proteins on artificial surface-induced thrombosis, *Trans. Am. Soc. Artif. Intern. Organs* **28:**498–503.

Zucker, M. B., Mosesson, M. W., Broekman, M. J., and Kaplan, K. L., 1979, Release of platelet fibronectin (cold-insoluble globulin) from alpha granules induced by thrombin or collagen; lack of requirement of plasma fibronectin in ADP-induced platelet aggregation, *Blood* **54:**8–12.

9

The Platelet–Fibrinogen Interaction

Joel S. Bennett

1. INTRODUCTION

Following the observation that adding ADP to platelet-rich plasma resulted in platelet aggregation (Gaarder *et al.*, 1961), it became apparent that factors in plasma were required for this process to proceed (Born and Cross, 1964). Experiments with washed platelets indicated that, at a minimum, calcium and fibrinogen were necessary (Cross, 1964; McLean *et al.*, 1964). Similarly, fibrinogen was needed for platelets to adhere to artificial surfaces such as glass (Zucker and Vroman, 1969). The relevance of these *in vitro* observations for physiologic platelet function became apparent from the study of individuals with congenital afibrinogenemia. Afibrinogenemic individuals have barely detectable concentrations of plasma fibrinogen and often have a prolonged bleeding time, indicating a concomitant defect in platelet function (Weiss and Rogers, 1971). The platelet defect is secondary to the deficiency of fibrinogen because the abnormal platelet function can be completely corrected by transfusions of fibrinogen. Further observations that platelets from individuals with Glanzmann's thrombasthenia do not aggregate (Caen, 1972), adhere to glass (Zucker and Vroman, 1969), or adsorb fibrinogen suggest that fibrinogen and the platelet surface undergo a specific interaction (Bang *et al.*, 1972). Until recently, an explanation for the fibrinogen requirement for platelet aggregation was not available. However, the demonstration that platelet stimulation exposes a limited number of fibrinogen receptors on the platelet surface provided the key to understanding the role of fibrinogen in platelet function.

Joel S. Bennett • Hematology-Oncology Section, Department of Medicine, University of Pennsylvania School of Medicine, Philadelphia, Pennsylvania 19104.

2. THE FIBRINOGEN MOLECULE

2.1. Plasma Fibrinogen

The dramatic conversion of soluble fibrinogen into an insoluble clot has provided the impetus for the intensive study of fibrinogen that has nearly solved its structure and explained its function. Plasma fibrinogen is a complex, dimeric glycoprotein with a molecular weight of 340,000 (Doolittle, 1973). It is composed of three pairs of homologous polypeptide chains designated Aα, Bβ, and γ held together by a series of disulfide bridges. The amino acid sequence of each polypeptide chain and the arrangement of the disulfide bridges have been determined (Doolittle, 1981). Carbohydrate is present on the β and γ chains, but its contribution to fibrinogen function is unclear. For example, removal of the sialic acid from fibrinogen does not adversely affect its ability to support platelet aggregation (Coller, 1979; Harfenist *et al.*, 1980b). The conversion of soluble fibrinogen to an insoluble fibrin clot is catalyzed by the enzyme thrombin. Thrombin cleaves the negatively charged A and B peptides from the amino terminal portions of the α and β chains, respectively. The resulting change in charge distribution over the molecule, now called fibrin monomer, and the exposure of two sets of polymerization sites permits the polymerization of fibrin into fibers and networks (Olexa and Budzynski, 1980).

Plasma fibrinogen is synthesized in the liver. The α, β, and γ chains appear to be coded by separate genes because it has been possible to isolate separate mRNA for each chain from rat liver (Crabtree and Kant, 1981). However, translation of the three genes appears to be coordinated since the hepatic levels of each mRNA species increase concomitantly after defibrination in the rat (Crabtree and Kant, 1982). Moreover, the α and γ genes in the rat (Kant and Crabtree, 1983) and the α and β genes in the human (Fowlkes *et al.*, 1983) appear to be physically linked, suggesting that the genetic material for plasma fibrinogen may exist as a large transcriptional unit.

Fibrin clots *in vivo* are degraded by the serine protease plasmin. Plasmin can also cleave fibrinogen (Marder *et al.*, 1982). Because the cleavage of fibrinogen by plasmin proceeds sequentially, the cleavage pattern has been used to investigate fibrinogen structure (Marder *et al.*, 1969; Marder and Budzynski, 1975; Takagi and Doolittle, 1975). The initial cleavages involve the removal of two thirds of the carboxy terminal end of the Aα chains and the first 42 amino acids of the amino terminal end of the Bβ chains, converting fibrinogen to the symmetrical, 250,000 molecular weight fragment X. The next series of cleavages split the X fragment asymmetrically into a 155,000 molecular weight fragment Y and a 100,000 mol. wt. fragment D. Finally, the fragment Y is cleaved into a second fragment D and a 45,000 mol. wt. fragment E. The model of fibrinogen (Figure 1) that best explains this pattern of plasmin degradation is that of a symmetrical molecule with a single central E domain containing the E fragment and two laterally disposed D domains containing the D fragments.

Other techniques have also been used to determine the three-dimensional structure of fibrinogen. Although it has not been possible to crystallize the intact fibrinogen molecule for x-ray diffraction analysis (Weisel *et al.*, 1981), consensus regarding the three-dimensional structure of fibrinogen has resulted from the synthesis of electron microscopic (Fowler and Erickson, 1979; Fowler *et al.*, 1980; Price *et al.*, 1981),

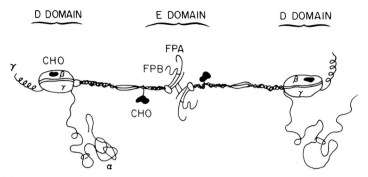

Figure 1. Schematic representation of the fibrinogen molecule: α, alpha chain; β, beta chain; γ, gamma chain; FPA, fibrinopeptide A; FPB, fibrinopeptide B; CHO, carbohydrate. Adapted from Marder *et al.* (1982); Hoeprich and Doolittle (1983).

hydrodynamic (Shulman, 1953), enzymatic fragmentation (Marder *et al.*, 1969; Takagi and Doolittle, 1975), primary sequence (Doolittle, 1981), and immunologic (Price *et al.*, 1981) data. The picture that emerges is an elongated molecule, 450 nm in length and 90 nm in diameter, formed by three nodules in a line connected by coiled strands (Doolittle, 1981). The strands have an α helical structure and contain 111–112 amino acids from each of the polypeptide chains. The central nodule, approximately 50 nm in diameter, contains the amino terminal segments of the six polypeptide chains and corresponds roughly to the fibrinogen fragment E (Takagi and Doolittle, 1975; Fowler and Erickson, 1979; Fowler *et al.*, 1980). The distal nodules, approximately 60 nm in diameter, contain the carboxy terminal two thirds of the β and γ chains (Takagi and Doolittle, 1975; Fowler and Erickson, 1979) and correspond roughly to the fibrinogen fragment D (Fowler *et al.*, 1980; Price *et al.*, 1981). The carboxy terminal two thirds of the α chain protrudes from the distal nodule and is cleaved by plasmin early in the conversion of fibrinogen to fragment X (Marder *et al.*, 1969)

2.2. Platelet Fibrinogen

Platelets contain fibrinogen in their α-storage granules (Holmsen and Weiss, 1979). This fibrinogen amounts to 120–150 μg/10^9 platelets and constitutes approximately 10% of the total platelet protein (Weiss and Kochwa, 1968). The origin of platelet fibrinogen and its identity with plasma fibrinogen remains an area of controversy. Castaldi and Caen (1965) found that radiolabeled plasma fibrinogen did not exchange with platelet fibrinogen, suggesting that the megakaryocyte is the site of platelet fibrinogen synthesis. Although this appears to be a reasonable supposition, because megakaryocytes synthesize another α-granule protein, von Willebrand factor (VWF) (Nachman *et al.*, 1977), there is no direct evidence to support this conclusion. In fact, studies of protein synthesis by guinea pig megakaryocytes using radioactive amino acids failed to demonstrate incorporation of label into platelet fibrinogen (Nachman *et al.*, 1978). Thus, a definitive answer to the source of α-granule fibrinogen may rest on the detection of mRNA directing fibrinogen synthesis in the megakaryocyte cytoplasm.

Another source of confusion regarding the origin of platelet fibrinogen is the question of the structural identity of platelet and plasma fibrinogen. Analyses of these fibrinogens have shown differences in their clottability (Nachman *et al.*, 1977), their intrinsic viscosity (Ganguly, 1972), their immunologic reactivity (Plow and Edgington, 1975), their subunit structure (Ganguly, 1972), and their susceptibility to cleavage by plasmin (James *et al.*, 1977). These differences, however, have not been found by all investigators. For example, Doolittle *et al.* (1974) found the two fibrinogens to be identical and suggested that they were products of the same gene. Furthermore, they suggested that the reported differences between plasma and platelet fibrinogen could be attributed to proteolysis during the isolation of the platelet species. Recent studies by Mosesson *et al.* (1984) demonstrated a difference between plasma and platelet fibrinogen chains. Whereas plasma fibrinogen is composed of 93% γ_A and 7% γ^1 chains, all platelet fibrinogen molecules contain only γ_A chains.

Fibrinogen has also been detected on the membrane of unstimulated platelets. Tangen *et al.* (1971), using gel filtration to separate platelets from plasma proteins, demonstrated with a fluorescent antifibrinogen antibody that fibrinogen remained associated with the platelet surface. Tollefsen and Majerus (1975), examining washed human platelets with antifibrinogen Fab fragments, also detected up to 50,000 molecules of fibrinogen per platelet. However, Nachman and Marcus (1968) found only a weak reaction between an agglutinating antifibrinogen antibody and platelets in plasma and no reaction between the antibody and washed platelets. Keenan (1972), using a tanned red cell hemagglutination inhibition immunoassay, could detect little or no fibrinogen associated with platelet membranes sedimented from platelet lysates. Finally, Zucker (1977), using an immunofluorescent technique and gel-filtered platelets, could not demonstrate fibrinogen on the platelet surface unless fibrinogen was added back to the platelet suspensions. In any case, if fibrinogen is normally adsorbed to the platelet surface, it is not available to support platelet aggregation, since fibrinogen must be added back to platelet suspensions prepared by gel filtration before ADP-induced aggregation will occur (Tangen *et al.*, 1971).

3. CHARACTERIZATION OF THE PLATELET FIBRINOGEN RECEPTOR

Unactivated platelets circulate in a milieu containing 2–4 mg/ml of fibrinogen. However, the mere presence of fibrinogen is unlikely to explain the fibrinogen requirement for platelet aggregation. Experiments by Mustard and co-workers (1978) demonstrated that radiolabeled fibrinogen associated with aggregating platelets stimulated by ADP and that fibrinogen dissociated from these platelets when they deaggregated. Therefore, they suggested that ADP induced changes in the platelet surface membrane that resulted in the transient association of fibrinogen with the stimulated platelets. The association of fibrinogen with stimulated platelets has now been intensively studied. Although many details of these studies remain controversial, there is general agreement that the platelet surface membrane contains a limited number of latent fibrinogen receptors that are exposed by platelet stimulation. Moreover, fibrinogen binding to these exposed receptors appears to explain the fibrinogen requirement for platelet aggregation.

3.1. Fibrinogen Binding to ADP-Stimulated Platelets

Of the various platelet agonists, the interaction of ADP-stimulated platelets with fibrinogen has been studied in greatest detail. The methodology for these studies has been adapted from that developed to study the polypeptide hormone receptors (Catt and Dufay, 1977). Purified radiolabeled fibrinogen is incubated with suspensions of washed or gel-filtered platelets. The platelets are then stimulated with ADP, and after a suitable period of time, the fibrinogen specifically associated with the platelets is quantitated. The presence of fibrinogen receptors is implied from the analysis of these binding data.

3.1.1. The Kinetics of Fibrinogen Binding to ADP-Stimulated Platelets

Radioiodinated fibrinogen rapidly associates with ADP-stimulated platelets (Bennett and Vilaire, 1979) (Figure 2). No fibrinogen binding occurs in the absence of platelet stimulation. At a fibrinogen concentration of 200 μg/ml and at 37 °C, fibrinogen binding to gel-filtered platelets stimulated with 10 μM ADP achieves a steady state within 60–90 sec. The measured association rate constant for this reaction (k_1) is equal to 8.49×10^6 M^{-1} min^{-1} (Table 1).

Although many investigators have observed a rate of fibrinogen binding similar to that depicted in Figure 1 (Hawiger *et al.*, 1980: Peerschke *et al.*, 1980b; Harfenist *et al.*, 1980a; Kornecki *et al.*, 1981; McEver *et al.*, 1983), Marguerie and colleagues (1979, 1980, 1981) have consistently measured a rate nearly tenfold slower. An explanation for the discrepancy in association rates is not readily apparent. However, the rapid rate of fibrinogen binding observed by most investigators is quite consistent with the rate of platelet aggregation observed in the aggregometer (Shattil and Bennett, 1981).

The initial platelet–fibrinogen interaction is reversible. Thus, platelet-bound

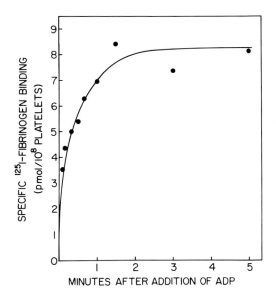

Figure 2. Time-course of [^{125}I]fibrinogen binding to ADP-stimulated platelets. Suspensions of 5×10^7 gel-filtered platelets in Tyrode's buffer, pH 7.4, were incubated at 37 °C with [^{125}I]fibrinogen (180 μg/ml) and 0.5 mM $CaCl_2$. At zero time, 10 μM ADP was added, and the incubations were continued without stirring for various time intervals. Incubations were terminated by centrifugation of the platelets through silicone oil. Nonspecific fibrinogen binding was measured by performing the binding assays in the presence of 4 mg/ml unlabeled fibrinogen and was subtracted from the total [^{125}I]fibrinogen associated with the centrifuges platelets. From Bennett and Vilaire (1979) with permission.

Table 1. Binding Parameters for the Specific Interaction
of [^{125}I]Fibrinogen with ADP-Stimulated Platelets[a]

Number of fibrinogen binding sites per platelet	44,000 ± 6,000 (S.E.)
K_d, equilibrium	8.1×10^{-8} M
K_d, kinetic	12.7×10^{-8} M
k_2	1.08 min^{-1}
k_1	8.49×10^6 M^{-1} min^{-1}

[a]Bennett and Vilaire (1979).

fibrinogen can completely dissociate from the platelet surface, at least during the first ten min following platelet stimulation (Bennett and Vilaire, 1979) (Figure 3). The rate constant for this dissociation (k_2) has been determined by measuring the rate at which excess unlabeled fibrinogen can displace labeled fibrinogen from the platelet surface and is equal to 1.08 min^{-1}.

Studies of the reversibility of fibrinogen binding at time points later than 10 min show that added fibrinogen progressively loses its ability to displace bound fibrinogen with time. Marguerie *et al.* (1980) found that after a 30 min incubation, only 10% of platelet-bound fibrinogen could be displaced. Similar results were reported by DiMinno *et al.* (1983). The mechanism for the irreversibility of fibrinogen binding at later time points remains to be explained. It apparently does not involve the formation of covalent bonds between fibrinogen and its receptor because sodium dodecyl sulfate–polyacrylamide gel electrophoresis (SDS–PAGE) reveals no change in the molecular weight of the platelet-bound fibrinogen (Marguerie *et al.*, 1979). The possibility that

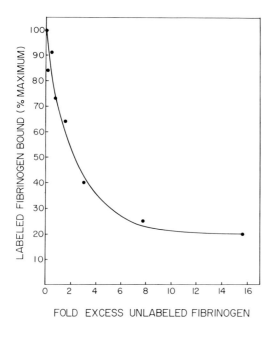

Figure 3. Displacement of platelet-bound [^{125}I]fibrinogen by unlabeled fibrinogen. Suspensions of 5×10^7 gel-filtered platelets were incubated with [^{125}I]fibrinogen (200 μg/ml), 0.5 mM CaCl$_2$, and 10 μM ADP for 3 min at 37 °C without stirring. Unlabeled fibrinogen in increasing concentrations was then added and the incubations were continued for 5 additional min. The incubations were terminated as described in Figure 2. In this experiment, nonspecific binding was determined by performing the entire binding assay in the presence of 4 mg/ml unlabeled fibrinogen and amounted to 20% of the maximum fibrinogen bound. From Bennett and Vilaire (1979) with permission.

platelet-bound fibrinogen is internalized with time has not been tested, but when experiments were performed at 10, 22, or 37 °C, no difference in the fraction of irreversibly bound fibrinogen could be detected (Marguerie and Plow, 1981). These data suggest that the affinity of the exposed fibrinogen receptors increases with time. However, since the formation of a hemostatic platelet plug occurs within minutes, the physiologic significance of this observation remains to be determined.

3.1.2. Examination of Fibrinogen Binding by Equilibrium Binding Methods

Because the initial binding of fibrinogen to ADP-stimulated platelets is reversible, it can be further characterized using equilibrium binding methods. Thus, when fibrinogen binding is measured at equilibrium as a function of fibrinogen concentration, the fibrinogen receptors are saturated at a fibrinogen concentration of approximately 100 μg/ml (Bennett and Vilaire, 1979) (Figure 4). Scatchard analysis of the binding data yields a straight line, consistent with the presence of a single class of fibrinogen receptors. Studies using platelets from multiple donors indicate that there are approximately 44,000 potential fibrinogen receptors per platelet with a dissociation constant (K_d) of 81 nM or 27 μg/ml fibrinogen (Table 1).

The validity of the dissociation constant derived from these equilibrium binding experiments can be tested by comparing it with the same constant calculated from the association and dissociation rate constants ($K_d = k_2/k_1$). The kinetic value is 127 nM or 43 μg/ml fibrinogen, in excellent agreement with the equilibrium K_d (Bennett and Vilaire, 1979) (Table 1).

Although the dissociation constants calculated from the kinetic and equilibrium binding studies are internally consistent, the relevance of these experiments to physiologic platelet function requires independent confirmation. As mentioned, the rate of fibrinogen binding is compatible with the rate of platelet aggregation *in vitro*. It is also sufficiently rapid to account for the rate of platelet plug formation *in vivo* as measured by the bleeding time (Bowie and Owen, 1974). Moreover, the fibrinogen concentration

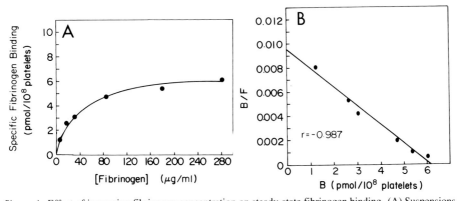

Figure 4. Effect of increasing fibrinogen concentration on steady-state fibrinogen binding. (A) Suspensions of 5 × 10⁷ gel-filtered platelets were incubated with increasing concentrations of [¹²⁵I]fibrinogen, 0.5 mM CaCl₂, and 10 μM ADP for 3 min at 37 °C. Specific fibrinogen binding were determined as described in Figure 2. (B) Scatchard analysis of the data from (A). From Bennett and Vilaire (1979) with permission.

needed for maximal platelet aggregation in afibrinogenemic plasma (Weiss and Rogers, 1971) or in washed platelet suspensions (Niewiarowski *et al.*, 1977) is 100–200 μg/ml, the same fibrinogen concentration required for maximal fibrinogen binding. Thus, the biochemical parameters for fibrinogen binding are quite consistent with the known fibrinogen requirements for platelet function.

Although these data indicate that the surface of ADP-stimulated platelets contains physiologically relevant fibrinogen receptors, the number of classes of receptors remains controversial. The equilibrium binding data in most reports (Bennett and Vilaire, 1979; Marguerie *et al.*, 1979; Hawiger *et al.*, 1980; Harfenist *et al.*, 1980a; McEver *et al.*, 1983; Pasqua and Pizzo, 1983) are consistent with a single class of fibrinogen receptors. However, equilibrium binding studies by Peerschke and coworkers (1980b) resulted in curvilinear Scatchard plots. They assumed these plots resulted from two classes of fibrinogen receptors—3500 sites per platelet with a dissociation constant of 130 nm and 9000 sites per platelet with a dissociation constant of 450 nm. To differentiate between two classes of noninteracting sites or a single class of sites with negative cooperativity, Peerschke (1982b) measured the rate of fibrinogen dissociation from platelets as a function of receptor occupancy. She reported that the apparent dissociation rate constant for bound fibrinogen increased as the number of occupied fibrinogen receptors increased, consistent with the presence of negative cooperativity. Similar Scatchard analyses were reported by Niewiarowski *et al.* (1981b) and Kornecki *et al.* (1981), who studies fibrinogen binding to chymotrypsin-treated and ADP-stimulated platelets and by DiMinno and colleagues (1983). In general, the interpretation of curvilinear Scatchard plots can be difficult because they can be artificially produced by an underestimation of nonspecific binding at low ligand concentrations or by the unrecognized presence of ligand with a lower binding affinity (Boeynaems and Dumont, 1977; Molinoff *et al.*, 1981). The interpretation of these curvilinear plots must also take into account the compelling evidence to be discussed that a single platelet surface structure is the fibrinogen receptor.

3.1.3. Effect of ADP Concentration on Fibrinogen Binding

The fibrinogen concentration in plasma is 20- to 30-fold greater than the minimal fibrinogen concentration required to saturate the platelet fibrinogen receptors (Bennett

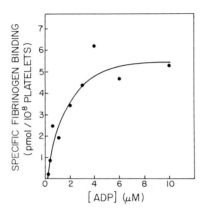

Figure 5. Effect of ADP concentration on the extent of [125I]fibrinogen binding. Suspensions of 5 × 10^7 gel-filtered platelets were incubated with [125I]fibrinogen (200 μg/ml), 0.5 mM CaCl_2, and increasing concentrations of ADP for 3 min at 37 °C. Specific fibrinogen binding was then determined. From Bennett and Vilaire (1979) with permission.

and Vilaire, 1979). Thus, factors other than the fibrinogen concentration, such as the strength of the platelet stimulus, must dictate the extent of fibrinogen binding. This is clearly the case when fibrinogen binding is measured as a function of the ADP concentration. Fibrinogen binding increases in a linear fashion as the ADP concentration is increased from 0.1 μM and reaches a maximum at ADP concentrations > 2 μM (Bennett and Vilaire, 1979) (Figure 5). These experiments also demonstrate that fibrinogen receptor exposure occurs within the concentration range that ADP acts as a platelet agonist (Weiss, 1975). Furthermore, since the extent of ADP-stimulated platelet aggregation increases within this same concentration range, these experiments indicate a general correspondence between the extent of fibrinogen binding and the extent of platelet aggregation.

3.2. Fibrinogen Binding Stimulated by Agonists other than ADP

3.2.1. Epinephrine

Epinephrine stimulates platelets by binding to 200–300 α_2-adrenergic receptors on the platelet surface (Hoffman et al., 1980). However, exogenous fibrinogen must be present before this stimulus can lead to platelet aggregation. To determine if platelet stimulation by epinephrine exposes fibrinogen receptors, equilibrium binding studies have been performed using this agonist. They demonstrate that epinephrine can expose approximately 48,000 fibrinogen receptors per platelet with a dissociation constant of 178 nm (Bennett and Vilaire, 1979). These values are similar to those determined with ADP. However, it is likely that additional factors are involved in the platelet response to epinephrine. The number of receptors exposed by epinephrine is decreased by 50% when platelets are pretreated with aspirin or indomethacin (Bennett et al., 1981). This suggests that prostaglandin or thromboxane synthesis is required for a portion of the response to this agonist. Moreover, the ability of ADP scavengers, such as apyrase or creatine phosphate/creatine phosphokinase, to partially (Peerschke, 1982a) or completely (Plow and Marguerie, 1980b) inhibit the fibrinogen binding stimulated by epinephrine suggests that platelet-derived ADP may also be involved. On the other hand, epinephrine clearly initiates fibrinogen receptor exposure because the entire response to this agonist can be prevented by the α-adrenergic antagonist phentolamine (Peerschke, 1982a).

3.2.2. Thrombin

In contrast to ADP and epinephrine, thrombin can stimulate platelet aggregation in the absence of exogenous fibrinogen. Nonetheless, there is considerable indirect evidence that fibrinogen is required for thrombin-stimulated aggregation. Tollefsen and Majerus (1975) were able to inhibit the thrombin-stimulated aggregation of washed platelets with anti-fibrinogen antibody Fab fragments. Similar data were reported by Miller et al. (1975), who used plasmin to degrade secreted platelet fibrinogen. In addition, several well-characterized antiplatelet monoclonal antibodies that inhibit fibrinogen binding to ADP-stimulated platelets also inhibit thrombin-stimulated platelet aggregation (McEver et al., 1983; Bennett et al., 1983; Coller et al., 1983). Thus, it is likely that fibrinogen binding to exposed receptors is required for thrombin-stimulated platelet aggregation and that the fibrinogen source can be the platelet α-granules.

The measurement of fibrinogen binding to platelets stimulated by thrombin is complicated by the formation of fibrin clots. To avoid this problem, the addition of thrombin to platelet suspensions has been quickly followed by the addition of hirudin to quench thrombin's enzymatic activity (Tam *et al.,* 1979; Holmsen *et al.,* 1981). Under these conditions, thrombin stimulates the exposure of fibrinogen receptors similar in number and affinity to those exposed by ADP (Hawiger *et al.,* 1980; Plow and Marguerie, 1980a). The reported ability of ADP scavengers to inhibit the fibrinogen binding stimulated by thrombin again suggests that platelet-derived ADP may be involved in this response (Plow and Marguerie, 1980a).

3.2.3. Prostaglandin Endoperoxides

The prostaglandin endoperoxide PGH_2 (Bennett *et al.,* 1981), two stable PGH_2 analogues (9, 11-aso PGH_2 and U46619) (Morinelli *et al.,* 1983), and arachidonic acid (DiMinno *et al.,* 1983) will stimulate fibrinogen binding by washed or gel-filtered platelets. The maximal number of fibrinogen receptors exposed by these agonists is identical to the number exposed by ADP (Bennett and Vilaire, 1979). However, the agent ultimately responsible for this response remains uncertain. Because prostaglandin endoperoxides and thromboxanes are potent stimuli for the platelet release reaction (Charo *et al.,* 1977), secreted ADP may be the responsible agent. Unfortunately, experiments to resolve this issued have yielded conflicting results. On the one hand, Bennett *et al.* (1981) reported that 20 μM ATP, a competitive antagonist of ADP, had little effect on the fibrinogen binding stimulated by 2–4 μM PGH_2, but inhibited the response to 1–2 μM ADP by 90%. On the other hand, Morinelli *et al.* (1983) reported that ADP scavengers or 1 mM ATP markedly inhibited fibrinogen binding stimulated by both ADP and the prostaglandin endoperoxide analogues. Thus, the question as to whether prostaglandins (and/or thromboxanes) can independently expose fibrinogen receptors cannot be answered with certainty at present.

3.2.4. Other Agonists

A limited number of studies have been performed with other platelet agonists. Collagen (DiMinno *et al.,* 1983) and cold (0–2 °C) (Peerschke and Zucker, 1981) appear to be capable of stimulating fibrinogen binding. Furthermore, the calcium ionophore A23187 is also capable of initiating this platelet response (J. S. Bennett and G. Vilaire, unpublished results).

3.3. Divalent Cation Requirements for Fibrinogen Binding

Like platelet aggregation, fibrinogen binding requires the presence of divalent cations and will not occur in the presence of EDTA (Bennett and Vilaire, 1979; Marguerie *et al.,* 1980). Fibrinogen binding is maximal at a Ca^{2+} concentration of 0.5 mM or at a Mg^{2+} concentration of 2.5 mM and decreases above and below these values (Bennett and Vilaire, 1979) (Figure 6). However, the effect of Ca^{2+} and Mg^{2+} is not additive, since fibrinogen binding in the presence of both cations was found to be

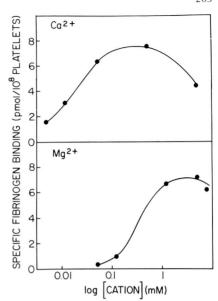

Figure 6. Effect of Ca^{2+} and Mg^{2+} concentration on the extent of [^{125}I]fibrinogen binding. Suspensions of 5 × 10^7 gel-filtered platelets were incubated with [^{125}I]fibrinogen (200 μg/ml), 10 μM ADP, and increasing concentrations of either $CaCl_2$ or $MgCl_2$ for 3 min at 37 °C. Specific fibrinogen binding was then determined. From Bennett and Vilaire (1979) with permission.

intermediate between that observed for either alone (Marguerie *et al.*, 1980). This suggests that these cations may bind to a limited number of specific sites. Manganese, the only other divalent cation studied, supports fibrinogen binding poorly (Marguerie *et al.*, 1980).

The role of divalent cations in fibrinogen binding is unknown. Whether the divalent cation requirement for fibrinogen binding is fulfilled by cation binding to the platelet, to fibrinogen, or to both, is not yet clear. The platelet surface avidly binds divalent cations (Brass and Shattil, 1982). Therefore, it is of considerable interest that thrombasthenic platelets, which fail to bind fibrinogen, also have a deficit in surface-bound calcium (Peerschke *et al.*, 1980a; Brass and Shattil, 1983). Calcium also binds to fibrinogen at three high-affinity sites and several low-affinity sites (Marguerie *et al.*, 1977). The dissociation constant for the higher affinity sites is in the micromolar range, well below the Ca^{2+} concentrations required for fibrinogen binding, and these sites only bind Ca^{2+}. On the other hand, the lower affinity sites have dissociation constants in the millimolar range and will bind either Ca^{2+} or Mg^{2+}. These sites could be involved in the interaction of fibrinogen with the fibrinogen receptor. However, whether divalent cations directly link fibrinogen to the fibrinogen receptor through metal-ion bridges or whether they are required to maintain the proper conformation of fibrinogen and/or the fibrinogen receptor remains to be determined.

3.4. Control Mechanisms for Fibrinogen Receptor Exposure

The fact that platelets must be stimulated before fibrinogen binding will occur implies that the exposure of fibrinogen receptors is an active metabolic process. As such, it should be subject to those influences that control other aspects of platelet function. Inhibition of platelet metabolism with agents such as antimycin A combined with 2-deoxyglucose (Mustard *et al.*, 1978; Peerschke and Zucker, 1981) or sodium azide (Coller, 1980) inhibits the ability of stimulated platelets to bind fibrinogen. Moreover, inhibiting platelet responsiveness by increasing the platelet concentration of cyclic AMP with agents such as PGE_1 (Coller, 1980; Peerschke and Zucker, 1981), PGI_2 (Coller, 1980), forskolin (Graber and Hawiger, 1982), or 1-methyl-3-isobutyl-xanthine (Graber and Hawiger, 1982) also inhibits fibrinogen receptor exposure. The nature of the intracellular processes subject to this inhibition, processes that presumably couple agonist stimulation to fibrinogen receptor exposure, is not yet known.

3.5. Sites on the Fibrinogen Molecule Interacting with the Fibrinogen Receptor

The unique ability of fibrinogen to support platelet aggregation has generated considerable interest in the features of the fibrinogen molecule responsible for this function. Experiments by Niewiarowski *et al.* (1977), performed before the description of the fibrinogen receptor, compared the ability of intact fibrinogen and the plasmin-generated cleavage products of fibrinogen to support platelet aggregation. They reported that cleavage of fibrinogen into the asymmetric fragment Y abolished the ability of fibrinogen to support platelet aggregation and suggested that the portions of the Aα-chain and of the Bβ-chain removed by plasmin were important for this function. The discovery that fibrinogen binds to specific platelet receptors stimulated further examination of this question. Hawiger and co-workers (1982) demonstrated that polymers of isolated fibrinogen γ-chains, and to a lesser extent polymers of α-chains, but not polymers of β-chains, could support platelet aggregation. They also demonstrated that anti-γ-chain antibody Fab fragments inhibited the binding of intact fibrinogen to ADP-stimulated platelets. Thus, sites on the γ-chain appear to interact with the fibrinogen receptor. To localize the sites on the γ-chain, these investigators isolated a 27-residue fragment from the carboxy-terminal segment of the γ-chain that inhibited both platelet aggregation and fibrinogen binding (Kloczewiak *et al.*, 1982). The region was further restricted to the carboxy-terminal 15 amino acids, a finding confirmed by showing that the plasmin-generated fibrinogen fragment D_1 containing these residues inhibited fibrinogen binding, whereas the fragment D_3, which lacks these residues, could not (Kloczewiak *et al.*, 1983). Marguerie and colleagues (1982) have reported similar results. They found that the larger D-fragments inhibited both fibrinogen binding and platelet aggregation, whereas the smaller D-fragments and fragment E were without effect. Recent experiments by Plow *et al.* (1984) have demonstrated inhibition of fibrinogen binding to ADP and thrombin-stimulated platelets by the synthetic peptides corresponding to the carboxy terminal 10, 11, or 12 amino acids of the γ-chain, whereas a peptide equal to the carboxy terminal 9 amino acids of the γ-chain was without effect. They have also reported that the tetrapeptide glycl-L-prolyl-L-arginyl-L-proline is inhibitory (Plow and Marguerie, 1982). This tetrapeptide is an analogue of the amino-terminal sequence of the α-chain following removal of the A peptide. As such, it interacts with a polymerization site in the D-domain of intact fibrinogen (Laudano and Doolittle, 1978, 1980). The interpretation of these latter experiments is ambiguous. The same result could have been produced by tetrapeptide binding to the fibrinogen receptor or to the D-domain of fibrinogen. There are similar difficulties in the interpretation of studies by Tomikawa *et al.* (1980), who reported that regions of the E-domain inhibited platelet aggregation. Furthermore, Pasqua and Pizzo (1983) have reported that although fibrinogen degradation fragments inhibit the binding of intact fibrinogen to platelets, the fragments themselves cannot bind to the platelet. They also reported that these fragments bind to intact fibrinogen, suggesting that the fragments inhibit fibrinogen binding to platelets by a fragment–fibrinogen interaction rather than a fragment–receptor interaction.

At present, the weight of available evidence suggests that the carboxy-terminal portions of the γ-chain of fibrinogen, and to a lesser extent, the carboxy-terminal regions of the α-chain, mediate the binding of fibrinogen to its platelet receptor. Other factors, such as molecular size and dimeric symmetry, then become important in the ability of the bound fibrinogen to support platelet aggregation.

3.6. Interaction of Platelets with Fibrin

Because fibrinogen binding to its platelet receptor appears to be mediated by sites in the D-domain, fibrin monomer should also bind to the fibrinogen receptor. This has not been specifically studied. However, the interaction of unactivated platelets with "polymerizing fibrin" has been investigated (Niewiarowski *et al.*, 1972; Brown *et al.*, 1977). In these experiments, washed platelets were incubated with fibrin monomer in solution and the number of platelets associated with the developing fibrin strands was quantitated. Although platelets did associate with the fibrin strands, there were a variety of differences between this interaction and the binding of intact fibrinogen to the fibrinogen receptor. In contrast to the latter reaction, the interaction of platelets with polymerizing fibrin occurred in the absence of platelet stimulation, occurred despite prior exposure of platelets to PGE_1, occurred after platelets were fixed with formalin, and occurred after platelets were depleted of metabolic ATP with deoxyglucose and antimycin A. Moreover, thrombasthenic platelets interacted with polymerizing fibrin just as well as normal platelets, although these platelets cannot bind fibrinogen (Niewiarowski *et al.*, 1981a). Thus, the physiologic significance of the interaction of platelets with polymerizing fibrin is unclear. Because thrombasthenic platelets are unable to aggregate yet interact with fibrin, this process is unlikely to be involved in platelet aggregation. Moreover, because thrombasthenic platelets also fail to support clot retraction, the interaction of platelets with polymerizing fibrin cannot account for this aspect of platelet function.

4. IDENTIFICATION OF THE PLATELET FIBRINOGEN RECEPTOR

The description of the platelet fibrinogen receptor has kindled intense interest in its identity. Although several different approaches have been taken to answer this question, each has reached a similar conclusion. Taken together, the data strongly suggest that the calcium-dependent complex formed by the platelet membrane glycoproteins IIb and IIIa contains the fibrinogen receptor.

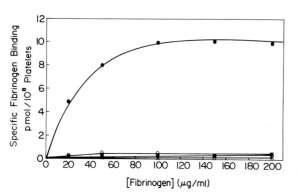

Figure 7. Comparison of ADP-stimulated fibrinogen binding to normal and thrombasthenic platelets. Gel-filtered platelets from one normal subject and from three patients with Glanzmann's thrombasthenia were incubated with increasing concentrations of [^{125}I]-fibrinogen, 0.5 mM CaCl$_2$ and 10 μM ADP for 3 min at 37 °C. Specific fibrinogen binding was determined as described in Figure 2. ●, Normal subject; ○, patient 1; ×, patient 2; ■, patient 3. From Bennett and Vilaire (1979) with permission.

4.1. Studies of Fibrinogen Binding to Thrombasthenic Platelets

Glanzmann's thrombasthenia is a hereditary bleeding disorder that is due to the inability of stimulated platelets to aggregate and to support clot retraction (Caen, 1972). Not only are thrombasthenic platelets functionally defective, but their surface membrane is abnormal, containing reduced concentrations of the glycoproteins IIb and IIIa (see Chapter 16). As seen in Figure 7, ADP-stimulated thrombasthenic platelets bind little if any fibrinogen (Bennett and Vilaire, 1979). Thus, the inability of thrombasthenic platelets to aggregate may be due to their inability to bind fibrinogen. Moreover, these data suggest that the glycoprotein IIb-IIIa complex may contain the fibrinogen receptor.

The inability of intact thrombasthenic platelets to bind fibrinogen has been confirmed (Mustard *et al.,* 1979; Coller, 1980; Peerschke *et al.,* 1980b). However, two groups have reported slightly different results. Lee and co-workers (1981) measured fibrinogen binding to platelets from four individuals with type I thrombasthenia and two individuals with type II thrombasthenia. The type I thrombasthenic platelets contained virtually no GP IIb and GP IIIa, whereas the type II platelets contained 13 to 15% of the normal quantities of each. As expected, neither the type I nor the type II platelets aggregated in response to ADP and the type I platelets also failed to bind fibrinogen. However, fibrinogen binding to the type II platelets was only slightly less than the lowest value of the normal range. The failure of the type II platelets to aggregate, in spite of their ability to bind fibrinogen, requires explanation. It does suggest that factors in addition to fibrinogen binding may be involved in platelet aggregation. Kornecki and co-workers (1981) reported that although fibrinogen did not bind to ADP-stimulated thrombasthenic platelets, a small amount of fibrinogen did bind when these platelets were treated with chymotrypsin. However, the platelets they studied were similar to type II platelets because they contained approximately 6% of the normal quantity of GP IIIa and retained a 66,000 mol. wt. fragment of GP IIIa following chymotrypsin treatment (Kornecki *et al.,* 1983).

4.2. Photoaffinity Labeling

Fibrinogen binding to its receptor is reversible. Consequently, isolation of the receptor with fibrinogen as a single complex may not be possible. One approach to

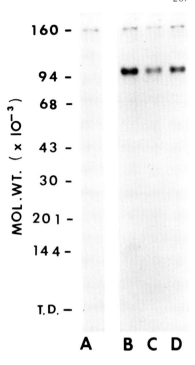

Figure 8. Interaction of photoactivated MABI–fibrinogen with ADP-stimulated platelets. Nonradiolabeled MABI-fibrinogen was prepared and added in the dark to suspensions of nonradiolabeled gel-filtered human platelets. The platelets were then stimulated with ADP, irradiated with ultraviolet light, sedimented through silicone oil, and dissolved in SDS. Fifty micrograms of dissolved platelet protein were electrophoresed on 0.1% SDS, 7.5% polyacrylamide slab gels. Covalent complexes composed of MABI–fibrinogen and cross-linked platelet proteins will not enter this gel. Following electrophoresis, the separated platelet proteins were transferred to strips of nitrocellulose paper. The strips were then incubated with either anti-Pl^{A1} antiserum or with nonimmune human serum, followed by an incubation with ^{125}I-labeled protein A. Autoradiography was used to visualize the sites of [^{125}I]protein A deposition. The nitrocellulose strip in lane A was incubated with nonimmune serum. Lanes B, C, and D were incubated with anti-Pl^{A1} antiserum; lane B, control platelets; lane C, platelets incubated with 200 μg/ml of MABI–fibrinogen; lane D, platelets incubated with 200 μg/ml of MABI–fibrinogen plus 2 mg/ml of native fibrinogen. The interaction of fibrinogen with GP IIIa is seen as a decreased intensity of the anti-Pl^{A1} reaction in lane C.

overcoming this difficulty is to chemically cross-link fibrinogen to the receptor (Peters and Richards, 1977). To this end, Bennett *et al.* (1982) modified fibrinogen with the photoreactive, heterobifunctional cross-linking reagent, methyl-r-azidobenzoimidate (MABI). In the dark, the imidoester of MABI reacts spontaneously with free amino groups in fibrinogen leaving the photoreactive azido group available to form crosslinks with nearby structures. When MABI–fibrinogen was bound to the surface of ADP-stimulated platelets in the dark, and the platelets were irradiated with ultraviolet light, the MABI–fibrinogen was specifically incorporated into a 105,000 mol. wt. polypeptide that contained the Pl^{A1} antigen (Figure 8). Glycoprotein IIIa contains the Pl^{A1} antigen (Kunicki and Aster, 1979), and the interaction of MABI–fibrinogen with GP IIIa to form high-molecular-weight complexes unable to enter the gel is seen as a decreased intensity of the GP IIIa band in lane C of Figure 8. Thus, MABI–fibrinogen appears to bind to the platelet surface on or near GP IIIa, providing additional evidence that this glycoprotein is at least one component of the platelet fibrinogen receptor.

4.3. Platelet-Specific Monoclonal Antibodies

The exquisite specificity of monoclonal antibodies makes them excellent reagents with which to study specific biologic questions, such as the identification of the fibrinogen receptor (Kohler and Milstein, 1975). Five groups of investigators have reported the generation of platelet-specific monoclonal antibodies that they used for this purpose (DiMinno *et al.*, 1983; McEver *et al.*, 1983; Bennett *et al.*, 1983; Coller *et al.*, 1983; Pidard *et al.*, 1983). Each antibody inhibited platelet aggregation stimu-

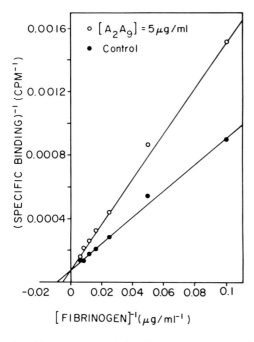

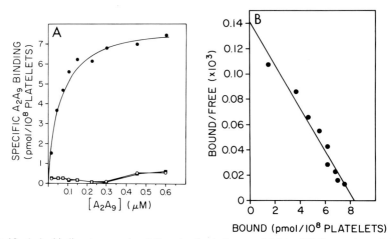

Figure 9. Double-reciprocal plot: Inhibition of fibrinogen binding to ADP-stimulated platelets by the monoclonal antiplatelet antibody A_2A_9. Suspensions of 5×10^7 gel-filtered platelets in Tyrode's/albumin buffer, pH 7.4, containing 1 mM $MgCl_2$ were incubated with a constant concentration of A_2A_9 and various concentrations of [^{125}I]fibrinogen. ●, [^{125}I]Fibrinogen binding in the absence of A_2A_9; ○, [^{125}I]fibrinogen binding in the presence of A_2A_9 at 5 μg/ml. From Bennett *et al.* (1983) with permission.

lated by a variety of platelet agonists and each inhibited, to a variable degree, fibrinogen binding to ADP-stimulated platelets. For example, the antibody reported by Bennett *et al.* (1983) behaved as a competitive inhibitor of fibrinogen binding with 50% inhibition of fibrinogen binding occurring at an antibody concentration of 65 nm (Figure 9). The antibody also bound to a single class of 47,000 sites on the surface of

Figure 10. A_2A_9 binding to normal and thrombasthenic human platelets. (A) Suspensions of 5×10^7 normal or thrombasthenic human platelets in Tyrode's/albumin buffer, pH 7.4, containing 1 mM $MgCl_2$ were incubated with various concentrations of ^{125}I-labeled A_2A_9 for 2 min at 37 °C. Specific A_2A_9 binding was determined after sedimenting the platelets through silicone oil. ●, Normal subject; ○, ■, thrombasthenic subjects. (B) Scatchard analysis of $A_2 A_9$ binding to the normal platelets in (A). From Bennett *et al.* (1984) with permission.

normal platelets with a K_d of 60 nm (Figure 10). Binding to thrombasthenic platelets was minimal. Other monoclonal antibodies among this group can also completely inhibit both ADP-induced fibrinogen binding and ADP-induced platelet aggregation (McEver *et al.*, 1983; Pidard *et al.*, 1983). Thus, there is a close correspondence between the ability of these antibodies to inhibit platelet fibrinogen binding and platelet aggregation and their ability to bind to the platelet surface. Moreover, the number of antibody-binding sites is equal to the number of fibrinogen receptors. When the antigen recognized by each of the five reported antibodies was determined, each was found to recognize an epitope on the GP IIb-IIIa complex. These experiments confirm the role of receptor-bound fibrinogen in platelet aggregation and provide a third piece of evidence that the GP IIb-IIIa complex contains the fibrinogen receptor.

4.4. Experiments Using Platelet Extracts

For these experiments, platelet proteins were extracted with the nonionic detergent Triton X-100, were separated and identified, and were reacted with ^{125}I-labeled fibrinogen. The ability of specific platelet proteins to interact directly with fibrinogen was then examined.

Nachman and Leung (1982) devised an enzyme-linked immunospecific assay (ELISA) to study the interaction of fibrinogen with GP IIb-IIIa. Fibrinogen, fibronectin, VWF, and plasminogen were each fixed to the wells of microtiter plates. Purified proteins, including GP IIb-IIIa, platelet membrane GP Ib, a mixture of erythrocyte glycoproteins, and human albumin were added to the wells and allowed to react with the fixed proteins overnight. The formation of specific complexes between the fixed and added proteins was detected with antisera against the added protein. Using this assay system, the only complex that could be detected was one that formed between an intact GP IIb-IIIa complex and fibrinogen.

Gogstad *et al.* (1982a) made similar observations using a different assay system. Dissolved platelets were subjected to crossed immunoelectrophoresis to separate the solubilized platelet proteins. The immunoelectrophoresis plates were then overlayed with ^{125}I-labeled fibrinogen and the proteins reacting with the fibrinogen were identified by autoradiography. Four immunoprecipitates reacted with the labeled fibrinogen. However, only one of the precipitates, that formed by GP IIb-IIIa, represented a platelet surface structure. When the crossed immunoelectrophoresis was performed in the presence of EDTA to dissociate the GP IIb-IIIa complexes (Kunicki *et al.*, 1981; Gogstad *et al.*, 1982b; Howard *et al.*, 1982), neither GP IIb nor IIIa alone reacted with fibrinogen. These data support the view that GP IIb-IIIa contains the fibrinogen receptor and suggest that an intact complex is required for the receptor activity to be expressed.

5. SUMMARY AND CONCLUSION

The search for an explanation of the fibrinogen requirement for platelet aggregation has resulted in our present concept of this process (Figure 11). Platelet stimulation exposes fibrinogen-binding sites on a complex formed by the membrane GP IIb and

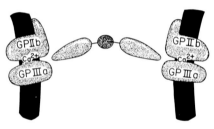

Figure 11. Diagrammatic representation of the interaction of fibrinogen with the surface of activated platelets. The calcium-mediated GP IIb-IIIa complex is represented with its fibrinogen-binding site exposed by an activation step, and is interacting with the D-domains of each end of the fibrinogen molecule.

IIIa. In the presence of calcium or magnesium, fibrinogen derived from the plasma or from the platelet α-granules binds to these sites and platelet aggregation ensues. This relatively straightforward construct has required the diligent efforts of numerous investigators and represents a major advance in our understanding of platelet physiology. However, like many advances, it has also raised many more unforeseen questions. How does platelet stimulation expose fibrinogen binding sites on GP IIb-IIIa? How can the suggestions by Fujimoto *et al.* (1982a,b) and Ruggeri *et al.* (1982) that VWF also binds to GP IIb-IIIa on stimulated platelets be reconciled with the array of fibrinogen-binding data? How does receptor-bound fibrinogen lead to the formation of platelet aggregates? Are other plasma or platelet granule proteins required? What is the relationship between the calcium needed to form the GP IIb-IIIa complex and the calcium needed for fibrinogen binding? Although these questions will not be easy to answer, they do provide a clear direction for future research.

REFERENCES

Bang, N. U., Heidenreich, R. O., and Trygstad, C. W., 1972, Plasma protein requirements for human platelet aggregation, *Ann. N.Y. Acad. Sci.* **201:**280–299.

Bennett, J. S., and Vilaire, G., 1979, Exposure of platelet fibrinogen receptors by ADP and epinephrine, *J. Clin. Invest.* **64:**1393–1401.

Bennett, J. S., Vilaire, G., and Burch, J. W., 1981, A role for prostaglandins and thromboxanes in the exposure of platelet fibrinogen receptors, *J. Clin. Invest.* **68:**981–987.

Bennett, J. S., Vilaire, G., and Cines, D. B., 1982, Identification of the fibrinogen receptor on human platelets by photoaffinity labeling, *J. Biol. Chem.* **257:**8049–8054.

Bennett, J. S., Hoxie, J. A., Leitman, S. F., Vilaire, G., and Cines, D. B., 1983, Inhibition of fibrinogen binding to stimulated human platelets by a monoclonal antibody, *Proc. Natl. Acad. Sci. U.S.A.* **80:**2417–2421.

Boeynaems, J. M., and Dumont, J. E., 1977, The two-step model of ligand-receptor interaction, *Mol. Cell. Endocrinol.* **7:**33–47.

Born, G. V. R., and Cross, M. J., 1964, Effects of inorganic ions and plasma proteins on the aggregation of blood platelets by adenosine diphosphate, *J. Physiol.* (*London*) **170:**397–414.

Bowie, E. J. W., and Owen, C. A., Jr., 1974, The bleeding time, *Prog. Hemostas. Thrombos.* **2:**249–271.

Brass, L. F., and Shattil, S. J., 1982, Changes in surface-bound and exchangeable calcium during platelet activation, *J. Biol. Chem.* **257:**14000–14005.

Brass, L. F., and Shattil, S. J., 1983, Identification of the saturable binding sites for Ca^{2+} on the surface of human platelets, *Thromb. Haemostasis* **50:**327.

Brown, R. S., Niewiarowski, S., Stewart, G. J., and Millman, M., 1977, A double-isotope study on incorporation of platelets and red cells into fibrin, *J. Lab. Clin. Med.* **90:**130–140.

Caen, J., 1972, Glanzmann thrombasthenia, *Clin. Haematol.* **1:**383–392.

Castaldi, P. A., and Caen, J., 1965, Platelet fibrinogen, *J. Clin. Pathol.* **18:**579–585.

Catt, K. J., and DuFau, M. L., 1977, Peptide hormone receptors, *Annu. Rev. Physiol.* **39:**529–557.

Charo, I. F., Feinman, R. D., Detwiler, T. C., Smith, J. B., Ingerman, C. M., and Silver, M. J., 1977, Prostaglandin endoperoxides and thromboxane A$_2$ can induce platelet aggregation in the absence of secretion, *Nature (London)* **269:**66–68.

Coller, B. S., 1979, Asialofibrinogen supports platelet aggregation and adhesion to glass, *Blood* **53:**325–332.

Coller, B. S., 1980, Interaction of normal, thrombasthenic, and Bernard-Soulier platelets with immobilized fibrinogen: Defective platelet-fibrinogen interaction in thrombasthenia, *Blood* **55:**169–178.

Coller, B. S., Peerschke, E. I., Scudder, L. E., and Sullivan, C. A., 1983, A murine monoclonal antibody that completely blocks the binding of fibrinogen to platelets produces a thrombasthenic-like state in normal platelets and binds to glycoproteins IIb and IIIa, *J. Clin. Invest.* **72:**325–338.

Crabtree, G. R., and Kant, J. A., 1981, Molecular cloning of cDNA for the α, β, and γ chains of rat fibrinogen, *J. Biol. Chem.* **256:**9718–9723.

Crabtree, G. R., and Kant, J. A., 1982, Coordinate accumulation of the mRNAs for the α, β, and γ chains of rat fibrinogen following defibrination, *J. Biol. Chem.* **257:**7277–7279.

Cross, M. J., 1964, Effect of fibrinogen on the aggregation of platelets by adenosine diphosphate, *Thrombos. Diathes. Haemorrh.* **12:**521–527.

DiMinno, G., Thiagarajan, P., Perussia, B., Martinez, J., Shapiro, S., Trinchieri, G., and Murphy, S., 1983, Exposure of platelet fibrinogen-binding sites by collagen, arachidonic acid, and ADP: Inhibition by a monoclonal antibody to the glycoprotein IIb-IIIa complex, *Blood* **61:**140–148.

Doolittle, R. F., 1973, Structural aspects of the fibrinogen to fibrin conversion, *Adv. Protein Chem.* **27:**1–109.

Doolittle, R. F., 1981, Fibrinogen and fibrin, in: *Haemostasis and Thrombosis* (A. L. Bloom, and D. P. Thomas, eds.), Churchill Livingstone, New York, pp. 163–191.

Doolittle, R. F., Takagi, T., and Cottrell, B. A., 1974, Platelet and plasma fibrinogens are identical gene products, *Science* **185:**368–369.

Fowler, W. E., and Erickson, H. P., 1979, Trinodular structure of fibrinogen, *J. Mol. Biol.* **134:**241–249.

Fowler, W. E., Fretto, L. J., Erickson, H. P., and McKee, P. A., 1980, Electron microscopy of plasmic fragments of human fibrinogen as related to trinodular structure of the intact molecule, *J. Clin. Invest.* **66:**50–56.

Fowlkes, D. M., Kant, J. A., Fornace, A. L., and Crabtree, G. R., 1983, Regulation and structure of rat and human fibrinogen genes, *Throm. Haemostasis* **50:**325.

Fujimoto, T., and Hawiger, J., 1982a, Adenosine diphosphate induces binding of von Willebrand factor to human platelets, *Nature (London)* **297:**154–156.

Fujimoto, T., Ohara, S., and Hawiger, J., 1982b, Thrombin-induced exposure and prostacyclin inhibition of the receptor for factor VIII/von Willebrand factor on human platelets, *J. Clin. Invest.* **69:**1212–1222.

Gaarder, A., Jonsen, J., Laland, S., Hellem, A., and Owren, P. A., 1961, Adenosine diphosphate in red cells as a factor in the adhesiveness of human blood platelets, *Nature (London)* **192:**531–532.

Ganguly, P., 1972, Isolation and some properties of fibrinogen from human blood platelets, *J. Biol. Chem.* **247:**1809–1816.

Gogstad, G. O., Brosstad, F., Krutnes, M-B., Hagen, I., and Solum, N. O., 1982a, Fibrinogen-binding properties of the human platelet glycoprotein IIb-IIIa complex: A study using crossed-radioimmunoelectrophoresis, *Blood* **60:**663–671.

Gogstad, G. O., Hagen, I., Krutnes, M.-B., and Solum, N. O., 1982b, Dissociation of the glycoprotein IIb-IIIa complex in isolated human platelet membranes, *Biochim. Biophys. Acta* **689:**21–30.

Graber, S. E., and Hawiger, J., 1982, Evidence that changes in platelet cyclic AMP levels regulate the fibrinogen receptor on human platelets, *J. Biol. Chem.* **257:**14606–14609.

Harfenist, E. J., Packham, M. A., and Mustard, J. F., 1980a, Reversibility of the association of fibrinogen with rabbit platelets exposed to ADP, *Blood* **56:**189–198.

Harfenist, E. J., Packham, M. A., and Mustard, J. F., 1980b, Identical behavior of fibrinogen and asialofibrinogen in reactions with platelets during ADP-induced aggregation, *Throm. Res.* **20:**353–358.

Harfenist, E. J., Packham, M. A., Kinlough-Rathbone, R. L., and Mustard, J. F., 1981, Inhibitors of ADP-induced platelet aggregation prevent fibrinogen binding to rabbit platelets and cause rapid deaggregation and dissociation of bound fibrinogen, *J. Lab. Clin. Med.* **97:**680–688.

Hawiger, J., Parkinson, S., and Timmons, S., 1980, Prostacyclin inhibits mobilisation of fibrinogen binding sites on human ADP- and thrombin-treated platelets, *Nature (London)* **283:**195–197.

Hawiger, J., Timmons, S., Kloczewiak, M., Strong, D. D., and Doolittle, R. F., 1982, γ and α chains of human fibrinogen possess sites reactive with human platelet receptors, *Proc. Natl. Acad. Sci. U.S.A.* **79:**2068–2071.

Hoeprich, P. D., and Doolittle, R. F., 1983, Dimeric half-molecules of human fibrinogen are joined through disulfide bonds in an antiparallel orientation, *Biochemistry* **22:**2049–2055.

Hoffman, B. B., Mullikin-Kilpatrick, D., and Lefkowitz, R. J., 1980, Heterogeneity of radioligand binding to α-adrenergic receptors, *J. Biol. Chem.* **255:**4645–4652.

Holmsen, H., and Weiss, H. J., 1979, Secretable storage pools in platelets, *Annu. Rev. Med.* **30:**119–134.

Holmsen, H., Dangelmaier, C. A., and Holmsen, H-K., 1981, Thrombin-induced platelet responses differ in requirement for receptor occupancy, *J. Biol. Chem.* **256:**9393–9396.

Howard, L., Shulman, S., Sadanandan, S., and Karpatkin, S., 1982, Crossed immunoelectrophoresis of human platelet membranes, *J. Biol. Chem.* **257:**8331–8336.

James, H. L., Ganguly, P., and Jackson, C. W., 1977, Characterization and origin of fibrinogen in blood platelets, *Throm. Haemostasis* **38:**939–954.

Kant, J. A., and Crabtree, G. R., 1983, The rat fibrinogen genes, *J. Biol. Chem.* **258:**4666–4667

Keenan, J. P., 1972, Platelet fibrinogen. I Quantitation using fibrinogen sensitized tanned red cells, *Med. Lab. Technol.* **29:**71–79.

Kloczewiak, M., Timmons, S., and Hawiger, J., 1982, Localization of a site interacting with human platelet receptor on carboxy-terminal segment of human fibrinogen γ-chain, *Biochem. Biophys. Res. Commun.* **107:**181–187.

Kloczewiak, M., Timmons, S., and Hawiger, J., 1983, Recognition site for the platelet receptor is present on the 15-residue carboxyl-terminal fragment of the γ-chain of human fibrinogen and is not involved in the fibrin polymerization reaction, *Throm. Res.* **29:**249–255.

Kohler, G., and Milstein, C., 1975, Continuous cultures of fused cells secreting antibody of predetermined specificity, *Nature (London)* **256:**495–497.

Kornecki, E., Niewiarowski, S., Morinelli, T. A., and Kloczwiak, M., 1981, Effects of chymotrypsin and adenosine diphosphate on the exposure of fibrinogen receptors on normal human and Glanzmann's thrombasthenic platelets, *J. Biol. Chem.* **256:**5696–5701.

Kornecki, E., Tuszynski, G. P., and Niewiarowski, S., 1983, Inhibition of fibrinogen receptor-mediated platelet aggregation by heterologous anti-human platelet membrane antibody, *J. Biol. Chem.* **258:**9349–9356.

Kunicki, T. J., and Aster, R. H., 1979, Isolation and immunologic characterization of the human platelet alloantigen, Pl[A1], *Mol. Immunol.* **16:**353–360.

Kunicki, T. J., Pidard, D., Rosa, J-P, and Nurden, A. T., 1981, The formation of Ca^{++}-dependent complexes of platelet membrane glycoproteins IIb and IIIa in solution as determined by crossed immunoelectrophoresis, *Blood* **58:**268–278.

Laudano, A. P., and Doolittle, R. F., 1978, Synthetic peptide derivatives that bind to fibrinogen and prevent the polymerization of fibrin monomers, *Proc. Natl. Acad. Sci. U.S.A.* **75:**3085–3089.

Laudano, A. P., and Doolittle, R. F., 1980, Studies on synthetic peptides that bind to fibrinogen and prevent fibrin polymerization. Structural requirements, number of binding sites, and species differences, *Biochemistry* **19;**1013–1019.

Lee, H., Nurden, A. T., Thomaidis, A., and Caen, J. P., 1981, Relationship between fibrinogen binding and the platelet glycoprotein deficiencies in Glanzmann's thrombasthenia type I and type II, *Br. J. Haematol.* **48:**47–57.

Marder, V. J., and Budzynski, A. Z., 1975, Data for defining fibrinogen and its plasmic degradation products, *Thrombos. Diathes. Haemorrh.* **33:**199–207.

Marder, V. J., Shulman, N. R., and Carroll, W. R., 1969, High molecular weight derivatives of human fibrinogen produced by plasmin, *J. Biol. Chem.* **244:**2111–2119.

Marder, V. J., Francis, C. W., and Doolittle, R. F., 1982, Fibrinogen structure and physiology, in: *Hemostasis and Thrombosis* (R. W. Colman, J. Hirsh, V. J. Marder, and E. W. Salzman, eds.), J. B. Lippincott Co., Philadelphia, pp. 145–163.

Marguerie, G. A., and Plow, E. F., 1981, Interaction of fibrinogen with its platelet receptor: Kinetics and effect of pH and temperature, *Biochemistry* **20:**1074–1080.

Marguerie, G., Chagniel, G., and Suscillon, M., 1977, The binding of calcium to bovine fibrinogen, *Biochim. Biophys. Acta* **490:**94–103.

Marguerie, G. A., Plow, E. F., and Edgington, T. S., 1979, Human platelets possess an inducible and saturable receptor specific for fibrinogen, *J. Biol. Chem.* **254:**5357–5363.

Marguerie, G. A., Edgington, T. S., and Plow, E. F., 1980, Interaction of fibrinogen with its platelet receptor as a part of a multistep reaction in ADP-induced platelet aggregation, *J. Biol. Chem.* **255:**154–161.

Marguerie, G. A., Ardaillou, N., Cherel, G., and Plow, E. F., 1982, The binding of fibrinogen to its platelet receptor, *J. Biol. Chem.* **257:**11872–11875.

McEver, R. P., Bennett, E. M., and Martin, M. N., 1983, Identification of two structurally and functionally distinct sites on human platelet membrane glycoprotein IIb-IIIa using monoclonal antibodies, *J. Biol. Chem.* **258:**5269–5275.

McLean, J. R., Maxwell, R. E., and Hertler, D., 1964, Fibrinogen and adenosine diphosphate-induced aggregation of platelets, *Nature (London)* **202:**605–606.

Miller, J. L., Katz, A. J., and Feinstein, M. B., 1975, Plasmin inhibition of thrombin-induced platelet aggregation, *Thrombos. Diathes. Haemorrh.* **33:**286–309.

Molinoff, P. B., Wolfe, B. B., and Weiland, G. A., 1981, Quantitative analysis of drug-receptor interactions: II: Determination of the properties of receptor subtypes, *Life Sci.* **29:**427–443.

Morinelli, T. A., Niewiarowski, S., Kornecki, E., Figures, W. R., Watchfogel, Y., and Colman, R. W., 1983, Platelet aggregation and exposure of fibrinogen receptors by prostaglandin endoperoxide analogues, *Blood* **61:**41–49.

Mosesson, M. W., Homandberg, G. A., and Amrani, D. L., 1984, Human platelet fibrinogen gamma chain structure, *Blood* **63:**990–995.

Mustard, J. F., Packham, M. A., Kinlough-Rathbone, R. L., Perry, D. W., and Regoeczi, E., 1978, Fibrinogen and ADP-induced platelet aggregation, *Blood* **52:**453–466.

Mustard, J. F., Kinlough-Rathbone, R. L., Packham, M. A., Perry, D. W., Harfenist, E. J., and Pai, K. R. M., 1979, Comparison of fibrinogen association with normal and thrombasthenic platelets on exposure to ADP and chrmotrypsin, *Blood* **54:**983–987.

Nachman, R. L., and Leung, L. L., 1982, Complex formation of platelet membrane glycoproteins IIb and IIIa with fibrinogen, *J. Clin. Invest.* **69:**263–269.

Nachman, R. L., and Marcus, A. J., 1968, Immunological studies of proteins associated with the subcellular fractions of thrombasthenic and afibrinogenaemic platelets, *Br. J. Haematol.* **15:**181–189.

Nachman, R., Levine, R., and Jaffe, E. A., 1977, Synthesis of Factor VIII antigen by cultured guinea pig megakaryocytes, *J. Clin. Invest.* **60:**914–921.

Nachman, R., Levine, R., and Jaffe, E., 1978, Synthesis of actin by cultured guinea pig megakaryocytes, complex formation with fibrin, *Biochim. Biophys. Acta* **543:**91–105.

Niewiarowski, S., Regoeczi, E., Stewart, G. J., Senyi, A. F., and Mustard, J. F., 1972, Platelet interaction with polymerizing fibrin, *J. Clin. Invest.* **51:**685–700.

Niewiarowski, S., Budzynski, A. Z., and Lipinski, B., 1977, Significance of the intact polypeptide chains of human fibrinogen in ADP-induced platelet aggregation, *Blood* **49:**635–644.

Niewiarowski, S., Levy-Toledano, S., and Caen, J. P., 1981a, Platelet interaction with polymerizing fibrin in Glanzmann's thrombasthenia, *Thrombos. Res.* **23:**457–463.

Niewiarowski, S., Budzynski, A. Z., Morinelli, T. A., Budzynski, T. M., and Stewart, G. J., 1981b, Exposure of fibrinogen receptor on human platelets by proteolytic enzymes, *J. Biol. Chem.* **256:**917–925.

Olexa, S., and Budzynski, A. Z., 1980, Evidence for four different polymerization sites involved in human fibrin formation, *Proc. Natl. Acad. Sci. U.S.A.* **77:**1374–1378.

Pasqua, J. J., and Pizzo, S. U., 1983, The role of ligand–ligand interactions in competition by fibrinogen and fibrin degradation products for fibrinogen binding to human platelets, *Biochim. Biophys. Acta* **757:**282–287.

Peerschke, E. I., 1982a, Induction of human platelet fibrinogen receptors by epinephrine in the absence of released ADP, *Blood* **60:**71–77.

Peerschke, E. I., 1982b, Evidence for interaction between platelet fibrinogen receptors, *Blood* **60:**973–978.

Peerschke, E. I., and Zucker, M. B., 1981, Fibrinogen receptor exposure and aggregation of human platelets produced by ADP and chilling, *Blood* **57:**663–670.

Peerschke, E. I., Grant, R. A., and Zucker, M. B., 1980a, Decreased association of ^{45}calcium with platelets unable to aggregate due to thrombasthenia or prolonged calcium deprivation, *Br. J. Haematol.* **46:**247–256.

Peerschke, E. I., Zucker, M. B., Grant, R. A., Egan, J. J., and Johnson, M. M., 1980b, Correlation between fibrinogen binding to human platelets and platelet aggregability, *Blood* **55:**841–847.

Peters, K., and Richards, F. M., 1977, Chemical cross-linking: Reagents and problems in studies of membrane structure, *Annu. Rev. Biochem.* **46:**523–551.

Pidard, D., Montgomery, R. R., Bennett, J. S., and Kunicki, T. J., 1983, Interaction of AP-2, a monoclonal antibody specific for the human platelet membrane glycoprotein IIb-IIIa complex, with intact platelets, *J. Biol. Chem.* **258:**12582–12586.

Plow, E. F., and Edgington, T. S., 1975, Unique immunochemical features and intracellular stability of platelet fibrinogen, *Thromb. Res.* **7:**729–742.

Plow, E. F., and Marguerie, G. A., 1980a, Participation of ADP in the binding of fibrinogen to thrombin-stimulated platelets, *Blood* **56:**553–555.

Plow, E. F., and Marguerie, G. A., 1980b, Induction of the fibrinogen receptor on human platelets by epinephrine and the combination of epinephrine and ADP, *J. Biol. Chem.* **255:**10971–10977.

Plow, E. F., and Marguerie, G., 1982, Inhibition of fibrinogen binding to human platelets by the tetrapeptide glycl-L-prolyl-L-arginyl-L-proline, *Proc. Natl. Acac. Sci. U.S.A.* **79:**3711–3715.

Plow, E. F., Srouji, A. H., Meyer, D., Marguerie, G., and Ginsberg, M. E., 1984, Evidence that three adhesive proteins interact with a common recognition site on activated platelets, *J. Biol. Chem.* **259:**5388–5391.

Price, T. M., Strong, D. D., Rudee, M. L., and Doolittle, R. F., 1981, Shadow-cast electron microscopy of fibrinogen with antibody fragments bound to specific regions, *Proc. Natl. Acad. Sci. U.S.A.* **78:**200–204.

Ruggeri, Z. M., Bader, R., and deMarco, L., 1982, Glanzmann's thrombasthenia: Deficient binding of von Willebrand factor to thrombin-stimulated platelets, *Proc. Natl. Acad. Sci. U.S.A.* **79:**6038–6041.

Shattil, S. J., and Bennett, J. S., 1981, Platelets and their membranes in hemostasis: Physiology and pathophysiology, *Ann. Intern. Med.* **94:**108–118.

Shulman, S., 1953, The size and shape of bovine fibrinogen, studies of sedimentation, diffusion and viscosity, *J. Am. Chem. Soc.* **75:**5846–5852.

Takagi, T., and Doolittle, R. F., 1975, Amino acid sequence studies on plasmin-derived fragments of human fibrinogen: Amino-terminal sequences of intermediate and terminal fragments, *Biochemistry* **14:**940–946.

Tam, S. W., Fenton, II, J. W., and Detwiler, T. C., 1979, Dissociation of thrombin from platelets by hirudin, *J. Biol. Chem.* **254:**8723–8725.

Tangen, O., Berman, H. J., and Marfey, P., 1971, Gel filtration, a new technique for separation of blood platelets from plasma, *Thrombos. Diath. Haemorrh.* **25:**268–278.

Tollefsen, D. M., and Majerus, P. W., 1975, Inhibition of human platelet aggregation by monovalent antifibrinogen antibody fragments, *J. Clin. Invest.* **55:**1259–1268.

Tomikawa, M., Iwamoto, M., Soderman, S., and Blomback, B., 1980, Effect of fibrinogen on ADP-induced platelet aggregation, *Thrombos. Res.* **19:**841–855.

Weisel, J. W., Phillips, Jr., G. N., and Cohen, C., 1981, A model from electron microscopy for the molecular structure of fibrinogen and fibrin, *Nature (London)* **289:**263–267.

Weiss, H. J., 1975, Platelet physiology and abnormalities of platelet function, *N. Engl. J. Med.* **293:**531–541, 580–588.

Weiss, H. J., and Kochwa, S., 1968, Studies of platelet function and proteins in 3 patients with Glanzmann's thrombasthenia, *J. Lab. Clin. Med.* **71:**153–165.

Weiss, H. J., and Rogers, J., 1971, Fibrinogen and platelets in the primary arrest of bleeding, *N. Engl. J. Med.* **285:**369–374.

Zucker, M. B., 1977, Relationship of plasma clotting factors to platelets, in: *Topics in Hematology,* (S. Seno, F. Takaku, and S. Irino, eds.), Excerpta Medica, Amsterdam-Oxford, pp. 280–282.

Zucker, M. B., and Vroman, L., 1969, Platelet adhesion induced by fibrinogen adsorbed onto glass, *Proc. Soc. Exp. Biol. Med.* **131:**318–320.

10

Platelet–von Willebrand Factor Interactions

Barry S. Coller

1. INTRODUCTION

Clinical observations indicate that the interaction between von Willebrand factor
(VWF)* and platelets plays a central role in platelet physiology. The most convincing
evidence is that patients with quantitative and/or qualitative abnormalities of their
VWF protein have a hemorrhagic diathesis characterized by abnormal platelet function
(Zimmerman and Ruggeri, 1983; Coller, 1984a). Additional support for the importance of this interaction comes from observations of patients whose platelets cannot
bind VWF to their surface as a result of abnormalities in the platelet membrane
receptor mechanisms; however, conclusions from these patients must be tempered by
an appreciation of the complexity of the disorders. Thus, patients with the Bernard-
Soulier syndrome, whose platelets do not bind human VWF in the presence of
ristocetin (Zucker *et al.*, 1977; Moake *et al.*, 1980) and cannot be aggregated by
bovine VWF (Bithell *et al.*, 1972; Howard *et al.*, 1973), have a serious bleeding
disorder (George *et al.*, 1984; Coller, 1984b). As discussed in Chapter 16, the platelets
from these patients lack normal glycoprotein (GP) Ib (Nurden and Caen, 1975), the
glycoprotein that several lines of evidence indicate is the receptor for both human VWF

---

*Abbreviations and definitions: (1) VIII:C (factor VIII coagulant activity; antihemophilic factor): The factor
  missing in the plasma of patients with hemophilia A required for a normal clotting time. (2) VIII:CAg (factor
  VIII coagulant antigen): The antigen corresponding to VIII:C. (3) VWF (von Willebrand factor): The factor
  missing from the plasma of patients with von Willebrand's disease (VWD) required for a normal bleeding
  time. (4) VIIIR:Ag (VWF antigen): The antigen corresponding to VWF. (5) VIII:RCo (ristocetin cofactor
  activity): The factor required for normal ristocetin-induced platelet agglutination. (6) F.VIII/VWF: The
  complex of VIII:C and VWF that copurifies under certain conditions.

Barry S. Coller • Division of Hematology, Departments of Medicine and Pathology, State University of
New York at Stony Brook, Stony Brook, New York 11794.

(when platelets are incubated with ristocetin) and bovine VWF (Kirby, 1977; Phillips, 1980; Coller *et al.*, 1983). In addition to the GP Ib abnormality, platelets from these patients also appear to lack GP V and a glycoprotein of M_r 17–22,000 (Clemetson *et al.*, 1982; Berndt *et al.*, 1983; Nurden *et al.*, 1983) that may be noncovalently associated with GP Ib (Coller *et al.*, 1983; Berndt *et al.*, 1983; Chong *et al.*, 1983; see Chapter 4). Patients with Glanzmann's thrombasthenia, whose washed platelets fail to bind VWF when stimulated with either thrombin (Ruggeri *et al.*, 1982a) or ADP (Gralnick *et al.*, 1984), also suffer from a severe bleeding disorder (George *et al.*, 1984; Coller, 1984b). It is impossible, however, to know how much of a contribution, if any, this defect in VWF binding contributes to their hemorrhagic diathesis since their platelets have major abnormalities of aggregation that appear to be related to their failure to bind fibrinogen (see Chapter 16). The platelets from these patients lack normal GP IIb-IIIa and there is considerable evidence that this abnormality is the cause of their functional defects, including their inability to bind VWF when stimulated with thrombin or ADP (Nachman and Leung, 1982; Coller, 1983; Ruggeri *et al.*, 1983b; Gralnick *et al.*, 1984).

In this chapter, I will review first the biosynthesis and structure of VWF and then discuss the *in vitro* data on VWF-dependent platelet function. Against that background, the more limited *ex vivo* and *in vivo* studies will be analyzed with a view toward identifying the possible mechanisms by which the VWF–platelet interaction contributes to normal hemostasis and pathological thrombosis.

2. BIOSYNTHESIS, LOCALIZATION, AND STRUCTURE OF VON WILLEBRAND FACTOR

2.1. Tissue Distribution of von Willebrand Factor

Endothelial cells and megakaryocytes have been reported to synthesize VWF (Jaffe *et al.*, 1974; Nachman *et al.*, 1977b). In fact, endothelial cells grown in culture release VWF into the medium and thus these cells are presumed to be the major source of plasma VWF. Endothelial cells also deposit VWF on their abluminal surface in close association with the internal elastic lamina (Rand *et al.*, 1980; Sussman and Rand, 1982). Although there is *in vitro* evidence that VWF interacts specifically with fibrillar collagen (Nyman, 1977; Legrand *et al.*, 1978; Santoro and Cowan, 1982; Santoro, 1983), treatment of the subendothelium with collagenase did not decrease the VWF detected by immunofluorescence (Rand *et al.*, 1980). Another component of the subendothelium that may function as a binding site for VWF is the elastin-associated microfibril, since extracts of this material cause platelets to aggregate only in the presence of VWF and the process requires that the platelet have an intact GP Ib (Legrand *et al.*, 1980; Fauvel *et al.*, 1983). In an *ex vivo* system, VWF in solution was found to bind to the subendothelial surface of human renal arteries and the amount bound correlated with the number of platelets that adhered (Sakariassen *et al.*, 1979; Sixma *et al.*, 1984). The contribution of plasma VWF to vessel-wall VWF *in vivo* has been questioned, however, since after de-endothelialization of rabbit aorta, there was little, if any, VWF present on the denuded neointima (which was in contact with

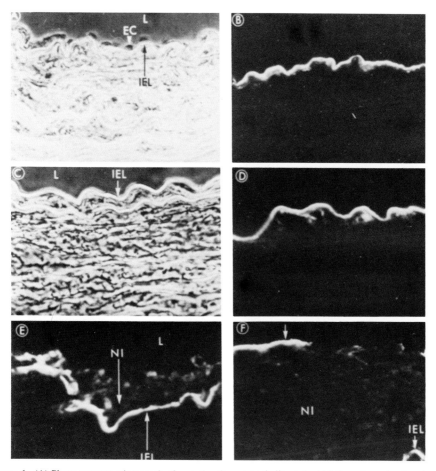

Figure 1. (A) Phase-contrast micrograph of a section from a nonballooned rabbit thoracic aorta, showing a single layer of endothelial cells (EC) luminal (L) to the internal elastic lamina (IEL) (magnification 330×). (B) Fluorescent micrograph of a section similar to (A), showing intensely positive immunofluorescent staining for VWF on the luminal surface. Small amounts of VWF are also seen in the vessel wall (magnification 330×). (C) Phase micrograph of a section of the same rabbit aorta, from a balloon de-endothelialized abdominal region, showing the absence of endothelial cells (magnification 330×). (D) Fluorescence micrograph of section similar to (C), again showing intense immunofluorescent staining for VWF along the de-endothelialized IEL (magnification 330×). (E) Fluorescence micrograph of a rabbit abdominal aorta 7 days after de-endothelialization, revealing intense staining for VWF along the IEL. Small amounts of VWF are also seen associated with the neointima (NI) cells luminal to the IEL (magnification 330×). (F) Abdominal aorta 2 weeks after de-endothelialization. This section of a blue/white junction reveals marked staining for VWF at the re-endothelialized surface (arrow), whereas no staining is observed in the non-re-endothelialized region to the right. The IEL still stains for VWF (magnification 330×). From Sussman and Rand, 1982, with permission.

plasma VWF), whereas there were significant amounts of VWF in areas of neointima that were covered by endothelial cells, which presumably deposited it (Sussman and Rand, 1982) (Figure 1). Thus, the relative contributions of VWF directly deposited by endothelial cells and VWF adsorbed from plasma at the sites of injury remain uncertain.

2.2. Platelet-Associated von Willebrand Factor

The VWF associated with platelets (Howard *et al.*, 1974; Coller *et al.*, 1975) is thought to be synthesized by megakaryocytes (Nachman *et al.*, 1977b). It is found primarily in the α-granules (Nachman and Jaffe, 1975), where, on appropriate stimulation, it can be released (Koutts *et al.*, 1978; Zucker *et al.*, 1979). When thrombin is the stimulus, VWF becomes exposed on the platelet membrane in a reaction that requires the presence of calcium (George and Onofre, 1982, Fernanda Lopez Fernandez *et al.*, 1982) and is probably dependent on the presence of GP IIb-IIIa (Ruggeri *et al.*, 1983a). There may also be a smaller pool of VWF associated with the membrane of unstimulated platelets (George and Onofre, 1982; Ruggeri *et al.*, 1983a). This pool appears to be independent of the presence of platelet GP IIb-IIIa (Ruggeri *et al.*, 1983a). The multimeric structure (see Section 2.3) of platelet VWF differs slightly from plasma VWF, containing a subpopulation of multimers of higher M_r (Fernanda Lopez Fernandez *et al.*, 1982). Minor differences in the peptide maps of platelet and plasma VWF have also been reported (Nachman *et al.*, 1980).

2.3. Structure of von Willebrand Factor

Human VWF has been purified by several different laboratories and there is general agreement about the fundamental aspects of its structure (Marchesi *et al.*, 1972; Legaz *et al.*, 1973; Shapiro *et al.*, 1973). As purified *in vitro*, it shows the unusual characteristic of being composed of a series of macromolecules of extremely high M_r (estimated to be between 1×10^6 and 15×10^6) (Zimmerman *et al.*, 1975; Hoyer and Shainoff 1980; Ruggeri and Zimmermann, 1980) (Figure 2). When disulfide bonds are fully reduced, only a single protein band of $M_r \sim 230,000$ can be seen in sodium dodecyl sulfate (SDS) gels, suggesting that this is the basic subunit from which all of the high M_r multimers are formed. It has also been suggested that either two (Counts *et al.*, 1978; Perret *et al.*, 1979; Weinstein and Deykin, 1979) or four (Hoyer and Shainoff, 1980; Meyer *et al.*, 1980; Ruggeri and Zimmerman, 1980) subunits join together to form a basic protomeric unit and that the bands seen in the large-pore gels (Figure 2; 1% agarose) represent quantum increments in the number of protomers joined together. However, with newer gel techniques, the earlier bands can be resolved into three to six closely spaced bands of varying intensity (Zimmerman *et al.*, 1984; Ruggeri and Zimmerman, 1981) (Figure 2; 2% agarose). Thus, the organization of the basic units into the multimers appears to be more complex than originally thought. Approximtely 5% of the VWF's weight is composed of carbohydrate, some of which is exposed on the surface of the molecule (Legaz *et al.*, 1973; Gralnick, 1978; Sodetz *et al.*, 1977, 1978). It is important to emphasize the enormous size of the VWF multimers, with the higher M_r ones being larger than some viruses. Electron microscopy of the molecule after rotary shadowing revealed a flexible filamentous structure with irregularly spaced nodules (Ohmori *et al.*, 1982) (Figure 3). The filaments were $\sim$20–30 nm wide and ranged in length from 500 to 11,500 nm, with an average of 4780 nm (compared with approximately 475 nm for fibrinogen). Another recent report confirms this basic pattern, but suggests a more compact molecule (Loscalzo *et al.*, 1983). Since the multimers are made up of subunits that appear to be identical, it is

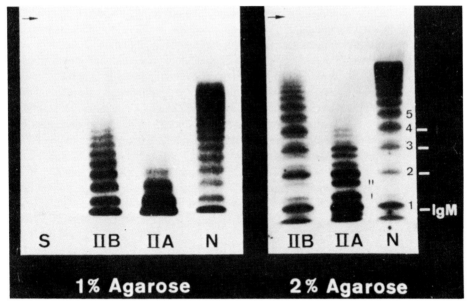

Figure 2. Autoradiograph of radioimmunodetectable VWF. SDS–electrophoresis in 1% (left) and 2% (right) agarose gels of normal plasma (N); plasma from a patient with VWD type IIA (IIA); a patient with VWD type IIB (IIB); and a patient with severe VWD(S). The arrow indicates the interface between stacking and running gel. The cathode was at the top and the anode at the bottom of the gel. Numbers from 1 to 5 indicate the smallest major bands in normal plasma, while the intervening bands are indicated by primes. The position of IgM and its cross-linked oligomers is shown by white marks. From Ruggeri and Zimmerman (1981), with permission.

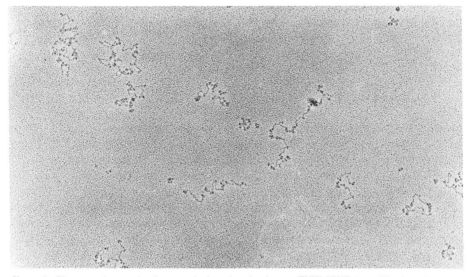

Figure 3. Electron micrograph of rotary shadowed native human FVIII/VWF (magnification 10,000×). From Ohmori *et al.* (1982) with permission. Reproduced at 75%.

likely that multiple platelet binding sites are always present on the surface of the molecule. The combination of large size and multiple-binding sites makes VWF an ideal bridging molecule for holding platelets together. One can liken these molecular features to another well-known agglutinating molecule, IgM. Thus, it is not surprising that the largest multimers are the most effective in causing platelet agglutination and that they seem to bind to platelets more readily than the smaller ones (Doucet-de-Bruine *et al.*, 1978; Gralnick *et al.*, 1981; Fernanda Lopez Fernandez *et al.*, 1982). Decreasing the size of the multimers by partial reduction of disulfide bonds results in a dramatic decrease in the platelet agglutinating activity of the VWF (Counts *et al.*, 1978; Gralnick *et al.*, 1981; Ohmori *et al.*, 1982). The larger and smaller M_r VWF

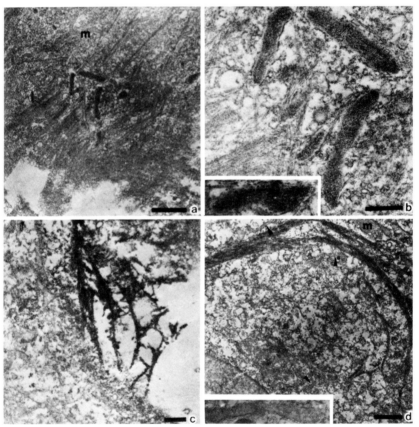

Figure 4. Electron micrograph of peroxidase–antiperoxidase-stained endothelial cells. In (A, B, and C), anti-von Willebrand protein antibody was used. (A) A cluster of positively stained Weibel-Palade bodies; unstained mitochondria (m) are also visible. Bar, 2 μm. (B) A higher magnification of the photograph in (A), showing electron dense deposits located parallel to the longitudinal axes of the organelles. Bar, 0.5 μm. The inset shows an unusually shaped Weibel-Palade body from a different cell (magnification 51,000×). (C) Strongly decorated extracellular filaments containing von Willebrand protein. Bar, 1 μm. (D) Control antisera, the large micrograph shows staining with antivimentin antibody, with positively stained intermediate filaments forming an arc around the nucleus. Mitochondria (m) and Weibel-Palade bodies (arrowheads) are unstained. Bar, 2 μm. The inset shows an unstained Weibel-Palade body from a cell treated with antifibronectin antiserum (magnification 28,600×). From Wagner *et al.* (1983) with permission.

multimers also differ in the ease with which they undergo cryoprecipitation and *in vivo* survival (Over *et al.*, 1978a,b).

2.4. Biosynthesis of von Willebrand Factor

Studies of VWF synthesis by endothelial cells in culture indicate that the basic subunit of M_r 230,000 is derived from a higher M_r precursor (Wagner and Marder, 1983; Lynch *et al.*, 1983). Incubation of the endothelial cells with tunicamycin to inhibit N-linked carbohydrate attachment did not alter precursor processing or secretion (Lynch *et al.*, 1983). It appears that multimer formation occurs before release from the surface of the endothelial cell (Wagner and Marder, 1983), but the possibility of multimer interconversion in plasma remains moot. Kinetic studies indicate that endothelial cells do not have large stores of VWF (Lynch *et al.*, 1983), but immunologic studies showed concentration of VWF in Weibel-Palade bodies where, it was postulated, the protein may be stored or processed (Figure 4) (Wagner *et al.*, 1982). Figure 5 schemat-

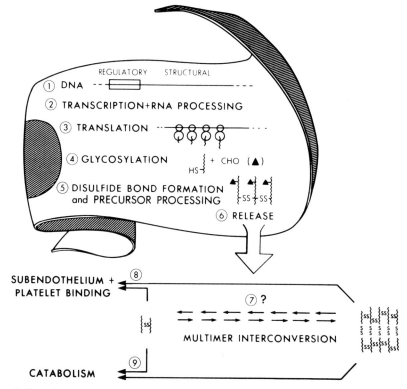

Figure 5. Steps in the synthesis, release, and disposition of VWF derived from endothelial cells. (1) Regulatory and structural genetic information is coded in DNA; rearrangements of DNA may occur when the genes are activated. (2) The structural gene for the precursor of the VWF subunit is transcribed and the resulting mRNA may undergo further processing. (3) The mRNA is translated and undergoes glycosylation (4), disulfide bond formation (5), and excision of the M_r ~20,000 peptide (5). (4) The mature VWF is released into the circulation where it may or may not undergo further multimer interconversion (7). (8) Under the appropriate circumstances, the circulating VWF can bind to platelets or subendothelium. (9) A catabolic mechanism independent of platelet or subendothelial cell binding may also exist. From Coller (1984a) with permission.

ically depicts the steps in VWF systhesis and catabolism, indicating the potential sites for abnormalities.

3.   VON WILLEBRAND'S DISEASE

A large number of quantitative and/or qualitative abnormalities of VWF have been described, leading to variable deficiencies in platelet function as judged by prolongation of the bleeding time, decreased platelet retention in columns of glass beads, abnormal platelet agglutination induced by either ristocetin or *Bothrops* species snake venoms, and abnormal behavior in *ex vivo* models (Zimmerman and Ruggeri, 1983; Coller, 1984a). A variety of classifications have been proposed and the one suggested by Zimmerman and Ruggeri (1983) is given in Table 1. It is sobering to note that none of the abnormalities currently recognized have been unequivocally localized to a given step in the scheme depicted in Figure 5. Several themes do, however, emerge from these observations as different types of VWD have been clinically recognized.

3.1.   Type I von Willebrand's Disease

If the amount of VWF decreases below a certain level, even if it is of apparently normal structure (as in patients with type I disease), the patient will have abnormal hemostasis. This type has parallel deficiencies in all VWF functions including the decrease in factor VIII coagulant activity, as outlined in Table 1.

3.2.   Type IIA von Willebrand's Disease

Minor reductions in VWF protein (as judged by total VWF antigen) can be associated with severe clinical symptoms if the protein is selectively deficient in the high M_r multimers (as in type IIA). In this case, the ability of the VWF to function as a cofactor in ristocetin-induced platelet agglutination correlates well with the clinical abnormality, whereas its ability to support *Bothrops* species venom-induced agglutination may correlate less well (Howard *et al.*, 1982; Brinkhous *et al.*, 1983). Although infusion of the vasopressin analogue, des-amino-D-arginine vasopressin (DDAVP), can increase the amount of VWF antigen in the circulation, the ristocetin cofactor activity increases very little in patients with type IIA, presumably because the additional VWF released from endothelial cells has the same deficiency in high M_r multimers (Ruggeri *et al.*, 1982b) (Figure 6). Although these observations tend to emphasize the importance of the size of the VWF, it should be noted that plasmin digestion fragments of VWF have been reported to retain considerable amounts of ristocetin cofactor activity (Guisasola *et al.*, 1978; Martin *et al.*, 1980). It is possible, of course, that such fragments aggregate by noncovalent association in the assays and reach a critical size, but there is no direct evidence for this. The type IIB VWF (Section 3.3) also behaves anomalously with regard to the size–activity hypothesis.

Table 1. Classification of von Willebrand's Disease [a]

	Type I	Type IIA	Type IIB	Type IIC	Type III
Genetic transmission	Autosomal dominant	Autosomal dominant	Autosomal dominant	Autosomal recessive	Autosomal recessive
Bleeding time	Prolonged	Prolonged	Prolonged	Prolonged	Prolonged
Crossed immunoelectrophoresis	Normal	Abnormal	Abnormal	Abnormal	Variable (mostly abnormal)
VIII:C [b]	Decreased	Decreased or normal	Decreased or normal	Normal	Markedly decreased
VIIIR:Ag	Decreased	Decreased or normal	Decreased or normal	Normal	Minute amounts
VIIIR:RCo	Decreased	Markedly decreased	Decreased or normal	Decreased	Absent
RIPA	Decreased or normal	Absent or decreased	Increased	Decreased	Absent
Multimeric structure	Normal in plasma and platelets	Absence of large and intermediate multimers from plasma and platelets	Absence of only larger multimers from plasma; normal in platelets	Absence of larger multimers from plasma and platelets; triplet structure is aberrant	Variable

[a] From Zimmerman and Ruggeri (1983) with permission.
[b] Abbreviations: VIII:C, factor VIII coagulant activity; VIIIR:Ag, factor VIII-related antigen (VWF antigen); VIIIR:RCo, ristocetin cofactor activity using normal platelets; RIPA, ristocetin-induced platelet agglutination/aggregation of patient's platelet-rich plasma.

3.3. Type IIB von Willebrand's Disease

The poorly understood type IIB variant is associated with a plasma VWF molecule that binds more readily to platelets in the presence of ristocetin than normal VWF, despite its deficiency of the high M_r multimers. The total amount of ristocetin cofactor activity (measured at high ristocetin concentrations) is usually mildly decreased and VIIIR:Ag values parallel the ristocetin cofactor levels. In contrast to type IIA disease, where the platelet and plasma VWF show essentially the same defect, the multimeric structure of platelet VWF in type IIB is essentially normal (Figure 7). Current speculation centers on there being a qualitative abnormality of the VWF in this disorder that permits ready interaction with platelets such that there is depletion of the high M_r multimers (which would be expected to bind most readily) from the plasma onto the platelets (Figure 2). The smaller multimers that remain in the plasma apparently do not function well *in vivo*, despite their increased interaction with ristocetin-treated platelets. Support for there being a process that results in the depletion of high M_r multimers comes from studies showing rapid removal of the high M_r multimers released into the circulation by DDAVP (Ruggeri *et al.*, 1982b) (Figure 8), and the recent report that DDAVP therapy can actually induce thrombocytopenia in type IIB

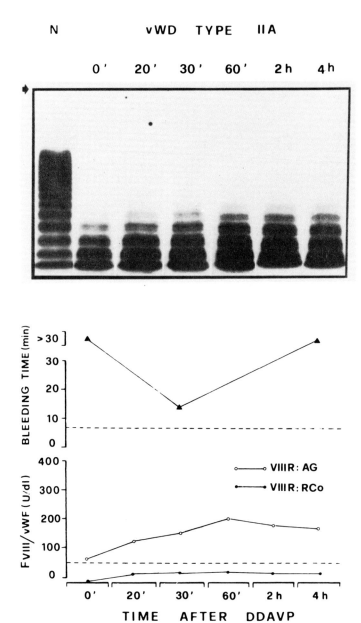

Figure 6. (Top) Autoradiograph pattern of plasma factor VIII/VWF from a patient with VWD type IIA, before (time 0) and at various times after infusion of DDAVP. The pattern of normal plasma (N) is shown for comparison on the extreme left lane. (Bottom) Quantitative changes of factor VIII/VWF and of the bleeding time observed at the same times. From Ruggeri *et al.* (1982) with permission.

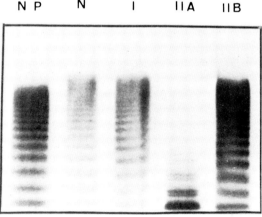

Figure 7. Autoradiograph pattern of factor VIII/VWF electrophoresed in 1.4% agarose in the presence of SDS and detected by reaction with ^{125}I-labeled affinity-purified antibody. Plasma factor VIII/VWF from a normal individual (NP) is shown on the left and compared with platelet factor VIII/VWF from a normal (N) and from patients with VWD type I, IIA, and IIB. The arrow indicates the origin of the running gel and the anode is at the bottom of the gel. From Ruggeri *et al.* (1982) with permission.

patients, presumably as a result of *in vivo* agglutination of platelets caused by the high M_r multimers (Holmberg *et al.*, 1983). Autologous platelet survival was normal in one patient studied under basal conditions (Holmberg *et al.*, 1984).

3.4. Abnormalities of VWF Carbohydrate Composition in von Willebrand's Disease

The role of carbohydrate in VWF platelet cofactor function is controversial. Gralnick *et al.* (1976) reported on three patients with a variant of VWD who had nearly normal levels of VWF antigen, but very low levels of ristocetin cofactor activity. Their purified VWF contained a molecular subunit that was indistinguishable from normal by protein stain, but the subunit did not stain with the carbohydrate-specific, periodic acid-Schiff (PAS) stain. In another study, this same group (Gralnick *et al.*, 1977) found a 60% decrease in the sialic acid associated with the VWF of two patients with type IIA VWD. Indirect support for a carbohydrate abnormality came from the observation that concanavalin A did not precipitate the VWF from some patients with variant VWD as well as normal VWF (Peake and Bloom, 1977; Howard *et al.*, 1979). De Marco and Shapiro (1981) reported that the VWF from one of two patients with type IIA VWD had a 50% decrease in sialic acid, whereas Zimmerman *et al.* (1979) could only find a decrease in PAS staining of VWF in 1 of 13 patients studied, and that abnormal patient was one previously reported to be abnormal by Gralnick. *In vitro* studies of the effect of carbohydrate removal have been equally controversial. Sodetz *et al.* (1977) reported that desialylation of VWF resulted in a 65% decrease in ristocetin cofactor activity, whereas Vermylen *et al.* (1976), Gralnick (1978), and De Marco and Shapiro (1981) found no effect on ristocetin cofactor activity. Desialylated VWF has the interesting property of being able to cause aggregation of platelets directly in the presence of fibrinogen and calcium (Vermylen *et al.*, 1976; De Marco and Shapiro, 1981; Gralnick *et al.*, 1985). Since the platelets of patients with Bernard-Soulier

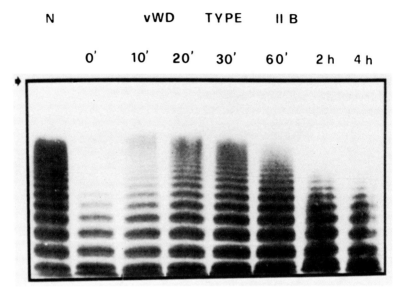

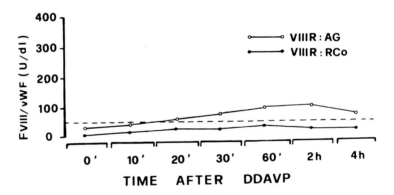

Figure 8. (Top) Autoradiograph pattern of plasma factor VIII/VWF from a patient with VWD type IIB, before (0 time) and at various times after infusion of DDAVP. The pattern of normal plasma (N) is shown for comparison on the extreme left lane. (Bottom) Quantitative changes of factor VIII/VWF observed at the same times. The bleeding time was not measured in this patient because of a concomitant thrombocytopenia. From Ruggeri *et al.* (1982) with permission.

syndrome do not aggregate normally with asialo-VWF (De Marco and Shapiro, 1981), it is likely that the asialo-VWF binds to GPIb. The requirements for calcium and fibrinogen suggest that the mechanism of platelet aggregation involves asialo-VWF-induced exposure of receptors for fibrinogen, perhaps via induction of the release reaction (Vermylen *et al.*, 1976). Studies with monoclonal antibodies confirm these observations (Gralnick *et al.*, 1985). The physiologic role for asialo-VWF remains undefined. The penultimate (but not ultimate) galactose residues have been reported to be crucial in VWF's ristocetin cofactor activity (Gralnick, 1978; Sodetz *et al.*, 1978;

Furlan *et al.*, 1979; Goudemand *et al.*, 1983), and preliminary evidence (Gralnick *et al.*, 1983) indicates that the penultimate galactose may be required for the maintenance of the normal multimeric structure. On the other hand, in a recent study using a potent endoglycosidase, there was no significant decrease in ristocetin cofactor activity or selective loss of high M_r multimers with the removal of more than 80% of the total VWF carbohydrate (Federici *et al.*, 1983).

3.5. Abnormalities of VWF Multimer Formation

The increased resolution of the fine structure of the VWF multimers will probably result in subdivision of the already-recognized categories. Thus, the one patient with type IIC VWD (Ruggeri *et al.*, 1982c) would have been included in the type IIA category if the high-resolution multimer technique were not available, and preliminary reports of subclassification of patients who might otherwise have been classified as type IIA have already appeared (Weiss *et al.*, 1983; Enayat and Hill, 1983; Kinoshita *et al.*, 1983; Mannucci *et al.*, 1983). Until the molecular basis for the fine structure is better understood, it will be difficult to evaluate the significance of this subdivision.

3.6. Pseudo-von Willebrand's Disease

In addition to the disorders considered above, in which the abnormal clinical hemorrhage is ascribed to an abnormal VWF protein, there have been several reports of hemorrhagic diatheses in which a platelet abnormality, characterized by excessive binding of the high M_r multimers of VWF, resulted in both reduced plasma levels of VWF and a predominance of the low M_r multimers (Takahashi, 1980; Weiss *et al.*, 1982; Gralnick *et al.*, 1982; Miller and Castella, 1982; Sakariassen *et al.*, 1983). This disorder has been termed "platelet-type" VWD or "pseudo-VWD." The patients have increased ristocetin-induced platelet agglutination at low concentration of ristocetin, along with variable degrees of thrombocytopenia and increased platelet size. In some of these patients, the addition of cryoprecipitate or purified VWF containing the high M_r multimers to platelet-rich plasma can actually induce platelet aggregation, a test that is useful in distinguishing this disorder from type IIB VWD (Weiss *et al.*, 1902; Miller *et al.*, 1983). This aggregation correlates with the binding of normal unmodified VWF to unstimulated platelets from patients with pseudo-VWD (Miller *et al.*, 1983). Infusion of cryoprecipitate into one such patient was followed by rapid depletion of the highest M_r multimers of VWF and the coincident development of thrombocytopenia (Miller *et al.*, 1984). Similarly, an infusion of DDAVP can cause spontaneous platelet aggregation and thrombocytopenia in these patients (Takahashi *et al.*, 1984).

4. VON WILLEBRAND FACTOR-DEPENDENT PLATELET FUNCTION

Platelet function is conveniently separated into adhesion, defined as the attachment of a platelet to something other than a platelet, and agglutination/aggregation (cohesion), defined as the attachment of a platelet to another platelet. *In vitro* and *ex vivo,* one can show a role for VWF in each of these processes. Based on data from *ex vivo* models, most current reviews emphasize VWF's role in platelet adhesion, but I do not believe that the available evidence excludes the possibility that VWF's major *in vivo* role is in platelet–platelet interactions (agglutination/aggregation).

4.1. Platelet Retention

In 1963, Zucker and Salzman independently reported that platelets from patients with VWD are retained in columns of glass beads less well than normal platelets. O'Brien (1967) showed that the blood had to be passed through the column rapidly in order to observe the defect in VWD, an observation that presaged the shear-rate-dependent abnormality subsequently found in the *ex vivo* systems (see Section 4.3). Other studies showed that cryoprecipitate and purified VWF could correct the defect (Bouma *et al.,* 1972; Weiss and Rogers, 1972). Although originally called a test of adhesiveness, in fact, this assay involves both adhesion of platelets to the glass and subsequent platelet aggregation. McPherson and Zucker (1976) dissected the retention test into its adhesion and aggregation components with platelet and plasma reconstitution experiments, and concluded that the VWF played a role in the platelet–platelet interactions and not in platelet adhesion. This is consistent with the observation that the direct adhesion to glass of platelets in VWD plasma is normal (George, 1972).

4.2. Platelet Agglutination and Aggregation

4.2.1. General Considerations

As a first approximation, one can consider platelets as any other particle in suspension and apply the theory of colloidal stability to predict their behavior (Pethica, 1961; Brooks *et al.,* 1967). In brief, this theory emphasizes the balance between attractive and repulsive forces. The attractive forces tend to operate only over a short range and are of diverse kinds (e.g., van der Waals dipole and dispersion forces, ion pair or triplet formation, etc.), whereas the repulsive forces operate over a longer range, are electrostatic in nature, and reflect the particle's surface charge. Thus, when particles stay apart, it is because the long-range repulsive forces are sufficiently strong to prevent the particles from getting close enough for the short-range attractive forces to operate. In fact, platelets do have an excess of negative charges on their surface as judged by their electrophoretic behavior (Seaman, 1976; Coller, 1983), and this is probably responsible for the generation of sufficient electrostatic repulsion to keep platelets from spontaneously agglutinating. Colloidal stability theory identifies two major mechanisms for altering the balance of forces in favor of attraction, namely, (1) reducing the electrostatic repulsion, or (2) interposing a large molecule that can bridge between particles that are being held at a distance by the repulsive forces. As indicated in Section 4.2.2, there is evidence that VWF-dependent platelet agglutination/aggregation involves both of these mechanisms.

A note on terminology is in order. I use the word agglutination to describe the attachment of platelets to each other by a process that is independent of platelet metabolism. The term emphasizes a more passive role for the platelet, but the process is platelet specific in that it requires an intact platelet membrane receptor mechanism. Operationally, this is often investigated by gently fixing platelets with low concentrations of formaldehyde. The word aggregation implies that the platelet must actively change the character of its membrane after agonist stimulation in a process that involves energy metabolism. Although it risks being an oversimplification, in general, GP Ib appears to be responsible for VWF-dependent agglutination phenomena, whereas GP IIb-IIIa appears to be responsible for aggregation phenomena that involve either VWF or fibrinogen. Each process may be going on simultaneously, as for example when ristocetin is added to platelet-rich plasma. In this case, the initial interaction between VWF and the platelet is an agglutination-type reaction, but this agglutination can itself trigger arachidonic acid metabolism and release of ADP. The latter, in turn, can induce a true aggregation response by exposing the receptor on GP IIb-IIIa that binds fibrinogen and, under the appropriate circumstance, VWF.

4.2.2. Ristocetin-Induced Agglutination

In 1971, Howard and Firkin observed that the antibiotic ristocetin induced platelet agglutination/aggregation in normal platelet-rich plasma, but much less or no agglutination/aggregation in the platelet-rich plasma of some patients with VWD. The defect could be corrected with normal plasma or purified VWF (Weiss and Rogers, 1972), thus permitting it to form the basis of an assay for this ristocetin cofactor function of VWF. Observations on patients with Bernard-Soulier syndrome, whose platelets cannot bind VWF when stimulated with ristocetin (Zucker et al., 1977; Moake et al., 1980) and platelets treated with proteases (Phillips, 1980; Yoshida et al., 1983) pointed to GP Ib as the binding site for VWF when platelets are stimulated with ristocetin. Immunologic studies with alloimmune (Tobelem et al., 1976; Degos et al., 1977), heterologous (Nachman et al., 1977a), and most recently, monoclonal (Mc-Michael et al., 1981; Ruan et al., 1981; Coller et al., 1983) antibodies to GP Ib add considerable support to this identification (Figure 9).

The mechanism by which ristocetin induces VWF-dependent platelet agglutination is only partially understood. Multiple lines of evidence have pointed to a primary role for electrostatic interactions (Evensen et al., 1974; Howard, 1975; Coller et al., 1976; Coller and Gralnick, 1977; Muraki, 1977; Coller, 1978). My current working hypothesis can be summarized as follows (Figure 10):

1. Platelets are normally held apart by electrostatic repulsion due to an excess of free carboxyl groups.
2. The platelet has a membrane receptor for VWF, but significant binding does not occur unless ristocetin is present.
3. Ristocetin is positively charged at neutral pH and can selectively bind to the platelet surface; this decreases the electrostatic repulsion between platelets and appears to permit the binding of VWF to its receptor (Coller et al., 1975; Coller and Gralnick, 1977).
4. The bound VWF facilitates agglutination by both acting as a bridge between platelets and decreasing the platelet's surface charge (Evensen et al., 1974; Coller, 1978).

A

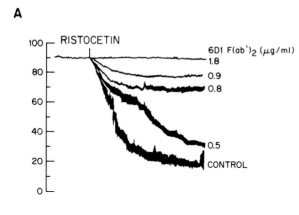

B

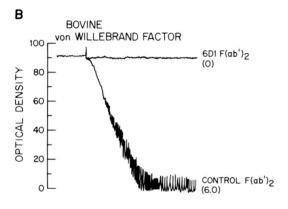

C

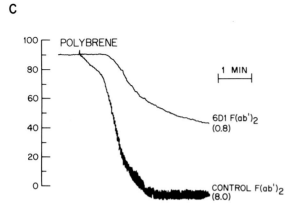

Figure 9. Effect of F(ab′)₂ fragments of a monoclonal antibody to GP Ib, 6DI, and a control monoclonal antibody on platelet aggregation. (A) Ristocetin-induced platelet aggregation. Platelet-rich plasma (0.4 ml, 3×10^{11} platelets/liter) was incubated with buffer (control) or 6DI F(ab′)₂ fragments (1–10 1) for 5 min at 22 °C, after which the cuvette was placed in the aggregometer and a baseline established at 90 optical density units (the zero baseline being set with platelet-poor plasma). At the indicated point, ristocetin (1.8 mg/ml final concentration) was added. In other experiments it was shown that control F(ab′)₂ fragments did not inhibit ristocetin-induced aggregation. (B) Bovine VWF-induced platelet aggregation. Platelet-rich plasma (0.45 ml, 3×10^{11} platelets/liter) and the ATP luminescence reagent (Chronolume, 50 μl) were incubated with 16 μl of control F(ab′)₂ (10 μg/ml final concentration) or 6DI F(ab′)₂ (2.5 μg/ml) for 1 min. At the indicated point, a crude preparation of bovine VWF (1 mg/ml), was added and both aggregation and ATP release were monitored. After the aggregation and release responses in each sample remained unchanged for 2 min, 1 μl of 1 mM ATP was added as an internal standard. The numbers in parentheses indicate the peak release of ATP in nmole/10^9 platelets. For comparison, thrombin (1 U/ml) produces peak release of ~13 nmole of ATP/10^9 platelets under similar circumstances. (C) Polybrene-induced aggregation. Conditions were as indicated in (B) with Polybrene (1.2 mg/ml final concentration) added at the indicated time. The numbers in parentheses indicate the peak release of ATP in nmole/10^9 platelets. From Coller *et al.* (1983) with permission.

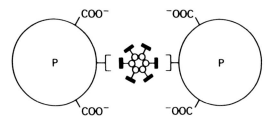

Normal VWF — No Ristocetin

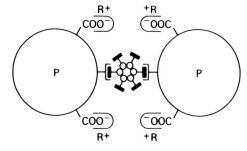

Figure 10. A hypothetical model of ristocetin-induced platelet agglutination. Abbreviations: P, platelet; VWF, von Willebrand factor; R, ristocetin. From Coller (1978) with permission.

Normal VWF + Ristocetin

A series of biochemical studies on ristocetin and vancomycin, an antibiotic similar to ristocetin that behaves as an inhibitor of ristocetin-induced agglutination, supported the requirements for ristocetin to have a positive charge and an intact phenolic group (Coller and Gralnick, 1977; Moake *et al.,* 1977; Boda *et al.,* 1979). Further support for ristocetin's mechanism of action relying on its cationic properties was derived from studies showing that polycations such as Polybrene (hexadimethrine bromide), polylysine, and polyornithine, which have no structural similarity to ristocetin, produce more extensive agglutination of platelets in the presence of VWF than in its absence (Rosborough and Swaim, 1978; Coller, 1980; Rosborough, 1980). Polybrene was also shown to affect platelet electrophoretic mobility in a manner similar to ristocetin and the agglutination Polybrene induces could be inhibited by vancomycin, lending further support to the similarity between the ristocetin and Polybrene phenomena (Coller, 1980). In fact, it has been suggested that any process causing the surface membranes of platelets to become spatially close can allow VWF binding (Senogles and Nelsestuen, 1983). However, not all maneuvers that decrease platelet surface charge result in binding of VWF and platelet agglutination (Coller, 1983). Thus, there must be some common specificity in the effects of the various compounds.

If one assumes that the charge neutralization model reflects the basic mechanism whereby platelets bind VWF to GP Ib, one can speculate on how such a mechanism may operate *in vivo.* Among the possibilities are: (1) release of a cationic substance from damaged blood vessels or hematopoietic cells, and (2) alteration in surface charge produced by the adhesion of a platelet to a surface by charge redistribution or activation of an enzyme that affects surface charge. Alternatively, VWF already bound to a solid surface (e.g., vessel wall or dialysis membrane) may be altered to allow direct interaction with the GP Ib of unstimulated platelets (Sakariassen *et al.*, 1979; Olson *et al.*, 1983; see Chapter 1).

The binding of VWF to ristocetin-stimulated platelets has been quantified by several groups with techniques that use either radiolabeled VWF or immunologic assays (Zucker *et al.*, 1977; Kao *et al.*, 1979; Morisato and Gralnick, 1980; Moake *et al.*, 1980). There is agreement that little or no VWF binds in the absence of ristocetin and that increasing amounts bind with increasing concentrations of ristocetin, reaching a saturation value of ~6.2–8.5 $\times$ g/10^8 platelets (reviewed by Kirby, 1982). Although it is tempting to convert this into molecules of VWF bound, uncertainties in the M_r of the VWF make only a rough estimate possible. Assuming an average M_r of 4×10^6, approximately 10,000 molecules are bound per platelet. This can be compared with the ~25,000 GP Ib sites recognized by monoclonal antibodies (McMichael *et al.*, 1981; Coller *et al.*, 1983). If, as the electron microscopy studies suggest, VWF is a flexible molecule, it is possible that a single VWF molecule could bind to several GP Ib molecules. It should be appreciated, however, that agglutination can occur when platelets bind only a fraction of the saturating amount, perhaps as little as 5–10%. The lack of uniformity of M_r and the different binding behavior of the VWF multimers makes it impossible to calculate a reliable binding affinity.

It is clear that alterations of the platelet membrane can modify the agglutination response mediated by VWF. Thus, Grant *et al.* (1976) found that ADP altered the platelet membrane in a manner that decreased VWF-dependent platelet agglutination. Agents that increase platelet cAMP also decrease VWF-dependent platelet agglutination by an effect on the platelet membrane (Coller, 1981; Moake *et al.*, 1981). This inhibition could be prevented by preincubation of platelets with agents that disrupt the platelet's cytoskeleton (Coller, 1981). Interestingly, even though the agents that increase cAMP prevented the decrease in platelet electrophoretic mobility associated with VWF binding (Coller, 1981), a study in which VWF binding was quantified did not find these agents to decrease VWF binding (Moake *et al.*, 1981). Brief incubation of platelets with tertiary amine local anesthetics, agents known to have diverse effects on platelets, including inhibition of the release reaction and induction of a spheroidal shape, caused an increase in VWF-dependent platelet agglutination despite decreasing the binding of VWF (Coller, 1982). More prolonged incubation with these agents led to decreased VWF-dependent agglutination, and this was found to correlate with digestion of GP Ib by a calcium-dependent protease(s) (Coller, 1982; Solum *et al.*, 1983). Thus, VWF-dependent agglutination is a complex phenomenon that is affected by both the amount of VWF bound to the platelet and the intrinsic agglutinability of the platelet. The latter is a function of membrane factors that are still poorly understand, but which may be affected by the level of platelet cAMP operating through a change in cytoskeletal elements.

4.2.3. Thrombin- and ADP-Induced Aggregation

Recent studies demonstrate that VWF binds to washed platelets stimulated with thrombin or ADP (Fujimoto *et al.*, 1982; Fujimoto and Hawiger, 1982). Unlike the ristocetin phenomenon, however, these are true aggregation responses, requiring metabolically intact platelets and calcium ions. Studies using platelets from patients with thrombasthenia and monoclonal antibodies specific for GP Ib and GP IIb-IIIa established that the VWF was not binding to GP Ib when platelets were stimulated with these agonists. In fact, virtually all of the ADP-induced VWF binding and the majority of the thrombin-induced VWF binding involves GP IIb-IIIa (Ruggeri *et al.*, 1982a, 1983b; Gralnick *et al.*, 1984). Although the initial report indicated that fibrinogen did not compete with VWF for binding to GP IIb-IIIa when platelets were stimulated with thrombin (Fujimoto *et al.*, 1982), subsequent studies have indicated that fibrinogen can compete effectively (Gralnick *et al.*, 1984) and that VWF does not bind to ADP or epinephrine-stimulated platelets in a plasma milieu (Schullek *et al.*, 1984). In fact, VWF and fibrinogen, as well as thrombospondin and fibronectin, appear to share the identical receptor on thrombin-activated platelets (Plow *et al.*, 1984; Ginsberg *et al.*, 1984). These observations are vital in assessing the physiologic significance of the phenomenon since in plasma there is much more fibrinogen than VWF.

4.2.4. Venom-Induced Platelet Agglutination

A variety of *Bothrops* species venoms contain a factor that causes VWF-dependent platelet agglutination of fresh or fixed platelets (Brinkhous *et al.*, 1983). As indicated above, the agglutination appears to differ from that induced by ristocetin in that multimers of lower M_r can support venom-induced agglutination, but not ristocetin-induced agglutination. Thus, the agglutination correlates better with total VWF antigen (Brinkhous *et al.*, 1983), but it may not correlate as well with clinical hemorrhage (Howard *et al.*, 1983). Bernard-Soulier platelets agglutinate only about one third as well as control platelets when stimulated with the purified venom factor, whereas thrombasthenic platelets agglutinate normally (Howard *et al.*, 1983). Thus, it appears that the VWF binds in part to GP Ib and in part to some other site on the platelet membrane (but not GP IIb-IIIa). The venom factor differs from ristocetin in two other ways: (1) vancomycin does not inhibit venom-induced agglutination, and (2) an association between the venom factor and VWF can be readily demonstrated by gel exclusion chromotography (Howard *et al.*, 1983). Thus, the venom factor may induce agglutination by altering the VWF, whereas ristocetin appears to act by altering the platelet.

4.2.5. Bovine and Porcine von Willebrand Factor-Induced Agglutination

In the 1950s, it was reported that preparations of bovine factor VIII produced thrombocytopenia when infused into patients and could aggregate platelets *in vitro* (Macfarlane *et al.*, 1954; Sharp and Bidwell, 1957). After some initial confusion about whether the agglutinating activity was bovine fibrinogen, bovine factor VIII, or bovine VWF (Caen *et al.*, 1966; Solum 1968; Forbes *et al.*, 1972), the identification of the activity as bovine VWF was established (Schmer *et al.*, 1972; Griggs *et al.*, 1973; Kirby and Mills, 1975). Biochemically, bovine VWF is very similar to human VWF except in having approximately four times as much carbohydrate on a weight basis (Schmer *et al.*, 1972; Bottecchia and Vermylen, 1976). Bovine VWF induces agglutination of fresh or fixed platelets in a manner that closely resembles the agglutination induced by ristocetin in the presence of human VWF, but ristocetin is not required. An interesting phenomenon induced by high concentrations of bovine VWF is "super-aggregation," wherein fresh but not fixed platelets form huge clumps (Kirby and Mills, 1975; Kirby et al., 1982). Paradoxically, platelets undergo more release of dense body contents when agglutinated/aggregated with lower concentrations of bovine VWF than when superaggregated with larger doses (Kirby and Mills, 1975; Kirby et al., 1982). As with human VWF, the high M_r multimers of bovine VWF are able to cause platelet agglutination much more readily than the lower M_r multimers (Kirby, 1981). Studies with the platelets of patients with thrombasthenia (Caen et al., 1966), Bernard-Soulier syndrome (Bithell et al., 1972; Howard et al., 1973), and with platelets treated with proteases (Kirby, 1977) suggested that bovine VWF binds to GP Ib. Additional evidence came from studies of the inhibition of ristocetin-induced binding of human VWF to platelets by bovine VWF (Kao et al., 1979) and monoclonal antibodies to GP Ib (Ruan et al., 1981; Coller et al., 1983).

Porcine VWF has many of the features of bovine VWF, but it is less potent than bovine VWF in inducing agglutination of human platelets (Forbes et al., 1972). It also has the multimeric structure found in bovine and human VWF (Fass et al., 1978).

4.3. Ex Vivo Models of Platelet Adhesion to the Vessel Wall

In an attempt to define better the interaction of platelets with the complex suben-dothelial surface of blood vessels, Baumgartner (1973) devised a technique wherein a rabbit's aorta is de-endothelialized in vivo with the aid of a balloon catheter after which it is removed and everted segments are mounted in a flow chamber. Blood is pumped through the chamber at a predetermined rate for a period of time, after which the segment is fixed and the extent of adhesion and aggregation determined morphometri-cally (Baumgartner, 1973). This technique can demonstrate a defect in the adhesion of platelets from the citrated blood of patients with VWD, but only when the shear rate in the system is relatively high (> 650 s^{-1}), corresponding to shear rates found in arteries and small blood vessels (Tschopp et al., 1974). When native blood is used, even higher shear rates (> 1300 s^{-1}) are required to demonstrate the defect and these rates correspond to those found only in small blood vessels (Weiss et al., 1978; Baumgartner et al., 1980). The shear-rate dependence of the defect suggests that at low shear rates, where the residence time between platelets and the subendothelium is relatively long, a VWF-independent adhesion mechanism can operate, most likely related to collagen. Although the idea that VWF's primary effect is on platelet adhe-sion was derived, in large part, from this model, more recent studies using essentially the same technique demonstrated a defect in platelet–platelet interactions with VWD blood (or normal blood treated with an antibody to VWF) in addition to the adhesion abnormality (Baumgartner et al., 1980; Weiss et al., 1981). Sakariassen et al. (1979) modified the technique by substituting human renal artery for rabbit aorta and recon-stituted blood with radiolabeled components for native blood. They concluded that the binding of fluid-phase VWF to the subendothelium was a prerequisite for normal adhesion, although only minute amounts of the VWF actually bound. In a subsequent study from the same group, it was demonstrated that fluid-phase VWF also binds to platelets that have adhered to the subendothelium and that platelet spreading is enhanced by the presence of VWF (Bolhuis *et al.*, 1981), an observation in accord with the morphologic studies of Barnhart and Chen (1978). Although these studies stressed the role of fluid-phase VWF, Turitto *et al.* (1981) showed that pretreatment of denuded vessel wall segments with antibody to VWF decreased the subsequent adhesion of normal platelets. Thus, it would appear that both fluid-phase and vessel-wall VWF contribute to platelet adhesion in this model. Interestingly, a report by Sixma *et al.* (1984) suggests that the multimeric composition of VWF is of less importance in binding to subendothelium and enhancing platelet adhesion than it is in ristocetin cofactor activity. In one of the few *in vivo* studies, Reddick *et al.* (1982) found no difference between normal and VWF-deficient bleeder swine in the adhesion of platelets to the subendothelial surface of coronary arteries that were denuded *in vivo* with a balloon catheter. They did, however, find that the adherent platelets from the bleeder swine contained fewer pseudopods than the control platelets and concluded that in the coronary arteries, where the shear rate is not great, VWF makes more of a contribution to platelet activation than adhesion. Recently, Aihara *et al.* (1984) demonstrated that VWF accelerates the rate of platelet adhesion to collagen, but does not affect the total number of platelets that bind (Aihara *et al.*, 1984).

4.4. Correlation among von Willebrand Factor-Dependent Platelet Functions

Although there is reason to believe that all of the functions discussed above are carried out by VWF, it is not at all clear that they rely on the same properties of the molecule; in fact, inconsistencies, as between ristocetin- and *Bothrops* species venom-induced platelet agglutination, make this unlikely. Substantial support for the idea that different sites on the VWF molecule subserve different functions comes from studies using monoclonal antibodies to VWF. Both in the bovine (Bowie *et al.*, 1983) and human (Meyer *et al.*, 1983) systems, there was little correlation between an antibody's ability to inhibit one function and its ability to inhibit other functions.

5. VON WILLEBRAND FACTOR IN NONHEMORRHAGIC DISEASES

Historically, research on VWF has centered around its role in normal hemostasis, with the description and analysis of the deficiency states associated with hemorrhage. More recently, there has been an appreciation of the role of platelets in acute thrombosis and microembolization. Moreover, the discovery of the potent mitogen, platelet-derived growth factor, that is stored in platelets and can be released during adhesion and aggregation (Ross *et al.*, 1974) sparked interest in a possible role of platelets in chronic atherosclerosis. It was logical, therefore, to look for differences in the platelet cofactor, VWF, in these processes.

Studies on pigs with severe deficiencies of VWF have shown a significant decrease in the amount of spontaneous and diet-induced aortic atherosclerosis (Fuster *et al.*, 1978; Bowie and Fuster, 1980; Griggs *et al.*, 1981), but the protection was far from complete. When spontaneous or balloon catheter-induced coronary atherosclerosis was studied in these animals, no differences between control and bleeder swine were noted (Griggs *et al.*, 1981). Thus, although VWF appears to play some role in the development of atherosclerosis, it is not a dominant one.

Studies of patients with diseases that predispose to the development of thrombosis, such as diabetes, have shown increases in plasma levels of VWF (Coller *et al.*, 1978), but the interpretation of such data is seriously complicated by the observation that VWF antigen is consistently increased in many diseases, including solid tumors, leukemia, liver disease, uremia, myocardial infarction, burns, and venous thrombosis, as well as in the postoperative state (Holmberg and Nilsson, 1979). Thus, although one could argue that many or all of these conditions are associated with an increased risk of thrombosis, the ubiquity of the increase means that other factors are likely to be of greater import. Since VWF appears to be made in endothelial cells, it is, perhaps, not surprising that any disease characterized by endothelial damage could lead to elevated plasma levels. Although only prospective studies can help to decide whether increases in VWF cause or simply reflect thrombotic disease, it seems reasonable to suggest that VWF may be involved in the development of a vicious circle wherein tissue injury results in increases in VWF, which, in turn, may contribute to further vascular disease.

The recent demonstration of significant homology between the platelet-derived growth factor and a simian oncogene (Waterfield *et al.*, 1983; Doolittle *et al.*, 1983) raises the intriguing possibility that a platelet cofactor such as VWF may have a modulating effect on malignancy. This exciting hypothesis, and each of the other hypotheses discussed in this brief review, awaits the availability of additional data.

ACKNOWLEDGMENTS.   Supported, in part, by grants from the National Heart, Lung and Blood Institute (HL19278) and by a Grant-in-Aid from the American Heart Association and with funds contributed in part by the American Heart Association, Suffolk County Chapter, New York.

REFERENCES

Aihara, M., Cooper, H. A., and Wagner, R. H., 1984, Platelet-collagen interactions: Increase in rate of adhesion of fixed washed platelets by factor VIII-related antigen, *Blood* **63**:495–501.

Barnhart, M. I., and Chen, S-T., 1978, Vessel wall models for studying interaction capabilities with blood platelets, *Semin. Thromb. Hemost.* **5**:112–155.

Baumgartner, H. R., 1973, The role of blood flow in platelet adhesion, fibrin deposition and formation of mural thrombi, *Microvasc. Res.* **5**:167–179.

Baumgartner, H. R., Turitto, V. T., and Weiss, H. J., 1980, Effect of shear rate on platelet interaction with subendothelium in citrated and native blood. II. Relationships among platelet adhesion, thrombus dimensions, and fibrin formation, *J. Lab. Clin. Med.* **95**:208–221.

Berndt, M. C., Gregory, C., Castaldi, P. A., and Zola, H., 1983, Purification of human platelet membrane Ib complex using a monoclonal antibody, *Thromb. Haemostasis* **50**:361.

Bithell, T. C., Parekh, S. J., and Strong, R. R., 1972, Platelet-function studies in the Bernard-Soulier syndrome, *Ann. N.Y. Acad. Sci.* **201**:145–160.

Boda, A., Solum, N. O., Sztaricska, R., and Kalman, R., 1979, Study of platelet agglutination induced by the antibiotics of the vancomycin group: Ristocetin, ristomycin, actinoidin and vancomycin, *Thromb. Haemostasis* **42**:1164–1180.

Bolhuis, P. A., Sakariassen, K. S., Sander, H. J., Bouma, B. N., and Sixma, J. J., 1981, Binding of factor VIII-von Willebrand factor to human arterial subendothelium precedes increased adhesion and enhances platelet spreading, *J. Lab. Clin. Med.* **97**:568–576.

Bottecchia, D., and Vermylen, J., 1976, Factor VIII and human platelet aggregation. I. Evidence that the platelet aggregating activity of bovine factor VIII is a property of its "carrier protein" subunit, *Br. J. Haematol.* **34:**303–320.

Bouma, B. N., Wiegernich, Y., and Sixma, J. J., 1972, Immunological characterization of purified anti-haemophilic factor (factor VIII) which corrects abnormal platelet retention in von Willebrand's disease, *Nature (New Biol.)* **236:**104–106.

Bowie, E. J. W., and Fuster, V., 1980, Resistance to atherosclerosis in pigs with von Willebrand's disease, *Acta Med. Scand.* (Suppl.) **642:**121–130.

Bowie, E. J. W., Fass, D. N. and Katzman, J. A., 1983, Functional studies of Willebrand factor using monoclonal antibodies, *Blood* **62:**146–151.

Brinkhous, K. M., Read, M. S., Fricke, W. A., and Wagner, R. H., 1983, Botrocetin (venom coagglutinin): Reaction with a broad spectrum of multimeric forms of factor VIII macromolecular complex, *Proc. Natl. Acad. Sci. U.S.A.* **80:**1463–1466.

Brooks, D. E., Millar, J. S., Seaman, G. V. F., and Vassar, P. S., 1967, Some physicochemical factors relevant to cellular interactions, *J. Cell Physiol.* **69:**155–168.

Caen, J. P., Castaldi, P. A., Leclerc, J. C., Inceman, S., Larrieu, M-J., Probst, M., and Bernard, J., 1966, Congenital bleeding disorders with long bleeding time and normal platelet count. I. Glanzmann's thrombasthenia (report of fifteen patients), *Am. J. Med.* **41:**4–26.

Chong, B. H., Berndt, M. C., Koutts, J., Castaldi, P. A., 1983, Quinidine-induced thrombocytopenia and leukopenia: Demonstration and characterization of the quinidine-dependent antiplatelet and anti-leukocyte antibodies, *Blood* **62:**1218–1223.

Clemetson, K. J., McGregor, J. L., James, E., Dechavanne, M., and Luscher, E. F., 1982, Characterization of the platelet membrane glycoprotein abnormalities in Bernard-Soulier syndrome and comparison with normal by surface-labeling techniques and high-resolution two-dimensional gel electrophoresis, *J. Clin. Invest.* **70:**304–311.

Coller, B. W., 1978, The effects of ristocetin and von Willebrand factor on platelet electrophoretic mobility, *J. Clin. Invest.* **61:**1168–1175.

Coller, B. S., 1980, Polybrene-induced platelet agglutination and reduction in electrophoretic mobility: Enhancement by von Willebrand factor and inhibition by vancomycin, *Blood* **55:**276–281.

Coller, B. S., 1981, Inhibition of von Willebrand factor-dependent platelet function by increased platelet cyclic AMP and its prevention by cytoskeleton disrupting agents, *Blood* **57:**846–855.

Coller, B. S., 1982, Effects of tertiary amine local anesthetics on von Willebrand factor-dependent platelet function: Alteration of membrane reactivity and degradation of GP Ib by a calcium-dependent protease(s), *Blood* **60:**731–743.

Coller, B. S., 1983, Biochemical and electrostatic considerations in primary platelet aggregation, *Ann. N.Y. Acad. Sci.* **416:**693–708.

Coller, B. S., 1984a, Von Willebrand's disease, in: *Disorders of Hemostasis* (O. D. Ratnoff and C. D. Forbes, eds.), Grune and Stratton, New York, pp. 241–269.

Coller, B. S., 1984b, Platelet disorders, in: *Disorders of Hemostasis* (O. D. Ratnoff and C. D. Forbes, eds.), Grune and Stratton, New York, pp. 73–176.

Coller, B. S., and Gralnick, H. R., 1977, Studies on the mechanism of ristocetin-induced platelet agglutination: Effects of structural modification of ristocetin and vancomycin, *J. Clin. Invest.* **60:**302–312.

Coller, B. S., Hirschman, R. J., and Gralnick, H. R., 1975, Studies on the factor VIII/von Willebrand factor antigen on the platelet surface, *Thromb. Res.* **6:**469–480.

Coller, B. S., Franza, B. R., Jr., and Gralnick, H. R., 1976, The pH dependence of quantitative ristocetin-induced platelet aggregation: Theoretical and practical implications. A new device for maintenance of platelet-rich plasma pH, *Blood* **47:**841–854.

Coller, B. S., Frank, R. N., Milton, R. C., and Gralnick, H. R., 1978, Plasma cofactors of platelet function: Correlation with diabetic retinopathy and hemoglobial A_{1a-c}. Studies in diabetic patients and normal persons, *Ann. Intern. Med.* **88:**311–316.

Coller, B. S., Peerschke, E. I., Scudder, L. E., and Sullivan, C. A., 1983, Studies with a murine nomoclonal antibody that abolishes ristocetin-induced binding of von Willebrand factor to platelets: Additional evidence in support of GP Ib as a platelet receptor for von Willebrand factor, *Blood* **61:**99–110.

Counts, R. B., Paskell, S. L., and Elgee, S. K., 1978, Disulfide bonds and the quaternary structure of factor VIII/von Willebrand factor, *J. Clin. Invest.* **62:**702–709.

De Marco, L., and Shapiro, S. S., 1981, Properties of human asialo-factor VIII. A ristocetin-independent platelet-aggregating agent, *J. Clin. Invest.* **68:**321–328.

Degos, L., Tobelem, G., Lethielleaux, P., Levy-Toledano, S., Caen, J., and Colombani, J., 1977, Molecular defect in platelets from patients with Bernard-Soulier syndrome, *Blood* **50:**899–903.

Doolittle, R. F., Humkapiller, M. W., Hood, L. E., Devare, S. G., Robbins, K. C., Aaronson, S. A., and Antoniades, H. N., 1983, Simian sarcoma virus *onc* gene, *v-sis,* is derived from the gene (or genes) encoding a platelet-derived growth factor, *Science* **221:**275–277.

Doucet-de Bruine, M. H. M., Sixma, J. J., Over, J., and Beeser-Visser, N. H., 1978, Heterogencity of human factor VIII. II. Characterization of forms of factor VIII binding to platelets in the presence of ristocetin, *J. Lab. Clin. Med.* **92:**96–107.

Enayat, S. M., and Hill, F. G. H., 1983, Subclassification of Type I$_A$ von Willebrand's disease patients using multimeric analysis, *Thromb. Haemostasis* **50:**32.

Evensen, S. A., Solum, N. O., Grottum, K. A., and Hovig, T., 1974, Familial bleeding disorder with moderate thrombocytopenia and giant blood platelet, *Scand. J. Haematol.* **13:**203–214.

Fass, D. N., Knutson, G. E., and Bowie, E. J. W., 1978, Porcine Willebrand factor: A population of multimers, *J. Lab. Clin. Med.* **91:**307–320.

Fauvel, F., Grant, M. E., Legrand, Y. J., Souchon, H., Tobelem, G., Jackson, D. S., and Caen, J. P., 1983, Interaction of blood platelets with a microfibriller extract from adult bovine aorta: Requirement for von Willebrand factor, *Proc. Natl. Acad. Sci. U.S.A.* **80:**551–554.

Federici, A. B., Elder, J. H., Ruggeri, Z. M., and Zimmerman, T. S., 1983, Role of carbohydrate in von Willebrand factor structure and function: Studies with a new specific endoglycosidase, *Thromb. Haemostasis* **50:**318.

Fernanda Lopez Fernandez, M., Ginsberg, M. H., Ruggeri, Z. M., Battlle, F. J., and Zimmerman, T. S., 1982, Multimeric structure of platelet factor VIII/von Willebrand factor: The presence of larger multimers and their reassociation with thrombin-stimulated platelets, *Blood* **60:**1132–1138.

Forbes, C. D., Ban, R. D., McNicol, G. P., and Douglas, A. S., 1972, Aggregation of human platelets by commercial preparations of bovine and porcine antihaemophilic globulin, *J. Clin. Pathol.* **25:**210–217.

Fujimoto, T., and Hawiger, J., 1982, Adenosine diphosphate induces binding of von Willebrand Factor to human platelets, *Nature (London)* **297:**154–156.

Fujimoto, T., Ohara, S., and Hawiger, J., 1982, Thrombin-induced exposure and prostacyclin inhibition of the receptor for factor VIII/von Willebrand Factor on human platelets, *J. Clin. Invest.* **69:**1212–1222.

Furlan, M., Perret, B. S., and Beck, E. A., 1979, Besteht ein zusammenhang zwischen zuckergehalt, plattchen aggregierender akrivitat und molekulgrosse von factor VIII? *Schweiz. Med. Wschr.* **109:**1369–1370.

Fuster, V., Bowie, E. J. W., Lewis, J. C., Fass, D. N., Owen, C. A., Jr., and Brown, A. L., Jr., 1978, Resistance to artherosclerosis in pigs with von Willebrand's disease: Spontaneous and high cholesterol diet-induced artherosclerosis, *J. Clin. Invest.* **61:**722–730.

George, J. N., 1972, Direct assessment of platelet adhesion to glass: A study of the forces of interaction and the effects of plasma and serum factors, platelet function, and modification of the glass surface, *Blood* **40:**862–874.

George, J. N., and Onofre, A. R., 1982, Human platelet surface binding of endogenous secreted factor VIII-von Willebrand factor and platelet factor 4, *Blood* **59:**194–197.

George, J. N., Nurden, A. T., and Phillips, D. R., 1984, Molecular defects that cause abnormalities of platelet-vessel wall interactions, *N. Engl. J. Med.* **311:**1084–1098.

Ginsberg, M. H., Wolff, R., Marguerie, G., Coller, B., McEver, R., and Plow, E. F., 1984, Thrombospondin binding to thrombin-stimulated platelets: Evidence for a common adhesive protein binding mechanism, *Clin Res.* **32:**308A.

Goudemand, J., Samor, B., Mazurier, C., and Goudemand, M., 1983, The role of the carbohydrate moiety of factor VIII/VWF on binding to human platelets, *Thromb. Haemostasis* **50:**191.

Gralnick, H. R., 1978, Factor VIII/von Willebrand factor protein. Galactose, a cryptic determinant of von Willebrand factor activity, *J. Clin. Invest.* **62:**496–499.

Gralnick, H. R., Coller, B. S., and Sultan, Y., 1976, Carbohydrate deficiency of the factor VIII/von Willebrand factor protein in von Willebrand's disease variants, *Science* **192:**56–59.

Gralnick, H. R., Sultan, Y., and Coller, B. S., 1977, Von Willebrand's disease: Combined qualitative and quantitative abnormalities, *N. Engl. J. Med.* **296**:1024–1030.

Gralnick, H. R., Williams, S. B., and Morisato, D. K., 1981, Effect of the multimeric structure of the factor VIII/von Willebrand factor protein on binding to platelets, *Blood* **58**:387–397.

Gralnick, H. R., Williams, S. B., Shafer, B. C., and Corash, L., 1982, Factor VIII/von Willebrand factor binding to von Willebrand's disease platelets, *Blood* **60**:328–332.

Gralnick, H. R., Williams, S., and Rick, M., 1983, The role of carbohydrate in the maintenance of the multimeric structure of the factor VIII/VWF protein, *Thromb. Haemostasis* **50**:318.

Gralnick, H. R., Williams, S. B., and Coller, B. S., 1984, Fibrinogen competes with von Willebrand factor for binding to the glycoprotein IIb/IIIa complex when platelets are stimulated with thrombin, *Blood* **64**:797–800.

Gralnick, H. R., Williams, S. B., and Coller, B. S., 1985, Asialo von Willebrand Factor interactions with platelets: Interdependence of glycoproteins Ib and Ib/IIIa for binding and aggregation, *J. Clin. Invest.* (in press).

Grant, R. A., Zucker, M. B., and McPherson, J., 1976, ADP-induced inhibition of von Willebrand factor-mediated platelet agglutination, *Am. J. Physiol.* **230**:1406–1410.

Griggs, T. R., Cooper, H. A., Webster, W. P., Wagner, R. H., and Brinkhous, K. M., 1973, Plasma aggregating factor (bovine) for human platelets: A marker for study of antihemophilic and von Willebrand factors, *Proc. Natl. Acad. Sci. U.S.A.* **70**:2814–2818.

Griggs, T. R., Reddick, R. L., Saltzer, D., and Brinkhous, K. M., 1981, Susceptibility to atherosclerosis in aortas and coronary arteries of swine with von Willebrand's disease, *Am. J. Pathol.* **102**:137–145.

Guisasola, J. A., Cockburn, C. G., and Hardisty, R. M., 1978, Plasmin digestion of factor VIII: Characterization of the breakdown products with respect to antigenicity and von Willebrand activity, *Thromb. Haemostasis* **40**:302–315.

Holmberg, L., and Nilsson, I. M., 1979, AHF related protein in clinical praxis, *Scand. J. Haematol.* **12**:221–231

Holmberg, L., Nilsson, I. M., Borge, L., Gunnarsson, M., and Sjörin, E., 1983, Platelet aggregation induced by DDAVP in type IIB von Willebrand's disease, *N. Engl. J. Med.* **309**:816–821.

Holmberg, L., Nilsson, I. M., and Lamme, S., 1984, Letter to the editor, *N. Engl. J. Med.* **310**:723.

Howard, M. A., 1975, Inhibition and reversal of ristocetin-induced platelet aggregation, *Thromb. Res.* **6**:489–499.

Howard, M. A., and Firkin, B. G., 1971, Ristocetin—A new tool in the investigation of platelet aggregation, *Thromb. Diath. Haemorrh.* **26**:362–369.

Howard, M. A., Hutton, R. A., and Hardisty, R. M., 1973, Hereditary giant platelet syndrome: A disorder of a new aspect of platelet function, *Br. Med. J.* **4**:586–588.

Howard, M. A., Montgomery, D. C., and Hardisty, R. M., 1974, Factor-VIII-related antigen in platelets, *Thromb. Res.* **4**:617–624.

Howard, M. A., Hendrix, L., and Firkin, B. G., 1979, Further studies on the factor VIII of a patient with a variant form of von Willebrand's disease, *Thromb. Res.* **14**:609–619.

Howard, M. A., Salem, H. H., Thomas, K. B., Hau, L., Perkin, J., Coghlan, M., and Firkin, B., 1982, Variant von Willebrand's disease type B-revisited, *Blood* **60**:1420–1428.

Howard, M. A., Perkin, J., Salem, H. H., and Firkin, B. S., 1984, The agglutination of human platelets by botrocetin: Evidence that botrocetin and ristocetin act at different sites on the factor VIII molecule and platelet membrane, *Br. J. Haematol.* **57**:25–35.

Hoyer, L. W., and Shainoff, J. R., 1980, Factor VIII related protein circulates in normal human plasma as high molecular weight multimers, *Blood* **55**: 1056–1059.

Jaffe, E. A., Hoyer, L. W., and Nachman, R. L., 1974, Synthesis of von Willebrand factor by cultured human endothelial cells, *Proc. Natl. Acad. Sci. U.S.A.* **71**:1906–1909.

Kao, K. J., Pizzo, S. V., and McKee, P. A., 1979, Demonstration and characterization of specific binding sites for factor VIII/von Willebrand Factor on human platelets, *J. Clin. Invest.* **63**:656–664.

Konoshita, S., Harrison, J., Lazerson, J., and Abildgaard, C., 1983, A new variant of dominant Type II von Willebrand disease with aberrant multimeric structure of factor VIII-related antigen, *Thromb. Haemostasis* **50**:78.

Kirby, E. P., 1977, Factor VIII-associated platelet aggregation, *Thromb. Haemostasis* **38**:1054–1072.

Kirby, E. P., 1981, Characterization of a form in bovine factor VIII which does not aggregate human platelets, *Thromb. Haemostasis* **46;**479–484.

Kirby, E. P., 1982, The agglutination of human platelets by bovine factor VIII, *R. J. Lab. Clin. Med.* **100:**963–976.

Kirby, E. P., and Mills, D. B., 1975, The interaction of bovine factor VIII with human platelets, *J. Clin. Invest.* **56:**491–502.

Kirby, E. P., Mills, D. C. B., Holmsen, H., and Russo, M., 1982, Factor VIII-induced superaggregation of human platelets, *Blood* **60:**1359–1369.

Koutts, J., Walsh, P. N., Plow, E. F., Fenton, J. W., Bouma, B. N., and Zimmerman, T. S., 1978, Active release of human platelet factor VIII related antigen by adenosine diphosphate, collagen and thrombin, *J. Clin. Invest.* **62:**1255–1263.

Legaz, M. E., Schmer, G., Counts, R. B., and Davie, E. W., 1973, Isolation and characterization of human factor VIII (antihemophilic factor), *J. Biol. Chem.* **248:**3946–3955.

Legrand, Y. J., Rodriguez-Zeballos, A., Kartalis, G., Fauvel, F., and Caen, J. P., 1978, Adsorption of factor VIII antigen-activity complex by collagen, *Thromb. Res.* **13:**909–911.

Legrand, Y. J., Fauvel, F., Gutman, N., Muh, J. P., Tobelem, G., Souchon, H., Karniguian, A., and Caen, J. P., 1980, Microfibrils platelet interaction: Requirement for von Willebrand factor, *Thromb. Res.* **19:**737–739.

Loscalzo, J., Slayter, H., and Handin, I., 1983, The native conformation of human von Willebrand protein: Electron microscopy and quasielastic light scattering analyses, *Blood* **62**(Suppl. 1):289a.

Lynch, D. C., Williams, R., Zimmerman, T. S., Kirby, E. P., and Livingston, D. M., 1983, Biosynthesis of the subunits of factor VIIIR by bovine aortic endothelial cells, *Proc. Natl. Acad. Sci. U.S.A.,* **80:**2738–2742.

Macfarlane, R. G., Biggs, R., and Bidwell, E., 1954, Bovine anti-haemophilic globulin in the treatment of hemophilia, *Lancet* **1:**1316–1319.

Mannucci, P. M., Lombardi, R., Pareti, F. I., Solinas, S., Mazzaconi, G., and Mariani, G., 1983, A variant of von Willebrand's disease characterized by recessive inheritance and missing triplet structure of von Willebrand factor multimers, *Thromb. Haemostasis* **50:**78.

Marchesi, S. L., Shulman, N. R., and Gralnick, H. R., 1972, Studies on the purification and characterization of human factor VIII, *J. Clin. Invest.* **51:**2151–2161.

Martin, S. W., Marder, V. J., Francis, C. W., Loftus, L. S., and Barlow, G. H., 1980, Enzymatic degradation of the factor VIII-von Willebrand protein: A unique tryptic fragment with ristocetin cofactor activity, *Blood* **55:**847–858.

McMichael, A. J., Rust, N. A., Pilch, J. R., Sochynsky, R., Morton, J., Mason, D. Y., Ruan, C., Tobelem, G., and Caen, J., 1981, Monoclonal antibody to human platelet glycoprotein I. I. Immunoglobulin studies, *Br. J. Haematol.* **49:**501–509.

McPherson, J., and Zucker, M. B., 1976, Platelet retention in glass bead columns: Adhesion to glass and subsequent platelet–platelet interactions, *Blood* **47:**55–67.

Meyer, D., Obert, B., Peitu, G., Lavergne, J. M., and Zimmerman, T. S., 1980, Multimeric structure of factor VIII/von Willebrand factor in von Willebrand's disease, *J. Lab. Clin. Med.* **95:**590–602.

Meyer, D., Lavergne, J. M., Baumgartner, H. R., Tobelem, G., Pietu, G., and Edgington, T. S., 1983, Monoclonal antibodies to human von Willebrand factor: Role of intramolecular loci in mediation of platelet adhesion to the subendothelium, *Thromb. Haemostasis* **50:**191.

Miller, J. L., and Castella, A., 1982, Platelet-type von Willebrand's disease: Characterization of a new bleeding disorder, *Blood* **60:**790–794.

Miller, J. L., Kupinski, J. M., Castella, A., and Ruggeri, Z. M., 1983, von Willebrand factor binds to platelets and induces aggregation in platelet-type but not type IIB von Willebrand's disease, *J. Clin. Invest.* **72:**1532–1542.

Miller, J. L., Boselli, B. D., and Kupinski, J. M., 1984, *In vivo* interaction of von Willebrand factor with platelets following cryoprecipitate transfusion of platelet-type von Willebrand's disease, *Blood* **63:**226–230.

Moake, J. L., Cimo, P. L., Peterson, D. M., Roper, P., and Natelson, E. A., 1977, Inhibition of ristocetin-induced platelet agglutination by vancomycin, *Blood* **50:**397–401.

Moake, J. L., Olson, J. D., Troll, J. H., Tang, S. S., Funicella, T., and Peterson, D. M., 1980, Binding of radioiodinated human von Willebrand factor to Bernard-Soulier, thrombasthenic and von Willebrand's disease platelets, *Thromb. Res.* **19:**21–27.

Moake, J. L., Tang, S. S., Olson, J. D., Troll, J. H., Cimo, P. L., and Davies, P. J. A., 1981, PGI_2 and AET inhibit F.VIII/VWF-mediated platelet cohesion, *Am. J. Physiol.* **241:**454–459.

Morisato, D. K., and Gralnick, H. R., 1980, Selective binding of the factor VIII/von Willebrand factor protein to human platelets, *Blood* **55:**9–15.

Muraki, H., 1977, Changes in electrokinetic properties of human platelets during their aggregation by ristocetin, *J. Keio. Med. Soc.* **54:**87–96.

Nachman, R. I., and Jaffe, E. A., 1975, Subcellular platelet factor VIII antigen and von Willebrand factor, *J. Exp. Med.* **141:**1101–1113.

Nachman, R. L., and Leung, L. L. K., 1982, Complex formation of platelet membrane glycoproteins IIb and IIIa with fibrinogen, *J. Clin. Invest.* **69:**263–269.

Nachman, R. L., Jaffe, E. A., and Weksler, B. B., 1977a, Immunoinhibition of ristocetin-induced platelet aggregation, *J. Clin. Invest.* **59:**143–148.

Nachman, R., Levine, R., and Jaffe, E. A., 1977b, Synthesis of factor VIII antigen by cultured guinea pig megakaryocytes, *J. Clin. Invest.* **60:**914–921.

Nachman, R. L., Jaffe, E. A., Ferris, B., 1980, Peptide map analysis of normal plasma and platelet factor VIII antigen, *Biochem. Biophys. Res. Commun.* **92:**1208–1214.

Nurden, A. T., and Caen, J. P., 1975, Specific roles for platelet surface glycoproteins in platelet function, *Nature (London)* **255:**720–722.

Nurden, A. T., Didry, D., and Rosa, J. P., 1983, Molecular defects of platelets in Bernard-Soulier syndrome, *Blood Cells* **9:**333–353.

Nyman, D., 1977, Interaction of collagen with the factor VIII antigen-activity-von Willebrand factor complex, *Thromb. Res.* **11:**433–438.

O'Brien, J., and Heywood, J. B., 1967, Some interactions between human platelets and glass: Von Willebrand's disease compared with normal, *J. Clin. Pathol.* **20:**56–64.

Ohmori, K., Fretto, L. J., Harrison, R. L., Switzer, M. E. P., Erickson, H. P., and McKee, P. A., 1982, Electron microscopy of human factor VIII/von Willebrand glycoprotein: Effect of reducing reagents on structure and function, *J. Cell. Biol.* **95:**632–640.

Olson, J. D., Moake, J. L., Collins, M. F., and Michael, B. S., 1983, Adhesion of human platelets to purified solid-phase von Willebrand factor: Studies of normal and Bernard-Soulier platelets, *Thromb. Res.* **32:**115–122.

Over, J., Bouma, B. N., and van Mourik, J. A., 1978a, Heterogeneity of human factor VIII. I. Characterization of factor VIII present in the supernatant of cryoprecipitate, *J. Lab. Clin. Med.* **91:**32–46.

Over, J., Sixma, J. J., Doucet-de Bruine, M. H. M., Trieschnigg, A. M. C., Vlooswijk, R. A. A., Beeser-Visser, N. H., and Bouma, B. N., 1978b, Survival of [125]iodine-labelled factor VIII in normals and patients with classic hemophilia. Observations on the heterogeneity of human factor VIII, *J. Clin. Invest.* **62:**223–234.

Peake, I. R., and Bloom, A. L., 1977, Abnormal factor VIII related antigen in von Willebrand's disease: Decreased precipitation by concanavalin A, *Thromb. Haemostasis* **37:**361–362.

Perret, B. A., Furlan, M., and Beck, E. A., 1979, Studies in factor VIII-related protein. II. Estimation of molecular size differences between factor VIII oligomers, *Biochim. Biophys. Acta* **578:**164–174.

Pethica, B. A., 1961, The physical chemistry of cell adhesion, *Exp. Cell Res.* **8:**123–140.

Phillips, D. R., 1980, An evaluation of membrane glycoproteins in platelet adhesion and aggregation, *Prog. Hemostas. Thromb.* **5:**81–109.

Plow, E. F., Srouji, A. H., Meyer, D., Marguerie, G., and Ginsberg, M. H., 1984, Evidence that three adhesive proteins interact with a common recognition site on activated platelets, *J. Biol. Chem.* **259:**5388–5391.

Rand, J. H., Sussman, I. I., Gordon, S. V., Chu, S. V., and Solomon, V., 1980, Localization of factor VIII related antigen in human vascular subendothelium, *Blood* **55:**752–756.

Reddick, R. L., Griggs, T. R., Lamb, M. A., and Brinkhous, K. M., 1982, Platelet adhesion to damaged coronary arteries: Comparison in normal and von Willebrand disease swine, *Proc. Natl. Acad. Sci. U.S.A.* **79:**5076–5079.

Rosborough, T. K., 1980, Von Willebrand factor, polycations, and platelet agglutination, *Thromb. Res.* **17:**481–490.

Rosborough, T. K., and Swaim, W. R., 1978, Abnormal polybrene-induced platelet agglutination in von Willebrand's disease, *Thromb. Res.* **12:**937–942.

Ross, R., Glomset, J., Kaviya, N., and Harker, I., 1974, A platelet-dependent serum factor that stimulates the proliferation of arterial smooth muscle cells in vitro, *Proc. Natl. Acad. Sci. U.S.A.* **71**:1207–1210.

Ruan, C., Tobelem, G., McMichael, A. J., Drouet, L., Legrand, Y., Degos, L., Kieffer, N., Lee, H., and Caen, J. P., 1981, Monoclonal antibody to human platelet glycoprotein I. II. Effects on human platelet function, *Br. J. Haematol.* **49**:511–519.

Ruggeri, Z. M., and Zimmerman, T. S., 1980, Variant von Willebrand's disease. Characterization of two subtypes by analysis of multimeric composition of factor VIII/von Willbrand factor in plasma and platelets, *J. Clin. Invest.* **65**:1318–1325.

Ruggeri, A. M., and Zimmerman, T. A., 1981, The complex multimeric composition of factor VIII/von Willebrand factor, *Blood* **57**:1140–1143.

Ruggeri, Z. M., Bader, R., and DeMarco, L., 1982a, Glanzmann thrombasthenia: Deficient binding of von Willebrand factor to thrombin-stimulated platelets, *Proc. Natl. Acad. Sci. U.S.A.* **79**:6038–6041.

Ruggeri, Z. M., Mannucci, P. M., Lombardi, R., Federici, A. B., and Zimmerman, T. S., 982b, Multimeric composition of Factor VIII/von Willebrand factor following administration of DDAVP: Implications for pathophysiology and therapy of von Willebrand's disease subtypes, *Blood* **59**:1272–1278.

Ruggeri, Z. M., Nilsson, I. M., Lombardi, R., Holmberg, L., and Zimmerman, T. S., 1982c, Aberrant multimeric structure of von Willebrand factor in a new variant of von Willebrand's disease (Type IIc), *J. Clin. Invest.* **70**:1124–1127.

Ruggeri, Z. M., Bader, R., Pareti, F. I., Mannucci, L., and Zimmerman, T. S., 1983a, High affinity interaction of platelet von Willebrand factor with distinct platelet membrane sites, *Thromb. Res.* **50**:35.

Ruggeri, Z. M., DeMarco, L., Montgomery, R. R., 1983b, Platelets have more than one binding site for von Willebrand factor, *J. Clin. Invest.* **72**:1–12.

Sakariassen, K. S., Bolhuis, P. A., and Sixma, J. J., 1979, Human blood platelet adhesion to artery subendothelium is mediated by factor VIII-von Willebrand factor bound to the subendothelium, *Nature (London)* **279**:636–638.

Sakariassen, K. S., Nieuwenhuis, H. K., Ottenhof-Rovers, and Sixma, J. J., 1983, Thrombocytopenia with large platelets and increased platelet binding of factor VIII-von Willebrand factor without aggregation with exogenous fVIII-VWF: A new entity? *Thromb. Haemostasis* **50**:79.

Salzman, E. W., 1963, Measurement of platelet adhesiveness. A simple *in vitro* technique demonstrating an abnormality in von Willebrand's disease, *J. Lab. Clin. Med.* **62**:724–735.

Santoro, S. A., 1983, Preferential binding of high molecular weight forms of von Willebrand factor to fibrillar collagen. I. *Biochim. Biophys. Acta* **756**:123–126.

Santoro, S. A., and Cowan, J. F., 1982, Adsorption of von Willebrand factor by fibrillar collagen—Implications concerning the adhesion of platelets to collagen, *Collagen Relat. Res.* **2**:31–43.

Schmer, G., Kirby, E. P., Teller, D. C., and Davie, E. W., 1972, The isolation and characterization of bovine factor VIII (antihemophilic factor), *J. Biol. Chem.* **247**:2512–2521.

Schullek, J., Jordan, J., and Montgomery, R. R., 1984, Interaction of von Willebrand factor with platelets in a plasma milieu, *J. Clin. Invest.* **73**:421–428.

Seaman, G. V. F., 1976, Electrochemical features of platelet interactions, *Thromb. Res.* (Suppl. II) **8**:235–246.

Senogles, S. E., and Nelsestuen, G. L., 1983, von Willebrand factor: A protein which binds at the cell surface interface between platelets, *J. Biol. Chem.* **258**:12327–12333.

Shapiro, G. A., Anderson, J. C., Pizzo, S. V., and McKee, P. A., 1973, The subunit structure of normal and hemophilic factor VIII, *J. Clin. Invest.* **52**:2198–2210.

Sharp, A. A., and Bidwell, E., 1957, The toxicity and fate of injected animal antihaemophilic globulin, *Lancet* **2**:359–362.

Sixma, J. J., Sakariassen, K. S., Beeser-Visser, N. H., Ottenhof-Rovers, M., and Bolhuis, P. A., 1984, Adhesion of platelets to human artery subendothelium. Effect of factor VIII-von Willebrand factor of various multimeric composition, *Blood* **63**:128–139.

Sodetz, J. M., Pizzo, S. V., and McKee, P. A., 1977, Relationship of sialic-acid to function and *in vivo* survival of human factor VIII-von Willebrand factor protein, *J. Biol. Chem.* **252**:5538–5546.

Sodetz, J. M., Paulson, M. C., Pizzo, S. V., and McKee, P. A., 1978, Carbohydrate on human factor VIII/von Willebrand factor. Impairment of function by removal of specific galactose residues, *J. Biol. Chem.* **253**:7202–7206.

Solum, N. O., 1968, Aggregation of human platelets by bovine platelet fibrinogen, *Scand. J. Haematol.* **5**:474–485.

Solum, N. O., Olsen, T. M., Gogstad, G. O., Hagen, I., and Brosstad, F., 1983, Demonstration of a new glycoprotein Ib-related component in platelet extracts prepared in the presence of leupeptin, *Biochim. Biophys. Acta* **729**:53–61.

Sussman, I. I., and Rand, J. H., 1982, Subendothelial deposition of von Willebrand's factor requires the presence of endothelial cells, *J. Lab. Clin. Med.* **100**:526–532.

Takahashi, H., 1980, Studies on the pathophysiology and treatment of von Willebrand's disease. IV. Mechanism of increased ristocetin-induced platelet aggregation in von Willebrand's disease, *Thromb. Res.* **19**:857–867.

Takahashi, H., Nagayama, R., Hattori, A., Shibata, A., 1984, Platelet aggregation induced DDAVP in platelet-type von Willebrand's disease, *N. Engl. J. Med.* **310**:722.

Tobelem, G., Levy-Toledano, S., Bredoux, R., Michel, H., Nurden, A., Degos, L., and Caen, J., 1976, New approach to determination of specific functions of platelet membrane sites, *Nature (London)* **263**:427–429.

Tschopp, T. B., Weiss, H. J., and Baumgartner, H. R., 1974, Decreased adhesion of platelets to subendothelium in von Willebrand's disease, *J. Lab. Clin. Med.* **83**:296–300.

Turitto, V., Weiss, H., Sussman, I., and Zimmerman, T., 1981, Factor VIII in vessel wall influences platelet interaction with subendothelium, *Thromb. Haemostasis* **46**:199.

Vermylen, J., Bottecchia, D., and Szpilman, H., 1976, Factor VIII and human platelet aggregation. III. Further studies on aggregation of human platelets by neuraminidase-treated human factor VIII, *Br. J. Haematol.* **34**:321–330.

Wagner, D. D., and Marder, V. J., 1983, Biosynthesis of von Willebrand protein by human endothelial cells. Identification of a large precursor polypeptide chain, *J. Biol. Chem.* **258**:2065–2067.

Wagner, D. D., Olmstead, J. B., and Marder, V. J., 1982, Immunolocalization of von Willebrand protein in Weibel-Palade bodies of human endothelial cells, *J. Cell. Biol.* **95**:355–360.

Waterfield, M. D., Scrace, G. T., Whittlen, N., Stroobant, P., Johnsson, A., Wasteson, A., Westermark, B., Heldin, C-H., Huang, J. S., and Deuel, T. F., 1983, Platelet-derived growth factor is structurally related to the putative transforming protein p28[sis] of simian sarcoma virus, *Nature (London)* **304**:35–39.

Weinstein, M., and Deykin, D., 1979, Comparison of factor VIII-related von Willebrand factor proteins prepared from human cryoprecipitate and FVIII concentrate, *Blood* **53**:1095–1105.

Weiss, H. J., and Rogers, J., 1972, Correction of the platelet abnormality in von Willebrand's disease by cryoprecipitate, *Am. J. Med.* **53**:734–738.

Weiss, H. J., Turitto, V. T., and Baumgartner, H. R., 1978, Effect of shear rate on platelet interaction with subendothelium in citrated and native blood. I. Shear dependent decrease of adhesion in von Willebrand's disease and the Bernard-Soulier syndrome, *J. Lab. Clin. Med.* **92**:750–764.

Weiss, H. J., Turitto, V. T., and Baumgartner, H. R., 1981, Platelet-fibrin deposition on subendothelium in congential bleeding disorders, *Thromb. Haemostasis* **46**:249.

Weiss, H. J., Meyer, D., Rabinowitz, R., Pietu, G., Girma, J-P., Vicic, W. J., and Rogers, J., 1982, Pseudo-von Willebrand's disease. An intrinsic platelet defect with aggregation by unmodified human factor VIII/von Willebrand factor and enhanced adsorption of its high-molecular-weight multimers, *N. Engl. J. Med.* **306**:326–333.

Weiss, H. J., Pietu, G., Rabinowitz, R., Girma, J-P., Rogers, J., and Meyer, D., 1983, Heterogeneous abnormalities in the multimeric structure, antigenic properties, and plasma–platelet content of factor VIII/von Willebrand factor in subtypes of classic (type I) and variant (type IIA) von Willebrand's disease, *J. Lab. Clin. Med.* **101**:411–415.

Yoshida, N., Weksler, B., and Nachman, R. L., 1983, Purification of human platelet calcium activated protease: Effect on platelet and endothelial cell function, *J. Biol. Chem.* **258**:7168–7174.

Zimmerman, T. S., and Ruggeri, Z. M., 1983, Von Willebrand's disease, *Clin. Haematol.* **12**:175–200.

Zimmerman, T. S., Roberts, J., and Edgington, T. S., 1975, Factor-VIII-related antigen: Multiple molecular forms in human plasma, *Proc. Natl. Acad. Sci. U.S.A.* **72**:5121–5125.

Zimmerman, T. S., Voss, R., and Edgington, T. S., 1979, Carbohydrate of factor VIII/von Willebrand factor in von Willebrand's disease, *J. Clin. Invest.* **64**:1298–1302.

Zimmerman, T. S., Dent, J. A., Federici, A. B., Ruggeri, Z. M., Nilsson, I. M., Holmberg, L., Abildgaard, C. A., and Nannini, L. H., 1984, High resolution Na Dod SO$_4$ - agarose electrophoresis identifies new molecular abnormalities in von Willebrand's disease, *Clin. Res.* **32:**501A.

Zucker, M. B., 1963, *In vitro* abnormality of the blood in von Willebrand's disease correctable by normal plasma, *Nature (London)* **197:**601–602.

Zucker, M. B., Kim, S. J., McPherson, J., and Grant, R. A., 1977, Binding of factor VIII to platelets in the presence of ristocetin, *Br. J. Haematol.* **35:**535–549.

Zucker, M. B., Broehman, F. J., and Kaplan, K. L., 1979, Factor VIII related antigen in human blood platelets. Localization and release by thrombin and collagen, *J. Lab. Clin. Med.* **94:**675–682.

11

Molecular Mechanisms of Platelet Adhesion and Platelet Aggregation

Ralph L. Nachman, Lawrence L. K. Leung, and Margaret J. Polley

1. PRIMITIVE CELL SYSTEMS

Certain primitive cell systems, such as cellular slime molds, exist either in unicellular vegetative forms or in a differentiated state in which they become adhesive and aggregate into a multicellular structure. An example of this mechanism of cellular behavior is shown in Figure 1, which illustrates the life cycle of *Acrasis rosea,* an acrasid cellular slime mold. The uninucleate ameboid cells move, divide by binary fission, deplete the local environment of the food supply, aggregate, and form fruiting bodies. The mechanisms associated with the development of the adhesive state have been studied most extensively in the dictyostelid cellular slime molds. When the cells of this species become adhesive, they synthesize surface polyvalent carbohydrate-binding proteins or lectins that mediate cell to cell adhesion (Frasier and Glaser, 1979; Barondes, 1981). This stage of the cell cycle is associated with activation of a large number of new genes (Blumberg *et al.*, 1982), presumably coding for cell-surface proteins that mediate the conversion to the adhesive state. Cell adhesion culminating in aggregation takes place via the interaction of carbohydrate-binding sites of a cell-surface lectin with oligosaccharides on membrane glycoprotein receptors. The initial interaction of one lectin molecule with one receptor oligosaccharide followed by binding at multiple sites leads to rapid and stable cohesion (aggregation). Discoidin I and II, two closely related galactose-binding lectins synthesized by *Dictyostelium dis-*

Ralph L. Nachman, Lawrence L. K. Leung, and Margaret J. Polley • Division of Hematology-Oncology, Department of Medicine, The New York Hospital-Cornell Medical Center, New York, New York 10021.

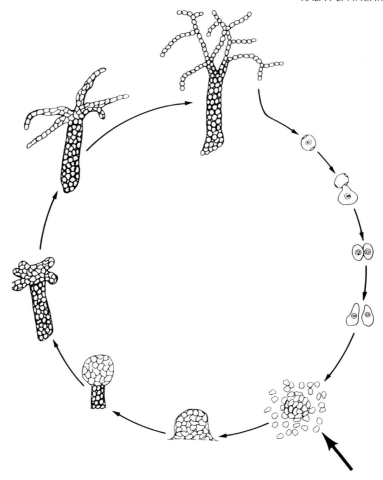

Figure 1. Life cycle of the acrasid cellular slime mold, *Acrasis rosea.* The aggregating cell stage (arrow) precedes the development of the fruiting body. From Olive (1975), with permission of Dr. J. Bonner.

coideum during the conversion from the noncohesive to the aggregating cohesive stage, may function at different stages of the aggregation process by binding to different oligosaccharide-containing receptors at the cell surface (Berger and Armant, 1982). Thus, at least in primitive systems, the aggregation process appears to involve the sequential exposure of a family of cell-surface recognition molecules, some of which serve as lectins. It is probable that human platelets may recapitulate some of these primitive cellular responses during the process of agonist-induced aggregation.

2. PLATELET DISORDERS AND INSIGHTS INTO FUNCTIONAL MEMBRANE DOMAINS

As discussed more extensively in Chapter 16, elucidation of the nature of two major congenital platelet disorders has provided evidence suggesting the presence of separate functional domains in the platelet membrane. For example, in the congenital hemorrhagic disorder, Bernard-Soulier syndrome, the platelets do not agglutinate in the presence of ristocetin and normal von Willebrand factor (VWF) and do not adhere to exposed vascular subendothelium under conditions simulating rapid blood flow. Platelet aggregation in response to ADP, epinephrine, and collagen is normal. The major defect in the platelet membrane is an absent or defective glycoprotein, GP Ib. GP V and GP IX are also diminished (Clemetson et al., 1982; Berndt et al, 1983). In contrast, in another congenital hemorrhagic disorder, Glanzmann's thrombasthenia, there is markedly diminished fibrinogen binding to the platelets and absent, decreased, or functionally abnormal GP IIb-IIIa associated with absent aggregation. It is of interest that recent studies suggest that thrombin-induced fibronectin and VWF binding are also decreased in thrombasthenia (Ginsberg et al., 1983; Ruggeri et al., 1982). The correlation of specific physiologic defects—adhesion to vessel wall and cohesion (aggregation) with adjacent platelets—with specific membrane glycoprotein abnormalities in these two disorders strongly suggests separate functional domains in the platelet membrane.

3. ADHESION: THE PLATELET VWF-SUBENDOTHELIAL AXIS

3.1. Von Willebrand Factor

Von Willebrand factor is a protein synthesized by endothelial cells (Jaffe et al., 1973) that circulates as a series of high-molecular-weight disulfide-linked multimers (Counts et al., 1978). This is the plasma factor necessary for adhesion (Weiss et al., 1973). The molecular dispersion of VWF represents polymeric combinations of apparently identical subunits (Nachman et al., 1980a). The higher molecular weight subsets provide the major adhesive supporting role and bind to a specific receptor(s) on the platelet membrane by a GP Ib-mediated mechanism (Ruan et al., 1981). Transfusion experiments using radiolabeled VWF in human volunteers demonstrate that the higher molecular weight forms disappear from the circulation faster than those with lower molecular weights. These observations raise the possibility that there may be a continuous conversion of the high-molecular-weight forms (possessing the adhesion-supporting functions of VWF) into lower molecular weight inactive forms (Over et al., 1980; Nachman et al., 1980a). Transfusion experiments in patients with von Willebrand's disease (VWD) suggest a similar in vivo conversion of high molecular weight VWF into lower molecular weight forms.

Comparison of VWF isolated from endothelial cell postculture medium with intracellular, nonsecreted endothelial VWF supports the idea that intracellular, low-molecular-weight monomers are converted into extracellular, higher molecular weight forms (Nachman, 1982; Wagner and Marder, 1983). This raises the possibility that an endothelial cell polymerase is necessary for full expression of VWF function. Von Willebrand factor exists in different molecular weight forms depending on whether it is within endothelial cells, secreted, or circulating. The higher molecular weight VWF subset seen in the postculture medium is not detected in circulating plasma and is strikingly absent from the intracellular compartment. It seems reasonable to speculate that factors, yet unknown, may regulate the flux of different molecular weight forms

A   VWF Synthesis

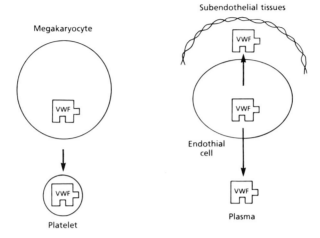

B   Platelet - VWF Interactions

Platelet adhesion to subendothelial VWF

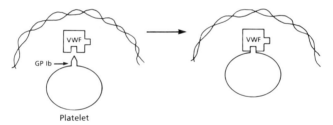

Agonist-induced binding of VWF to platelets

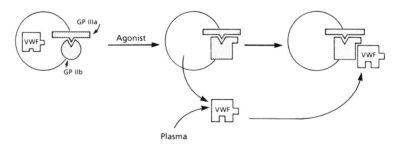

Figure 2. Multifunctional role of VWF in platelet physiology. (A) Two sites of synthesis of VWF are illustrated: megakaryocytes, the source of platelet alpha granule VWF, and endothelial cells, the source of plasma and subendothelial VWF. (B) Platelet adhesion to subendothelium is dependent on the interaction of subendothelial-bound VWF and platelet GP Ib that has been conformationally altered. Agonist-induced binding is dependent on the platelet alpha granule VWF as well as plasma VWF binding to a conformationally activated GP IIb-IIIa.

that may be crucial in determining the activation of the adhesion-supporting function of VWF. Recent studies suggest that endothelial cells secrete VWF in a bidirectional manner, i.e., luminally into circulating blood and abluminally onto the subendothelium (Rand et al., 1982). It is possible that, in fact, the most physiologically relevant VWF may be that fraction bound to the subendothelium. The multimeric nature of subendothelial VWF remains to be clarified. The synthesis and functional sites of VWF are diagrammed in Figure 2.

3.2. Von Willebrand Factor Receptor on Platelets

The nature of the VWF-platelet membrane interaction is under intensive study. Platelets from patients with Bernard-Soulier syndrome lack normal GP Ib (as well as GP V and GP IX), show decreased adhesion to the subendothelium, and do not bind VWF in the presence of ristocetin. Immunoinhibitory probes using polyclonal as well as monoclonal GP Ib antisera suggest that GP Ib on the platelet membrane is the VWF receptor (Ruan et al., 1981; Nachman et al., 1977). Recent studies, however, suggest that platelets have more than one binding site for VWF. Thrombin induces platelet binding of [^{125}I]-labelled VWF. This binding reaction is deficient in Glanzmann's thrombasthenia and normal to increased in Bernard-Soulier syndrome (Ruggeri et al., 1983b). In addition, a monoclonal antibody specific for the GP IIb-IIIa complex inhibits thrombin and ADP-induced platelet binding of VWF (Gralnick and Coller, 1983). These studies, taken together, support the concept of a two-site receptor system for the VWF. One involves GP Ib and binds VWF in the presence of ristocetin and mediates adhesion to subendothelial-bound VWF. The other involves GP IIb-IIIa and binds VWF when platelets are stimulated with thrombin or ADP (Figure 2). It is of interest that addition of ADP to EDTA-treated normal platelets or thrombasthenic platelets in the presence of sodium citrate interferes with, rather than augments, ristocetin-induced platelet agglutination (Grant et al., 1976; Cohen et al., 1975), suggesting that there may be some interaction between these two receptor systems.

The marked hemorrhagic diathesis in Bernard-Soulier syndrome makes it probable that GP Ib is a vital component in mediating the platelet adhesion process. Perhaps, agonist-induced binding of VWF by the GP IIb-IIIa complex reflects a role in aggregation for the VWF-derived from platelet alpha granules. Platelet VWF is functionally (Ruggeri et al., 1983b) and may be structurally (Nachman et al., 1980b) different from circulating VWF in plasma in that it binds more rapidly and with a higher affinity to thrombin-stimulated platelets. We speculate that plasma and subendothelial VWF support adhesion, whereas platelet alpha granule VWF (synthesized by megakaryocytes) supports, but is not necessary for, aggregation (Figure 2).

4. AGGREGATION: GLYCOPROTEIN IIb-IIIa AND FIBRINOGEN

Fibrinogen is necessary for platelet aggregation. The plasma protein and possibly its alpha granule counterpart bind to specific receptor sites on the membrane that are induced following stimulation of the resting platelet by physiologic stimuli such as ADP, thrombin, or epinephrine (Bennett and Vilaire, 1979; Marguerie et al., 1979). The binding reaction exhibits the characteristics of a saturable receptor system and is

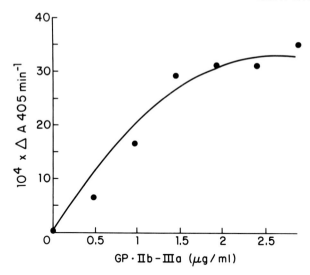

Figure 3. GP IIb-IIIa complex formation with absorbed fibrinogen as measured by ELISA. Fibrinogen (3 μg/ml) in coating buffer was added to plastic wells for 18 hr at 4 °C. Following extensive washing, varying amounts of the undenatured GP IIb-IIIa complex were added for 24 hr at 4 °C. The extent of specific GP IIb-IIIa complex formation with adsorbed fibrinogen was detected by attachment of specific monospecific antisera to GP IIb and GP IIIa to the plastic plate, monitored by an alkaline phosphatase labeled protein A probe as described in Table 1. Color formation indicative of bound GP IIb-IIIa is expressed on the ordinate by the change in absorbance at 405 nm. From Nachman and Leung (1982).

divalent cation dependent. Glycoprotein IIb and IIIa are present as major components in the membrane as a calcium-binding macromolecular complex. These membrane glycoproteins are markedly diminished to absent in Glanzmann's thrombasthenia, a hemorrhagic disorder characterized by absent platelet aggregation (see Chapter 16). Glycoprotein IIb and IIIa, as isolated from the platelet membrane, can be separated into structurally distinct and antigenically different components (Leung *et al.*, 1981). As discussed more extensively in Chapter 9, several lines of evidence strongly suggest that the GP IIb-IIIa complex is the fibrinogen receptor. Mixtures of monospecific, polyclonal rabbit antisera to the individual proteins have been used to demonstrate direct binding of the GP IIb-IIIa complex to fibrinogen using an enzyme-linked immunospecific assay (ELISA) system (Figure 3). This binding is specific, saturable, and calcium dependent. The receptor-binding site appears to require the native macromolecular complex isolated in an undenatured condition as evidenced by the fact that mixtures of individually purified and partially denatured GP IIb and GP IIIa do not bind to fibrinogen (Table 1).

Both GP IIb and GP IIIa are normally exposed as a complex on the unstimulated platelet surface which does not bind fibrinogen (McEver *et al.*, 1983; see also Chapter 4). Membrane stimulation by agonists may cause changes or new spatial interactions within the macromolecular membrane complex that are necessary for the generation of the receptor-binding site. Recent studies suggest that the binding of fibrinogen to the platelet surface is a necessary but insufficient condition for the platelet aggregation process, and that membrane fluidity and receptor clustering may be required (Peers-

Table 1. Complex Formation of Isolated Purified GP IIb and GP IIIa with Fibrinogen [a]

Glycoproteins	$10^4 \times \Delta A405$ min^{-1}
GP IIb-IIIa, purified and denatured	7.1 ± 0.6
GP IIb-IIIa, undenatured mixture	112.3 ± 2.6
Control	8.1 ± 1.4

[a]Fibrinogen (6 μg/ml) in coating buffer was applied to the plastic wells for 18 hr at 4 °C and the plates washed. Isolated, purified, and partially denatured GP IIb and GP IIIa (added together, 1.44 μg/ml each) or the GP IIb-IIIa complex mixture (2.88 μg/ml) then added for 24 hr at 4 °C. Controls included purified human albumin. After washing, a mixture of anti-GP IIb and anti-GP IIIa γ-globulin (10 μg/ml each) was added for 24 hr at 4 °C. After further washing, alkaline phosphatase-labeled protein A was added to 3 hr at room temperature. The substrate, *p*-nitrophenyl phosphate, was added and color development was followed in a spectrophotometer at 405 nm. The reaction was expressed as the enzymatic activity of the bound alkaline phosphatase. The undenatured GP IIb-IIIa complex mixture was isolated by lentil lectin affinity chromatography. The individual subunits, GP IIb and GP IIIa, were further purified by electrophoretic elution of sodium dodecyl sulfate-polyacrylamide gels. From Nachman and Leung (1982).

chke and Zucker 1981; Lee *et al.*, 1981). Topographic analyses of GP IIb and GP IIIa in the platelet membrane utilizing immunoelectron microscopy demonstrate random distribution of the two glycoproteins in the unstimulated state. Following thrombin stimulation, large clusters of GP IIb-IIIa are demonstrated (Polley *et al.*, 1981). In the absence of exogenously added fibrinogen, thrombin stimulation of the platelet induced secretion of endogenous fibrinogen that co-clustered on the platelet surface with GP IIb-IIIa complexes. Figure 4 indicates co-clustering of GP IIIa and fibrinogen when platelets were stimulated with thrombin in the absence of exogenous fibrinogen. Thus intracellular alpha granule fibrinogen may serve as the binding ligand across adjacent aggregated platelets.

Molecular models relating fibrinogen binding to the GP IIb-IIIa complex in platelet aggregation are shown in Figure 5. In the resting state, the receptor-binding site for fibrinogen is not expressed. Receptor induction subsequent to membrane stimulation and activation is a prerequisite for fibrinogen binding and cell aggregation. The precise nature of the receptor induction is not fully known. Figure 5 illustrates two hypotheses. One is that conformational modification of the GP IIb-IIIa membranous complex results in the acquisition of a binding function. The alternative hypothesis is that GP IIb and GP IIIa are discrete, individual glycoproteins and that complex formation only results from agonist stimulation. Data suggesting the presence of approximately 40,000 fibrinogen-binding sites per platelet and a similar number of GP IIb-IIIa complexes per platelet support these models and suggest a 1:1 stoichiometry. Recent experimental data demonstrate that the major platelet-reactive sites on the fibrinogen molecule reside on the carboxy terminal region of the γ-chain with the D-domains (Hawiger *et al.*, 1982, Marguerie *et al.*, 1982). Two sites per fibrinogen molecule thus allow the symmetrical fibrinogen to "bridge" adjacent activated platelets across identical receptor complexes.

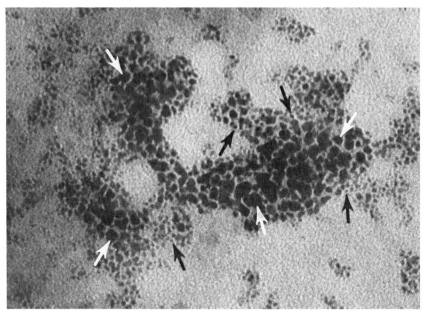

Figure 4. Thrombin-treated platelets reacted with antibody to platelet membrane GP IIIa followed by ferritin-conjugated goat anti-rabbit γ-globulin and then followed by gold-conjugated antifibrinogen. Stained with bismuth subnitrate. Mixed clusters of ferritin (black arrows) and gold (white arrows) are shown (magnification, 278,400×). No significant difference in the uptake and spatial organization of gold was seen when the platelets were stimulated with thrombin in the presence or absence of exogenous fibrinogen.

5. SECONDARY AGGREGATION MECHANISMS

From an operational point of view, as delineated by aggregometer studies, platelet aggregation can be viewed as occurring in two sequences (Figure 6). Primary aggregation reflects the initial responses of the stimulated platelet associated with fibrinogen receptor induction and fibrinogen binding, and secondary aggregation reflects the consequences of secretion and the participation of alpha-granule proteins in membrane-mediated events. As discussed more extensively in Chapter 8, recent studies demonstrate specific binding of fibronectin and thrombospondin, in addition to fibrinogen and VWF, to thrombin-activated platelets.

An interesting example of alteration of the platelet membrane following the initiation of secretion is the development of a membrane-bound lectin activity that may be important in aggregation by binding to a specific receptor on other platelets (Gartner *et al.*, 1981; see Chapter 12). Thrombospondin, also termed thrombin-sensitive protein and GP G, is an alpha-granule constituent that has a mol. wt. of 450,000 (Lawler *et al.*, 1978), binds to platelet membrane in a calcium-dependent manner (Phillips, 1980), and is the endogenous lectin of human platelets (Jaffe *et al.*, 1982). Thus purified thrombospondin agglutinates formalin-fixed, trypsin-treated sheep erythrocytes, an activity blocked by glucosamine and mannosamine, but not by their N-acetyl derivatives. In addition, thrombospondin blocked hemagglutination induced by for-

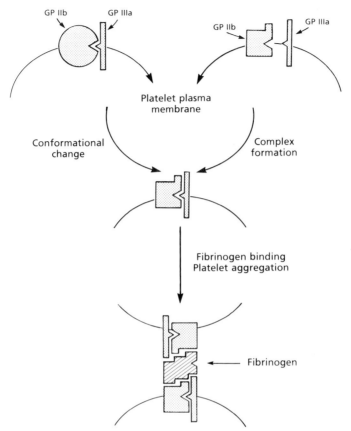

Figure 5. Molecular models of agonist-induced receptor induction and fibrinogen binding in platelets. The figure illustrates two hypotheses. On the left, GP IIb and GP IIIa are represented in a naturally occurring complex in resting platelets. Upon agonist stimulation, this complex undergoes a conformational change to be the fibrinogen receptor. On the right, GP IIb and GP IIIa are represented as separated individual glycoproteins that form a complex only after agonist stimulation, then becoming the fibrinogen receptor.

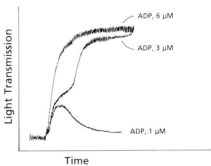

Figure 6. Platelet aggregation patterns in platelet rich plasma. 1. Reversible primary aggregation, ADP, 1 μM. 2. Aggregation without detectable separate primary and secondary waves; ADP, 6 μM. 3. Biphasic aggregation showing distinct primary and secondary waves; ADP, 3 μM.

Table 2. Complex Formation of
Thrombospondin with Fluid-Phase Proteins [a]

Proteins	$10^4 \times \Delta A405$ min^{-1}
Fibrinogen	129.0 ± 6.4
Fibronectin	22.0 ± 0.6
Albumin	2.0 ± 0.6
γ-Globulin	4.0 ± 0.8
VWF	5.0 ± 1.4
α_2-Plasmin inhibitor	3.0 ± 1.9
Platelet GP IIb-IIIa	4.0 ± 1.1
Antithrombin III	1.4 ± 0.6

[a]Thrombospondin (4 μg/ml) in coating buffer was applied to plastic wells for 18 hr at 4 °C. After washing, the indicated fluid-phase proteins (4 μg/ml) were added for 24 hr at 4 °C. Complex formation was detected by attachment of the corresponding monospecific antisera to the plastic plate, monitored by an alkaline phosphatase labeled protein A probe as described in Table 1. Modified from Leung and Nachman (1982).

malin-fixed thrombin-treated platelets (Jaffe *et al.*, 1982). Using an ELISA system, it has been possible to demonstrate specific and saturable binding of fibrinogen to purified thrombospondin. Thrombospondin was also capable of binding to isolated fibronectin; however, no complex formation was noted with a number of other proteins including albumin, γ-globulin, VWF, α_2-plasmin inhibitor, GP IIb-IIIa, and antithrombin III (Table 2) (Leung and Nachman, 1982; Lahav *et al.*, 1982).

These observations suggest that fibrinogen plays a unique role in platelet aggregation. If primary aggregation depends on the initial reaction of fibrinogen with GP IIb-IIIa, how is this reaction related to the subsequent interactions of endogenous-secreted thrombospondin with the platelet surface and fibrinogen? In order to evaluate these relationships, binding studies were performed using the ELISA system to determine whether GP IIb-IIIa in the fluid phase competed with solid-phase thrombospondin for

Table 3. Influence on Fibrinogen-Thrombospondin Complex Formation by GP IIb-IIIa [a]

Protein coat	Fluid phase	Antibody probe	$10^4 \times \Delta A405$ min^{-1}
Thrombospondin	Fibrinogen	Antifibrinogen	102 ± 4
	Fibrinogen + GPIIb-IIIa	Antifibrinogen	106 ± 2
Fibrinogen	Thrombospondin	Antithrombospondin	58 ± 6
	Thrombospondin + GPIIb-IIIa	Antithrombospondin	51 ± 11
	GPIIb-IIIa	Anti-GPIIb + anti-GIIIa	64 ± 1
	Thrombospondin + GPIIb-IIIa	Anti-GPIIb + anti-GPIIIa	70 ± 9

[a]For experiments with thrombospondin-coated wells, the protein (4 μg/ml) was applied to plastic wells for 18 hr at 4 °C. After washing, fibrinogen (4 μg/ml in Ca^{2+} buffer) was added in the absence or presence of GP IIb-IIIa (4 μg/ml). For experiments with fibrinogen-coated wells, the protein (4 μg/ml) was applied to plastic wells. After washing, thrombospondin or GP IIb-IIIa were added separately or as a mixture. Complex formation was detected by the attachment of the indicated monospecific antiserum followed by a protein A probe as described in Table 1. From Leung and Nachman (1982).

Figure 7. Hypothetical models of platelet aggregation. Primary reversible aggregation is shown on the
left with the interaction between fibrinogen (φ) and
GP IIb-IIIa. The molecular model of GP IIb and GP
IIIa is the same as in Figures 2 and 5. Secondary,
irreversible aggregation is shown on the right, dia-
gramming the additional interaction of endogenous secreted thrombospondin (TSP) with the platelet surface
and fibrinogen to stabilize the platelet aggregate.

fibrinogen-complex formation. No inhibition of fibrinogen-complex formation with
thrombospondin was noted in the presence of the platelet membrane glycoproteins
(Table 3). In an alternative fashion, thrombospondin did not inhibit the binding of GP
IIb-IIIa to solid-phase fibrinogen nor did GP IIb-IIIa interfere with the binding of
thrombospondin to fibrinogen. These studies suggest that the fibrinogen binding sites
for thrombospondin are different from the fibrinogen binding sites for GP IIb-IIIa. In
view of the fact that the D-domains of fibrinogen γ-chains contain the recognition sites
for GP IIb-IIIa, it is attractive to postulate that the E-domain of fibrinogen may contain
the recognition site for thrombospondin. In recent studies using immunoelectron microscopy, we have shown that thrombospondin specifically co-clusters with fibrinogen
in the stimulated platelet membrane (A. Asch, manuscript in preparation). The precise
molecular assembly of thrombospondin, fibrinogen, and GP IIb-IIIa on the activated
platelet membrane remains to be fully clarified. A hypothetical model illustrating the
molecular events occurring on the membrane during secondary aggregation compared
with the previously described sequence of events in primary aggregation is shown in
Figure 7. The symmetrical fibrinogen molecule linking GP IIb-IIIa receptor sites on
adjacent platelets is associated with primary aggregation (Figure 7, left). The bridging
complex is reinforced and eventually converted into an irreversible state following
secretion by the attachment of thrombospondin to a different site on the intercellular
fibrinogen bridge (Figure 7, right). The functional roles of the other alpha-granule
proteins in the platelet aggregation process remain to be determined.

REFERENCES

Barondes, S. H., 1981, Lectins: Their multiple endogenous cellular functions, *Annu. Rev. Biochem* **50:**207–
    231.
Bennett, J. S., and Vilare, G., 1979, Exposure of platelet fibrinogen receptors by ADP and epinephrine, *J.
    Clin. Invest.* **64:**1393–1401.
Berger, E. A., and Armant Randall, E., 1982, Discoidins I and II: Common and unique regions on two
    lectins implicated in cell-cell cohesion in *Dictyostelium discoideum, Proc. Natl. Acad. Sci. U.S.A.*
    79:2162–2166.
Berndt, M. C., Gregory, C., Chong, B. H., Zola, H., and Castaldi, P. A., 1983, Additional glycoprotein
    defects in Bernard-Soulier's syndrome: Confirmation of genetic basis by parental analysis, *Blood*
    62:800–807.
Blumberg, D. D., Margolskee, J. P., Barklis, E., Chung, S. N., Cohen, N. S., and Lodish, H. F., 1982,
    Specific cell–cell contacts are essential for induction of gene expression during differentiation of
    Dictyostelium discoideum, *Proc. Natl. Acad. Sci. U.S.A.* **79:**127–131.

Clemetson, K. J., McGregor, J. L., James, E., Dechavanne, M., and Luscher, E., 1982, Characterization of the platelet membrane glycoprotein abnormalities in Bernard-Soulier syndrome and comparison with normal by surface-labeling techniques and high resolution two-dimensional gel electrophoresis, *J. Clin. Invest.* **70:**304–311.

Cohen, I., Glaser, T., and Seligsohn, U., 1975, Effect of ADT and ATP on bovine fibrinogen and ristocetin-induced platelet aggregation in Glanzman's thrombasthenia, *Br. J. Haematol.* **31:**343–347.

Counts, R. B., Poskell, S. L., and Elgee, S. K., 1978, Disulfide bonds and the quarternary structure of Factor VIII/von Willebrand factor, *J. Clin. Invest.* **62:**702–709.

Frazier, W., and Glaser, L., 1979, Surface components and cell recognition, *Annu. Rev. Biochem.* **48:**491–523.

Gartner, K. T., Gerrard, J. M., White, J. G., and Williams, D. C., 1981, Fibrinogen is the receptor for the endogenous lectin of human platelets, *Nature (London)* **289:**688–690.

Ginsberg, M. H., Forsyth, J., Lightsey, A., Chediak, J., and Plow, E. F., 1983, Reduced surface expression and binding of fibronectin by thrombin-stimulated thrombasthenic platelets, *J. Clin. Invest.* **71:**619–624.

Gralnick, H., and Coller, B., 1983, Platelets stimulated with thrombin and ADP bind von Willebrand factor to different sites than platelets stimulated with ristocetin, *Clin. Res.* **31:**482A.

Grant, R. A., Zucker, M. B., and McPherson, J., 1976, ADP-induced inhibition of von Willebrand factor-mediated platelet agglutination, *Am. J. Physiol.* **230:**1406–1410.

Hawiger, J., Timmons, S., Kloczewiak, M., Strong, D. D., and Doolittle, R. R., 1982, $\gamma \alpha$ chains of human fibrinogen possess sites reactive with human platelet receptors, *Proc. Natl. Acad. Sci. U.S.A.* **79:**2068–2071

Jaffe, E. A., Hoyer, L., and Nachman, R. L., 1973, Synthesis of AHF antigens by cultured human endothelial cells, *J. Clin. Invest.* **52:**2757–2764.

Jaffe, E. A., Leung, L. L. K., Nachman, R. L., Levin, R., and Mosher, D. F., 1982, Thrombospondin is the endogenous lectin of human platelets, *Nature (London)* **295:**246–248.

Lehav, J., Schwartz, M. A., and Hynes, R. O., 1982, Analysis of platelet adhesion with a radioactive chemical crosslinking reagent: Interaction of thrombospondin with fibronectin and collagen, *Cell* **31:**253–262.

Lawler, J. W., Slater, H. S., and Coligan, J. E., 1978, Isolation and characterization of a high molecular weight glycoprotein from human platelets, *J. Biol. Chem.* **273:**8609–8616.

Lee, H., Nurden A., and Caen, J. P., 1981, Relationship between fibrinogen binding and the platelet glycoprotein deficiencies in Glanzmann's thrombasthenia Type I and Type II, *Br. J. Haematol.* **48:**47–57.

Leung, L. L. K., and Nachman, R. L., 1982, Complex formation of platelet thrombospondin with fibrinogen, *J. Clin. Invest.* **70:**542–549.

Leung, L. L. K., Kinoshita, T., and Nachman, R. L., 1981, Isolation, purification, and partial characterization of platelet membrane glycoproteins IIb and IIIa, *J. Biol. Chem.* **256:**1994–1997.

Marguerie, G. A., Plow, E. F., and Edgington, T. S., 1979, Human platelets possess an inducible and saturable receptor specific for fibrinogens, *J. Biol. Chem.* **254:**5357–5363.

Marguerie, G. A., Ardaillon, N., Cherel, G., and Plow, E. F., 1982, The binding of fibrinogen to its platelet receptor, *J. Biol. Chem.* **257:**11872–11875.

McEver, R. P., Bennett, E. M., and Martin, M. N., 1983, Identification of two structurally and functionally distinct sites on human platelet membrane glycoprotein IIb-IIIa using monoclonal antibodies, *J. Biol. Chem.* **258:**5264–5275.

Nachman, R. L., 1982, von Willebrand's disease: A clinical and molecular enigma, *West. J. Med.* **136:**318–325.

Nachman, R. L., and Leung, L. L. K., 1982, Complex formation of platelet membrane glycoproteins IIb and IIIa with fibrinogens, *J. Clin. Invest.* **69:**263–269.

Nachman, R. L., Jaffe, E. A., and Weksler, B. W., 1977, Immunoinhibition of ristocetin induced platelet aggregation, *J. Clin. Invest.* **59:**143–148.

Nachman, R. L., Jaffe, E. A., Miller, C., and Brown, W. T., 1980a, Structural analysis of factor VIII antigen in von Willebrand's disease, *Proc. Natl. Acad. Sci. U.S.A.* **77:**6832–6836.

Nachman, R. L., Jaffe, E. A., and Ferris, B., 1980b, Peptide map analysis of normal plasma and platelet factor VIII antigen, *Biocehm. Biophys. Res. Commun.* **92:**1208–1214.

Olive, L. S., 1975, *The Mycetozoans,* Academic Press, New York, p. 163.

Over, J., Bouma, B. N., Sixma, J. J., Bolhuis, P. A., and Vlooswijk, R. A. A., 1980, Heterogeneity of human factor VIII-III. Transition between forms of factor VIII present in cryoprecipitate and in cryosupernatant plasma, *J. Lab. Clin. Med.* **95:**323–334.

Peerschke, E. I., and Zucker, M. B., 1981, Fibrinogen receptor exposure and aggregation of human blood platelets produced by ADP and chilling, *Blood* **57:**663–668.

Phillips, D. R., Jennings, L. K., and Prasanna, H. R., 1980, Ca^{2+}-mediated association of glycoprotein G (thrombin sensitive protein, thrombospondin) with human platelets, *J. Biol. Chem.* **255:**11629–11632.

Polley, M. J., Leung, L. L. K., Clark, F., and Nachman, R. L., 1981, Thrombin induced platelet membrane glycoprotein IIb and IIIa complex formation: An electron microscope study, *J. Exp. Med.* **154:**1058–1068.

Rand, J. H., Gordon, R. E., Sussman, I. I., Chu, S. V., and Solomon, V., 1982, Electron microscopic localization of a factor-VIII related antigen in adult human blood vessels, *Blood* **60:**627–634.

Ruan, C., Tobelem, G., McMichael, A. J., Drouet, L., Legrand, Y., Degos, L., Kieffer, N., Lee, H., and Caen, J. P., 1981, Monoclonal antibody to human platelet glycoprotein I, *Br. J. Haematol.* **49:**511–519.

Ruggeri, Z. M., Bader, R., and De Marco, L., 1982, Glanzmann's thrombasthenia: Deficient binding of von Willebrand factor to thrombin-stimulated platelets, *Proc. Natl. Acad. Sci. U.S.A.* **79:**6038–6041.

Ruggeri, Z. M., Bader, R., and Zimmerman, T. S., 1983a, High affinity interaction of platelet von Willebrand factor with distinct platelet membrane sites, *Clin. Res.* **31:**322A.

Ruggeri, Z. M., De Marco, L., and Montgomery, R. R., 1983b, Platelets have more than one binding site for von Willebrand factor, *Clin. Res.* **31:**322A.

Wagner, D. D., and Marder, V. J., 1983, Biosynthesis of von Willebrand protein by human endothelial cells, *J. Biol. Chem.* **258:**2065–2067.

Weiss, H. J., Hoyer, L. W., Rickles, F. R., Varma, A., and Rogers, J., 1973, Quantitative assay of a plasma factor deficient in von Willebrand's disease that is necessary for platelet aggregation. Relationship to factor VIII procoagulant activity and antigen content, *J. Clin. Invest.* **52:**2708–2716.

12

Lectin–Carbohydrate Binding as a Model for Platelet Contact Interactions

T. Kent Gartner

1. INTRODUCTION

Lectins have been defined as sugar-binding proteins or glycoproteins of nonimmune origin that can agglutinate cells and/or glycoconjugates (Goldstein *et al.*, 1980). Lectins occur in plants and animals, and are devoid of enzymatic activity for the sugars to which they bind and do not require glycosidic hydroxyl groups on these sugars for their binding (Kocoureck and Horejsi, 1981; Barondes, 1981). One of the functions suggested for lectins is that of mediating direct cell–cell interactions (reviewed in Phillips and Gartner, 1980; Barondes, 1981). The first data suggesting this functional activity was published by Rosen *et al.* (1973), who showed that a lectin expressed by the cellular slime mold *Dictyostelium discoideum* may mediate the cohesiveness of these cells. Since these initial studies, numerous studies have appeared concerning the role of carbohydrate-binding proteins in aggregation of slime mold cells (reviewed in Bartles *et al.*, 1982; Barondes, 1981). The concept developed from these studies is that lectins, bound to carbohydrate-containing receptors on adjacent cells, mediate cell–cell interaction. This concept has been utilized to explain the interactions of a wide variety of cells, including sea urchin egg–sperm interactions (Glabe *et al.*, 1982), the attachment of *Rhizobium trifolii* to the roots of clover (Dazzo, 1981), attachment of *Escherichia coli* to mucosal cells (Ofek *et al.*, 1977), the attachment of the Sendai virus to host cells (Markwell *et al.*, 1981), and phagocytosis by human neutrophils (Doolittle *et al.*, 1983).

T. Kent Gartner • Department of Biology, Memphis State University, Memphis, Tennessee 38152.

2. DISCOVERY OF THE ENDOGENOUS PLATELET LECTIN

The studies of the role of lectins in the aggregation of slime mold amebae resulted in an experimental approach that underlies many of the studies done to determine if lectin–carbohydrate binding mediates mammalian cell–cell interactions. This approach is based on the comparative abilities of cohesive and noncohesive cells to agglutinate formalin-fixed erythrocytes. The cells of interest are screened for expression of hemagglutination activity to see if the ability to agglutinate red cells correlates with the acquisition of cohesiveness. The assay used to measure the hemagglutinin activity is usually a modification of the microtiter plate assay described by Rosen *et al.* (1973). In this assay, the agglutinin present on the membrane surface of the test cell binds to receptors on the fixed erythrocytes and cross-links them. As in the case with most hemagglutination assays (Sharon and Lis, 1972), the erythrocytes used to detect cellular agglutinins must be treated with trypsin before fixation with formaldehyde. If hemagglutination is observed, the sensitivity of the agglutinin activity to inhibition by various carbohydrate compounds is determined. The agglutinin is identified as a lectin if it is inhibited by specific carbohydrates. As a preliminary test of the role of the lectin in mediating cell–cell interactions, the carbohydrate that inhibits the hemagglutination activity of the lectin is then tested for its ability to directly block aggregation (cohesion) of the test cells. Inhibition by the carbohydrate in the absence of apparent physiologically deleterious effects may indicate that the lectin mediates cell–cell interaction.

Platelet aggregation is a complicated process that depends on metabolic responses of the platelet to various stimuli and culminates in an altered plasma membrane surface. Unstimulated discoid platelets have a low affinity for each other, whereas the membrane surfaces of activated platelets are mutually cohesive, permitting platelet aggregation to occur. This feature makes platelets an ideal system to determine whether lectin–carbohydrate binding mediates the aggregation of these cells. In our initial studies (Gartner *et al.*, 1978), we compared the ability of control and α-thrombin-stimulated human platelets to agglutinate trypsinized, formaldehyde-fixed erythrocytes from several species. It was found that bovine and ovine erythrocytes were more extensively agglutinated by thrombin-activated platelets than control platelets (Figure 1). Erythrocytes from other sources were agglutinated similarly by control and activated platelets. These results suggested that an agglutinin was expressed on the surface of activated platelets that was not present on unstimulated platelets.

As an initial attempt to characterize the agglutinin on activated platelets, a variety of substances was tested for their ability to inhibit the agglutinin activity. The amino sugars galactosamine, glucosamine, and mannosamine inhibited the hemagglutination activity of thrombin-stimulated platelets, but not that of resting platelets. This is illustrated for galactosamine in Figure 1. Mannosamine was a more potent inhibitor than either galactosamine or glucosamine (Gartner *et al.*, 1978). In contrast, the *N*-acetylated derivatives of the amino sugars and other neutral sugars were without effect. The agglutinin was termed a lectin because this hierarchy of potency for inhibition by amino sugars was assumed to reflect sugar specificity in the function of the agglutinin. However, the agglutinin activity of thrombin-stimulated platelets was also inhibited by arginine and partially inhibited by lysine. This apparent broad specificity indicates that

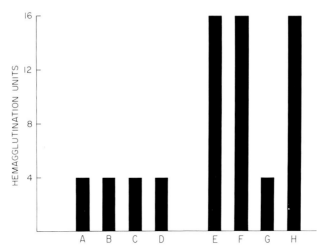

Figure 1. Expression of the endogenous platelet lectin. Agglutinin activity of control platelets (A–D) using formaldehyde-fixed bovine erythrocytes in: buffer alone (A), buffer plus galactose (B), buffer plus galactosamine (C), or buffer plus *N*-acetylgalactosamine (D). (E–H) Thrombin-treated platelets assayed in: buffer alone (E), buffer plus galactose (F), buffer plus galactosamine (G), or buffer plus *N*-acetylgalactosamine (H). All sugars were at 30 mM final concentration. This figure shows that thrombin activation of platelets causes the expression of a lectin capable of agglutinating bovine erythrocytes and that this lectin is inhibited by galactosamine but not by galactose or *N*-acetylgalactosamine. Published with permission from Gartner *et al.* (1978).

it may be more correct to term this activity a ''lectinlike agglutinin'' rather than a lectin. For consistency in nomenclature, however, the term ''lectin'' is preferred and used throughout this chapter.

The functional role of the endogenous platelet lectin was established by identifying the effects of lectin inhibitors on platelet function. It was found that the same sugars that inhibited the lectin hemagglutination activity (i.e., amino sugars) also inhibited thrombin-induced platelet aggregation. This inhibition occurred without significant modification of thrombin-induced secretion of serotonin (Gartner *et al.*, 1978). Sugars without effect on the agglutinin activity were also without effect on platelet aggregation. It is important to note that greater concentrations of amino sugars, relative to α-thrombin, blocked both aggregation and secretion (Gartner *et al.*, 1978). Although the reason for this inhibition was not investigated further, it is believed to be caused by a mechanism other than inhibition of lectin function.

3. SURFACE-BOUND VERSUS SOLUBLE LECTIN

The finding that the lectin activity and aggregation were inhibited by similar sugars suggested that the platelet–platelet interactions may be mediated by lectin–lectin receptor interactions (Gartner *et al.*, 1978). Two modes of expression of this activity were envisioned: one was that the lectin cross-linked erythrocytes as a soluble agent; the other was that the lectin was expressed on the surface of activated platelets.

One approach used to distinguish between these possibilities was to compare the lectin activities from platelets stimulated by different agonists, including α-thrombin, γ-thrombin, thrombocytin, and the Ca^{2+} ionophore, A23187 (Gartner et al., 1980a). The agonist γ-thrombin is a derivative of α-thrombin that causes platelets to aggregate and undergo the release reaction, but does not convert fibrinogen to fibrin at a significant rate (Fenton et al., 1977; Charo et al., 1977; Gartner and Walz, 1983). Thrombocytin also is a serine protease that causes platelets to aggregate and undergo the release reaction, but does not convert fibrinogen to fibrin (Kirby et al., 1979; Niewiarowski et al., 1979). The Ca^{2+} ionophore A23187 is an antibiotic that causes platelets

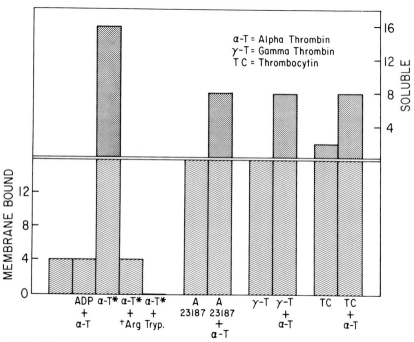

Figure 2. Platelet-bound and soluble hemagglutination activities of washed platelets. The platelets were tested directly (first bar on the left) or after the treatments described. Washed platelets plus 5 μM ADP were in Tyrode's solution without fibrinogen (bar 2). Washed platelets represented by bars 3–5 were exposed to 0.5 μ/ml of human α-thrombin for 1 min before testing or fractionation. Washed platelets represented by bars 6 and 7 were activated with a 2 μM concentration of A23187; bars 8 and 9, by 5 μg/ml of γ-thrombin, and bars 10 and 11, by 5 μg/ml of thrombocytin. The platelet-free supernatant fractions were derived from duplicate samples of these preparations and were activated with 0.5 μg/ml of α-thrombin. In one experiment (bar 5), α-thrombin-activated platelets and the platelet-free supernatant fraction of the activated platelets were treated with 20 μg/ml (final concentration) of trypsin and incubated at room temperature for 15 min, then the reaction was stopped by 200 μg/ml (final concentration) of soybean trypsin inhibitor. Control samples were treated with a trypsin preparation previously incubated with soy bean trypsin inhibitor.[†] All enhanced hemagglutination activities presented in Figure 2 were inhibited by 30 mM arginine, galactosamine, glucosamine, and mannosamine, but not by glutamine, galactose, glucose, mannose, and the N-acetylated derivatives of the amino sugars at the same concentration. Reprinted with permission from Gartner et al. (1980a).

to secrete the contents of their storage granules and aggregate (Feinman and Detwiler, 1974; White *et al.*, 1974). In each case, the soluble and platelet-bound hemagglutinin activities were assayed. Platelets stimulated with α-thrombin expressed both soluble and platelet-bound hemagglutinin activities (Figure 2). Both were inhibited by arginine and amino sugars. That both activities were protein or glycoprotein in nature was shown by their susceptibility to trypsin digestion. Platelets treated with A23187, γ-thrombin, or thrombocytin behaved differently than α-thrombin-treated platelets in that they expressed only platelet-bound lectin activity (Figure 2). Thrombocytin-stimulated platelets also expressed a low level of soluble agglutinin activity, but this was caused by thrombolectin, a lectin present in slight amounts in the thrombocytin preparation (Gartner *et al.*, 1980b). Thrombolectin could be distinguished from the platelet-derived lectin in that the former was blocked by lactose.

Stimulation of platelets with A23187, γ-thrombin, or thrombocytin resulted in the secretion of a latent form of the lectin. Treatment of the supernatant fractions of such preparations with α-thrombin caused the appearance of lectin activity. The common feature of the agonists A23187, α-thrombin, and thrombocytin is that they can, under appropriate conditions, induce the release reaction and subsequent aggregation (Feinman and Detwiler, 1974; White *et al.*, 1974; Charo *et al.*, 1977; Gartner and Walz, 1983; Niewiarowski *et al.*, 1979). However, since they cannot convert secreted fibrinogen to fibrin, as can α-thrombin, these results suggested that fibrin was a plausible candidate for the soluble agglutinin. As a test of this idea, ADP, γ-thrombin, thrombocytin, and α-thrombin were tested for their effect on the hemagglutinin activity of fibrinogen. Only α-thrombin converted fibrinogen into an active hemagglutinin. The hemagglutinin activity of fibrin was inhibited by the same concentrations of the same compounds that inhibit the soluble hemagglutinin activities of α-thrombin-stimulated platelets. Thus, the soluble agglutinin activity of α-thrombin-stimulated platelets is fibrin dependent. Although the agglutinin activities of the platelet-bound agglutinin and fibrin were inhibited by similar sugars, our experiments suggest that the platelet-bound hemagglutination activity is not dependent on trace amounts of fibrin. Platelets stimulated with A23187 in the presence or absence of 10 U/ml of the thrombin inhibitor, hirudin, were found to have identical amounts of the platelet-bound agglutinin activity (T. K. Gartner, unpublished observations). Thus two agglutinin activities can be measured, depending on the platelet stimulus used.

4. REGULATION OF LECTIN EXPRESSION

The soluble, fibrin-mediated activity was assumed to result from secretion of fibrinogen from α-granules of platelets followed by proteolysis; however, the source of the platelet-bound agglutinin was unknown. Two hypotheses were considered for the origin of this platelet-bound activity. One stated that the platelet-bound agglutinin was a consequence of platelet secretion, possibly representing the expression of granule-associated proteins on the platelet membrane surface. The other stated that the agglutinin resulted from a structural change in plasma membrane proteins and did not require secretion. To distinguish between these possibilities, the agglutinin activities of platelets treated by different agonists and antagonists were compared. Washed platelets in Tyrode's solution, treated with ADP, underwent a shape change, but did not aggregate because no exogenous fibrinogen was present. No soluble hemagglutination activity was observed in such preparations and the platelet-bound agglutinin activity was indistinguishable from that present in control platelets (Gartner et al., 1980a). Since platelets treated with ADP under these conditions do not undergo secretion (Mustard et al., 1975), these results suggested that expression of the platelet-bound agglutinin was secretion dependent. Platelets treated with α-thrombin in the presence of prostaglandin E_1 to inhibit secretion also had a level of agglutinin activity indistinguishable from resting platelets (Gartner et al., 1980a). Platelets exposed to α-thrombin inactivated with phenylmethylsulfonyl fluoride failed to express either form of the agglutinin activity (Gartner et al., 1980a). These results demonstrated that neither shape change alone, proteolysis of the platelet surface by thrombin, nor binding of proteolytically inactive thrombin is sufficient for expression of either form of the agglutinin activity.

Two approaches were used to evaluate the hypothesis that expression of the endogenous platelet lectin is controlled by secretion (Gartner and Walz, 1983). The first was based on the premise that primary aggregation induced by ADP or epinephrine in the presence of fibrinogen does not cause secretion of dense body and α-granule constituents (Charo et al., 1977). Platelets that had undergone primary aggregation were found to have a level of hemagglutination activity characteristic of resting platelets, suggesting that activation of platelets is insufficient to cause expression of the endogenous platelet lectin. The other approach used to determine whether expression of the agglutinin is controlled by secretion was based on studies using trifluoperazine, an inhibitor of calmodulin-mediated reactions. Trifluoperazine, under the appropriate conditions, was found to inhibit the expression of the platelet-bound agglutinin (Figure 3). Since this drug does not inhibit primary aggregation, but does inhibit secretion of serotonin, α-granule contents, and secondary aggregation

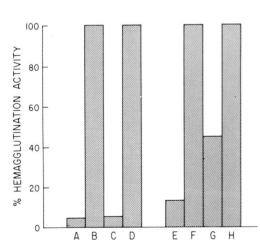

Figure 3. Effect of trifluoperazine on the expression of the platelet-bound lectin. Washed platelets were tested for hemagglutination activity directly or after exposure to 1.5 μM (final concentration) of A23187 or 5 μg/ml of γ-thrombin in the presence or absence of 100 μM final concentration of trifluoperazine. The trifluoperazine was added 1 min before or 1 min after stimulation of the platelets. (A, E) washed platelets; (B) washed platelets stimulated with A23187; (C) washed platelets exposed to trifluoperazine before exposure to A23187; (D) washed platelets stimulated with A23187 before exposure to trifluoperazine; (F) washed platelets stimulated with γ-thrombin; (G) washed platelets treated with trifluoperazine before exposure to γ-thrombin; (H) washed platelets treated with trifluoperazine after exposure to γ-thrombin. These data show that inhibition of secretion with trifluoperazine inhibits expression of the platelet-bound lectin.

(Kindness *et al.*, 1980; White and Raynor, 1980; Gartner and Walz, 1983), this finding supports the hypothesis that expression of the endogenous platelet lectin is controlled by secretion.

5.   LECTIN ACTIVITY IN INHERITED BLEEDING DISORDERS

The agglutinin activities of platelets from patients with well-characterized bleeding disorders were studied in an effort to correlate a platelet compositional deficiency with absent or diminished expression of the agglutinin activity. In these studies the platelet-bound agglutinin activities of normal and abnormal platelets were compared (Gartner *et al.*, 1981b). Platelets from patients with congenital afibrinogenemia, Glanzmann's thrombasthenia, and the Hermansky-Pudlak syndrome had a normal amount of the platelet-bound agglutinin activity (Figure 4). This suggested that this activity was not dependent on platelet fibrinogen, platelet dense bodies, or platelet membrane GP IIb-IIIa, respectively (see Table 1). In contrast, platelets from patients with the gray platelet syndrome had a low level of activity (Figure 4), indicating that

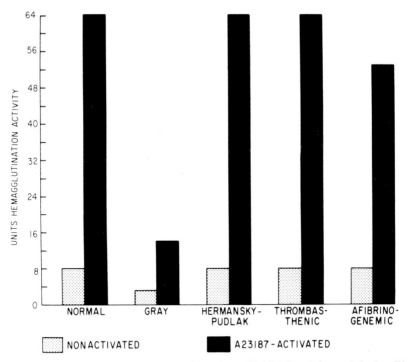

Figure 4. Platelet-bound lectin agglutinin activity of ionophore A23187-activated abnormal platelets. Washed platelets were tested directly or 1 min after exposure to 2 μM A23187. These studies show that the lectin activity of activated platelets from patients with Hermansky-Pudlak syndrome, Glanzmann's thrombasthenia, and afibrinogenemia is similar to controls, whereas that from patients with the gray platelet syndrome is markedly lower. This suggests that the agglutinin is of α-granule origin.

Table 1. Abnormal Platelets Used in Studies to Characterize Agglutinin Activity[a]

Platelet source	Compositional deficiency	Platelet fibrinogen	Other α-granule proteins
Hermansky-Pudlak syndrome	Dense bodies ($\approx$1% of normal)[b]	100%	100%
Gray platelet syndrome	α-Granules (<15% of normal)[c]	<10% (0.3 μg/10^8 platelets)	<10%
Glanzmann's thrombasthenia	Glycoproteins IIb, IIIa (membrane)[d]	$\approx$10%[f]	100%
Congenital afibrinogenemia	Fibrinogen[e]	<10% (<0.6 μg/10^8 platelets)	100%

[a]Reprinted with permission from Gartner *et al.* (1981b). Further information on these disorders is to be found in Chapter 15.
[b]Gerrard *et al.*, 1975.
[c]Gerrard *et al.*, 1980; White, 1972.
[d]Phillips and Agin, 1977.
[e]Shapiro *et al.*, 1980.
[f]Gartner *et al.*, 1981b.

the platelet-bound agglutinin originates from α-granules. This finding provides an explanation for the decreased aggregation response of gray platelets (Gerrard *et al.*, 1980; Levy-Toledano *et al.*, 1981).

6. IDENTIFICATION OF THE LECTIN

α-Granule proteins have been tested to determine whether they have activity similar to that of the platelet-bound agglutinin. Several proteins have been found to lack activity, including platelet-derived growth factor, platelet factor 4, and β-thromboglobulin (Phillips and Gartner, 1980). Fibronectin, another α-granule protein (Zucker *et al.*, 1979), was also examined because it has agglutinin activity that is inhibited by compounds similar to those which inhibit the agglutinin expressed on platelets (Gartner *et al.*, 1977, 1978; Yamada *et al.*, 1975). It was found, however, that plasma fibronectin was not active in the conditions used to measure the platelet agglutinin (Gartner *et al.*, 1981a; Jaffe *et al.*, 1982). Also, cell-surface fibronectin was without activity and monospecific anti-cell-surface fibronectin antibodies (kindly supplied by K. M. Yamada) had no effect on platelet hemagglutination activity (T. K. Gartner, unpublished observations). In view of these observations, it appears that fibronectin is not the endogenous platelet lectin.

Thrombospondin, an α-granule protein secreted from platelets (Hagen, 1975), is a trimeric glycoprotein consisting of three disulfide-linked subunits with mol. wt. of 150,000 (Lawler *et al.*, 1978). After secretion, this glycoprotein binds to the platelet surface (George *et al.*, 1980; Phillips *et al.*, 1980). In 1982, Jaffe *et al.* reported that thrombospondin was the endogenous platelet agglutinin. They found that thrombospondin agglutinated trypsinized, formaldehyde-fixed sheep erythrocytes and that this agglutination was inhibited by amino sugars, but not by the *N*-acetylated forms of the

Figure 5. Effect of antithrombospondin serum on the platelet-bound lectin activity. The platelet-bound lectin activity of control and activated platelets was determined according to our standard procedures. (A) Control platelets. (B) Platelets stimulated with γ-thrombin. (C) Platelets stimulated with γ-thrombin, but tested for lectin activity in the presence of antithrombospondin serum. (D) Control platelets. (E) Platelets stimulated with A23187. (F) Platelets stimulated with A23187, but tested for lectin activity in the presence of preimmune serum. (G) Platelets stimulated with A23187 and tested for lectin activity in the presence of antithrombospondin serum. (H) Platelets stimulated with A23187 and tested against a 1:1 dilution of the antithrombospondin serum. This figure shows that antithrombospondin serum inhibits the platelet-bound lectin activity.

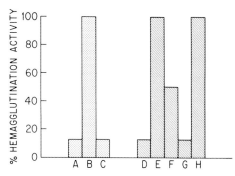

amino sugars. The agglutinin activity of thrombospondin has also been examined in my laboratory using purified thrombospondin and antithrombospondin sera (Figure 5) (Gartner *et al.*, 1983). Although these data demonstrate that the expression of the platelet-bound agglutinin is dependent on thrombospondin, it does not establish that thrombospondin, but itself, is entirely responsible for the agglutinin activity expressed on the platelet membrane surface.

The binding characteristics of secreted thrombospondin and the behavior of the platelet-bound lectin support the conclusion that platelet thrombospondin is the endogenous platelet lectin. Most secreted thrombospondin binds to the platelet surface in a Ca^{2+}-dependent manner (Phillips *et al.*, 1980). However, a small, but easily demonstrable, amount of secreted thrombospondin binds to the platelet plasma membrane in the absence of free Ca^{2+} (George *et al.*, 1980; Gartner and Dockter, 1983). It is this thrombospondin that is the platelet-bound lectin measured in the hemagglutination assay (Gartner *et al.*, 1983; Gartner and Dockter, 1983). Interestingly, the thrombospondin causing the platelet-bound lectin activity of platelets stimulated with A23187 is not removed by extensive washing in the presence of excess EDTA (Gartner and Dockter, 1983). Thus, the thrombospondin that is the platelet-bound agglutinin binds to the platelet surface in a manner that is independent of free Ca^{2+}.

7. RECEPTORS FOR THE PLATELET-DERIVED LECTIN

A variety of experimental evidence suggests that fibrinogen is a receptor for the platelet-bound agglutinin. This was suggested by studies showing that platelets fixed in normal plasma, but not in afibrinogenemic plasma, were agglutinated by thrombin-stimulated platelets (Gartner *et al.*, 1981a). Also, fibrinogen, in the absence of Ca^{2+}, was found to inhibit the platelet-bound agglutinin activity expressed by platelets stimulated with A23187, γ-thrombin, or thrombocytin. Albumin and plasma fibronectin were without effect. Two additional studies support the concept that fibrinogen is a receptor for the endogenous platelet agglutinin. Leung and Nachman (1982) found that human fibrinogen formed a complex with platelet thrombospondin in an enzyme-linked immunosorbent assay. The formation of this complex was partially inhibited by amino sugars, but not by N-acetylated amino sugars. In addition, Agam and Livne (1983) demonstrated that formaldehyde-fixed platelets were incorporated into platelet aggregates formed in response to either A23187 or α-thrombin but only when the fixed platelets had fibrinogen covalently attached to their membrane surface. These studies suggest that the platelet-bound agglutinin binds to fibrinogen on the platelet membrane surface, but do not exclude other receptors, which are suggested by the presence of lectin activity on thrombin-activated thrombasthenic platelets.

8. ROLE OF THE LECTIN IN PLATELET AGGREGATION

A role for the endogenous platelet agglutinin in platelet aggregation is the subject of current experimentation. The finding that the same compounds that inhibited the function of the agglutinin also inhibited platelet aggregation suggested that the lectin mediates secretion-dependent platelet aggregation (Gartner *et al.*, 1978). However, these same compounds also inhibit the binding of fibrinogen to ADP-stimulated platelets (Kinlough-Rathbone *et al.*, 1979). Thus, these compounds may inhibit aggregation by inhibiting the binding of fibrinogen to its receptors. This does not exclude a role for the platelet lectin in secretion-dependent aggregation, however, and Nurden *et al.* (1983) have recently demonstrated that monovalent Fab fragments of mono-specific antithrombospondin IgG completely inhibited the aggregation of platelets stimulated with α-thrombin, whereas the same antibodies had little or no effect on the primary aggregation of washed platelets in response to fibrinogen and ADP. These observations, coupled with what is known about the agglutinin, indicate that the endogenous platelet lectin mediates the aggregation of platelets stimulated by α-thrombin. Further work is required to determine if the lectin also mediates secretion-dependent aggregation of platelets in response to other agonists. Finally, these data do not exclude a role for other α-granule proteins in enhancing the function of thrombospondin in aggregation.

Acknowledgments. Research support for the author's unpublished data included in this chapter was provided by USPHS Grant HL 23010.

REFERENCES

Agam, G., and Livne, A., 1983, Passive participation of fixed platelets in aggregation facilitated by covalently bound fibrinogen, *Blood* **61**:186–191.

Barondes, S. H., 1981, Lectins: Their multiple endogenous cellular functions, *Annu. Rev. Biochem.* **50**:207–231.

Bartles, J. R., Frazier, W. A., and Rosen, S. D., 1982, Slime mold lectins, *Int. Rev. Cytol.* **75**:61–99.

Charo, I. F., Feinman, R. D., and Detwiler, T. C., 1977, Interrelations of platelet aggregation and secretion, *J. Clin. Invest.* **60**:866–873.

Dazzo, F. B., 1981, Bacterial attachment as related to cellular recognition in the rhizobium-legume symbiosis, *J. Supramol. Struct.* **16**:29–41.

Doolittle, R. L., Packman, C. H., and Lichtman, M. A., 1983, Amino-sugars enhance recognition and phagocytosis of particles by human neutrophils, *Blood* **62**:697–701.

Feinman, R. D., and Detwiler, T. C., 1974, Platelet secretion induced by divalent cation ionophores, *Nature (London)* **249**:172–173.

Fenton, J. W. II, Landis, B. H., Walz, D. A., and Finlayson, J. S., 1977, Human thrombins, in: *Chemistry and Biology of Thrombin* (R. L. Lundblade, J. W. Fenton II, and K. G. Mann, eds.), Ann Arbor Science Publishers, Ann Arbor, Michigan, pp. 43–70.

Gartner, T. K., and Dockter, M. E., 1983, Secreted platelet thrombospondin binds monovalently to platelets and erythrocytes in the absence of free Ca^{2+}, *Thromb. Res.* **33**:19–30.

Gartner, T. K., and Walz, D. A., 1983, TSP inhibits the expression of the endogenous platelet lectin and secretion of α-granules, *Thromb. Res.* **29**:63–74.

Gartner, T. K., Williams, D. C., and Phillips, D. R., 1977, Platelet plasma membrane lectin activity, *Biochem. Biophys. Res. Commun.* **79**:592–599.

Gartner, T. K., Williams, D. C., Minion, F. C., and Phillips, D. R., 1978, Thrombin-induced platelet aggregation is mediated by a platelet plasma membrane-bound lectin, *Science* **200**:1281–1283.

Gartner, T. K., Phillips, D. R., and Williams, D. C., 1980a, Expression of thrombin-enhanced platelet lectin activity is controlled by secretion, *Febs Lett.* **113**:196–200.

Gartner, T. K., Stocker, K., and Williams, D. C., 1980b, Thrombolectin: A lectin isolated from *Bothrops atrox* venom, *Febs Lett.* **117**:13–16.

Gartner, T. K., Gerrard, J. M., White, J. G., and Williams, D. C., 1981a, Fibrinogen is the receptor for the endogenous lectin of human platelets, *Nature (London)* **289**:688–690.

Gartner, T. K., Gerrard, J. M., White, J. G., and Williams, D. C., 1981b, The endogenous lectin of human platelets is an α-granule component, *Blood* **58**:153–157.

Gartner, T. K., Doyle, M. J., and Mosher, D. F., 1983, Anti-TSP serum inhibits the endogenous lectin activity of A23187 activated and γ-thrombin activated human platelets, *Thromb. Haemostasis* **50**:124a.

George, J. N., Lyons, R. M., and Morgan, R. K., 1980, Membrane changes associated with exposure of actin on the platelet surface after thrombin-induced secretion, *J. Clin. Invest.* **66**:1–9.

Gerrard, J. M., White, J. G., Roa, G. H. R., Krivit, W., and Witkop, C. J., Jr., 1975, Labile aggregation stimulating substance (LASS): The factor from storage pool deficient platelets correcting defective aggregation and release of aspirin-treated normal platelets, *Br. J. Haematol.* **29**:657–665.

Gerrard, J. M., Phillips, D. R., Rao, G. H. R., Plow, E. F., Walz, D. A., Ross, R., Harker, L. A., and White, J. G., 1980, Biochemical studies of two patients with the gray platelet syndrome-selective deficiency of platelet alpha granules, *J. Clin. Invest.* **66**:102–109.

Glabe, C. G., Grabel, L. B., Vacquier, V. D., and Rosen, S. D., 1982, Carbohydrate specificity of sea urchin sperm binding: a cell surface lectin mediating sperm-egg adhesion, *J. Cell Biol.* **94**:123–128.

Goldstein, I. J., Hughes, R. C., Monsigny, M., Osawa, T., and Sharon, N., 1980, What should be called a lectin?, *Nature (London)* **285**:66.

Hagan, I., 1975, Effects of thrombin on washed, human platelets: Changes in subcellular fractions, *Biochim. Biophys. Acta* **392**:242–254.

Jaffe, E. A., Leung, L. L. K., Nachman, R. L., Levin, R. I., and Mosher, D. F., 1982, Thrombospondin is the endogenous lectin of human platelets, *Nature (London)* **295**:246–248.

Kindness, G., Williamson, F. B., and Long, W. F., 1980, Inhibitory effect of trifluoperazine on aggregation of human platelets, *Thromb. Res.* **17**:549–554.

Kinlough-Rathbone, R. L., Chahil, A., Perry, D. W., Packham, M. A., and Mustard, J. F., 1979, Effect of amino sugars that block platelet lectin activity on fibrinogen binding to washed rabbit or human platelets, *Blood* **54**:249a.

Kirby, E. P., Niewiarowski, S., Stocker, K., Kettner, C., Shaw, E., and Brudzynski, T. M., 1979, Thrombocytin, a serine protease from *Bothrops atrox* venom. 1. Purification and characterization of the enzyme, *Biochemistry* **18**:3564–3570.

Kocoureck, J., and Horejsi, V., 1981, Defining a lectin, *Nature (London)* **290**:188.

Lawler, J. W., Slayter, H. S., and Coligan, J. E., 1978, Isolation and characterization of a high molecular weight glycoprotein from human blood platelets, *J. Biol. Chem.* **253**:8609–8616.

Leung, L. L. K., and Nachman, R. L., 1982, Complex formation of platelet thrombospondin with fibrinogen, *J. Clin. Invest.* **70**:542–549.

Levy-Toledano, S., Caen, J. P., Gorius-Breton, J., Rendu, F., Golenzer-Cywiner, C., Dupuy, E., Legrand,

Y., and Maclouf, J., 1981, Gray platelet syndrome: α-Granule deficiency, its influence on platelet function, *J. Lab. Clin. Med.* **98**:831–848.

Markwell, M. A. K., Svennerholm, L., and Paulson, J. C., 1981, Specific gangliosides function as host cell receptors for Sendai virus, *Proc. Natl. Acad. Sci. U.S.A.* **78**:5406–5410.

Mustard, J. F., Perry, D. W., Rathbone-Kinlough, R. L., and Packham, M. A., 1975, Factors responsible for ADP-induced release reaction of human platelets, *Am. J. Physiol.* **228**:1757–1765.

Niewiarowski, S., Kirby, E. P., Brudzynski, T. M., and Stocker, K., 1979, Thrombocytin, a serine protease from *Bothrops atrox* venom. 2. Interaction with platelets and plasma clotting factors, *Biochemistry* **18**:3570–3577.

Nurden, A. T., Hasitz, M., and Rosa, J-P., 1983, Inhibition of thrombin-induced platelet aggregation by a rabbit antibody against human platelet thrombospondin, *Thromb. Haemostasis* **50**:132a.

Ofek, I., Mirelman, D., and Sharon, N., 1977, Adherence of *Escherichia coli* to human mucosal cells mediated by mannose receptors, *Nature (London)* **265**:623–625.

Phillips, D. R., and Agin, P. P., 1977, Platelet membrane defects in Glanzmann's thrombasthenia, *J. Clin. Invest.* **60**:535–545.

Phillips, D. R., and Gartner, T. K., 1980, Cell recognition systems in eukaryotic cells, in: *Bacterial Adherence (Receptors and Recognition,* Series B, Volume 6) (E. H. Beachy, ed.), Chapman and Hall, London, pp. 401–438.

Phillips, D. R., Jennings, L. K., and Prasanna, H. R., 1980, Ca^{2+}-mediated association of glycoprotein G (thrombin sensitive protein, thrombospondin) with human platelets, *J. Biol. Chem.* **255**:11629–11632.

Rosen, S. D., Kafka, J. A., Simpson, D. L., and Barondes, S. L., 1973, Developmentally regulated, carbohydrate-binding protein in *Dictyostelium discoidium, Proc. Natl. Acad. Sci. U.S.A.* **70**:2554–2557.

Shapiro, R. S., Gerrard, J. M., Ramsay, N. K. C., Nesbit, M. E., Coccia, P. F., Stoddard, S. F., Plow, E. F., White, J. G., and Krivit, W., 1980, Selective deficiencies in collagen-induced platelet aggregation during L'asparaginase therapy, *Am. J. Pediatr. Hematol. Oncol.* **2**:207–212.

Sharon, N., and Lis, H., 1972, Lectins: Cell-agglutinating and sugar specific proteins, *Science* **177**:949–959.

White, G. C. II, and Raynor, S. T., 1980, The effects of trifluoperazine, an inhibitor of calmodulin, in platelet function, *Thromb. Res.* **18**:279–284.

White, J. G., 1972, Interaction of membrane systems in blood platelets, *Am J. Pathol.* **66**:295–312.

White, J. G., Rao, G. H., and Gerrard, J. M., 1974, Effects of ionophore A23187 in blood platelets, *Am. J. Pathol.* **77**:135–149.

Yamada, K. M., Yamada, S. S., and Pastan I., 1975, The major cell surface glycoprotein of chick embryo fibroblasts is an agglutinin, *Proc. Natl. Acad. Sci. U.S.A.* **72**:3158–3162.

Zucker, M. B., Mosesson, M. W., Broekman, M. J., and Kaplan, K. L., 1979, Release of platelet fibronectin (cold insoluble globulin) from alpha granules induced by thrombin or collagen; lack of requirement for plasma fibronectin in ADP-induced platelet aggregation, *Blood* **54**:8–12.

IV

Interactions of Platelet Membrane Glycoproteins with the Intracellular Cytoskeleton

13

The Organization of Platelet Contractile Proteins

Joan E. B. Fox

1. INTRODUCTION

Ever since 1959, when Bettex-Galland and Lüscher first demonstrated the presence of the contractile proteins actin and myosin in platelets (Bettex-Galland and Lüscher, 1959), it has been assumed that these proteins function in the responses of platelets to stimulation. The observation that the filopodia of stimulated platelets contain bundles of actin filaments (Nachmias, 1980; White, 1968; Zucker-Franklin *et al.*, 1967) suggested a role for contractile proteins in filopodia extension, whereas studies showing a ring of microfilaments surrounding the granules in activated platelets (White, 1974) indicated that secretion involves a contractile process. Other responses, such as the condensation of platelet aggregates and the retraction of fibrin clots, are even more obviously contractile in nature (Cohen *et al.*, 1982; van Deurs and Behnke, 1980).

Because contractile proteins appear to be involved in the responses of platelets to stimulation, detailed information about the organization and regulation of these proteins in unstimulated and stimulated platelets is essential for a full understanding of the mechanisms involved in platelet activation. This review will summarize the properties of the contractile proteins and describe changes that occur in the organization of these proteins upon platelet activation. Ways in which these changes may be regulated will then be discussed. Finally, the current knowledge concerning the manner in which these proteins are attached to platelet membranes will be reviewed.

2. PROPERTIES OF PLATELET CONTRACTILE PROTEINS

2.1. Actin

Actin is the most abundant protein in platelets. It has been estimated that it comprises 15–20% of the total platelet protein. Actin has been purified from platelets

Joan E. B. Fox • The Gladstone Foundation Laboratories for Cardiovascular Disease, Cardiovascular Research Institute, University of California, San Francisco, California 94140.

by several methods (Gordon *et al.*, 1977; Landon *et al.*, 1977; Rosenberg *et al.*, 1981b), and has been shown to exist in the β and γ isomeric forms found in other nonmuscle cells (Landon *et al.*, 1977). These isomeric forms have slightly different amino acid sequences compared with the α form found in skeletal muscle (Gordon *et al.*, 1977; Landon *et al.*, 1977). However, the three forms appear to have very similar functional properties (Korn, 1978, 1982). Actin monomers (otherwise known as globular or G-actin) have a $M_r = 42,000$ and are slightly asymmetrical (5.5×3.5 nm).

Under favorable conditions, monomers polymerize almost completely into filaments (also known as F-actin). Polymerization occurs in a two-step process consisting of nucleation and elongation. The rate-limiting step is the formation of "nuclei," which are thought to include at least three actin monomers (Korn, 1982; Pollard and Craig, 1982). Although nucleation is a slow process, it can be stimulated by certain ionic conditions such as the addition of 1 mM $MgCl_2$ or 0.1 M KCl. Nucleation is also induced by some proteins that complex more than one actin monomer or by an increase in the concentration of actin monomers. Once nuclei have formed, elongation continues by the rapid addition of monomers to either end of the growing filament.

Because the monomers are not symmetrical, the filaments that form have a defined polarity, which can be detected by electron microscopy of filaments that have been exposed to heavy meromyosin. Heavy meromyosin "decorates" the actin filaments with a characteristic arrowhead pattern (Ishikawa *et al.*, 1969) so that the two ends of a filament, termed the "barbed end" and the "pointed end," can be identified by the direction of the arrowheads. Although actin monomers can add onto either end of an actin filament, they do so much more rapidly at the barbed end than at the pointed end (Pollard and Mooseker, 1981). The addition of monomers continues until the concentration of actin monomers remaining in solution decreases to the "critical concentration." This concentration is affected both by temperature and ionic strength (Korn, 1982), and polymerization of actin will not occur unless it is exceeded. The critical concentration of platelet actin is similar to that of other nonmuscle actins and is ~30 μM at 25 °C and physiological ionic strengths (Gordon *et al.*, 1977).

Measurement of the rate constants for polymerization and depolymerization of actin at the two ends of heavy meromyosin-decorated filaments have shown that under certain ionic conditions (75 mM KCl and 1 mM $MgCl_2$) the critical concentration for polymerization at the barbed ends is lower than that for polymerization at the pointed ends (Pollard and Mooseker, 1981). Thus, under these conditions and at steady state, the concentration of free actin monomers is higher than the critical concentration at the barbed end of the filament and lower than the critical concentration at the pointed end. Actin monomers, therefore, can add onto the barbed end of the filament, traverse through the filament, and dissociate from the pointed end in a process that has been termed "treadmilling" (Wegner, 1976).

2.2. Myosin

Platelet myosin consists of six polypeptide chains, two $M_r = 200,000$ heavy chains, two $M_r = 20,000$ light chains, and two $M_r = 16,000$ light chains (Adelstein and Conti, 1972; Pollard *et al.*, 1974). Although there are no reported amino acid sequence comparisons, the lack of cross-reactivity between antibodies against muscle myosin and platelet myosin (Pollard *et al.*, 1977) indicates differences between the proteins from the two sources. As in muscle myosin, the two heavy chains of platelet myosin are coiled in an α-helical tail at one end, whereas at the other end they form two globular heads with which the light chains are associated. Under certain defined *in vitro* conditions, the tails can polymerize to form bipolar filaments. The filaments formed by platelet myosin are much shorter than those formed by muscle myosin, containing only about 28 myosin molecules (Niederman and Pollard, 1975).

Myosin filaments have not yet been observed within intact platelets, possibly because the concentration of myosin in platelets is very low (estimated at 1–2% of the platelet protein) (Adelstein and Conti, 1972). It is also possible that myosin filaments do not exist within platelets, since there is no evidence that contractile forces in platelets are generated by the sliding filament mechanism that functions in muscle. Whatever the mechanism, the observation that force can be generated by isolated threads of platelet actomyosin (Lebowitz and Cooke, 1978) has provided compelling evidence that force is generated in some way by the interaction of platelet myosin with actin filaments. Support for this comes from the observation that the globular heads of platelet myosin, like those of muscle myosin, contain the actin-binding sites and actin-activatable ATPase activity required for generation of such force (Adelstein *et al.*, 1971).

The $M_r = 20,000$ light chains of myosin can be phosphorylated by myosin light-chain kinase. This kinase contains calmodulin as part of its structure and is dependent on the presence of micromolar concentrations of Ca^{2+} (Daniel and Adelstein, 1976; Hathaway and Adelstein, 1979). Adelstein and Conti (1975) showed that phosphorylation of the $M_r = 20,000$ light chain results in the stimulation of actin-activatable ATPase activity, indicating that phosphorylated myosin interacts differently with actin filaments than does unphosphorylated myosin. Phosphorylation of the light chain has also been shown to cause myosin to assemble into filaments at physiological ionic strength and Mg–ATP concentrations (Scholey *et al.*, 1980). Several groups have shown that the light chain becomes phosphorylated during the stimulation of platelets (Daniel *et al.*, 1977; Haslam and Lynham, 1977; Haslam *et al.*, 1979; Lyons *et al.*, 1975), and there is one report indicating that this phosphorylation causes myosin to associate with actin filaments in stimulated platelets (Fox and Phillips, 1982) (see Section 5.1), although there is as yet no indication of whether or not stimulation causes myosin filaments to form within intact platelets.

2.3. Other Actin-Associated Proteins

Table 1 summarizes the other actin-associated proteins known to be present in platelets. Tropomyosin, which comprises about 1.6% of the total platelet protein, was first isolated from platelets by Cohen and Cohen in 1972; it has subsequently been obtained in a highly purified form (Coté and Smillie, 1981b). Like the tropomyosin from muscle, it consists of two very similar, although not identical subunits, α and β. These subunits, however, are shorter than those from skeletal muscle ($M_r = 28,500$ vs. 33,000), and amino acid analysis (Coté *et al.*, 1978a) has shown that the NH_2-terminal region of the platelet protein is lacking 37 amino acids present in the muscle protein. In addition, the COOH-terminal sequence of platelet tropomyosin is markedly different

Table 1. Some Actin-Associated Proteins in Platelets

	Molecular weight		Number of subunits	Reference
Myosin	200,000		2	Adelstein and Conti, 1972
	20,000		2	
	16,000		2	Pollard et al., 1977
Tropomyosin	28,500	α	1	Coté and Smillie, 1981b
	28,500	β	1	
α-Actinin	100,000		1	Landon and Olomucki, 1983
	102,000		1	
Actin-binding protein	250,000		2	Schollmeyer et al., 1978
Gelsolin	90,000		1	Lind and Stossel, 1982
Profilin	16,000		1	Markey et al., 1978
P235	235,000		2	Collier and Wang, 1982a
Vinculin	—		—	Jenkins et al., 1982
				Langer et al., 1982a
Spectrin	—		—	Fox et al.[a]

[a]J. E. B. Fox, C. G. Wong, and D. R. Phillips, unpublished observations.

from that of the muscle protein. Although the remaining amino acid sequences of the molecules appear to be very similar (Coté et al., 1978b), the differences in terminal sequences have marked effects on the properties of the molecules. Unlike skeletal muscle tropomyosin, platelet tropomyosin does not polymerize (Coté et al., 1978a), it binds only weakly to purified actin filaments (Coté and Smillie, 1981a), and it has a low affinity for skeletal muscle troponin (Coté et al., 1978b). The function of tropomyosin in muscle is to bind to actin filaments and, by acting in concert with the troponin complex, to regulate the interaction of myosin with actin filaments. Since platelet tropomyosin has a low affinity for both actin filaments and the troponin complex, this mechanism is unlikely to operate in platelets. Furthermore, troponin has not been detected in platelets where the binding of myosin to actin filaments appears to be regulated by the level of the phosphorylation of the myosin light chain (see Section 5.1). Since no other function for tropomyosin has been described, its role in the regulation of platelet contractile activity is not presently understood.

α-Actinin has been purified from platelets as a highly asymmetric, rodlike structure (Landon and Olomucki, 1983). Both morphological and immunological studies have shown it to be very similar to α-actinin from muscle. However, amino acid analysis (Landon and Olomucki, 1983) and one-dimensional peptide mapping (Rosenberg et al., 1981a) have shown that the two molecules are different. Unlike the subunits of the muscle protein, the two subunits of platelet α-actinin can be resolved on hydroxyapatite columns, and can be separated on sodium dodecyl sulfate (SDS) gels into subunits of $M_r = 100,000$ and 102,000 (Landon and Olomucki, 1983). Like muscle α-actinin, the platelet protein binds to F-actin, inducing the formation of actin filament networks (Landon and Olomucki, 1983: Rosenberg et al., 1981a). An important difference between the muscle and platelet proteins, however, is that the interaction of platelet α-actinin with actin is inhibited by calcium, whereas that of muscle α-actinin is not affected by this cation (Landon and Olomucki, 1983; Rosenberg et al., 1981a).

Actin-binding protein cross-links platelet actin filaments into networks. It is a long, semiflexible rod that exists in solution as a dimer formed by the "head-to-head" association of two subunits (each of M_r = 250,000) (Hartwig and Stossel, 1981). This protein was originally purified from macrophages (Hartwig and Stossel, 1975) and was subsequently observed in many nonmuscle cells including platelets (Schollmeyer et al., 1978). It is similar, if not identical, to filamin, a protein isolated from chicken gizzard (Wallach et al., 1978). Actin-binding protein binds to actin filaments in a ratio of one actin-binding protein dimer to 14 actin monomers, resulting in the cross-linking of filaments into networks. This interaction has been observed with the electron microscope, where bipolar actin-binding protein dimers, linked head-to-head, have been seen to have their tails associated with actin filaments (Hartwig and Stossel, 1981). Hartwig et al. (1980) have demonstrated that actin-binding protein also nucleates the polymerization of monomeric actin, shortening the lag time required for polymerization to begin.

Disruption of actin filament networks can be brought about by a protein called gelsolin. This protein was originally isolated from macrophages as a globular, monomeric protein of M_r = 91,000, which binds calcium at two specific sites (K_a = 1.09 × 10^6 M^{-1}) (Yin and Stossel, 1980). It has subsequently been shown to be immunologically and functionally similar to a M_r = 90,000 protein present in platelets (Lind and Stossel, 1982). Electron microscopy has revealed that gelsolin, when added to networks of actin filaments, severs the filaments into smaller pieces, thus loosening the networks (Yin et al., 1980, 1981). This activity is dependent on the presence of calcium at micromolar concentrations or above, and could therefore be involved in disrupting actin filament networks during platelet activation (see Section 3.2). Gelsolin has also been shown to nucleate the polymerization of purified actin by binding two actin monomers (Yin et al., 1981). The binding of gelsolin to monomeric actin is also dependent on the presence of micromolar calcium. As discussed in Section 3.2., however, it seems unlikely that gelsolin nucleates actin polymerization in this way in intact cells.

Profilin is another protein detected only in nonmuscle cells. It is a basic protein of M_r = 16,000 that was originally isolated from spleen by Carlsson et al. (1976) and was later identified in several other cells, including platelets (Markey et al., 1978; Reichstein and Korn, 1979). It forms a 1:1 complex with actin monomers and in this way prevents the formation of actin filaments (Carlsson et al., 1976). Because the affinity of profilin for actin monomers is relatively low (Korn, 1978), it may act as a buffer, preventing the formation of actin nuclei and therefore inhibiting filament formation. Evidence for such a mechanism comes from observations showing that profilin has a much greater inhibitory effect on the nucleation of purified actin than on filament elongation. Furthermore, the effect of profilin on purified actin can be readily reversed by the addition of nuclei such as short actin filaments (Korn, 1982; Markey et al., 1982).

The rate of actin polymerization is also inhibited under certain conditions by P235 (also known as band 2), a polypeptide of M_r = 235,000 that has been described only in platelets. This protein, which contributes 3–8% of the total platelet protein, has been purified from the soluble fraction of platelet lysates as a dimer of $M_r = 470,000$ (Collier and Wang, 1982a). The physical parameters of the purified protein suggest that it is nearly globular (frictional ratio of 1.3), a conclusion that is supported by electron microscopy. Functional assays (Collier and Wang, 1982b) have indicated that under certain ionic conditions, P235 inhibits the initial rate of actin polymerization and restricts the lengths of filaments produced, while having no effect on the extent of polymerization. This suggests that it may act as an additional ''profilinlike'' protein, inhibiting the nucleation of actin polymerization (Collier and Wang, 1983a). However, because P235 stimulates the initial rate of actin polymerization under other ionic conditions (Collier and Wang, 1983b), it is not clear at present how this protein acts in the intact cell.

Vinculin (Jenkins et al., 1982; Langer et al., 1982a) and spectrin (J. E. B. Fox, C. G. Wong, and D. R. Phillips, unpublished observations) are additional actin-associated proteins that have been identified immunologically in platelets, but have not yet been purified or characterized. In other cells, vinculin is a peripheral membrane protein of $M_r = 130,000$ (Geiger, 1979), which has been implicated as an attachment site for actin filaments (Geiger et al., 1981). Vinculin isolated from chicken gizzard has been shown to bind to actin filaments, possibly at the barbed ends, and to inhibit actin polymerization (Wilkins and Lin, 1982). Spectrin has best been characterized in red blood cells (Lux, 1979a), where it exists as a tetramer consisting of two subunits of $M_r = 260,000$ and two subunits of $M_r = 225,000$. Its function in red blood cells is to attach actin filaments to membranes. Spectrin has recently been purified from several other cell types (V. Bennett et al., 1979; Burridge et al., 1982; Glenney et al., 1982a, b). Although the subunit molecular weights and amino acid compositions differ widely, spectrin proteins from other sources appear to have the same function as red cell spectrin.

3. STIMULUS-DEPENDENT CHANGES IN CONTRACTILE PROTEINS

3.1. Polymerization of Actin during Platelet Stimulation

3.1.1. Measurement of Actin Filament Content

The actin filament content of platelets can be measured by two biochemical techniques. Both techniques were developed as a result of the observation of Bray and Thomas (1976) that actin filaments are insoluble in the nonionic detergent Triton X-100, whereas most other proteins, including monomeric actin, are soluble. In the first method, actin filaments in Triton-lysed platelets are separated from unpolymerized actin by centrifugation at 100,000g for 3 hr. The F-actin content of the sedimented filaments is quantitated by densitometry of the actin on Coomassie Blue-stained gels (Jennings et al., 1981). This procedure also allows for the simultaneous identification and quantitation of other proteins that are Triton insoluble as a result of their association with actin filaments. However, this method is time consuming and is not very useful when many samples must be processed. In a modification of this method, filaments are sedimented for shorter times by centrifugation at low gravity forces (typically 12,000g for 4 min) (Jennings *et al.*, 1981; Markey *et al.*, 1981; Rosenberg *et al.*, 1981b). Although this makes the rapid processing of many samples feasible, it can give artificially low values, because few of the filaments in unstimulated platelets are sufficiently cross-linked to sediment at these low speeds (Jennings *et al.*, 1981; Markey *et al.*, 1981; Fox *et al.*, 1984).

The second method for measuring actin filament content in platelet lysates was developed by Lindberg and co-workers (Blikstad *et al.*, 1978; Lazarides and Lindberg, 1974), who showed that monomeric actin binds to deoxyribonuclease I (DNase I) in a 1:1 complex, inhibiting enzyme activity. The amount of F-actin can be determined by comparing the DNase I inhibitory activity present in Triton lysates before and after depolymerization of actin filaments with guanidine hydrochloride. With this technique, samples can be processed rapidly. It has been widely used for measuring the actin filament content of many cell types in addition to platelets (Carlsson *et al.*, 1979; Fox *et al.*, 1981; Howell and Tyhurst, 1980; Swanston-Flatt *et al.*, 1980).

3.1.2. Actin Filament Content of Unstimulated Platelets

Both centrifugation analyses and DNase I inhibition studies using Triton X-100 lysates have indicated that unstimulated platelets contain ~40% of their actin in a filamentous form (Casella *et al.*, 1981; Fox *et al.*, 1981; Jennings *et al.*, 1981; Markey *et al.*, 1981). It has been demonstrated that these filaments are not induced by the procedures used to isolate the platelets from plasma. Similar levels of filament content have been observed both in discoid platelets isolated by gel filtration and in those isolated by centrifugation in the presence of either prostacyclin (which inhibits centrifugation-induced shape change) or cytochalasin D (which inhibits activation-induced polymerization of actin without affecting the filament content of unstimulated cells) (Fox *et al.*, 1984). Under the defined lysis conditions of these experiments, filaments are completely stable for at least 2 hr after lysis, indicating that the filament content measured in Triton lysates is the same as that present in intact cells (Fox *et al.*, 1981).

The actin filaments detected biochemically in extracts of unstimulated platelets have proved difficult to observe in intact platelets using electron microscopy because of problems in preserving the filaments during osmium tetroxide fixation. Boyles (1982) has recently developed a fixation procedure, however, that preserves actin filament morphology. Lysine is added to the glutaraldehyde fixative to prevent any damage to filaments that would otherwise occur during fixation with osmium tetroxide. The electron micrograph in Figure 1 was obtained with this procedure; it shows the dense filamentous network of actin filaments that exists in unstimulated, discoid platelets. Detergent and fixative were added simultaneously, allowing for the removal of detergent-soluble cellular components and the clear visualization of filaments, while preventing postlysis changes in filament organization. Thus, the filaments present in the cytoskeletal remnant retained the discoid shape of unextracted platelets. The visualization of extensive filament networks in unstimulated platelets is consistent with the biochemical data cited above indicating that ~40% of the actin in unstimulated platelets is filamentous. In contrast, Nachmias (1980) has suggested that filaments are not present in unstimulated platelets. Although the reason for this discrepancy is not known, it is possible that a postlysis reorganization of actin filaments might have occurred in the latter study.

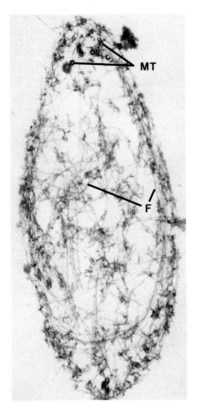

Figure 1. Electron micrograph of actin filaments in a Triton-solubilized platelet. Discoid platelets were isolated by the method of Nachmias (1980) and fixed simultaneously with lysis by the method of Boyles (1982). This method includes lysine in the glutaraldehyde fixative, which results in the preservation of actin filaments. This figure shows that un-stimulated, discoid platelets contain an extensive network of actin filaments (F) and a circumferential band of micro-tubules (MT) (magnification, 44,415×; reproduced at 70%). From J. K. Boyles, unpublished data.

3.1.3. Actin Filament Content of Activated Platelets

Measurement of the actin filament content of thrombin-activated platelets by either the DNase I inhibition assay (Carlsson *et al.,* 1979; Casella *et al.,* 1981; Fox and Phillips, 1981) or by sedimentation assays (Jennings *et al.,* 1981) has shown that 70–80% of the total actin in these platelets is filamentous. Similar values for F-actin content have been observed in platelets stimulated with ADP or the calcium ionophore A23187 (Fox and Phillips, 1981). Since only 40% of the actin is filamentous in unstimulated platelets, these findings indicate that there must be mechanisms for inducing the polymerization of actin during platelet activation. As shown in Figure 2, this stimulus-induced polymerization occurs very rapidly. After the addition of throm-bin, there is a slight lag of about 5 sec, which is followed by a rapid increase in the amount of F-actin; a plateau is reached by about 20 sec.

3.2. Structural Reorganization of Actin Filaments during Platelet Stimulation

As shown in Figure 1, morphological studies of F-actin in unstimulated, discoid platelets have indicated that the F-actin is organized in loose networks throughout the cytosol. In contrast, morphological studies of activated platelets have shown that the F-

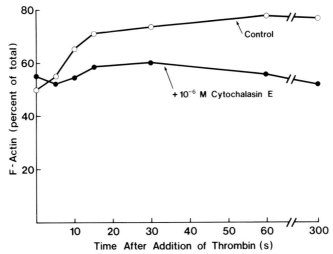

Figure 2. Thrombin-induced polymerization of actin within platelets and inhibition by cytochalasin E. Platelets were incubated with 0.2% (v/v) DMSO (○) or 10^{-6} M cytochalasin E and 0.2% DMSO (●), then stimulated with thrombin (0.1 U/ml) for the times indicated and lysed with Triton X-100. The actin filament content of lysates was measured by the DNase I inhibition assay. This figure shows that unstimulated platelets contain ~40–50% of their total actin as filaments and that this value increases to ~70% within 20 sec of stimulation by thrombin. The thrombin-induced polymerization of actin is inhibited by cytochalasin E. From Fox and Phillips (1981) with permission.

actin is highly organized as bundles of filaments in filopodia or in rings of filaments surrounding centralized granules. These observations suggest that there are mechanisms that induce the structural reorganization of F-actin during platelet activation, in addition to those that induce increased actin polymerization.

Electron microscopy of the actin filaments isolated by centrifugation from Triton X-100 lysates has confirmed that the F-actin in activated platelets is organized differently than that in unstimulated platelets. Filaments isolated from unstimulated platelets exist in a random array, whereas those isolated from thrombin-activated platelets have a much more compact or dense structure (Jennings *et al.*, 1981). Filaments in detergent-extracted cytoskeletons of platelets activated with thrombin for 15 sec are observed as "balls" of about 2.1 μm in diameter, which surround the centralized granules. If the platelets are incubated with thrombin for longer periods of time prior to extraction, these balls of filaments appear to be more condensed, with diameters of about 1.5 μm.

Examination of the polypeptide content of the isolated, filamentous structures has shown that actin filaments become associated with other proteins during platelet activation, and it is likely that these proteins are involved in the structural reorganization of actin filaments (Feinstein *et al.*, 1983; Jennings *et al.*, 1981). Figure 3 shows an SDS–polyacrylamide gel of the Triton-insoluble material isolated from unstimulated and thrombin-activated platelets at increasing periods of time after thrombin addition. Triton-insoluble material from unstimulated platelets consists mainly of actin, with small amounts of actin-binding protein, myosin, and a polypeptide of $M_r = 31,000$

(which has not been identified but has a molecular weight similar to that of the tropomyosin subunits). When platelets are stimulated with thrombin, increasing amounts of actin become Triton insoluble; this increase plateaus ~15 sec after the addition of thrombin. This is consistent with the increased polymerization of actin that has been measured by the DNase I inhibition assay. The Triton-insoluble material from thrombin-activated platelets also contains increasing amounts of actin-binding protein and myosin. The amount of Triton-insoluble actin-binding protein increases form ~5% of the total in unstimulated platelets to ~25% in thrombin-activated platelets. The time course of the increased Triton insolubility of this protein is very similar to that observed for the formation of actin filaments. The change in the amount of myosin that is Triton insoluble is more dramatic. In unstimulated platelets, ~14% of the total myosin is Triton insoluble; this increases very rapidly until ~30 sec after thrombin addition, when essentially all of the myosin is Triton insoluble. After longer periods of time, the amount of myosin that is Triton insoluble declines until only about 60% of it remains Triton insoluble. Additional polypeptides of $M_r = 50,000$–$60,000$ become Triton insoluble at longer periods of time after thrombin addition (~10 min). These polypeptides have recently been identified as the polypeptide chains of fibrin (Casella et al., 1983). Although the Triton insolubility of these fibrin polypeptides may result from the inherent Triton insolubility of fibrin, neither actin-binding protein nor myosin are themselves insoluble in Triton X-100. Evidence that the Triton insolubility of these two proteins results from their binding to actin filaments comes from the observation that myosin and actin-binding protein are no longer soluble if actin filaments in Triton lysates are depolymerized with Ca^{2+} or DNase I.

Several pieces of evidence suggest that the stimulus-induced association of myosin with actin filaments is involved in the contractile events necessary for secretion. This association between myosin and F-actin does not occur when platelets are stimulated with ADP (Fox and Phillips, 1982), which causes aggregation and shape change but not secretion. Similarly, platelets stimulated with phorbol 12-myristate 13-acetate change shape and aggregate, but myosin does not associate with the F-actin in these platelets and little secretion occurs (Carroll et al., 1982). Furthermore, immunofluorescence studies have localized myosin to the central region of activated platelets, where rings of actin filaments appear to be forcing granules to secrete their contents into the open canalicular system (Debus et al., 1981; Painter and Ginsberg, 1982; Pollard et al., 1977).

Evidence that actin-binding protein is involved in the structural reorganization of actin filaments comes from immunofluorescence studies showing that actin-binding protein is distributed throughout unstimulated platelets, but is concentrated in the filopodia of stimulated platelets (Debus et al., 1981). Purified actin-binding protein has been shown to bind to actin and affect the structural organization of filaments. Hartwig et al. (1980) have shown that actin-binding protein causes purified actin filaments to form networks. Rosenberg et al. (1981b; Rosenberg and Stracher, 1982) and Schollmeyer et al. (1978) found that purified platelet actin-binding protein causes actin filaments to gel.

Another protein that may be involved in the stimulus-induced reorganization of actin filaments is gelsolin. At the low concentrations of Ca^{2+} that exist in unstimulated platelets, this protein has no known affect on actin. At the micromolar levels existing in the cytosol of activated platelets, however, gelsolin can either nucleate actin polymerization or sever actin filaments (Yin *et al.,* 1980, 1981). As discussed in Section 4.2., it seems unlikely that gelsolin has a role in nucleating polymerization. It probably acts by cleaving filaments into shorter ones; such an action would result in the disruption of networks and therefore affect the reorganization of actin filaments (Lind and Stossel, 1982).

In summary, unstimulated platelets contain ~40% of their actin as filaments that exist in random networks. When platelets are activated, there is a rapid increase in actin polymerization until about 70–80% of the actin is filamentous. An activation-induced reorganization of actin filaments also occurs, which probably results from the interaction of actin filaments with myosin, actin-binding protein, and gelsolin.

4. REGULATION OF STIMULUS-INDUCED ACTIN POLYMERIZATION

4.1. Regulation of the Filament Content of Unstimulated Platelets

The concentration of actin in platelets has been estimated to be between 0.25 mM and 1.0 mM (Blikstad *et al.,* 1978; Pollard *et al.,* 1977). Because this is well above the critical concentration of about 30 μM, it would seem that most of the actin in platelets would exist in a polymerized form. The finding that only about 40% of the actin in unstimulated platelets is filamentous raises the interesting question of how the rest of the actin is prevented from polymerizing. It could be that (1) certain proteins bind to monomeric actin and lower the effective concentration of monomers, or (2) certain proteins bind to the ends of pre-existing filaments and prevent further polymerization at these ends. One protein that is known to inhibit polymerization by binding to actin monomers is profilin. As discussed in Section 2.3, profilin apparently acts as a buffer, preventing the formation of actin nuclei. Its effect can be readily reversed by the addition of other nuclei—short actin filaments for example (Korn, 1982; Markey *et al.,* 1982). Evidence that profilin inhibits polymerization in intact platelets comes from the work of Markey *et al.* (1981), who have shown that Triton X-100 lysates of unstimulated platelets contain profilin in a 1:1 complex with actin, whereas lysates from thrombin-activated platelets contain dissociated profilin. This observation demonstrates that profilin acts as an actin monomer buffer in unstimulated platelets and is released in activated platelets when new actin nuclei are presumably formed.

It has been suggested that P235 is another profilinlike protein that prevents the polymerization of actin monomers in unstimulated platelets (Collier and Wang, 1983a). Although there is no evidence that P235 acts in this way in intact cells, it does inhibit the initial rate of polymerization of purified actin under certain ionic conditions.

Proteins that inhibit actin polymerization by "capping" the ends of pre-existing filaments could potentially bind to either end of the filaments. However, because polymerization occurs much more rapidly at the barbed ends than at the pointed ends, proteins binding to the barbed ends would be more likely to have a regulatory role in inhibiting the net polymerization of filaments. Grumet and Lin (1980) first assayed for the presence of a platelet protein that would bind to the barbed ends of filaments by measuring the inhibition of binding of radioactively labeled cytochalasin by platelet extracts. They purified a $M_r = 90,000$ protein that competed with cytochalasin for binding to actin filaments and inhibited the salt-induced elongation of actin (Lin *et al.*, 1982). This protein, subsequently shown to be gelsolin (Lin *et al.*, 1982; Lind *et al.*, 1982), only binds to actin filaments in the presence of micromolar concentrations of calcium. It is therefore unlikely that it acts as a capping protein in unstimulated platelets, since the free calcium concentration in the cytoplasm is only about 100 nM (Rink *et al.*, 1982). It is more likely that gelsolin plays a role in regulating actin filament structure in activated platelets, where the calcium levels rise to about 3 μM.

Membrane-associated proteins may also regulate the polymerization of actin at the barbed ends of filaments, since membrane-associated actin filaments are attached to membranes by these barbed ends. Vinculin has been implicated as a membrane attachment site for actin filaments in other cell types on the basis of its colocalization with the ends of such filaments (Geiger, 1979; Geiger *et al.*, 1981). It remains to be determined whether it is a barbed end-capping protein in platelets.

Another protein that has been implicated as a membrane attachment site, and therefore as a potential barbed end-capping protein, is α-actinin (Geiger *et al.*, 1981). It was originally suggested that this protein might be the platelet membrane GP IIIa (Gerrard *et al.*, 1979). Although it has since been shown that these two proteins are distinct (Langer *et al.*, 1982b; Sixma *et al.*, 1982), α-actinin has been demonstrated to be at least partly membrane associated (Sixma *et al.*, 1982). Platelet α-actinin binds to actin filaments in a calcium-dependent manner (Landon and Olomucki, 1983; Rosenberg *et al.*, 1981a); binding only occurs at concentration of calcium lower than micromolar. Thus, if binding occurs at the barbed ends, α-actinin could be a potentially important regulator of polymerization in intact cells, since it would be released at the concentrations of calcium that exists in the cytosol of activated platelets (about 3 μM).

4.2. Regulation of the Filament Content of Activated Platelets

The increased polymerization of actin that occurs during platelet activation is more rapid and more extensive than that which has been reported for any other cell type. Platelets therefore offer a useful model system for studying the mechanisms regulating actin polymerization within cells. The mechanisms responsible for regulating this increased polymerization presumably involve the increasing cytoplasmic calcium concentrations that are reached during platelet activation (Feinstein, 1980; Le Breton *et al.*, 1976; Rink *et al.*, 1982). Evidence for such a conclusion comes from reports stating that actin polymerization is induced by the calcium ionophore A23187 (Fox and Phillips, 1981) and that polymerization can be inhibited by agents that elevate cyclic AMP (Feinstein *et al.*, 1983), which is thought to act by removing Ca^{2+} from the cytosol (Fox *et al.*, 1979; Käser-Glanzmann *et al.*, 1978). Two ways in which polymerization may be induced are by the generation of new actin nuclei, which would allow actin monomers from the profilin-bound pool to polymerize, or by the release of capping proteins from the ends of pre-existing filaments, thus allowing monomers to add onto these ends.

Cytochalasins have been used to identify the direction of polymerization from the stimulus-induced nuclei. These mold metabolites bind to the barbed end of actin filaments and inhibit further polymerization from this end (Brenner and Korn, 1979; Brown and Spudich, 1979; Flanagan and Lin, 1980; MacLean-Fletcher and Pollard, 1980). Figure 2 shows that cytochalasin E inhibits the thrombin-induced polymeriza- tion of actin in platelets. Cytochalasins D and B have been shown to have similar effects (Fox and Phillips, 1981). The relative order of their effectiveness in inhibiting polymerization (D > E > B) is consistent with the relative affinities of the three cytochalasins for binding to the barbed ends of actin filaments and inhibiting actin polymerization from these ends. These results show that stimulus-induced actin poly- merization in platelets is nucleated from the barbed ends of filaments. Thus, studies of actin polymerization in activated platelets should concentrate on those mechanisms that allow for polymerization at the barbed ends of filaments rather than those [like the polymerization induced by gelsolin (Yin *et al.*, 1981)] that involve polymerization at the pointed ends.

One protein that has been shown to nucleate polymerization from the barbed end of purified actin filaments is actin-binding protein. Hartwig *et al.* (1980) demonstrated that this protein shortens the lag time required for the initiation of polymerization of purified G-actin. In their experiments, the filaments that formed were highly cross- linked, and electron microscopy of filaments that had been decorated with heavy meromyosin showed that the actin-binding protein which remained at the branch points had nucleated the polymerization of actin from the barbed ends of the filaments. Actin- binding protein is associated with actin filaments within platelets (Figure 3). It is therefore possible that it is involved in nucleating the stimulus-induced polymerization of actin within platelets, although the way in which such an activity could be initiated is not known.

An alternative way in which actin polymerization might be induced during platelet activation is by the release of capping proteins from the ends of actin filaments. Since preformed actin filaments are efficient nuclei for polymerization, the release of such proteins could induce a rapid burst of polymerization. Because stimulus-induced poly- merization occurs at the barbed ends of filaments, such a mechanism would involve the release of barbed end-capping proteins such as those discussed above. Studies measur- ing the effects of cytochalasins on the actin filament content of unstimulated and stimulated platelets have provided indirect evidence that polymerization is induced in this way. These studies have shown that cytochalasins have no effect on the filament content of unstimulated platelets (Casella *et al.*, 1981; Fox and Phillips, 1981), but rapidly reverse the rise in actin filament content of thrombin-stimulated platelets to the levels found in unstimulated platelets (Casella *et al.*, 1981). One explanation for this may be that the filaments in unstimulated platelets are stable because their barbed ends are already capped by a protein that, like the cytochalasins, prevents addition of monomers to these ends; only the pointed ends would be in equilibrium with the pool of monomeric actin. Stimulation may release the barbed end-capping protein, allowing the filaments to polymerize and obtain a new steady state where filaments may tread- mill. The subsequent addition of cytochalasin would cap the barbed ends of the fila- ments again, preventing further addition of actin monomers at these ends, whereas depolymerization would continue at the pointed ends until the steady state originally present in unstimulated platelets is restored.

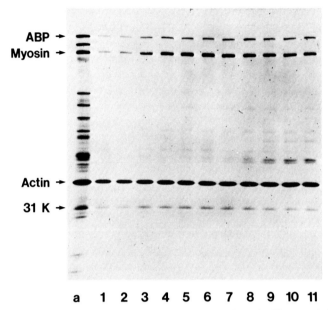

Figure 3. Time-dependent changes in the association of proteins with actin filaments during thrombin-stimulation of platelets. Platelets were lysed with Triton X-100 directly, lane 1, or after exposure to thrombin (0.1 U/ml) for 5 sec, lane 2; 10 sec, lane 3; 15 sec, lane 4; 30 sec, lane 5; 1 min, lane 6; 2 min, lane 7; 5 min, lane 8; 10 min, lane 9; 20 min, lane 10; or 30 min, lane 11. Insoluble material was isolated by centrifugation and electrophoresed on SDS–polyacrylamide gels. This figure shows that actin-binding protein and myosin become associated with actin filaments during platelet activation. From Jennings *et al.* (1981) with permission.

In summary, unstimulated platelets contain ~40% of their actin in a filamentous form. The barbed ends of these filaments may be capped by proteins that, together with proteins such as profilin that bind to monomeric actin, prevent the polymerization of the remaining actin into filaments. When platelets are stimulated, there is a rapid increase in the amount of F-actin. This probably results from the formation of new actin nuclei, to which actin monomers add on from the profilin-bound pool. These actin nuclei are probably proteins such as actin-binding protein. Alternatively, polymerization may result from the release of barbed end-capping proteins from pre-existing filaments.

5. REGULATION OF STIMULUS-INDUCED REORGANIZATION OF ACTIN FILAMENTS

It is likely that the structural reorganization of actin filaments that occurs during platelet activation is regulated by one or more of the several proteins present in platelets that are known to affect actin filament structure in isolated systems (Table 1). The most probable candidates are myosin and actin-binding protein, proteins known to become associated with filaments during platelet activation. The stimulus-induced modification of these proteins may affect their ability to bind to actin, and therefore affect their ability to direct the reorganization and functioning of actin filaments.

5.1. Modification of Myosin during Platelet Activation

Several groups have shown that $M_r = 20,000$ light chains of myosin become phosphorylated during platelet activation (Daniel *et al.*, 1977; Haslam *et al.*, 1979; Lyons *et al.*, 1975). Daniel *et al.* (1981) have quantitated the amount of light chain that is phosphorylated in platelets, showing that only about 10% of the $M_r = 20,000$ light chain is phosphorylated in unstimulated platelets, but that almost 100% is phosphorylated within 30 sec of the addition of thrombin. Phosphorylation of the $M_r = 20,000$ light chains of myosin is known to affect its actin-activatable ATPase activity and therefore, presumably, its interaction with actin. Phosphorylation is a way in which platelet stimulation may lead to the association of myosin with actin filaments. Evidence that this is so comes from the demonstration that the time course for phosphorylation of the myosin light chain is identical to that for the association of myosin with actin filaments (Fox and Phillips, 1982), a correlation that holds when platelets are stimulated by thrombin, collagen, ADP, or the calcium ionophore A23187, agents that induce phosphorylation of the light chain at very different rates. The correlation

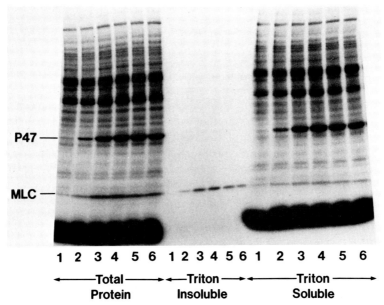

Figure 4. Selective association of phosphorylated myosin with actin filaments in platelets. ^{32}P-labeled platelets were exposed to thrombin (0.1 U/ml) for 0 sec, lane 1; 15 sec, lane 2; 30 sec, lane 3; 60 sec, lane 4; 2 min, lane 5; or 15 min, lane 6; and then solubilized in SDS or lysed by adding Triton X-100. Triton-insoluble material and Triton-soluble fractions were separated by centrifugation, and the protein was electrophoresed on SDS–polyacrylamide gels. Phosphopolypeptides were detected by autoradiography; P47, $M_r = 47,000$ phosphopolypeptide; MLC, myosin light chain. This figure shows that phosphorylation of the myosin light chain causes myosin to bind to actin filaments. From Fox and Phillips, 1982, with permission.

between the phosphorylation of myosin light chains and the association of myosin with actin filaments is also observed when phosphorylation is blocked by the calmodulin antagonist triflorperazine, which inhibits the myosin light-chain kinase, or with prostaglandin E_1 or prostacyclin, agents that probably act by removing calcium from the cytosol. As a direct test of whether phosphorylation of the myosin light chain causes the myosin molecule to associate with actin filaments, the distribution of phosphorylated and nonphosphorylated myosin has been examined in Triton soluble and insoluble fractions of Triton-lysed platelets. Figure 4 shows that the phosphorylated form was recovered with the Triton-insoluble actin filaments. Comparison of the amount of myosin and the amount of radioactivity recovered with the actin filaments confirmed that the phosphorylated form of myosin is selectively associated with actin filaments. Thus, the phosphorylation of the myosin light chain during platelet activation causes myosin to bind to actin filaments and, as discussed above, induces the contractile activity required for the secretion of granule contents.

5.2. Modification of Actin-Binding Protein during Platelet Activation

The stimulus-induced modification of actin-binding protein that accounts for its ability to cross-link actin filaments in activated platelets is not well understood. It has been suggested that the binding of actin-binding protein to actin filaments in activated platelets is regulated by the phosphorylation of actin-binding protein (Stracher et al., 1982). One group has detected an increased phosphorylation of actin-binding protein during platelet activation (Carroll and Gerrard, 1982), but since this increase was small and was only demonstrated with thrombin as a stimulus, the importance of this modification is not clear.

An additional mechanism by which actin-binding protein may regulate the organization of actin filaments during platelet stimulation has been described recently (Fox et al., 1983a). This mechanism involves the activation of the calcium-dependent protease in platelets. This protease is a neutral, thiol-dependent enzyme (Phillips and Jakábová, 1977; Tsujinaka et al., 1982), which has been isolated from many cell types (Murachi et al., 1981; Truglia and Stracher, 1981; Yoshida et al., 1983). Original reports suggested that the protease requires millimolar Ca^{2+} for activation (Dayton et al., 1976; Mellgren et al., 1982), indicating that its activation following platelet stimulation was unlikely. However, forms of the protease that are activated by micromolar levels of Ca^{2+}, which exist in the cytosol of activated platelets, have recently been isolated (DeMartino, 1981; Kishimoto et al., 1983; Mellgren, 1980; Tsujinaka et al., 1982; Zimmerman and Schlaepfer, 1982). Thus, activation of the calcium-dependent protease during platelet activation is feasible.

The proteins that have been identified as substrates for the calcium-dependent protease in platelets include actin-binding protein, P235, and protein kinase C (Fox et al., 1983a; Kishimoto et al., 1983; Phillips and Jakábová, 1977; White, 1980). In a recent study in our laboratory, two-dimensional gels were used to compare the proteins in unstimulated platelets to those present in activated platelets (Fox et al., 1983a). Actin-binding protein, P235, and a $M_r = 87,000$ polypeptide were selectively hydrolyzed during the activation of platelets by thrombin, collagen, or the divalent cation ionophore A23187. Decreases in the amounts of these polypeptides were accompanied

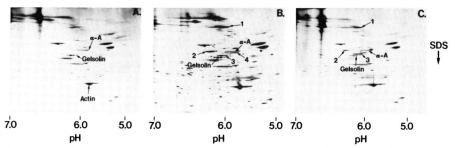

Figure 5. Stimulus-induced changes in the polypeptide content of platelets. Unstimulated platelets (A), or those stirred with thrombin (0.1 U/ml) for 30 sec (B), or collagen (20 μg/ml) for 15 min (C), were solubilized and their polypeptide contents were analyzed by two-dimensional gel electrophoresis. Gels were stained by a silver-staining method; α-A, α-actinin. Polypeptides 1–3 are major new polypeptides arising during platelet activation.

by the generation of at least three new polypeptides observed on two-dimensional gels. Figure 5 shows the generation of these three polypeptides during activation of platelets with thrombin or collagen. The polypeptides have been designated polypeptide 1 (M_r = 200,000), polypeptide 2 (M_r = 100,000), and polypeptide 3 (M_r = 91,000).

Several lines of evidence indicate that the three new polypeptides arose from the hydrolysis of actin-binding protein by the calcium-dependent protease: First, these polypeptides were produced in platelet lysates in the presence of calcium cation; their generation was inhibited when divalent cations were chelated with EGTA. Second, the generation of the new polypeptides in platelet lysates was inhibited by N-ethyl-maleimide, leupeptin, or mersalyl, known inhibitors of the calcium-dependent protease (Phillips and Jakábová, 1977; Rodemann et al., 1982; Toyo-Oka et al., 1978). Third, the generation of polypeptides 1–3 in either platelet lysates or intact cells was accompanied by a decrease in the concentration of actin-binding protein. Fourth, purified actin-binding protein was hydrolyzed by the calcium-dependent protease during the production of polypeptides 1–3. Fifth, polypeptides 1–3 cross-reacted with an affinity-purified antibody to actin-binding protein. These results show that actin-binding protein is hydrolyzed by the calcium-dependent protease during platelet activation. Hydrolysis was not inhibited by 5,5'-dithiobis (2-nitrobenzoic acid), an inhibitor of the calcium-dependent protease that cannot permeate membranes, confirming that the calcium-dependent protease hydrolyzes actin-binding protein within intact cells.

The function of the hydrolysis of actin-binding protein by the calcium-dependent protease may be to promote the reorganization of actin filaments during platelet activation. Purified actin-binding protein cross-links actin filaments, resulting in the formation of networks. The hydrolysis of actin-binding protein causes it to lose its ability to cross-link filaments (Davies et al., 1978; Truglia and Stracher, 1981). Thus, activation of the calcium-dependent protease during platelet activation may provide a mechanism by which actin-binding protein is hydrolyzed, leading to the disassembly of actin filament networks. In such a model, other proteins would then be able to bind to the filaments and cause further changes in their organization.

In summary, actin filaments are reorganized when platelets are activated. The initial reorganization required for secretion and shape change is probably mediated by the association of myosin and actin-binding protein with filaments. The subsequent hydrolysis of actin-binding protein, together with the activity of gelsolin, may result in the additional reorganization required for later events, such as clot retraction.

6. MEMBRANE ATTACHMENT SITES

It is assumed that actin filaments within platelets are attached to membranes, resulting in the transmission of force to accomplish shape change, filopodia extension, and clot retraction. Indirect evidence for such attachments comes from observations that actin co-isolates with platelet plasma membranes (Taylor *et al.*, 1975). More direct evidence for interactions between actin filaments and membranes comes from morphological studies, such as that shown in Figure 1 where filaments can be visualized as submembranous lattices. Other studies have shown that the barbed ends of actin filaments are attached to membranes in the filopodia of activated platelets (Zucker-Franklin and Grusky, 1972).

Platelets contain several proteins that may be involved in linking filaments to membranes. For example, platelets have been shown to contain ankyrin (Bennett, 1979), a protein that links actin filaments to membranes in red blood cells. In red blood cells (Lux, 1979a,b) actin filaments bind to spectrin in a process that is enhanced by a protein known as band 4.1. Spectrin binds to ankyrin, which binds to an integral membrane glycoprotein known as band 3. It was originally thought that this actin–spectrin–ankyrin system was unique to red cells; however, the observation that ankyrin is present in platelets raises the possibility that it might also function in these cells. Furthermore, spectrin has recently been demonstrated in several other cell types (V. Bennett *et al.*, 1979; Burridge *et al.*, 1982; Glenney *et al.*, 1982a,b). It has been isolated from some of these cells and has been shown to be structurally and functionally similar to red blood cell spectrin. In recent experiments (J. E. B. Fox, C. G. Wong, and D. R. Phillips, unpublished observations), spectrin has been detected immunologically in platelets, suggesting the possibility that the actin–spectrin–ankyrin system is involved in the attachment of actin filaments to an integral membrane protein in platelets.

Additional proteins that have been implicated in the linkage of actin filaments to integral membrane proteins in other cell types are vinculin and α-actinin. They are both peripheral membrane proteins that have been detected at the ends of actin filaments in immunofluorescence studies. Both α-actinin and vinculin have been identified immunologically in platelets. α-Actinin has been purified and shown to be at least partly associated with the platelet plasma membrane (Sixma *et al.*, 1982). Immunofluorescence studies suggest that it is concentrated on the cytoplasmic face of the plasma membrane (Debus *et al.*, 1981).

Three glycoproteins have been implicated as the integral membrane proteins to which actin filaments are linked either directly or indirectly through intermediate proteins such as ankyrin, spectrin, α-actinin, or vinculin. One of these is GP Ib, which is retained with the Triton-insoluble actin filaments isolated from unstimulated platelets (Solum *et al.*, 1983; Fox *et al.*, 1983b). The other two are GP IIb and IIIa, which are retained with Triton-insoluble actin filaments from platelets that have aggregated in response to stimulation by thrombin (Phillips *et al.,* 1980) or that have previously been exposed to concanavalin A (Painter and Ginsberg, 1982). Since a complex of GP IIb and IIIa is also the receptor site for fibrinogen (J. S. Bennett *et al.,* 1982; Gogstad *et al.,* 1982; Nachman and Leung, 1982), they may be responsible for connecting fibrin clots on the external surface of activated platelets to actin filaments within the cytosol. The polarity of actin filaments in filopodia is such that, if myosin interacts with actin as it does in skeletal muscle, the actin–myosin interactions would pull filaments into the body of the platelet. Thus, the attachment of actin filaments to the GP IIb-IIIa complex in activated platelets could result in the retraction of fibrin clots. Although previous models for clot retraction have not always suggested that a retraction of filopodia occurs, a recent publication (Cohen *et al.,* 1982) has provided morphological evidence that this may indeed be the case, and that the retraction of filopodia is dependent on platelet contractile proteins (see Chapter 14).

In summary, the actin filament networks in unstimulated platelets and the bundles of actin filaments in the filopodia of stimulated platelets appear to interact with the inner surface of plasma membranes. These filaments may be linked to integral membrane glycoproteins such as ankyrin, spectrin, vinculin, or α-actinin. Glycoprotein Ib may be an attachment site in unstimulated platelets, whereas GP IIb-IIIa complex may serve the same purpose in aggregated platelets. The interaction of the GP IIb-IIIa complex with fibrinogen on the outer surface of platelets and with actin filaments inside platelets may provide a mechanism whereby this glycoprotein complex mediates clot retraction.

7. CONCLUSIONS

Many platelet responses to stimulation involve contractile proteins. The studies summarized in this review have led to an understanding of how changes in these contractile proteins during platelet stimulation regulate the content and organization of actin filaments and, therefore, the functioning of platelets. Forty percent of the actin in unstimulated, discoid platelets is polymerized into filaments. In stimulated platelets, 70–80% of the actin is filamentous. In unstimulated platelets, the actin filaments are organized in loose networks, whereas those in activated platelets form bundles in the filopodia and a condensed shell around the centralized granules. These differences in the organization of actin filaments in unstimulated and stimulated platelets suggest that mechanisms exist in platelets to both regulate the polymerization of actin into filaments and reorganize existing actin filaments.

The changes that occur in actin and associated proteins as a result of thrombin-induced platelet stimulation are shown in Figure 6. The changes are explained in terms of the responses to the increased concentration of cytoplasmic Ca^{2+} that occurs during platelet stimulation. The increased polymerization of actin, which seems to be under Ca^{2+} control, may be regulated by proteins such as actin-binding protein, which nucleates polymerization, or by the Ca^{2+}-dependent release of barbed end-capping proteins. Proteins that associate with actin seem to directly affect the reorganization of actin filaments. Myosin binds to actin due to the Ca^{2+}-mediated phosphorylation of the myosin light chain. Actin-binding protein also binds to F-actin during platelet

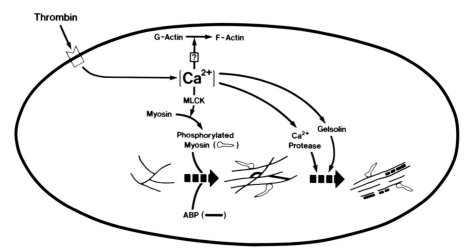

Figure 6. Mechanisms involved in the stimulation-dependent changes in actin filament content and organization. These changes are depicted as responses to the increased concentration of cytoplasmic Ca^{2+} that occurs during platelet stimulation. Calcium causes an increase in the amount of F-actin by an unknown mechanism. The organization of actin filaments is affected by proteins that associate with the actin. Calcium ions mediate the phosphorylation of the myosin light chain by the myosin light-chain kinase (MLCK), causing myosin to bind to actin filaments. Actin-binding protein (ABP) also binds to filaments during platelet stimulation. Subsequent to the initial reorganization of filaments, actin-binding protein is hydrolyzed by the calcium-dependent protease and gelsolin severs actin filaments, allowing further filament reorganization.

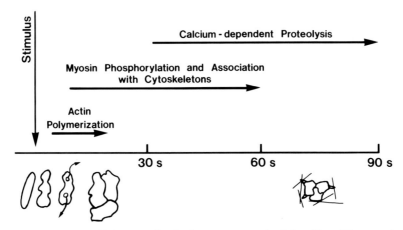

Figure 7. Temporal relationship between the platelet responses to stimulation. The solid arrows above the time axis point out the biochemical changes involved in actin filament reorganization. The diagrams below the time axis illustrate the physiological responses to stimulation—shape change, secretion, aggregation, and clot retraction.

stimulation, but the manner in which the binding is regulated is unknown. Subsequent to the initial reorganization of filaments, actin-binding protein is hydrolyzed by the calcium-dependent protease. This hydrolysis results in further filament reorganization, occurring at the same time as the retraction of fibrin clots. The Ca^{2+}-dependent severing of filaments by gelsolin is an additional mechanism regulating the reorganization of actin filaments.

Figure 7 illustrates the temporal relationship between these biochemical changes and platelet function. The stimulation of platelets with thrombin causes them to lose their discoid shape within milliseconds. The addition of newly formed actin filaments is not detected until about 5 sec after stimulation. Secretion and aggregation occur 5–30 sec after the addition of thrombin. During this time, actin polymerizes and myosin undergoes phosphorylation, causing it to bind to actin filaments. This association of myosin with actin filaments has been implicated in secretion, a finding consistent with the observed organization of myosin and the centralization of granules that can be seen at this time. At later points in time after thrombin addition, the calcium-dependent protease is activated and hydrolyzes actin-binding protein, thus causing further changes in filament structure. These changes occur at the same time that platelet aggregates condense and fibrin clots retract.

ACKNOWLEDGMENTS.   I would like to express my thanks to Dr. J. K. Boyles for the electron micrograph used in Figure 1, and to James Warger and Mark Sterne for graphics. I also greatly appreciate the editorial assistance of Russell Levine and the manuscript preparation of Debora Springer.

REFERENCES

Adelstein, R. S., and Conti, M. A., 1972, The characterization of contractile proteins from platelets and fibroblasts, *Cold Spring Harbor Symp. Quant. Biol.* **37**:599–606.

Adelstein, R. S., and Conti, M. A., 1975, Phosphorylation of platelet myosin increases actin-activated myosin ATPase activity, *Nature (London)* **256**:597–598.

Adelstein, R. S., Pollard, T. D., and Kuehl, W. M., 1971, Isolation and characterization of myosin and two myosin fragments from human blood platelets, *Proc. Natl. Acad. Sci. U.S.A.* **68**:2703–2707.

Bennett, J. S., Vilaire, G. T., and Cines, D. B., 1982, Identification of the fibrinogen receptor on human platelets by photoaffinity labeling, *J. Biol. Chem.* **257**:8049–8054.

Bennett, V., 1979, Immunoreactive forms of human erythrocyte ankyrin are present in diverse cells and tissues, *Nature (London)* **281**:597–599.

Bennett, V., Davis, J., and Fowler, W. E., 1982, Brain spectrin, a membrane-associated protein related in structure and function to erythrocyte spectrin, *Nature (London)* **299**:126–131.

Bettex-Galland, M., and Lüscher, E. F., 1959, Extraction of an actomyosin-like protein from human thrombocytes, *Nature (London)* **184**:276–277.

Blikstad, I., Markey, F., Carlsson, L., Persson, T., and Lindberg, U., 1978, Selective assay of monomeric and filamentous actin in cell extracts, using inhibition of deoxyribonuclease I, *Cell* **15**:935–943.

Blikstad, I., Sundkvist, I., and Eriksson, S., 1980, Isolation and characterization of profilactin and profilin from calf thymus and brain, *Eur. J. Biochem.* **105**:425–433.

Boyles, J. K., 1982, A modified fixation for the preservation of microfilaments in cells and isolated F-actin (abstract), *J. Cell Biol.* **95**:287a.

Bray, D., and Thomas, C., 1976, Unpolymerized actin in fibroblasts and brain, *J. Mol. Biol.* **105**:527–544.

Brenner, S. L., and Korn, E. D., 1979, Substoichiometric concentrations of cytochalasin D inhibit actin polymerization. Additional evidence for an F-actin treadmill, *J. Biol. Chem.* **254**:9982–9985.

Brown, S. S., and Spudich, J. A., 1979, Cytochalasin inhibits the rate of elongation of actin filament fragments, *J. Cell Biol.* **83:**657–662.

Burridge, K., Kelly, T., and Mangeat, P., 1982, Nonerythrocyte spectrins: Actin-membrane attachment proteins occurring in many cell types, *J. Cell Biol.* **95:**478–486.

Carlsson, L., Nyström, L. E., Lindberg, U., Kannan, K. K., Cid-Dresdner, H., Lovgren, S., and Jornvall, H., 1976, Crystallization of a non-muscle actin, *J. Mol. Biol.* **105:**353–366.

Carlsson, L., Markey, F., Blikstad, I., Persson, T., and Lindberg, U., 1979, Reorganization of actin in platelets stimulated by thrombin as measured by the DNase I inhibition assay, *Proc. Natl. Acad. Sci. U.S.A.* **76:**6376–6380.

Carroll, R. C., and Gerrard, J. M., 1982, Phosphorylation of platelet actin-binding protein during platelet activation, *Blood* **59:**466–471.

Carroll, R. C., Butler, R. G., and Morris, P. A., 1982, Separable assembly of platelet pseudopodal and contractile cytoskeletons, *Cell* **30:**385–393.

Casella, J. F., Flanagan, M. D., and Lin, S., 1981, Cytochalasin D inhibits actin polymerization and induces depolymerization of actin filaments formed during platelet shape change, *Nature (London)* **293:**302–305

Casella, J. F., Masiello, N. C., Lin, S., Bell, W., and Zucker, M. B., 1983, Identification of fibrinogen derivatives in the Triton-insoluble residue of human blood platelets, *Cell Motil.* **3:**21–30.

Cohen, I., and Cohen, C., 1972, A tropomyosin-like protein from human platelets, *J. Mol. Biol.* **68:**383–387.

Cohen, I., Gerrard, J. M., and White, J. G., 1982, Ultrastructure of clots during isometric contraction, *J. Cell Biol.* **93:**775–787.

Collier, N. C., and Wang, K., 1982a, Purification and properties of human platelet P235. A high molecular weight protein substrate of endogenous calcium-activated protease(s), *J. Biol. Chem.* **257:**6937–6943.

Collier, N. C., and Wang, K., 1982b, Human platelet P235: A high M_r protein which restricts the length of actin filaments, *FEBS Lett.* **143:**205–210.

Collier, N. C., and Wang, K., 1983a, Calcium sensitive modulation of actin polymerization by human platelet P235 (abstract), *J. Cell Biol.* **96:**289a.

Collier, N. C., and Wang, K., 1983b, Human platelet P235: A high M_r cytoplasmic protein which modulates actin polymerization (abstract), *Biophys. J.* **41:**86a.

Coté, G. P., and Smillie, L. B., 1981a, The interaction of equine platelet tropomyosin with skeletal muscle actin, *J. Biol. Chem.* **256:**7257–7261.

Coté, G. P., and Smillie, L. B., 1981b, Preparation and some properties of equine platelet tropomyosin, *J. Biol. Chem.* **256:**11004–11010.

Coté, G., Lewis, W. G., and Smillie, L. B., 1978a, Non-polymerizability of platelet tropomyosin and its NH_2- and COOH-terminal sequences, *FEBS Lett.* **91:**237–241.

Coté, G. P., Lewis, W. G., Pato, M. D., and Smillie, L. B., 1978b, Platelet tropomyosin: Lack of binding to skeletal muscle troponin and correlation with sequence, *FEBS Lett.* **94:**131–135.

Daniel, J. L., and Adelstein, R. S., 1976, Isolation and properties of platelet myosin light chain kinase, *Biochemistry* **15:**2370–2377.

Daniel, J. L., Holmsen, H., and Adelstein, R. S., 1977, Thrombin-stimulated myosin phosphorylation in intact platelets and its possible involvement in secretion, *Thromb. Haemost.* **38:**984–989.

Daniel, J. L., Molish, I. R., and Holmsen, H., 1981, Myosin phosphorylation in intact platelets, *J. Biol. Chem.* **256:**7510–7514.

Davies, P. J. A., Wallach, D., Willingham, M. C., Pastan, I., Yamaguchi, M., and Robson, R. M., 1978, Filamin-actin interaction. Dissociation of binding from gelation by Ca^{2+}-activated proteolysis, *J. Biol. Chem.* **253:**4036–4042.

Dayton, W. R., Reville, W. J., Goll, D. E., and Stromer, M. H., 1976, A Ca^{2+}-activated protease possibly involved in myofibrillar protein turnover. Partial characterization of the purified enzyme, *Biochemistry* **15:**2159–2167.

Debus, E., Weber, K., and Osborn, M., 1981, The cytoskeleton of blood platelets viewed by immunofluorescence microscopy, *Eur. J. Cell Biol.* **24:**45–52.

DeMartino, G. N., 1981, Calcium-dependent proteolytic activity in rat liver: Identification of two proteases with different calcium requirements, *Arch. Biochem. Biophys.* **211:**253–257.

Feinstein, M. B., 1980, Release of intracellular membrane-bound calcium precedes the onset of stimulus-induced exocytosis in platelets, *Biochem. Biophys. Res. Commun.* **93:**593–600.

Feinstein, M. B., Egan, J. J., and Opas, E. E., 1983, Reversal of thrombin-induced myosin phosphorylation and the assembly of cytoskeletal structures in platelets by the adenylate cyclase stimulants prostaglandin D_2 and forskolin, *J. Biol. Chem.* **258:**1260–1267.

Flanagan, M. D., and Lin, S., 1980, Cytochalasins block actin filament elongation by binding to high affinity sites associated with F-actin, *J. Biol. Chem.* **255:**835–838.

Fox, J. E. B., and Phillips, D. R., 1981, Inhibition of actin polymerization in blood platelets by cytochalasins, *Nature (London)* **292:**650–652.

Fox, J. E. B., and Phillips, D. R., 1982, Role of phosphorylation in mediating the association of myosin with the cytoskeletal structures of human platelets, *J. Biol. Chem.* **257:**4120–4126.

Fox, J. E. B., Say, A. K., and Haslam, R. J., 1979, Subcellular distribution of the different platelet proteins phosphorylated on exposure of intact platelets to ionophore A23187 or to prostaglandin E_1. Possible role of a membrane phosphopolypeptide in the regulation of calcium-ion transport, *Biochem. J.* **184:**651–661.

Fox, J. E. B., Dockter, M. E., and Phillips, D. R., 1981, An improved method for determining the actin filament content of nonmuscle cells by the DNase I inhibition assay, *Anal. Biochem.* **117:**170–177.

Fox, J. E. B., Reynolds, C. C., and Phillips, D. R., 1983a, Calcium-dependent proteolysis occurs during platelet aggregation, *J. Biol. Chem.* **258:**9973–9981.

Fox, J. E. B., Baughan, A. K., and Phillips, D. R., 1983b, Direct linkage of GP Ib to a $M_r = 250,000$ polypeptide in platelet cytoskeletons, *Blood* **62:**255a.

Fox, J. E. B., Boyles, J. K., Reynolds, C. C., and Phillips, D. R., 1984, Actin filament content and organization in unstimulated platelets, *J. Cell Biol.* **98:** 1985–1991.

Geiger, B., 1979, A 130K protein from chicken gizzard: Its localization at the termini of microfilament bundles in cultured chicken cells, *Cell* **18:**193–205.

Geiger, B., Dutton, A. H., Tokuyasu, K. T., and Singer, S. J., 1981, Immunoelectron microscope studies of membrane-microfilament interactions: Distributions of α-actinin, tropomyosin, and vinculin in intestinal epithelial brush border and chicken gizzard smooth muscle cells, *J. Cell Biol.* **91:**614–628.

Gerrard, J. M., Schollmeyer, J. V., Phillips, D. R., and White, J. G., 1979, α-Actinin deficiency in thrombasthenia. Possible identity of α-actinin and glycoprotein III, *Am. J. Pathol.* **94:**509–523.

Glenney, J. R., Jr., Glenney, P., and Weber, K., 1982a, F-Actin-binding and cross-linking properties of porcine brain fodrin, a spectrin-related molecule, *J. Biol. Chem.* **257:**9781–9787.

Glenney, J. R., Glenney, P., Osborn, M., and Weber, K., 1982b, An F-actin- and calmodulin-binding protein from isolated intestinal brush borders has a morphology related to spectrin, *Cell* **28:**843–854.

Gogstad, G. O., Brosstad, F., Krutnes, M-B., Hagen, I., and Solum, N. O., 1982, Fibrinogen-binding properties of the human platelet glycoprotein IIb–IIIa complex: A study using crossed-radioimmunoelectrophoresis, *Blood* **60:**663–671.

Gordon, D. J., Boyer, J. L., and Korn, E. D., 1977, Comparative biochemistry of non-muscle actins, *J. Biol. Chem.* **252:**8300–8309.

Grumet, M., and Lin, S., 1980, A platelet inhibitor protein with cytochalasin-like activity against actin polymerization *in vitro*, *Cell* **21:**439–444.

Hartwig, J. H., and Stossel, T. P., 1975, Isolation and properties of actin, myosin, and a new actin-binding protein in rabbit alveolar macrophages, *J. Biol. Chem.* **250:**5696–5705.

Hartwig, J. H., and Stossel, T. P., 1981, Structure of macrophage actin-binding protein molecules in solution and interacting with actin filaments, *J. Mol. Biol.* **145:**563–581.

Hartwig, J. H., Tyler, J., and Stossel, T. P., 1980, Actin-binding protein promotes the bipolar and perpendicular branching of actin filaments, *J. Cell Biol.* **87:** 841–848.

Haslam, R. J. and Lynham, J. A., 1977, Relationship between phosphorylation of blood platelet proteins and secretion of platelet granule constituents. I: Effects of different aggregating agents. *Biochem. Biophys. Res. Commun.* **77:**714–722.

Haslam, R. J., Lynham, J. A., and Fox, J. E. B., 1979, Effects of collagen, ionophore A23187 and prostaglandin E_1 on the phosphorylation of specific proteins in blood platelets, *Biochem. J.* **178:**397–406.

Hathaway, D. R., and Adelstein, R. S., 1979, Human platelet myosin light chain kinase requires the calcium-binding protein calmodulin for activity, *Proc. Natl. Acad. Sci. U.S.A.* **76:**1653–1657.

Howell, S. L., and Tyhurst, M., 1980, Regulation of actin polymerization in rat islets of Langerhans, *Biochem. J.* **192:**381–383.

Ishikawa, H., Bischoff, R., and Holzer, H., 1969, Formation of arrowhead complexes with heavy meromyosin in a variety of cell types, *J. Cell Biol.* **43:**312–328.

Jenkins, C. S. P., Gordon, P. B., Hatcher, V. B., and Puszkin, E. G., 1982, The presence of vinculin in cultured human endothelial cells and in platelets (abstract), *J. Cell Biol.* **95:**282a.

Jennings, L. K., Fox, J. E. B., Edwards, H. H., and Phillips, D. R., 1981, Changes in the cytoskeletal structure of human platelets following thrombin activation, *J. Biol. Chem.* **256:**6927–6932.

Käser-Glanzmann, R., Jakábová, M., George, J. N., and Lüscher, E. F., 1978, Further characterization of calcium-accumulating vesicles from human blood platelets, *Biochim. Biophys. Acta* **512:**1–12.

Kishimoto, A., Kajikawa, N., Shiota, M., and Nishizuka, Y., 1983, Proteolytic activation of calcium-activated, phospholipid-dependent protein kinase by calcium-dependent neutral protease, *J. Biol. Chem.* **258:**1156–1164.

Korn, E. D., 1978, Biochemistry of actomyosin-dependent cell motility (a review), *Proc. Natl. Acad. Sci. U.S.A.* **75:**588–599.

Korn, E. D., 1982, Actin polymerization and its regulation by proteins from nonmuscle cells, *Physiol. Rev.* **62:**672–737.

Landon, F., and Olomucki, A., 1983, Isolation and physico-chemical properties of blood platelet α-actinin, *Biochim. Biophys. Acta* **742:**129–134.

Landon, F., Huc, C., Thomé, F., Oriol, C., and Olomucki, A., 1977, Human platelet actin. Evidence of β and γ forms and similarity of properties with sarcomeric actin, *Eur. J. Biochem.* **81:**571–577.

Langer, B., Gonnella, P., Nachmias, V., Leung, L., and Siliciano, J., 1982a, Presence of α-actinin and vinculin in normal and thrombasthenic platelets (abstract), *J. Cell Biol.* **95:**296a.

Langer, B. G., Leung, L. L. K., Gonnella, P. A., Nachmias, V. T., Nachman, R. L., and Pepe, F. A., 1982b, α-Actinin and membrane glycoprotein IIIa are different proteins in human blood platelets, *Proc. Natl. Acad. Sci. U.S.A.* **79:**432–435.

Lazarides, E., and Lindberg, U., 1974, Actin is the naturally occurring inhibitor of deoxyribonuclease I, *Proc. Natl. Acad. Sci. U.S.A.* **71:**4742–4746.

Lebowitz, E. A., and Cooke, R., 1978, Contractile properties of actomyosin from human blood platelets, *J. Biol. Chem.* **253:**5443–5447.

Le Breton, G. C., Dinerstein, R. J., Roth, L. J., and Feinberg, H., 1976, Direct evidence for intracellular divalent cation redistribution associated with platelet shape change, *Biochem. Biophys. Res. Commun.* **71:**362–370.

Lin, S., Wilkins, J. A., Cribbs, D. H., Grumet, M., and Lin, D. C., 1982, Proteins and complexes that affect actin-filament assembly and interactions, *Cold Spring Harbor Symp. Quant. Biol.* **46:**625–632.

Lind, S. E., and Stossel, T. P., 1982, The microfilament network of the platelet, *Prog. Hemost. Thromb.* **6:**63–84.

Lind, S. E., Yin, H. L., and Stossel, T. P., 1982, Human platelets contain gelsolin. A regulator of actin filament length, *J. Clin. Invest.* **69:**1384–1387.

Lux, S. E., 1979a, Dissecting the red cell membrane skeleton, *Nature (London)* **281:**426–429.

Lux, S. E., 1979b, Spectrin-actin membrane skeleton of normal and abnormal red blood cells, *Semin. Hematol.* **16:**21–51.

Lyons, R. M., Stanford, N., and Majerus, P. W., 1975, Thrombin-induced protein phosphorylation in human platelets, *J. Clin. Invest.* **56:**924–936.

MacLean-Fletcher, S., and Pollard, T. D., 1980, Mechanism of action of cytochalasin B on actin, *Cell* **20:**329–341.

Markey, F., Lindberg, U., and Eriksson, L., 1978, Human platelets contain profilin, a potential regulator of actin polymerisability, *FEBS Lett.* **88:**75–79.

Markey, F., Persson, T., and Lindberg, U., 1981, Characterization of platelet extracts before and after stimulation with respect to the possible role of profilactin as microfilament precursor, *Cell* **23:**145–153.

Markey, F., Larsson, H., Weber, K., and Lindberg, U., 1982, Nucleation of actin polymerization from profilactin opposite effects of different nuclei, *Biochim. Biophys. Acta* **704:**43–51.

Mellgren, R. L., 1980, Canine cardiac calcium-dependent proteases: Resolution of two forms with different requirements for calcium, *FEBS Lett.* **109:**129–133.

Mellgren, R. L., Repetti, A., Muck, T. C., and Easly, J., 1982, Rabbit skeletal muscle calcium-dependent protease requiring millimolar Ca^{2+}. Purification, subunit structure, and Ca^{2+}-dependent autoproteolysis, *J. Biol. Chem.* **257:**7203–7209.

Murachi, T., Tanaka, K., Hatanaka, M., and Murakami, T., 1981, Intracellular Ca^{2+}-dependent protease (calpain) and its high molecular-weight endogenous inhibitor (calpastatin), *Adv. Enzyme Regul.* **19:**407–424.

Nachman, R. L., and Leung, L. K., 1982, Complex formation of platelet membrane glycoproteins IIb and IIIa with fibrinogen, *J. Clin. Invest.* **69:**263–269.

Nachmias, V. T., 1980, Cytoskeleton of human platelets at rest and after spreading, *J. Cell Biol.* **86:**795–802.

Niederman, R., and Pollard, T. D., 1975, Human platelet myosin. II. *In vitro* assembly and structure of myosin filaments, *J. Cell Biol.* **67:**72–92.

Painter, R. G., and Ginsberg, M., 1982, Concanavalin A induces interactions between surface glycoproteins and the platelet cytoskeleton, *J. Cell Biol.* **92:**565–573.

Phillips, D. R., and Jakábová, M., 1977, Ca^{2+}-Dependent protease in human platelets, *J. Biol. Chem.* **252:**5602–5605.

Phillips, D. R., Jennings, L. K., and Edwards, H. H., 1980, Identification of membrane proteins mediating the interaction of human platelets, *J. Cell Biol.* **86:**77–86.

Pollard, T. D., and Craig, S. W., 1982, Mechanism of actin polymerization, *Trends Biochem. Sci.* **7:**55–58.

Pollard, T. D., and Mooseker, M. S., 1981, Direct measurement of actin polymerization rate constants by electron microscopy of actin filaments nucleated by isolated microvillus cores, *J. Cell Biol.* **88:**654–659.

Pollard, T. D., Thomas, S. M., and Niederman, R., 1974, Human platelet myosin. I. Purification by a rapid method applicable to other nonmuscle cells, *Anal. Biochem.* **60:**258–266.

Pollard, T. D., Fujiwara, K., Handin, R., and Weiss, G., 1977, Contractile proteins in platelet activation and contraction, *Ann. N.Y. Acad. Sci.* **283:**218–236.

Reichstein, E., and Korn, E., 1979, Acanthamoeba profilin. A protein of low molecular weight from *Acanthamoeba castellanii* that inhibits actin nucleation, *J. Biol. Chem.* **254:**6174–6179.

Rink, T. J., Smith, S. W., and Tsien, R. Y., 1982, Cytoplasmic free Ca^{2+} in human platelets: Ca^{2+} thresholds and Ca-independent activation for shape-change and secretion, *FEBS Lett.* **148:**21–26.

Rodemann, H. P., Waxman, L., and Goldberg, A. L., 1982, The stimulation of protein degradation in muscle by Ca^{2+} is mediated by prostaglandin E_2 and does not require the calcium-activated protease, *J. Biol. Chem.* **257:**8716–8723.

Rosenberg, S., and Stracher, A., 1982, Effect of actin-binding protein on the sedimentation properties of actin, *J. Cell Biol.* **94:**51–55.

Rosenberg, S., Stracher, A., and Burridge, K., 1981a, Isolation and characterization of a calcium-sensitive α-actinin-like protein from human platelet cytoskeletons, *J. Biol. Chem.* **56:**12986–12991.

Rosenberg, S., Stracher, A., and Lucas, R. C., 1981b, Isolation and characterization of actin and actin-binding protein from human platelets, *J. Cell Biol.* **91:**201–211.

Scholey, J. M., Taylor, K. A., and Kendrick-Jones, J., 1980, Regulation of non-muscle myosin assembly by calmodulin-dependent light chain kinase, *Nature (London)* **287:**233–235.

Schollmeyer, J. V., Rao, G. H. R., and White, J. G., 1978, An actin-binding protein in human platelets. Interactions with α-actinin on gelatin of actin and the influence of cytochalasin B, *Am. J. Pathol.* **93:**433–446.

Sixma, J. J., Schiphorst, M. E., Verhoeckx, C., and Jockusch, B. M., 1982, Peripheral and integral proteins of human blood platelet membranes. α-Actinin is not identical to glycoprotein III, *Biochim. Biophys. Acta* **704:**333–344.

Solum, N. O., Olsen, T., and Gogstad, G., 1983, GP Ib in the Triton-soluble (cytoskeletal) fraction of platelets (abstract), *Thromb. Haemost.* **50:**372a.

Stracher, A., Qingqi, Z., Lawrence, J., and Rosenberg, S., 1982, Role of phosphorylation in the regulation of platelet cytoskeletal formation (abstract), *Fed. Proc.* **41:**657.

Swanston-Flatt, S. K., Carlsson, L., and Gylfe, E., 1980, Actin filament formation in pancreatic β-cells during glucose stimulation of insulin secretion, *FEBS Lett.* **117:**299–302.

Taylor, D. G., Mapp, R. J., and Crawford, N., 1975, The identification of actin associated with pig platelet membranes and granules, *Biochem. Soc. Trans.* **3:**161–164.

Toyo-Oka, T., Shimizu, T., and Masaki, T., 1978, Inhibition of proteolytic activity of calcium-activated neutral protease by leupeptin and antipain, *Biochem. Biophys. Res. Commun.* **82:**484–491.

Truglia, J. A., and Stracher, A., 1981, Purification and characterization of a calcium-dependent sulfhydryl protease from human platelets, *Biochem. Biophys. Res. Commun.* **100:**814–822.

Tsujinaka, T., Sakon, M., Kambayashi, J., and Kosaki, G., 1982, Cleavage of cytoskeletal proteins by two forms of Ca^{2+} activated neutral proteases in human platelets, *Thromb. Res.* **28**:149–156.

van Deurs, B., and Behnke, O., 1980, Membrane structure of nonactivated and activated human blood platelets as revealed by freeze-fracture: Evidence for particle redistribution during platelet contraction, *J. Cell Biol.* **87**:209–218.

Wallach, D., Davies, P. J. A., and Pastan, I., 1978, Purification of mammalian filamin. Similarity to high molecular weight actin-binding protein in macrophages, platelets, fibroblasts, and other tissues, *J. Biol. Chem.* **253**:3328–3335.

Wegner, A., 1976, Head to tail polymerization of actin, *J. Mol. Biol.* **108**:139–150.

White, G. C., 1980, Calcium-dependent proteins in platelets: Response of calcium-activated protease in normal and thrombasthenic platelets to aggregating agents, *Biochim. Biophys. Acta* **631**:130–138.

White, J. G., 1968, Fine structural alterations induced in platelets by adenosine diphosphate, *Blood* **31**:604–622.

White, J. G., 1974, Electron microscopic studies of platelet secretion, *Prog. Hemost. Thromb.* **2**:49–98.

Wilkins, J. A., and Lin, S., 1982, High-affinity interaction of vinculin with actin filaments in vitro, *Cell* **28**:83–90.

Yin, H. L., and Stossel, T. P., 1980, Purification and structural properties of gelsolin, a Ca^{2+}-activated regulatory protein of macrophages, *J. Biol. Chem.* **255**:9490–9493.

Yin, H. L., Zaner, K. S., and Stossel, T. P., 1980, Ca^{2+} control of actin gelation. Interaction of gelsolin with actin filaments and regulation of actin gelation, *J. Biol. Chem.* **255**:9494–9500.

Yin, H. L., Hartwig, J. H., Maruyama, K., and Stossel, T. P., 1981, Ca^{2+} control of actin filament length. Effects of macrophage gelsolin on actin polymerization, *J. Biol. Chem.* **256**:9693–9697.

Yoshida, N., Weksler, B., and Nachman, R., 1983, Purification of human platelet calcium-activated protease. Effect on platelet and endothelial function, *J. Biol. Chem.* **258**:7168–7174.

Zimmerman, U-J. P., and Schlaepfer, W. W., 1982, Characterization of a brain calcium-activated protease that degrades neurofilament proteins, *Biochemistry* **21**:3977–3983.

Zucker-Franklin, D., and Grusky, G., 1972, The actin and myosin filaments of human and bovine blood platelets, *J. Clin. Invest.* **51**:419–430.

Zucker-Franklin, D., Nachman, R. L., and Marcus, A. J., 1967, Ultrastructure of thrombosthenin, the contractile protein of human blood platelets, *Science* **157**:945–946.

14

The Mechanism of Clot Retraction

Isaac Cohen

1. INTRODUCTION

Nearly 2000 years elapsed between the first observation of blood clotting by Aristotle and that of Hewson (1772) on the subsequent retraction of the spontaneously formed blood clot. The recognition of the crucial role of platelets in the phenomenon of clot retraction had to await the promotion of platelets from the undignified status of being derived from red cells or chyle (Donné, 1842) to that of a glorified element with its own megakaryocytic parenthood (Wright, 1906). The early observation that clots formed by the blood of severely thrombocytopenic subjects or prepared from platelet-poor plasma of normal subjects do not retract, whereas clots containing normal platelets retract drastically to about 10% of their original volume, leaves little doubt as to the role of platelets in clot retraction. The first monograph on clot retraction published by Budtz-Olsen three decades ago (1951) still represents an excellent document on this subject and there have been several more recent reviews (Cohen and Lüscher, 1975; Behnke, 1976; Pollard *et al.*, 1977; Cohen, 1982).

The phenomenon of clot retraction is reminiscent of muscle contraction and this similarity was affirmed by Bettex-Galland and Lüscher (1959) with the discovery of thrombosthenin, the platelet counterpart to muscle actomyosin. Since thrombosthenin represents 15 to 20% of the cell protein content, it is reasonable to assume that the contractile system plays a major role in various platelet functions. The first manifestation of a contractile activity following platelet activation is platelet shape change. Actin binding to the cytoplasmic membrane appears to be essential for the shape change process. Similarly, fibrin binding to specific loci of the platelet surface is a *sine qua*

Isaac Cohen • Atherosclerosis Program, Rehabilitation Institute of Chicago, and Department of Molecular Biology, Northwestern University Medical School, Chicago, Illinois 60611.

non requirement for the transmembrane coupling of fibrin to the force-generating apparatus in platelets. Since clot retraction, similar to muscle contraction, represents a manifestation of an activated state of the platelet rather than just being a passive syneresis process, the term clot contraction is more appropriate. As spectacular as this phenomenon is, its physiological role is still a subject of controversy.

2. MEASUREMENT OF CONTRACTILE FORCE

Fibrinogen clotting and platelet activation are mediated by thrombin and are required for clot retraction. Clotting can be dissociated from clot retraction, however, with the use of reptilase, the *Bothrops atrox* enzyme that clots fibrinogen by releasing fibrinopeptide A (Blombäck, 1958), but does not activate platelets. Such reptilase-induced clots will retract only in the presence of a platelet agonist, such as ADP, added either immediately before or after clotting (de Gaetano *et al.*, 1973, 1974). In these cases, the platelets uniformly distributed in the clot will undergo activation resulting in the protrusion of long pseudopods and the transmission of the contractile force to the fibrin strands. The spatially even distribution of the platelets, prior to their activation in the fibrinogen–fibrin gel, is crucial for retracting the whole clot. Indeed, the addition of polymerizing fibrin to ADP-induced preaggregated platelets does not result in clot retraction (de Gaetano *et al.*, 1971).

The first semiquantitative method for assessing the extent of clot retraction included the measurement of the residual clot length or extruded serum following the spontaneous clotting of whole blood or thrombin-induced clotting of whole or diluted citrated platelet-rich plasma (Budtz-Olsen, 1951; Didisheim and Bunting, 1966). In order to prevent the adhesion of clots to glass, several methods were proposed, such as flaming red-hot or gelatinizing the test tubes prior to sample introduction (Lüscher, 1956; Massini *et al.*, 1982), or dislodging a cylindrical clot from a glass tubing and measuring its rate of shortening when freely floating in Tyrode buffer at 37 °C (Cohen *et al.*, 1975), the so-called "floating clot technique." The first attempt to quantitate the force generated by platelets in a contracting system involved a kymographic recording technique (Majno *et al.*, 1972). Since the contractile force of platelets *per se* could not be determined, the structures on which the bound platelets exerted their contractile effect consisted of either the native fibrin mesh (Cohen and de Vries, 1973; Bottecchia and Fantin, 1973; de Gaetano *et al.*, 1973, 1974) or a nylon mesh (Salganicoff *et al.*, 1977). With the fibrin system used in our laboratory, a thrombin-induced cylindrical clot obtained from platelet-rich plasma is tied at one end to a silver alloy wire holder and at the upper end to a force displacement transducer linked to a polygraph system. The clot is immersed into an isolated tissue bath containing Ca^{2+}-free Tyrode buffer maintained at 37 °C. A preload of 80 to 100 mg is applied to straighten the clot. Since the transducer developed a displacement rate of 0.02 mm/g, isometric contraction is measured in the longitudinal direction. With the nylon mesh system, platelets become activated while being pelleted at 10,000 g over a nylon mesh and the tension is measured at 37 °C on strips of this mesh using a kymographic system. Contractile force and viscoelastic properties of a platelet-rich plasma clot measured with a fluid rheometer or a Weissenberg rheogoniometer provide an accurate quantitative approach for correlating clot structure and contractile force (Glover *et al.*, 1975; Kuntamukkula *et al.*, 1978; Jen and McIntire, 1982).

3. ·MECHANICAL ASPECTS OF PLATELET–FIBRIN INTERACTION

Although it is clear that clot retraction results from the interaction of a static, nonmotile fibrin mesh and dynamic, actively motile blood platelets, the plethora of models to explain this phenomenon illustrates how poorly understood it is from a mechanistic point of view. The extent of clot shortening in the absence of an external load was shown to be inversely proportional to the concentration of fibrin (Budtz-Olsen, 1951; Lüscher, 1956). We measured the interaction of fibrin and blood platelets in the presence and absence of an external load in clots containing different ratios of these two components (Cohen *et al.*, 1975). Under isometric conditions the tension increased to a saturating value when plotted as a function of the concentration of either component at a constant concentration of the other. The maximal tension was found to increase linearly with the minimal platelet concentration (Ps) required in order to reach saturation at various different fibrin concentrations (Figure 1). A linear relationship was obtained when the logarithm of the fibrin concentrations used was plotted against the critical platelet concentration at saturation points (Ps). The exponential increase of fibrin concentration with platelet concentration to achieve saturation (Figure 2) suggests that the fibrin three-dimensional network is a function of the number of pseudopodia generated by platelets upon activation. Noteworthy is the low fibrin concentra-

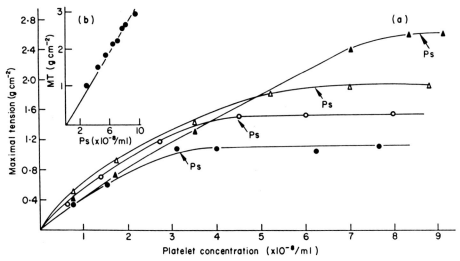

Figure 1. (A) Saturation of fibrin by platelets in isometric conditions. The maximal tension (MT) obtained for various constant fibrin concentrations is plotted against platelet concentration. Fibrin concentrations: ●, 0.05 mg/ml; 0, 0.01 mg/ml; △, 0.2 mg/ml; ▲, 0.75 mg/ml. Ps = Critical platelet concentration at saturation point, above which MT is constant. Only a few curves are presented for the sake of clarity. (B) MT obtainable for a given platelet number. Saturation MT values are plotted against Ps. From Cohen *et al.* (1975) with permission.

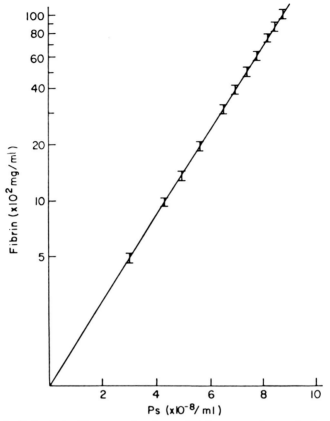

Figure 2. Platelet-fibrin relationship at saturation. Log$_{10}$ of fibrin concentration vs. platelet concentration at Ps (see Figure 1). From Cohen et al. (1975) with permission.

tion (0.05 mg/ml) necessary for the apparent saturation of physiological concentrations of platelets (3 × 10^8 platelets/ml) (Figures 1 and 2) for the development of maximal tension. At relatively high fibrin concentrations, part of the fibrin fibers may not participate in the contractile process because of steric hindrance. The importance of platelet–fibrin interaction over platelet–platelet interaction for force generation is further substantiated by the direct linear increase of the rate of isometric tension development with increasing platelet concentration (Figure 3). Indeed, if platelet–platelet interaction was the important component for force generation, small platelet numbers should be much less able to make pseudopod–pseudopod contact and tension should fall off exponentially with a decrease in platelet numbers. The observation that the rate of tension development decreases linearly with a decrease in platelet concentration argues strongly for the importance of platelet–fibrin interaction.

In the absence of an external load, the velocity of shortening normalized with respect to the momentary length, as measured by the floating clot technique, decreased in an hyperbolic manner with increasing fibrin concentrations. The curve in Figure 4 is reminiscent of the velocity–load relationship of a contracting muscle. The presence of fibrin, although essential for the formation of a continuous three-dimensional network,

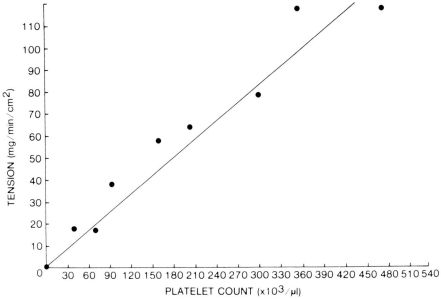

Figure 3. Variation in the rate of tension development in a plasma clot with increasing concentrations of washed platelets. From Cohen *et al.* (1982) with permission.

appears to impose concurrently an internal resistance that may be considered as the equivalent of an external load in the case of muscle.

4. ULTRASTRUCTURE AND TENSION GENERATION IN CONTRACTED CLOTS

4.1. Normal Clots

An examination of clots allowed to contract in the absence of an external load reveals platelets and fibrin organized into large clumps with fibrin concentrated cen-

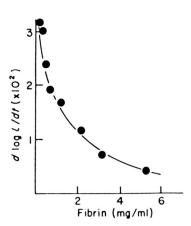

Figure 4. Velocity of clot shortening (d log l/dt) vs. fibrin concentrations. From Cohen *et al.* (1975) with permission.

trally and connecting these clumps (Sokal, 1960; Cohen *et al.,* 1982). A puzzling feature is the disproportion between the relatively small mass of platelets and the huge sea of gelled fibrin retracting under the influence of platelets. This could be explained by Sokal's observation (1960) that platelet pseudopodia emerge radially and cover a considerable distance from platelet masses while attaching to fibrin fibers. The protrusion of long pseudopodia, reaching a length at least five times the diameter of a platelet body, indeed increases dramatically the "sphere" of influence of platelets over the fibrin gel.

Whereas the ultrastructure of platelet–fibrin interaction is often difficult to interpret in clots contracted in the absence of an external load, clots contracted under isometric conditions reveal an orientation of fibrin strands and platelets in the direction of force generation that facilitates the interpretation of ultrastructural observations. In such clots of normal platelet-rich plasma, tension developed at an initial rate of 0.1 to 0.2 g/min/cm^2 (initial cross-sectional area). When fixed under tension after having attained a force of 1.6 g/cm^2, the longitudinal sections of platelet pseudopods were observed in close apposition to long, thin fibrin strands oriented like cables in the axis of tension (Figure 5). Spindle-shaped platelet bodies and pseudopods extended in the axis of tension and thereby contributed to the longitudinally oriented platelet–fibrin associations. Adjacent serial sections in the longitudinal plane through the same platelet showed that microfilaments as well as microtubules assumed an orientation in the direction of force generation (Figure 6B, C). This quasi-uniform orientation of the microtubules is seen best in cross-sections of the isometrically contracted clot (Figure 6A). Portions of platelets in close apposition to fibrin strands contained hollow-cored circular profiles of microtubules.

The role of platelets in the structure of the clot and subsequently to the generation of force is suggested by our results. This was confirmed by Jen and McIntire (1982) who, using a rheological technique, demonstrated a maximum elastic modulus (a measure of clot strength) of 6,000 dynes/cm^2 and a maximum contractile force of 1,500 dynes/cm^2 in clotted platelet-rich plasma.

4.2.   Thrombasthenic and Storage-Pool-Deficient Clots

Clots made from the platelet-rich plasma of patients with Glanzmann's thrombasthenia lacking platelet membrane glycoproteins (GP) IIb and IIIa neither generate tension nor show an orientation of the fibrin strands in the direction of isometric tension (Figure 7). The platelets, as expected, do not form clumps. These results suggest that a platelet membrane component, absent or defective in thrombasthenia, is required for clot contraction. On the other hand, clots made from plasma containing platelets with reduced number of dense bodies and α-granules have a normal ultrastructure appearance and generate tension normally, demonstrating that these platelet constituents are not essential for clot contraction (Cohen *et al.,* 1982).

4.3.   Factor XIII-Deficient Clots

Clots formed from normal washed platelets suspended in factor XIII-deficient plasma do not generate tension (Figure 8) and lack the typical orientation of the fibrin

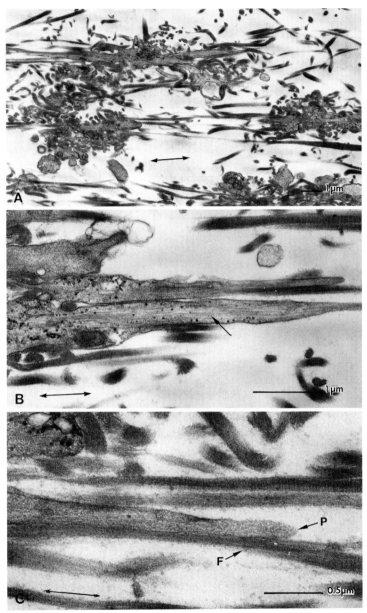

Figure 5. (A) A clot was formed using platelet-rich plasma and then allowed to undergo contraction isometrically until half-maximum tension was achieved. Platelet pseudopods and fibrin strands can be seen aligning in the direction of tension. Longitudinal section, uranyl acetate and lead citrate stain (magnification 7800×). (B) A higher power view of a sample of platelet-rich plasma contracted isometrically until maximum tension was achieved. A long platelet pseudopod with numerous microtubules (arrow) can be seen. Longitudinal section, uranyl acetate and lead citrate stain (magnification 27,000×). (C) Another view of the sample shown in (B) to show close interaction between a platelet pseudopod (P) and a fibrin strand (F). Longitudinal section, uranyl acetate and lead citrate stain (magnification 46,000×). From Cohen *et al.* (1982) with permission. Entire figure reproduced at 70%.

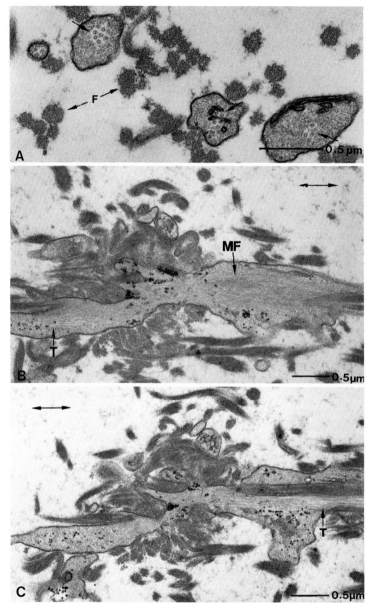

Figure 6. (A) A cross section of the clot formed in platelet-rich plasma and allowed to contract isometrically to reach maximal tension. Fibrin (F) strands are cut almost exclusively in cross section as are platelet pseudopods. Microtubules (arrows) are present within all cross sections of pseudopods. Cross section, uranyl acetate and lead citrate stain (magnification 50,000×). (B, C) Adjacent serial sections in the longitudinal plane though the same platelet. The body and pseudopods extend in the axis of tension. Microtubules (T) are prominent in one pseudopod and masses of microfilaments (MF) in the opposite extension in (B). The microfilaments are replaced by microtubules (T) in a deeper section of the same pseudopod in (C). Longitudinal sections, uranyl acetate and lead citrate stain. (B) magnification 30,500×; (C) magnification 30,000×. From Cohen *et al.* (1982) with permission. Entire figure reproduced at 70%.

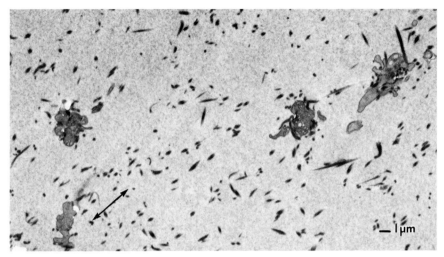

Figure 7. A clot was formed using a sample of platelet-rich plasma from a patient with Glanzmann's thrombasthenia and allowed to contract isometrically. No tension was produced. Platelets remained single and did not form platelet-fibrin clumps. No alignment of platelet pseudopods or fibrin strands in the direction of tension was seen. Longitudinal section, uranyl acetate and lead citrate stain (magnification 3750×). From Cohen *et al.* (1982) with permission. Reproduced at 85%.

fibers. Instead, they are randomly oriented and form only a few thick, striated fibers. Platelets clump within these clots, but the clumps remain widely separated. When highly purified factor XIII is added to the factor XIII-deficient system before clotting is induced with thrombin, tension is restored to normal (Figure 8) and thick fibrin strands oriented in the direction of tension are produced. Platelets also develop long pseudopods that extend along the fibrin strands. These results suggest that the structure of the clot and the generation of tension require cross-linked fibrin strands.

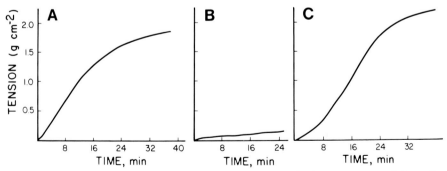

Figure 8. The role of factor XIII in tension development. Washed platelets were resuspended at a final concentration of 5.5×10^8 platelets/ml in normal plasma (A), in factor XIII-deficient plasma (B), or in factor XIII-deficient plasma supplemented with 24 μg pure factor XIII/ml (C). Factor XIII-deficient plasmas from two other patients gave the same result. Clots (A) and (C) were insoluble in 5 M urea or 1% monochloracetic acid whereas, clot (B) was soluble under the same conditions. From Cohen *et al.* (1982) with permission.

Log [Ca^{2+}]

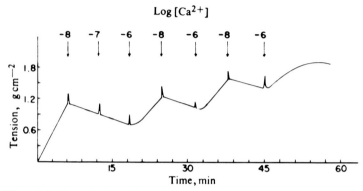

Figure 9. Effect of Ca^{2+} on the isometric tension developed in a thrombin-clotted platelet-rich plasma. From Cohen and de Vries (1973) with permission.

5.   ACTIVATORS AND INHIBITORS OF CLOT RETRACTION

5.1.   Divalent Cations

Ca^{2+} and Mg^{2+} are required for the development of tension in a platelet-rich plasma clot (Bottecchia and Fantin, 1973). To a certain extent, Mg^{2+} is able to substitute for Ca^{2+} (Massini *et al.*, 1982). Cycles of contraction and relaxation were recorded during the development of isometric tension when the free Ca^{2+} concentrations in Ca–EGTA buffers, added at half-maximal tension, was cycled between 10^{-8} and 10^{-6} M (Cohen and de Vries, 1973) (Figure 9). A concentration of 10^{-6} M Ca^{2+} is similar to that required to stimulate the Mg^{2+}-activated ATPase of platelet actomyosin (Cohen and Cohen, 1972). No relaxation was observed when EGTA was added after maximal tension. Whether Ca^{2+} is involved in platelet–fibrin interaction and/or in cytoskeletal interactions leading to force generation by the platelet actomyosin complex is uncertain. The loose association of platelets with fibrin responsible for the loss of tension and the randomness of fibrin orientation in the presence of EGTA (Cohen *et al.*, 1982) leave little doubt as to the requirement of external Ca^{2+} for the platelet pseudopod–fibrin interaction. The increase in the cytosolic Ca^{2+}, presumably from internal membrane stores, is crucial for platelet activation mechanisms (Le Breton *et al.*, 1976). Recently, Vickers *et al.* (1982) correlated the phospholipase C-mediated phosphodiesterase cleavage of phosphatidylinositol biphosphate (PI$_{4,5}$ P$_2$) in stimulated platelets (Agranoff *et al.*, 1983) with the release of Ca^{2+} into the cytosol from the PI$_{4,5}$ P$_2$, presumed to be located on the inner surface of the platelet membrane (Michell, 1975; Perret *et al.*, 1979). The uptake of Ca^{2+} following platelet activation has also been reported (Mürer and Holme, 1970; Massini and Lüscher, 1976; Massini *et al.*, 1978). Recently, Rink *et al.* (1982) using quin 2, a fluorescent Ca^{2+} indicator, showed that the shape change in thrombin-stimulated platelets was associated with a rapid rise in the cytosolic free Ca^{2+} from a basal level of 100 nM to 3 μM. This rise, which occurred in the presence of 1 mM CaCl$_2$, was due to Ca^{2+} influx, and only 10% could be attributable to internal release. It is therefore probable that external Ca^{2+} participates in platelet–fibrin association as well as in intracellular cytoskeletal processes resulting in the generation of force.

An attractive hypothesis for the need for extracellular Ca^{2+} is also provided by its binding to phospholipid heads of synthetic lipid bilayers, thereby neutralizing their charges and producing a compression wave that could be effective at long distances (Rubalcava et al., 1969; Vanderkooi and Martonosi, 1969; Singer, 1971). Such Ca^{2+}-generated compression waves might cause structural changes in proteins embedded in these lipid bilayers.

Although the Ca^{2+} metabolism in platelets is not the scope of this chapter, Figure 10 may assist in understanding the role of cytosolic Ca^{2+} in platelet intermediary metabolism. Whereas Ca^{2+} and Mg^{2+} can sustain clot retraction, manganese inhibits clot retraction or reverses it if added after the onset of tension development (Bottecchia and Fantin, 1973). This effect may be due to an uptake of Mn^{2+} competing with the internal Ca^{2+}, similarly to smooth muscle.

5.2. Prostaglandin Metabolites

Thrombin-induced platelet release reaction is not affected by cyclooxygenase inhibitors, and therefore a dissociation of the release reaction from clot retraction

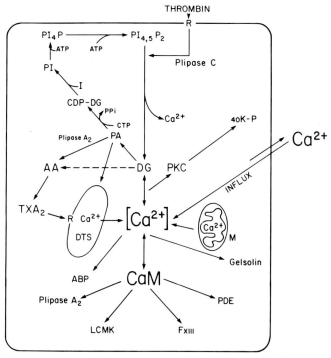

Figure 10. Calcium movements in stimulated platelets. $[Ca^{2+}]$, free cytosolic Ca^{2+}; R, receptor; $PI_{4,5} P_2$, phosphatidylinositol 4,5-biphosphate; PI_4P, phosphatidylinositol 4-phosphate; PI, phosphatidylinositol; DG, diacylglycerol; PA, phosphatidic acid; AA, arachidonic acid; TXA_2, thromboxane A_2; DTS, dense tubular system; M, mitochondrion; ABP, actin-binding protein; CaM, calmodulin; LCMK, light-chain myosin kinase; FXIII, factor XIII; PDE, phosphodiesterase; PKC, protein kinase C; 40 K-P, 40,000 daltons protein. For the sake of clarity, localization of the various processes in membrane or cytosolic compartments was not attempted.

mediated by nonsteroidal anti-inflammatory drugs is impossible with thrombin. On the other hand, reptilase, which clots fibrinogen without activating platelets, is useful for screening activators and inhibitors of the clot retraction process and has been widely used by de Gaetano *et al.* (1982). The addition of agonists such as ADP or epinephrine either before or after clotting stimulates the generation of tension, an effect that is not prevented by aspirin or indomethacin (de Gaetano *et al.*, 1973). The release reaction that is inhibited by these cyclooxygenase inhibitors is therefore not associated with clot retraction, nor is thromboxane A_2 when ADP or epinephrine are used as platelet agonists. On the other hand, the induction of clot retraction by arachidonic acid in a reptilase clot is completely blocked by aspirin. This suggests that shape change and primary aggregation, which are maintained in ADP- and epinephrine- but not in arachidonic acid-stimulated platelets treated with aspirin, are important for clot retraction. Clot retraction in "arachidonic acid-reptilase clots" is not affected by the thromboxane synthetase inhibitors 4-methyl-imidazole and UK 37,248-01, suggesting that the endoperoxides PGG_2 and PGH_2, potent platelet stimulants, are important in clot retraction. As expected, the induction of clot retraction by U-46619, the endoperoxide and thromboxane A_2 mimetic, is not affected by inhibitors of either cyclooxygenase or thromboxane synthetase activities. On the other hand, azoprostanoic acid derivatives, which compete with endoperoxide and thromboxane A_2 receptors on the platelet surface, do inhibit clot retraction induced by either arachidonic acid or U-46619. Adenosine diphosphate scavengers do not affect the arachidonic acid- or U-46619-induced clot retraction process, suggesting that release of ADP is not necessary for this process to occur. In conclusion, clot retraction proceeds normally whenever the first stages of platelet stimulation, comprising shape change and primary aggregation, are maintained.

Agents that increase the cytosolic cyclic AMP levels by either activating the platelet adenyl cyclase activity or inhibiting the platelet phosphodiesterase activity inhibit clot retraction induced by all known agonists. This effect of cAMP is probably due to its property to either prevent the release of Ca^{2+} from internal membrane structures into the cytosol or to trigger a reuptake of the cytosolic Ca^{2+} by membrane systems (Käser-Glanzmann *et al.*, 1977, 1978, 1979). The adenyl cyclase activators that inhibit clot retraction and that have receptors on the platelet membrane are PGI_2, PGE_1, and PGD_2 (Kuntamukkula *et al.*, 1978; de Gaetano *et al.*, 1982). Thromboxane A_2 prevents the stimulation of adenyl cyclase by these agents (Miller *et al.*, 1977). Phosphodiesterase inhibitors potentiate the effect of the adenyl cyclase activators by preventing the breakdown of cAMP and pyrimido-pyrimidine compounds (dipyridamole) as well as 3-methyl-xanthines (isobutyl-methyl-xanthine, caffeine, theophylline, papaverine) have indeed been reported to inhibit clot retraction (de Gaetano *et al.*, 1982).

5.3. Cytoskeletal Destabilizing Agents

Cytochalasin B and the more specific cytochalasin E either prevent the actin assembly or disassemble preformed actin filaments by preventing the treadmill mechanism of actin assembly (Lin *et al.*, 1980; Maruyama, 1981; Brown and Spudich, 1981). The cytochalasins prevent the assembly of actin triggered in platelets by thrombin (Fox and Phillips, 1981). When considering the interaction of actin and myosin in platelets to produce force, the inhibitory effect of cytochalasins on clot retraction (Majno *et al.*, 1972) becomes self-explanatory. Cytochalasins preserve the discoid shape of platelets and prevent pseudopod formation as well as tension generation (White, 1971; Cohen *et al.*, 1982; Jen and McIntire, 1982). Removal of cytochalasin B or E by washing the clot several times restores normal contractile activity and full isometric tension (Cohen *et al.*, 1982).

The role of platelet microtubules in clot retraction, studied with the use of de-stabilizing agents such as vinca alkaloids, which cause microtubule disassembly, has long been controversial. Whereas colchicine and vincristine were reported to prevent clot retraction when studied by standard or kymographic procedures (Shepro *et al.*, 1969; Chao *et al.*, 1976; Cohen *et al.*, 1982), these results could not be confirmed when rheological techniques were used (Kuntamukkula *et al.*, 1982; Jen and McIntire, 1982). The controversy may have been resolved by the use of taxol, a microtubule stabilizing agent. Taxol is known to prevent vincristine-induced microtubule disassem-bly in platelets (White and Rao, 1982). However, despite pretreatment with taxol, vincristine still prevented the development of tension or caused relaxation if added at half-maximal tension (I. Cohen and J. G. White, personal communication). This result, similar to the effect of vincristine on platelet aggregation (White and Rao, 1982), suggests that vincristine has a cytotoxic effect unrelated to its ability to dis-assemble microtubules.

5.4. Energy Metabolism Inhibitors

The platelet contractile elements, including actin, myosin, tropomyosin, actin-binding protein, gelsolin, and α-actinin represent about 50% of the total platelet protein content. Platelets are therefore really tiny muscles circulating in our blood-stream, and as such are heavily dependent on a highly developed energetic metabo-lism. Inhibitors of either the glycolytic pathway (2-deoxyglucose) or the oxidative phosphorylation pathway (cyanide, antimycin, oligomycin) do not inhibit significantly clot retraction. On the other hand, the combination of both types of inhibitors, which totally prevents ATP synthesis, does inhibit clot retraction (Mürer, 1969; Kirkpatrick *et al.*, 1980)

6. TRANSMEMBRANE LINKAGE OF CYTOSKELETAL COMPONENTS WITH FIBRIN

Since actin and myosin are the main building stones of the contractile apparatus in muscle and nonmuscle systems, and since motile events in cells require the attachment of actin filaments to membranes, it is plausible that clot retraction involves a direct or indirect linkage of actin to fibrin. Moreover, the development of an effective force presupposes the binding of actin and fibrin bundles at specific sites on the platelet membrane. This binding of the two filament types could occur either on the platelet surface following actin extrusion (cis-membrane binding) or on both surfaces with actin bound on the cytoplasmic surface and fibrin on the outer platelet surface (transmembrane linkage). There is no convincing evidence for the presence of either actin or myosin on the outer surface of nonactivated platelets. Fujiwara and Pollard (1976), using myosin-specific IgG coupled to fluorescent dyes and purified by affinity chromatography, did not detect any fluorescence on the surface of nonactivated platelets. Similarly, Gabbiani *et al.* (1973) reported an absence of binding of antiactin autoantibodies to nonactivated platelets. The reports of surface localization based on work with antibodies obtained from antigens possibly contaminated with membrane should be viewed with caution. Bennett *et al.* (1981) using the tritiated adenine nucleotide affinity label 5'-*p*-fluorosulfonylbenzoyl adenosine did not find any labeling of myosin and actin in platelet-rich plasma or gel-filtered platelets unless platelets were washed, which may have resulted in membrane damage. No labeling was observed when gel-filtered platelets were stimulated by either ADP or thrombin under nonstirring conditions. George *et al.* (1980) found no surface-labeled actin using washed platelets reversibly aggregated with ADP and then reacted with [^{125}I]diazotized diiodosulfanilic acid. However, platelets stimulated with thrombin in the presence of EDTA did demonstrate subsequent labeling of actin by this reagent. Bouvier *et al.* (1977) found binding of fluorescent antiactin autoantibodies to platelets reversibly or irreversibly aggregated with ADP in the presence or absence of EDTA. Stirring in the absence of platelet agonists caused only a small fluorescent staining. Antiactin autoantibodies also bind to platelets in thrombin-induced retracted platelet-rich plasma clots. In no case did the binding of the antibody affect either aggregation or clot retraction. Regardless of whether actin is or is not exposed on the platelet surface following platelet stimulation, the lack of functional effects of antiactin antibodies, such as on clot retraction, suggests that actin-to-fibrin binding on the activated platelet surface is not involved in clot retraction. A more plausible model is the transmembrane binding of plasmatic fibrin to actin bound to the cytoplasmic surface.

The identification of "effective" receptors for fibrin and actin on both sides of the membrane is essential for understanding the mechanism of force transduction involving the platelet membrane. The identification of fibrinogen and fibrin receptors exposed on the platelet surface following stimulation with platelet agonists is an attractive line of investigation. Chapter 9 presents the evidence for the role of GP IIb-IIIa as the fibrinogen receptor required for normal platelet aggregation. A recent report of an association between fibrin and the Triton-insoluble residue of thrombin-stimulated platelets (Tuszynski *et al.*, 1984) further adds support to the possible attachment of the fibrin clot to the platelet cytoskeleton in the mechanism of clot retraction. This report also presents evidence that the association between fibrin and the cytoskeleton is mediated by the membrane GP IIb-IIIa.

As reported by Polley *et al.* (1981), a prerequisite for fibrinogen (fibrin)-binding to platelets may involve the redistribution and clustering of the GP IIb-IIIa complexes found in thrombin-stimulated platelets. However, van Deurs and Behnke (1980), using freeze-fracture techniques, failed to find intramembrane particle clusters following either ADP or thrombin-induced stimulation of platelets. The membrane particles were as evenly distributed as in nonactivated platelets, showing 1500 particles/μm^2 of membrane with two to three times more particles on the external side (E face) than on the protoplasmic side (P face). The firm association of the particles with the E leaflet was thought to be due to some kind of interaction between the surface-exposed

Figure 11. Freeze-fracture of platelets in a retracting clot with both E and P faces seen, the latter exhibiting many clustered particles. From van Deurs and Behnke (1980) with permission.

glycoproteins and glycolipids. On the other hand, freeze-fracture following clot retraction showed the particles to be most frequent on the P face and they tended to form occasional clusters (Figure 11). This may be due to a stronger attachment of glycoproteins to clustered actin filaments either directly or via α-actinin (Figure 12). Since shape change alone, stimulated by ADP or thrombin, is not associated with the formation of membrane clusters, these may form only when force is to be transmitted to fibrin strands.

Although Chapter 13 deals extensively with the interaction of platelet membrane and cytoskeletal proteins, I felt that it was important to briefly mention the following data for a better understanding of transmembrane coupling occurring during the clot retraction process. The association of actin and a 95,000 mol. wt. protein with the platelet membrane has been elegantly demonstrated with the use of [^{125}I]iodonaphtyl azide (or ^{125}INA) as a hydrophobic probe for labeling membrane proteins embedded in the lipid bilayer (Rotman et al., 1982a). The interaction of GP IIb and IIIa with the actin-rich Triton residue of either thrombin-aggregated or concanavalin A-stimulated platelets has been recently reported (Phillips et al., 1980; Rotman et al., 1982a; Painter and Ginsberg, 1982). Glycoprotein Ia was found to bind to the Triton residue of resting

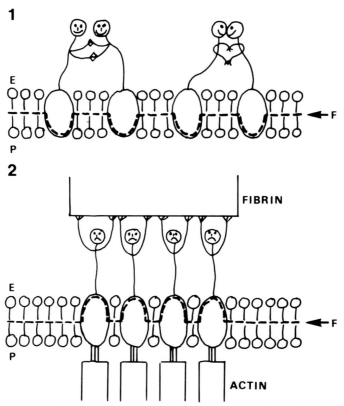

Figure 12. Interpretation of the freeze-fracture observations on membranes of nonactivated platelets (1) and of platelets in retracting clots (2). In (1) relatively strong bonds between the external segments of the evenly distributed integral glycoproteins are indicated, explaining the fracture plane (dashed line labeled F). In (2) bonds have been established between the actin filaments on the inside of the membrane and the integral proteins. This causes aggregations of the membrane proteins and may explain the fracture plane provided the bonds to the external fibrin network are weaker. From van Deurs and Behnke (1980) with permission.

as well as of ADP- or thrombin-aggregated platelets (Rotman *et al.,* 1982a). Zucker and Masiello (1983) caution on the interpretation of these results since the increase in the Triton-residue radioactivity of membrane-labeled platelets may result from an inadequate membrane lysis. Even if a controversy still persists as to the identity of the membrane components that interact with the platelet cytoskeleton, the data presented here show, however, compelling evidence for a transmembrane linkage of fibrin to platelet cytoskeletal components.

7. REGULATION OF THE CYTOSKELETAL APPARATUS AND MODEL OF CLOT RETRACTION

The use of a model helps to explain biological phenomena. I shall attempt to assemble all the facts known at present pertaining to (1) the platelet cytoskeleton and the related functions of cytoskeletal components in other biological systems, (2) the membrane glycoproteins, and (3) the platelet–fibrin interaction, mixing them together and extracting a 1984 model of clot retraction without referring to earlier proposed models. This model is essentially the same as that recently proposed (Cohen *et al.*, 1982).

There is overwhelming ultrastructural and biochemical evidence that the majority of the platelet actin content in nonactivated platelets is present in a depolymerized state (Behnke *et al.*, 1971; Behnke, 1976; Phillips *et al.*, 1980). These "globular" actin molecules with bound ATP–Mg may be associated with profilin (Markey and Lindberg, 1978; Harris and Weeds, 1978) into a stable profilactin complex. At the onset of thrombin-induced platelet activation, by mechanisms yet to be elucidated, the inhibitor profilin probably dissociates from actin, thereby enabling actin to polymerize and ATP to be hydrolyzed to ADP and Pi. Platelet α-actinin, upon "stimulation," could be a candidate for triggering actin polymerization since it has been shown to induce the polymerization of actin from profilactin (Blikstad *et al.*, 1980). The large influx of Ca^{2+} into the platelet cytoplasm at the onset of thrombin stimulation (Rink *et al.*, 1982) may be involved in triggering the platelet actin polymerization process. This probably involves a slow nucleation of three to four actin molecules and a rapid elongation process by the addition of monomers to the nuclei (Brenner and Korn, 1979, 1980; Simpson and Spudich, 1980). Recently, it was demonstrated that actin nuclei and F-actin oligomers can reverse the inhibitory action of profilin (Reichstein and Korn, 1979; Grumet and Lin, 1980a). This implies a higher affinity of actin for actin oligomers than for profilin, establishing therefore a mechanism for actin–profilin dissociation once actin nuclei are generated. An inhibitor has been found in platelets that reacts with actin nuclei in a cytochalasinlike fashion, and this inhibitor may prove to be the real regulator of actin polymerization (Grumet and Lin, 1980b). Polymerization of the globular actin is probably due to a head-to-tail process and elongation evolves at both ends of the filament, not necessarily at the same rate (Wegner, 1976). At steady state, the rate of elongation at one end is equal to the rate of shortening at the other end, resulting in a treadmilling of actin "beads" through the filament. The energy necessary for driving the constant flux of actin in a filament is provided by the hydrolysis of ATP. The polymerization process is an essential property, since only two-stranded superhelices of F-actin are able to transmit tension and activate the Mg^{2+}-stimulated ATPase activity of myosin during the mechanochemical coupling. As in other nonmuscle systems, there is an equilibrium in platelets between the G- and F-actin, with the equilibrium shifted in one direction or the other in accordance with the motile requirements. If actin filaments are needed in one part of the cell, filaments from other parts of the cell could be recruited as nucleation sites are formed. α-Actinin is probably involved in the formation of actin filament bundles. Cross-linking and branching of the actin filaments is due to the "hooking" effect of the actin-binding protein (Hartwig *et al.*, 1980). Actin-binding protein may also be involved in nucleating polymerization or by cross-linking newly formed oligomers (Rosenberg *et al.*, 1981; Rosenberg and Stracher, 1982). The effect of actin-binding protein may be mediated by a phosphorylation-dephosphorylation process. Indeed, thrombin-stimulated platelets catalyze phosphorylation of actin-binding protein. This is followed by a dephosphorylation process if platelets are aggregated (Carroll and Gerrard, 1982). The prevention of an overpolymerization of actin, or if needed, depolymerization is induced by at least three types of inhibitors:

1. The cytochalasinlike inhibitor (Grumet and Lin, 1980b); this inhibitor competes with cytochalasin for binding to actin complexes, and like cytochalasin, binds to actin nuclei probably at the barbed end,* thereby preventing the exchange reaction and increasing the G-actin pool at steady state. This effect of cytochalasin has been found in stimulated platelets (Fox and Phillips, 1981). Thus, this inhibitor blocks the nuclei-induced actin polymerization (the elongation process) and induces depolymerization of F-actin, whereas profilin inhibits the nucleation reaction.
2. Gelsolin, a Ca^{2+}-dependent 91,000 mol. wt. protein found in platelets, binds reversibly to actin and severs actin filaments. Gelsolin could fulfill the role of a dynamic regulator of actin filament length in the human platelet (Lind *et al.*, 1982). Regulation of the length or the degree of cross-linking of actin filaments determines the level of the contractile force. Although actin-binding protein amplifies actomyosin contraction, Ca^{2+}-dependent gelsolin can either decrease the force generated or confer a directionality of this force (Stendahl and Stossel, 1980).
3. The 235,000 mol. wt. protein, termed P235, restricts the length of the actin filaments in platelets (Collier and Wang, 1982).

A fine tuning is therefore present in platelets for controlling the polymerized–depolymerized state of actin.

If actin filaments represent the wheels of the contractile biological machine, myosin filaments represent the driving motor force for the wheels to roll forward. Under near-physiological conditions, platelet myosin assembly yields bipolar filaments of 320 nm long by 10.5 nm wide containing about 28 myosin molecules (Pollard, 1975); these are shorter than the skeletal muscle filaments that are 1500 nm long by 31 nm wide and contain about 400 myosin molecules. Multipolar filaments of 3 μm long and 100 nm wide and formed by side-wise or transverse assembly of myosin molecules have been reported by Hinsen *et al.* (1978). We postulate that the two kinds of filaments may be present in platelets in accordance with motile requirements. Short bipolar filaments could occur in regions where small tension is required, whereas long, multipolar filaments could form where substantial tensions are needed, for instance, in the vicinity of the membrane (Cohen, 1982). The difficulty in detecting myosin in platelets may be due to the high ratio of actin-to-myosin in platelets. Also, if long multipolar filaments were formed, very few of these would be present and it would be very difficult to locate them. The myosin driving force, possibly generated by Huxley's (1963) swinging cross-bridge mechanism, is triggered by the calmodulin-Ca^{2+}-dependent phosphorylation of the 20,000 mol. wt. myosin light chain (Sellers *et al.*, 1981). Stimulation of platelets by thrombin has indeed been shown to cause phosphorylation of the myosin light chain (Daniel *et al.*, 1981), which is accompanied by

---

*Defined by the orientation of arrowheads following decoration with the heavy meromyosin subfragment 1 (HMM-S1).

the interaction of myosin with platelet cytoskeletal structures insoluble in Triton X-100 (Fox and Phillips, 1982). A regulatory role of tropomyosin and troponin associated with the actin thin filaments cannot be ruled out (Cohen and Cohen, 1972; Cohen *et al.*, 1973). Phosphorylation–dephosphorylation mechanisms are intimately associated with platelet stimulation and the phosphorylation of a protein kinase C-dependent 40 K protein following thrombin stimulation may be of utmost importance (Sano *et al.*, 1983).

It has long been postulated that the emergence of the long pseudopodial spikes is due to a sliding contractile mechanism. This is very unlikely in view of Nachmias *et al.* (1980) finding that HMM-S1 decoration of actin filaments in pseudopods yields arrowheads pointing away from the microspikes tips. It is more probable that actin–myosin interaction around a "weak" membrane region could squeeze out pseudopodial protrusions, a process associated with intense and rapid actin polymerization. This concept is supported by the fact that the protein composition of the pseudopodial membrane is different from that of the platelet body (Rotman *et al.*, 1982b). The tremendous increase in the surface-to-volume ratio following pseudopodial formation may be due to unruffling of canaliculi from the surface-connected canalicular system resulting in their partial evagination (Cohen *et al.*, 1979). The platelet shape change may be due to the considerable contractile pressure at the rim of the flat disk forcing the cell to reduce its size in the equatorial plane and to deform into the only possible shape, that of a sphere. The marginal microtubules may break as a result of pressure from the contractile mesh or/and as a result of the cytosolic Ca^{2+} redistribution. At an advanced stage of platelet activation microtubules are mainly seen in pseudopods (White, 1974).

In the presence of fibrin, the development of platelet contractile activity is associated with platelet–fibrin interaction (Figure 13). External calcium is needed both for the generation of a maximal platelet contractile force and for platelet–fibrin attachment. Receptors for fibrin may be the clustered GP IIb-IIIa complexes. An effective transduction of the force requires factor XIII-induced cross-linking of the fibrin strands (Cohen *et al.*, 1982). This is not surprising since one could not conceive "flabby" fibrin strands to effectively transmit the platelet-derived tension. In view of the alignment of the long pseudopods with the cross-linked fibrin fibers, one could speculate that these pseudopods are able to crawl and elongate only on fibrin fibers strengthened by factor XIII-generated covalent cross-links. Several lines of evidence point to a continuous process of pseudopod extension and retraction. Using a microscope equipped with Nomarski differential interference contrast, washed platelets are seen to extend and retract pseudopodia at a relatively high speed (Allen *et al.*, 1979). Recently, Massini *et al.* (1982), using the same kind of microscopical technique on a retracting platelet-rich fibrin clot, observed platelets to change their shape continuously by forming and retracting protrusions at a speed of 3 nm/sec. This allows multiple sequential attachments of pseudopods to fibrin fibers. The gradual pulling of pseudopods bound to fibrin, perhaps even as they continue to extend at the tips, provides a mechanism for platelets to gather in fibrin strands so that strands remaining outside the developing platelet–fibrin clumps become stretched and aligned in the direction of tension (Figures 13 and 14). This process is therefore compatible with the picture seen at maximal tension, i.e., large platelet–fibrin clumps with tightly stretched fibrin in between. The irreversibility of the tension generated following achievement of max-

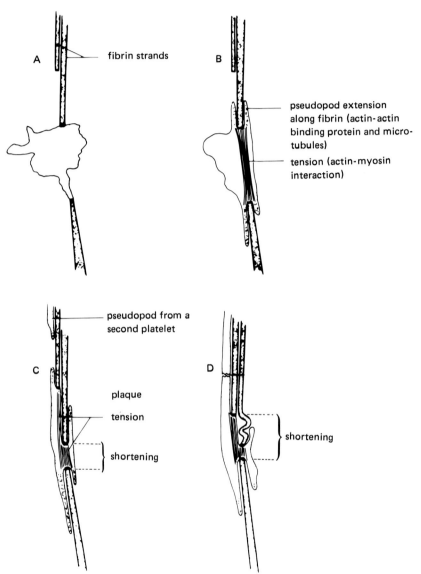

Figure 13. A proposed model for the interaction of platelets and fibrin in an isometrically contracting clot. The platelet initially forms attachments to fibrin strands (A). Pseudopods then start to crawl along the fibrin strands (B). We speculate that such pseudopod extension may involve the interaction of actin with actin binding protein and microtubules. In addition, we speculate that actin–myosin interaction within the body of the platelet (B, C, D) is responsible for drawing fibrin strands attached on opposite sides of the platelet closer together. Compaction of a fibrin strand in association with the platelet is beginning in (D). As more fibrin strands are compacted within such platelet–fibrin aggregates, the strands that remain outside such aggregates are pulled to create the tension measured and to align them in the direction of tension as shown in Figure 14. From Cohen *et al.* (1982) with permission.

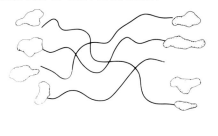

CLOT IMMEDIATELY
ON FORMATION

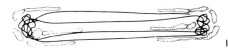

CLOT AT MAXIMAL
TENSION DURING
ISOMETRIC CONTRACTION

Figure 14. Alignment of fibrin strands during tension development as it would occur postulating the model of clot retraction shown in Figure 13. From Cohen *et al.* (1982) with permission.

imal tension may be due to internal factor XIII-dependent cross-linking, locking the ATP-depleted, contracted platelet in an irreversible rigorlike fashion (Lim and Cohen, 1982).

8. CONCLUDING REMARKS

Considerable progress has been achieved in the past few years in the field of platelet membrane glycoproteins as evidenced by the chapters of this volume summarizing the state of the art in this field. One is struck by the relative slow progress in understanding the mechanism of clot retraction at the molecular level. As much as this monograph may be premature at the moment, it still establishes the bases of working hypotheses for elucidating the many still obscure aspects of clot retraction. These include the unequivocal identification of platelet receptors for fibrin, the receptors for actin on the cytoplasmic side and whether actin binds directly to the membrane or via other cytoskeletal components such as α-actinin or a vinculinlike protein structure, and finally, the mechanism of regulation of platelet activation initiating the contractile process. The latter will entail the discovery of many more agonists and antagonists of the contractile mechanism. The physiological role of clot retraction is still not unequivocally established, although most of the evidence supports the view that it is not merely an unconsequential *in vitro* curiosity. Clot retraction may be responsible for the tightness of platelet–fibrin interaction, which in turn may be crucial for an effective hemostatic plug. The retraction of a clot may also be important for allowing the recanalization of a partially or totally obstructed vessel. And finally, as shown by Carroll *et al.* (1981), retraction of a clot, by tightening the proximity of fibrinolytic substrates and enzymes, may be necessary for effective fibrinolysis and eventual recanalization of a thrombosed vessel.

ACKNOWLEDGMENT.   I am grateful to Dr. David Green (Northwestern University) for careful review of the manuscript.

REFERENCES

Agranoff, B. W., Murthy, P., and Seguin, E. B., 1983, Thrombin-induced phosphodiesteratic cleavage of phosphatidylinositol biphosphate in human platelets, *J. Biol. Chem.* **258**:2076–2078.

Allen, R. D., Zacharski, L. R., Widirstky, S. T., Rosenstein, R., Zaitlin, L. M., and Burgess, D. R., 1979, Transformation and motility of human platelets, *J. Cell Biol.* **83**:126–142.

Behnke, O., 1976, The blood platelet, a potential smooth muscle cell, in: *Contractile Systems in Non-Muscle Tissue* (S. V. Perry, ed.), Elsevier/North Holland Biomedical Press, Amsterdam, pp. 105–115.

Behnke, O., Kristensen, B. I., and Engdahl-Nielsen, L., 1971, Electron microscopical identification of platelet contractile proteins, in: *Platelet Aggregation* (J. Caen, ed.), Masson, Paris, pp. 3–13.

Bennett, J. S., Vilaire, G., Colman, R. F., and Colman, R. W., 1981, Localization of human platelet membrane-associated actomyosin using the affinity label 5'-p-fluorosulfonylbenzoyl adenosine, *J. Biol. Chem.* **256**:1185–1190.

Bettex-Galland, M., and Lüscher, E. F., 1959, Extraction of an actomyosin-like protein from human thrombocytes, *Nature (London)* **184**:276–277.

Blikstad, I., Eriksson, S., and Carlsson, L., 1980, α-Actinin promotes polymerization of actin from profilactin, *Eur. J. Biochem.* **109**:317–323.

Blombäck, B., 1958, Studies on the action of thrombin enzymes on bovine fibrinogen as measured by N-terminal analysis, *Arkh. Kemi* **12**:321–325.

Bottecchia, D., and Fantin, G., 1973, Platelets and clot retraction—effect of divalent cations and several drugs, *Thromb. Diath. Haemorrh.* **30**: 567–576.

Bouvier, C. A., Gabbiani, G., Ryan, G. B., Badonnel, M. C., Majno, G., and Lüscher, E. F., 1977, Binding of anti-actin autoantibodies to platelets, *Thromb. Haemostasis* **37**:321–328.

Brenner, S. L., and Korn, E. D., 1979, Substoichiometric concentrations of cytochalasin D inhibit actin polymerization. Additional evidence for an F-actin treadmill, *J. Biol. Chem.* **254**:9982–9985.

Brenner, S. L., and Korn, E. D., 1980, Spectrin/actin complex isolated from sheep erythrocytes accelerates actin polymerization by simple nucleation, *J. Biol. Chem.* **255**:1670–1676.

Brown, S. S., and Spudich, J. A., 1981, Mechanism of action of cytochalasin: Evidence that it binds to actin filament ends, *J. Cell Biol.* **88**:487–491.

Budtz-Olsen, O. E. (ed.), 1951, *Clot Retraction,* Thomas, Springfield, Ill.

Carroll, R. C., and Gerrard, J. M., 1982, Phosphorylation of platelet actin-binding protein during platelet activation, *Blood* **59**:466–471.

Carroll, R. C., Gerrard, J. M., and Gilliam, J. M., 1981, Clot retraction facilitates clot lysis, *Blood* **57**:44–48.

Chao, F. C., Shepro, D., Tullis, J. L., Belamarich, F. A., and Curby, W. A., 1976, Similarities between platelet contraction and cellular motility during mitosis: Role of platelet microtubules in clot retraction, *J. Cell Sci.* **20**:569–588.

Cohen, I., 1982, Contractile platelet proteins, in: *Hemostasis and Thrombosis—Basic Principles and Clinical Practice* (R. W. Colman, J. Hirsh, V. J. Marder, and E. W. Salzman, eds.), J. B. Lippincott Company, Philadelphia and Toronto, pp. 459–471.

Cohen, I., and Cohen, C., 1972, A tropomyosin-like protein from human platelets, *J. Mol. Biol.* **68**:383–387.

Cohen, I., and de Vries, A., 1973, Platelet contractile regulation in an isometric system, *Nature (London)* **246**:36–37.

Cohen, I., and Lüscher, E. F., 1975, The blood platelet contractile system, *Haemostasis* **4**:125–243.

Cohen, I., Kaminski, E., and de Vries, A., 1973, Actin-linked regulation of the human platelet contractile system, *Febs Lett.* **34**:315–317.

Cohen, I., Gabbay, J., Glaser, T., and Oplatka, A., 1975, Fibrin-blood platelet interaction in a contracting clot, *Br. J. Haematol.* **31**:45–50.

Cohen, I., Gerrard, J. M., Bergman, R. N., and White, J. G., 1979, The role of contractile filaments in platelet activation, in: *Protides of the Biological Fluids* (H. Peeters, ed.), Pergamon Press, Oxford and New York, pp. 555–566.

Cohen, I., Gerrard, J. M., and White, J. G., 1982, Ultrastructure of clots during isometric contraction, *J. Cell Biol.* **93**:775–787.

Collier, N. C., and Wang, K., 1982, Human platelet P235: A high Mr protein which restricts the length of actin filaments, *Febs Lett.* **143**:205–210.

Daniel, J. L., Molish, I. R., and Holmsen, H., 1981, Myosin phosphorylation in intact platelets, *J. Biol. Chem.* **256**:7510–7514.

de Gaetano, G., Donati, M. B., Vermylen, J., and Verstraete, M., 1971, Inhibition of clot retraction by previous in vitro platelet aggregation, *Thromb. Diath. Haemorrh.* **26**:449–454.

de Gaetano, G., Bottecchia, D., and Vermylen, J., 1973, Retraction of Reptilase-clots in the presence of agents inducing or inhibiting the platelet adhesion-aggregation reaction, *Thromb. Res.* **2**:71–84.

de Gaetano, G., Franco, R., Donati, M. B., Bonaccorsi, A., and Garratini, S., 1974, Mechanical recording of Reptilase clot retraction—effect of adenosine-5'-diphosphate and prostaglandin E$_1$, *Thromb. Res.* **4**:189–192.

de Gaetano, G., Bertelé, V., Cerletti, C., and Di Minno, G., 1982, Prostaglandin effects on platelets, in: *Prostaglandins in Clinical Medicine—Cardiovascular and Thrombotic Disorders* (K. K. Wu and E. C. Rossi, eds.), Year Book Medical Publishers, Chicago, London, pp. 49–74.

Didisheim, P., and Bunting, D., 1966, Abnormal platelet function in myelofibrosis, *Am. J. Clin. Pathol.* **45**:566–573.

Donné, M. A., 1842, De l'origine des globules du sang, de leur mode de formation et de leur fin, *C.R. Seances Acad. Sci.* **14**:366–368.

Fox, J. E. B., and Phillips, D. R., 1981, Inhibition of actin polymerization in blood platelets by cytochalasins, *Nature (London)* **292**:650–652.

Fox, J. E. B., and Phillips, D. R., 1982, Role of phosphorylation in mediating the association of myosin with the cytoskeletal structures of myosin with the cytoskeletal structures of human platelets, *J. Biol. Chem.* **257**:4120–4126.

Fujiwara, K., and Pollard, T. D., 1976, Fluorescent antibody localization of myosin in the cytoplasm, cleavage furrow, and mitotic spindle of human cells, *J. Cell Biol.* **71**:848–875.

Gabbiani, G., Ryan, G. B., Lamelin, J. P., Vassali, P., Majno, G., Bouvier, C. A., Cruchaud, A., and Lüscher, E. F., 1973, Human smooth antibody. Its identification as antiactin antibody and a study of its binding to nonmuscular cells, *Am. J. Pathol.* **72**:473–484.

George, J. N., Lyons, R. M., and Morgan, R. K., 1980, Membranes changes associated with platelet activation. Exposure of actin on the platelet surface after thrombin-induced secretion, *J. Clin. Invest.* **66**:1–9.

Glover, C. J., McIntire, L. V., Brown III, C. H., and Natelson, E. A., 1975, Dynamic coagulation studies: Influence of normal and abnormal platelets on clot structure formation, *Thromb. Res.* **7**:185–198.

Grumet, M., and Lin, S., 1980a, Reversal of profilin inhibition of actin polymerization *in vitro* by erythrocyte cytochalasin-binding complexes and cross-linked actin nuclei, *Biochem. Biophys. Res. Commun.* **92**:1324–1334.

Grumet, M., and Lin, S., 1980b, A platelet inhibitor protein with cytochalasin-like activity against actin polymerization *in vitro*, *Cell* **21**:439–444.

Harris, H. E., and Weeds, A. G., 1978, Platelet actin: Subcellular distribution and association with profilin, *Febs Lett.* **90**:84–88.

Hartwig, J. H., Tyler, J., and Stossel, T. P., 1980, Actin-binding protein promotes the bipolar and perpendicular branching of actin filaments, *J. Cell Biol.* **87**:841–848.

Hewson, W. (ed.), 1772, *Experimental Inquiry into the Properties of Blood,* Cadell, London.

Hinsen, H., d'Haese, J., Small, J. V., and Sobieszek, A., 1978, Mode of filament assembly of myosins from muscle and nonmuscle cells, *J. Ultrastruct. Res.* **64**:282–302.

Huxley, H. E., 1963, Electron microscope studies of natural and synthetic protein filaments from striated muscle, *J. Mol. Biol.* **7**:281–308.

Jen, C. J., and McIntire, L. V., 1982, The structural properties and contractile force of a clot, *Cell Motil.* **2**:445–455.

Käser-Glanzmann, R., Jakábová, M., George, J. N., and Lüscher, E. F., 1977, Stimulation of calcium uptake in platelet membrane vesicles by adenosine 3', 5'-cyclic nonaphosphate and protein kinase, *Biochim. Biophys. Acta* **466**:429–440.

Käser-Glanzmann, R., Jakábová, M., George, J. N., and Lüscher, E. F., 1978, Further characterization of calcium-accumulating vesicles from human blood platelets, *Biochim. Biophys. Acta* **512**:1–12.

Käser-Glanzmann, R., Gerber, E., and Lüscher, E. F., 1979, Regulation of the intracellular calcium level in human blood platelets: Cyclic adenosine 3', 5'-monophosphate dependent phosphorylation of a 22,000 dalton component in isolated Ca^{2+}-accumulating vesicles, *Biochim. Biophys. Acta* **558**:344–347.

Kirkpatrick, J. P., McIntire, L. V., Moake, J. L., and Peterson, D. M., 1980, Metabolic requirements of contractile force generation in platelet-rich plasma—a rheological study, *Biorrheology* **17**:411–418.

Kuntamukkula, M. S., McIntire, L. V., Moake, J. L., Peterson, D. M., and Thompson, W. J., 1978, Rheological studies of the contractile force within platelet-fibrin clots: Effects of prostaglandin E₁, dibutyryl cAMP and dibutyrly cGMP, *Thromb. Res.* **13**:957–969.

Kuntamukkula, M. S., Moake, J. L., McIntire, L. V., and Cimo, P. L., 1982, Effects of colchicine and vinblastine on platelet contractility and release, *Thromb. Res.* **26**:329–339.

Le Breton, G. C., Dinerstein, R. J., Rothe, L. J., and Feinberg, H., 1976, Direct evidence for intracellular divalent cation redistribution associated with platelet shape change, *Biochem. Biophys. Res. Commun.* **71**:362–370.

Lim, C. T., and Cohen, I., 1982, Irreversible cross-linking process in thrombin-induced platelet aggregates (abstract), *Circulation* **66**(Part II):698.

Lin, D. C., Tobin, K. D., Grumet, M., and Lin, S., 1980, Cytochalasins inhibit nuclei-induced actin polymerization by blocking filament elongation, *J. Cell Biol.* **84**:455–460.

Lind, S. E., Yin, H. L., and Stossel, T. P., 1982, Human platelets contain gelsolin, a regulator of actin filament length, *J. Clin. Invest.* **69**:1384–1387.

Lüscher, E. F., 1956, Viscous metamorphoresis of blood platelets and clot retraction, *Vox Sang.* **1**:133–154.

Majno, G., Bouvier, C. A., Gabbiani, G., Ryan, G. B., and Statkov, P., 1972, Kymographic recording of clot retraction: Effects of papaverine, theophylline and cytochalasin B, *Thromb. Diath. Haemorrh.* **28**:49–53.

Markey, F., and Lindberg, U., 1978, Human platelets contain profilin, a potential regulator of actin polymerisability, *Febs Lett.* **88**:75–79.

Maruyama, K., 1981, Sonic vibration induces the nucleation of actin in the absence of magnesium ions and cytochalasins inhibit the elongation of the nuclei, *J. Biol. Chem.* **256**:1060–1062.

Massini, P., and Lüscher, E. F., 1976, On the significance of the influx of calcium ions into stimulated human blood platelets, *Biochim. Biophys. Acta* **436**:652–663.

Massini, P., Käser-Glanzmann, R., and Lüscher, E. F., 1978, Movement of calcium ions and their role in the activation of platelets, *Thromb. Haemostasis* **40**:212–218.

Massini, P., Näf, U., and Lüscher, E. F., 1982, Clot retraction does not require Ca ions and depends on continuous contractile activity, *Thromb. Res.* **27**:751–756.

Miller, O. V., Johnson, R. A., and Gorman, R. R., 1977, Inhibition of PGE₁-stimulated cAMP accumulation in human platelets by thromboxane A₂, *Prostaglandins* **13**:599–609.

Michell, R. H., 1975, Inositol phospholipids and cell surface receptor function, *Biochim. Biophys. Acta* **415**:81–147.

Mürer, E. H., 1969, Clot retraction and energy metabolism of platelets. Effect and mechanism of inhibitors, *Biochim. Biophys. Acta* **172**:266–276.

Mürer, E. H., and Holme, R., 1970, A study of the release of calcium from human blood platelets and its inhibition by metabolic inhibitors, N-ethylmaleimide and aspirin, *Biochim. Biophys. Acta* **222**:197–205.

Nachmias, V. T., Sullender, J., Fallon, J., and Asch, A., 1980, Observations on the cytoskeleton of human platelets, *Thromb. Haemostasis* **42**:1661–1666.

Painter, R. G., and Ginsberg, M., 1982, Concanavalin A induces interaction between surface glycoproteins and the platelet cytoskeleton, *J. Cell Biol.* **92**:565–573.

Perret, B., Chap, H. J., and Douste-Blazy, L., 1979, Asymmetric distribution of arachidonic acid in the plasma membrane of human platelets, *Biochim. Biophys. Acta* **556**:434–446.

Phillips, D. R., Jennings, L. K., and Edwards, H. H., 1980, Identification of membrane proteins mediating the interaction of human platelets, *J. Cell Biol.* **86**:77–86.

Pollard, T. D., 1975, Functional implications of the biochemical and structural properties of cytoplasmic contractile proteins, in: *Molecules and Cell Movement* (S. Inoue and R. E. Stephens, eds.), Raven Press, New York, pp. 259–285.

Pollard, T. D., Fujiwara, K., Handin, R., and Weiss, G., 1977, Contractile proteins in platelet activation and contraction, *Ann. N.Y. Acad. Sci.* **283**:218–236.

Polley, M. J., Leung, L. L. K., Clark, F. Y., and Nachman, R., 1981, Thrombin-induced platelet membrane glycoprotein IIb and IIIa complex formation, *J. Exp. Med.* **154**:1058–1068.

Reichstein, E., and Korn, E., 1979, Acanthamoeba profilin. A protein of low molecular weight from *Acanthamoeba castellanii* that inhibits actin nucleation, *J. Biol. Chem.* **254**:6174–6179.

Rink, T. J., Smith, S. W., and Tsien, R. Y., 1982, Cytoplasmic free Ca^{2+} in human platelets: Ca^{2+} thresholds and Ca-independent activation for shape change and secretion, *Febs Lett.* **148**:21–26.

Rosenberg, S., and Stracher, A., 1982, Effect of actin-binding protein on the sedimentation properties of actin, *J. Cell Biol.* **94**:51–55.

Rosenberg, S., Stracher, A., and Lucas, R. C., 1981, Isolation and characterization of actin and actin-binding protein from human platelets, *J. Cell Biol.* **91**:201–211.

Rotman, A., Heldman, J., and Linder, S., 1982a, Association of membrane and cytoplasmic proteins with the cytoskeleton in blood platelets, *Biochemistry* **21**:1713–1719.

Rotman, A., Makov, N., and Lüscher, E. F., 1982b, Isolation and partial characterization of proteins from platelet pseudopods, *Proc. Natl. Acad. Sci. U.S.A.* **79**:4357–4361.

Rubalcava, B., Martinez de Munoz, D., and Gitler, C., 1969, Interaction of fluorescent probes with membranes. I. Effect of ions on erythrocyte membranes, *Biochemistry* **8**:2742–2747.

Salganicoff, L., Russo, M., and Loughnane, M., 1977, The platelet strip: A new model for the study of the mechanochemical properties of a platelet aggregate, *Thromb. Haemostasis* **38**:155.

Sano, K., Takai, Y., Yamanishi, J., and Nishizuka, Y., 1983, A role of calcium-activated phospholipid-dependent protein kinase in human platelet activation, *J. Biol. Chem.* **258**:2010–2013.

Sellers, J. R., Pato, M. D., and Adelstein, R. S., 1981, Reversible phosphorylation of smooth muscle myosin, heavy meromyosin, and platelet myosin, *J. Biol. Chem.* **256**:13137–13142.

Shepro, D., Belamarich, F. A., and Chao, F. C., 1969, Retardation of clot retraction after incubation of platelets with colchicine and heavy water, *Nature (London)* **221**:563–565.

Simpson, P. A., and Spudich, J. A., 1980, ATP-driven steady-state exchange of monomeric and filamentous actin from *Dictyostelium discoideum, Proc. Natl. Acad. Sci. U.S.A.* **77**:4610–4613.

Singer, S. J., 1971, The molecular organization of biological membranes, in: *Structure and Function of Biological Membranes* (L. I. Rothfield, ed.), Academic Press, New York, pp. 145–222.

Sokal, G. (ed.), 1960, Etude morphologique de la metamorphose visqueuse et de la structuration du caillot en milieu plasmatique, in: *Plaquettes sanguines et structure du caillot,* Arscia, Brussels, pp. 17–59.

Stendahl, O. I., and Stossel, T. P., 1980, Actin-binding protein amplifies actomyosin contraction, and gelsolin confers calcium control on the direction of contraction, *Biochem. Biophys. Res. Commun.* **92**:675–681.

Tuszynski, G. P., Kornecki, E., Cierniewski, C., Knight, L., Koshy, A., Srivastava, S., Niewiarowski, S., and Walsh, P. N., 1984, Association of fibrin with the platelet cytoskeleton, *J. Biol. Chem.* **259**:5247–5254.

Vanderkooi, J., and Martonosi, A., 1969, Sarcoplasmic reticulum. VIII. Use of 8-amilino-1-naphtalene sulfonate as conformational probe on biological membranes, *Arch. Biochem. Biophys.* **133**:153–163.

Van Deurs, B., and Behnke, O., 1980, Membrane structure of nonactivated and activated human blood platelets as revealed by freeze-fracture: Evidence for particle redistribution during platelet contraction, *J. Cell Biol.* **87**:209–218.

Vickers, J. D., Kinlough-Rathbone, R. L., and Fraser Mustard, J., 1982, Changes in phosphatidyl inositol-4, 5-biphosphate 10 seconds after stimulation of washed rabbit platelets with ADP, *Blood* **60**:1247–1250.

Wegner, A., 1976, Head-to-tail polymerization of actin, *J. Mol. Biol.* **108**:139–150.

White, J. G., 1971, Platelet microtubules and microfilaments: Effects of cytochalasin B on structure and function, in: *Platelet Aggregation* (J. Caen, ed.), Masson, Paris, pp. 15–52.

White, J. G., 1974, Electron microscopic studies of platelet secretion, in: *Progress in Hemostasis and Thrombosis 2* (T. H. Spaet, ed.), Grune and Stratton, New York and London, pp. 49–98.

White, J. G., and Rao, G. H. R., 1982, Effects of a microtubule stabilizing agent on the response of platelets to vincristine, *Blood* **60**:474–483.

Wright, J. H., 1906, The origin and nature of the blood plates, *Boston Med. Surg. J.* **154**:643–645.

Zucker, M. B., and Masiello, N. C., 1983, The Triton X-100 insoluble residue ("cytoskeleton") of aggregated platelets contains increased lipid phosphorus as well as ^{125}I-labeled glycoproteins, *Blood* **61**:676–683.

V

*Platelet Membrane Glycoprotein
Immunology and Abnormalities*

Immunology of the Platelet Surface

Sharron L. Pfueller

1. INTRODUCTION

Although the platelet is not an effector cell of the immune system, it interacts with soluble products of the immune system in several ways. Its susceptibility to attack by antibodies arising either spontaneously or after blood transfusion, pregnancy, or ingestion of certain drugs has been known for a long time. The resulting platelet destruction causes major clinical problems. Platelets are also activated by products of antibody reaction with nonplatelet antigens. The exact role of platelet activation by these substances in normal body function or in disease is not yet fully understood.

2. METHODS FOR DETECTION OF PLATELET ANTIGENS AND THEIR ANTIBODIES

Characterization of platelet antigens has only been possible because of procedures that detect interaction of antibodies from patient serum with target platelets. Assay techniques form four main categories: (1) complement fixation as a result of antigen–antibody complex formation, (2) measurement of platelet stimulation or damage (aggregation or agglutination, increase in platelet factor 3 availability, serotonin release, complement-dependent platelet lysis), (3) inhibition of platelet function (clot retraction, migration, aggregation), and (4) binding of antibody immunoglobulin. The methods have been recently reviewed by McMillan (1981), Mueller-Eckhardt (1977), Kelton and Gibbons (1982), and Klein and Blajchman (1982). The first three types of

---

Sharron L. Pfueller • Department of Medicine, Monash University Medical School, Alfred Hospital, Prahran 3181, Victoria, Australia.

assay often do not detect so-called "blocking" antibodies. It is believed that these antibodies bind to the platelet surface, but because the antigens are not present in sufficient concentration or in the right configuration, no biological effects are observed. Another complication is that IgG antibodies are often complement-fixing but not agglutinating, whereas IgM agglutinins do not fix complement (reviewed by Svejgaard, 1969).

The most useful approach is immunoglobulin binding measured either by antiglobulin consumption or by using antiglobulins labeled with radioisotopes, fluorescent markers or by uptake of staphylococcal protein A. In recent years a plethora of new procedures has been developed to increase sensitivity, specificity, speed, and ease of performance (Myers *et al.*, 1981; Gudino and Miller, 1981; Faig and Karpatkin, 1982; Kelton *et al.*, 1981b; Mueller-Eckhardt *et al.*, 1982a; Schiffer and Young, 1983; Kickler *et al.*, 1983; von dem Borne *et al.*, 1980a).

3. PLATELET-ASSOCIATED ANTIGENS

Two classes of platelet antigen exist: those also found on other cells (blood group, histocompatibility, T and Tn antigens) and those unique to platelets. Although antibodies to either type of antigen may cause thrombocytopenia, the nature and relationship of these antigens to the platelet membrane are different.

3.1. Blood Group Antigens

3.1.1. ABO Antigens

Although ABO antigens are mainly located on the erythrocyte surface, they are found in other tissues, e.g., gastrointestinal mucosa, saliva of "secretors," and on the platelet surface (Moreau and Andre, 1954; Gurevitch and Nelken, 1954; Coombs and Bedford, 1955). These glycosidic antigenic determinants are present on both glycosphingolipids and glycoproteins (reviewed by Hakomori, 1981). The antigen-bearing molecules on platelets are, in contrast to erythrocytes, probably not an intrinsic part of the membrane structure, but rather absorbed from the plasma (Lewis *et al.*, 1960). Kelton *et al.* (1982a) found that the amount of anti-A antibody bound to platelets correlated directly with the amount of A antigen in plasma from the platelet donor. A antigen was also taken up by washed platelets after *in vitro* incubation with plasma.

ABO antigens associated with platelets have different properties from those on erythrocytes. They are more susceptible to proteolysis than those on erythrocytes (Lewis *et al.*, 1960). Transfusion of ABO-incompatible platelets causes only an initial reduction in platelet recovery, whereas the remaining platelets survive normally (Aster, 1965). The incompatible antigens, loosely adsorbed to platelets, may dissociate, allowing the platelets to avoid destruction.

3.1.2. Other Blood Group Antigens

Platelets probably do not carry any of the other blood group antigens so far examined (reviewed by Aster, 1984).

3.2. HLA Antigens

Antigens of the HLA system are the primary determinants of tissue graft acceptance. The products of the HLA-A,-B, and -C genes are expressed on most cells including platelets (reviewed by Perkins, 1979). HLA-D and -Dr are restricted to lymphocytes, monocytes, and endothelial cells. On lymphocytes the antigens consist of two subunits: the larger carries the A, B, or C specificity and the smaller is β_2 microglobulin which is common to all specificities. Platelets carry 73% of the total HLA-A and -B content of blood (Aster *et al.*, 1973). HLA-B has a variable expression in relation to lymphocytes (Liebert and Aster, 1977; Szatkowski and Aster, 1980), whereas HLA-C is only weakly expressed (Mueller-Eckhardt *et al.*, 1980a). The presence of HLA antigens on platelets has led to one of the more commonly used tests for tissue HLA typing (Shulman *et al.*, 1964). Antibodies to HLA determinants (Heinrich *et al.*, 1977) and to β_2 microglobulin (Falus *et al.*, 1982) cause platelet aggregation.

After solubilization of platelets with deoxycholate or papain, antigen-bearing glycopeptides of $M_r = 39,000–43,000$ are obtained (Trägårdh *et al.*, 1979; Gockerman and Jacob, 1979). They are associated with a polypeptide of 12,000 daltons, which is β_2 microglobulin (Bernier *et al.*, 1974). If the orientation of the HLA molecule in platelets is like that in lymphocytes (Walsh and Crumpton, 1977), the NH_2-terminal portions of the heavy chains containing the carbohydrate are on the outside of the membrane and the -COOH terminal is on the cytoplasmic side. How β_2 microglobulin is associated with heavy chains is not known.

There is, however, evidence that HLA antigens may occur in a loosely bound form. They can be taken up from plasma *in vitro* by platelets (Lalezari and Driscoll, 1982) and may be liberated from platelets (Aster *et al.*, 1973; Dautigny *et al.*, 1975). Since increased levels of intracellular cyclic AMP, which inhibit platelet aggregation, increase extractability of HLA antigens with salt solution (Pincus *et al.*, 1976), it is possible that the metabolic state of the platelet controls the strength with which HLA antigens are bound.

3.3. Tn and T (Thomsen-Friedenreich) Antigens

The cryptic Tn antigen has α-N-acetyl-D-galactosamine (GalNAc) residues as the immunodominant group and is responsible for polyagglutination of certain red cells by antibodies in normal sera. It is also present on platelets (Cartron and Nurden, 1979). Exposure of the Tn antigen forms the basis of the Tn syndrome and may also occur in hemolytic anemia, leukemia, and preleukemia (reviewed by Nurden *et al.*, 1982). Abnormal exposure of the Tn antigen may result from deficiency of UDP Gal: GalNAc-β-3-D-galactosyl transferase (T transferase) activity and is often associated with thrombocytopenia. In red cells, GalNAc residues are linked through an alkali-labile O-glycosidic bond to serine or threonine on glycophorins A and B and are usually substituted to form β-D-galactosyl (1–3)-N-GalNAc, the immunodominant group of the T (Thomsen-Friedenreich) antigen. The T antigen is also present in platelets and can be exposed by neuraminidase treatment (Glöckner *et al.*, 1978).

Platelets from three Tn individuals studied by Cartron and Nurden (1979) had abnormalities of glycoprotein (GP)Ib, which showed decreased staining with periodic

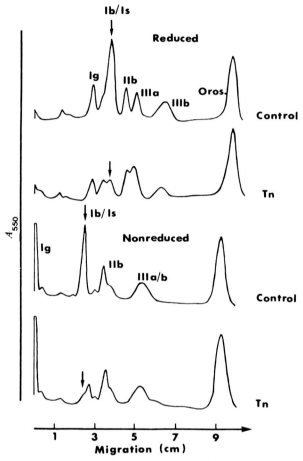

Figure 1. A comparison of the glycoprotein profiles of normal human and Tn platelets. Washed platelets were subjected to SDS–PAGE on 6% polyacrylamide gels after reduction with 2-mercaptoethanol (upper profiles) or on 7% polyacrylamide nonreduced (lower profiles). Glycoproteins were located by PAS staining and the gels scanned densitometrically at 550 nm. From Cartron and Nurden, 1979.

acid Schiff's reagent (PAS) and a slightly faster than normal electrophoretic mobility (Figure 1), but normal labeling by [^{125}I]lactoperoxidase. Labeling of carbohydrate of GP Ib, by the galactose oxidase and sodium [^{3}H]borohydride method, did not require prior treatment with neuraminidase (Nurden *et al.*, 1982), indicating decreased levels of sialic acid and increased amounts of galactose or *N*-acetylgalactosamine in the terminal position. Since on analysis of normal platelets depleted of sialic acid by neuraminidase treatment, mobility of GP Ib is decreased, it is likely that GP Ib of Tn platelets has more severe structural alterations than simple sialic acid deficiency. This altered GP Ib also showed unusual reactivity during crossed-immunoelectrophoresis. Since Tn platelets reacted normally with a quinidine-dependent antibody and with a monoclonal antibody (AN 51) directed toward the ristocetin-binding site of GP Ib (Ruan *et al.*, 1981), and also aggregated normally with ristocetin (Cartron *et al.*, 1980), the Tn antigen is not involved in any of these functions of GP Ib.

Fractionation of Tn platelets on affinity columns of *Helix pomatia* lectin yielded two platelet populations: one had a normal glycoprotein profile (Tn$^-$), whereas the other (Tn$^+$) had a more marked abnormality of GP Ib than the total platelet sample (Cartron and Nurden, 1979). The ratio of Tn$^+$ to Tn$^-$ platelets varied markedly between different patients. These two platelet populations are derived from separate megakaryocyte clones (Vainchenker *et al.*, 1982). Since separate clones of Tn$^+$ and Tn$^-$ erythroid and granulocyte lines were also observed, the mutation causing Tn transformation must have occurred in a single pluripotent stem cell followed by clonal expansion. Thus Tn activation may be equivalent to the preleukemic state.

4.   PLATELET-SPECIFIC ANTIGENS

4.1.   Alloantigens

The existence of distinct platelet-specific antigens was first recognized by Harrington *et al.* (1953) and Stefanini *et al.* (1953), who found that sera of normal mothers whose infants had thrombocytopenia, or of patients who had received multiple transfusions, caused agglutination of normal platelets. Six antigenic systems containing nine different antigens have now been described (Table 1). Current knowledge of the surface location of these antigens is summarized in Figure 2. All antigens are the products of autosomal dominant genes. Although the Pl^A (Zw), Pl^E, and Ko systems are diallelic, only single alleles of Bak^a, Lek^a, and DUZO have been detected. It is possible that as techniques for detecting platelet antigens become more specific and sensitive, more will be found. Their detection is often complicated by the coexistence, in sera from multiply-transfused patients, of antibodies of HLA or blood group antigens as well as to platelet-specific antigens. The early serological and genetic studies of platelet antibodies have been reviewed by Van de Wiel *et al.* (1961), Shulman *et al.* (1964), and Svejgaard (1969).

Table 1. Alloantigens of the Platelet Surface

Antigen system	Alleles	Phenotype frequency (%)	Approximate gene frequency (%)	References
Pl^A (Zw)	Pl^{A1} (Zw^a)	97	0.83	van Loghem *et al.*, 1959; van der
	Pl^{A2} (Zw^b)	27	0.17	Weerdt *et al.*, 1963; Shulman *et al.*, 1961; Mueller-Eckhardt *et al.*,1982
Pl^E	Pl^{E1}	99.9	0.975	Shulman *et al.*, 1962
	Pl^{E2}	5	0.025	
Ko	Ko^a	17–21	0.08	Dausset and Berg, 1963; van der
		15		Weerdt *et al.*, 1963
	Ko^b	99		van der Weerdt *et al.*, 1961
DUZO	DUZO	22	0.12	Moulinier, 1961
Bak	Bak^a	91	0.6	von dem Borne *et al.*, 1980
Lek	Lek^a	98		Boizard and Wautier, 1984; Kieffer *et al.*, 1984

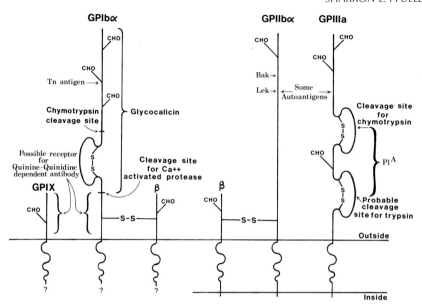

Figure 2. Probable antigenic sites on surface glycoproteins as adapted from the glycoprotein models shown in Chapter 3.

4.1.1. Pl^A (Zw)

4.1.1.a. Nomenclature.   The Zw antigen described by Van Loghem *et al.* (1959) is identical to an antigen termed Pl^{A1} by Shulman *et al.* (1961). Although by precedence the antigen should be termed Zw, the nomenclature of Shulman *et al.* (1964) to term platelet antigens Pl with a defining superscript will be used here. The Zw^b antigen (initially termed Hau^a by Van der Weerdt *et al.*, 1963) will be termed Pl^{A2}. Pl^A antigens are found on platelets of humans and other primates, dogs, and rabbits (Shulman *et al.*, 1964).

4.1.1.b. Absence in Glanzmann's Thrombasthenia.   The first information on the chemical structure of Pl^{A1} antigen came from observations that platelets of patients with Glanzmann's thrombasthenia show decreased reactivity with anti-Pl^{A1} antibodies (Kunicki and Aster, 1978; Muller *et al.*, 1978; van Leeuwen *et al.*, 1981). These platelets have several surface abnormalities, in particular a marked decrease in amounts of GP IIb and GP IIIa (described in Chapter 16). Since quinine- and quinidine-dependent antibodies and autoantibodies caused normal lysis of Glanzmann's thrombasthenia platelets, it is most likely that their defect is a result of antigen deletion rather than abnormal platelet responsiveness. Platelets that were attacked least by anti-Pl^{A1} antibodies showed the most marked decrease in GP IIb and GP III and carriers of the thrombasthenic trait had intermediate levels of Pl^{A1} activity (Kunicki *et al.*, 1981a). Platelets from Pl^{A1}-negative normal individuals had normal glycoprotein levels.

The Pl^{A2} (Zw^b) determinant is, as expected, also missing in patients with Glanzmann's thrombasthenia (van Leeuwen *et al.*, 1981). Of eleven patients with

Glanzmann's thrombasthenia, platelets from only three showed any reactivity with anti-Pl[A1] or anti-Pl[A2] sera in an immunofluorescence assay. Thus Glanzmann's thrombasthenia can be termed a Pl[A] (Zw) null disease.

4.1.1.c. Partial Purification. When solubilized platelet membranes were subjected either to affinity chromatography on *Lens culinaris* lectin and concanavalin A columns or to preparative sodium dodecyl sulfate–polyacrylanide gel electrophoresis (SDS–PAGE), material containing the Pl[A1] antigen with $M_r = 95,000 \pm 10,000$ was obtained, containing GP IIIa (Kunicki and Aster, 1979).

4.1.1.d. Immunochemical Identification. The Pl[A1] antigen has been examined by techniques in which platelets were solubilized in SDS and subjected to SDS–PAGE in slab gels. After electrophoresis, Pfueller *et al.* (1981b) incubated the gels with anti-

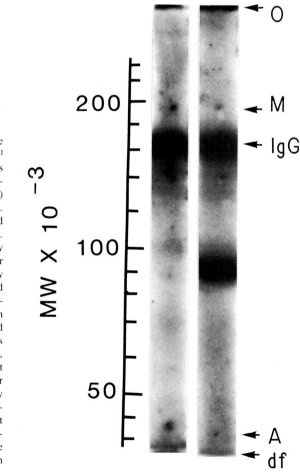

Figure 3. Demonstration of the platelet protein to which an anti-Pl[A1] antibody binds. Washed platelets were solubilized in SDS and subjected to SDS–PAGE (nonreduced) on 6.5% polyacrylamide slab gels. Lanes cut from the gels were washed extensively with buffered saline. They were incubated sequentially with normal human serum (N) or serum containing anti-Pl[A1] antibody (P), rabbit anti-human IgG, and [125I]protein A, with extensive washing in buffered saline between each incubation. The gel lanes were dried and exposed to x-ray film. Arrows indicate the position of gel origin (o), dye front (df) and molecular weight markers electrophoresed in other lanes on the gel and visualized by staining with Coomasssie Blue: myosin (M) and actin (A) in platelet lysates, and IgG. The [125I]autoradiograph band at M_r 95,000 identifies the location of the antigen reacting with the anti-Pl[A1]antibody.

Pl^{A1} serum, rabbit antihuman IgG, and then [^{125}I]protein A. Autoradiography then visualized both IgG present in platelet lysates and a band with $M_r = 95,000 \pm 1,000$ (Figure 3). McMillan *et al.* (1982) used similar procedures in which electrophoresed platelet proteins were transferred to nitrocellulose paper by Western blotting. Although platelet IgG was not visualized, a band of $M_r = 100,000$ was produced by anti-Pl^{A1}. No bands were obtained if platelets were examined after disulfide bond reduction. When anti-Pl^{A1} was first concentrated by adsorption and elution from normal platelets, several extra radioactivity bands ($M_r = 86,000–175,000$) appeared. With platelets from a patient with Glanzmann's thrombasthenia, the intensity of the band of $M_r = 100,000$ was decreased and an additional high-molecular-weight band was found.

Similar characteristics of the Pl^{A1}-carrying protein have been obtained by immunoprecipitation of radiolabeled platelets. Solubilized ^{125}I-labeled platelet proteins were incubated with anti-Pl^{A1} and then with either rabbit antihuman IgG (Kunicki and Aster, 1979) or formalin-fixed *Staphylococcus aureus* (Lane *et al.*, 1982). When Kunicki and Aster (1979) analyzed the precipitated material by SDS–PAGE, they found a labeled protein with $M_r = 100,000$ under nonreducing conditions and 120,000 after reduction (Figure 4). Similar Pl^{A1}-containing material ($M_r = 90,000$ unreduced) from both human and dog platelets was found by Lane *et al.* (1982). One anti-Pl^{A1} serum examined also precipitated a slightly higher molecular weight component. Thus, allowing for slight differences in molecular size, possibly resulting from different electrophoretic conditions, the molecule with Pl^{A1} activity has the characteristics of GP IIIa. Since anti-Pl^{A1} sera react with megakaryocytes (Kanz *et al.*, 1982), Pl^{A1} determinants are present on GP IIIa before platelets are liberated from their progenitor cells.

4.1.1.e. Biochemical Characteristics. The susceptibility of the antigen to destruction by heating at 100° for 10 min, to trypsin hydrolysis, and to reducing agents as well as the failure of any of 11 sugars commonly present in cell-surface carbohydrates to inhibit its reactivity with anti-Pl^{A1} suggest that the antigen is part of the polypeptide portion of GP IIIa. Its configuration is dependent on one or both of the two intrachain disulfide bonds (Figure 2) and is located between the sites of chymotrypsin and trypsin cleavage (Kunicki and Aster, 1979).

4.1.1.f. Inheritance and Gene–Dosage Effects. Pl^{A1} and Pl^{A2} antigens are inherited as codominant alleles with phenotype frequencies of 97% and 27%, respectively (van Loghem *et al.*, 1959; Shulman *et al.*, 1961; van der Weerdt *et al.*, 1963). Using complement fixation, Shulman *et al.* (1961) and Mueller-Eckhardt *et al.* (1982b) found the amount of antigen was proportional to the genetic composition of the individual, i.e., homozygotes Pl^A (1,1) showed double the reactivity of heterozygotes Pl^A (1,2) with anti-Pl^{A1}, whereas homozygotes Pl^A (2,2) did not react at all. Such a gene–dosage effect was not found with less quantitative immunofluorescent techniques (van Leeuwen *et al.*, 1981). Although the Pl^A determinants are missing in Glanzmann's thrombasthenia, the genes controlling these two traits are independent (Kunicki and Aster, 1979; Kunicki *et al.*, 1981a; van Leeuwen *et al.*, 1981).

4.1.1.g. In vitro Effects of Anti-Pl^A Antibodies on Platelets. Anti-Pl^{A1} antibodies may cause complement-dependent aggregation and lysis (Cines and Schreiber, 1979) and agglutination (van der Weerdt *et al.*, 1963) of platelets. They may also inhibit aggregation (van Leeuwen *et al.*, 1979) or fibrinogen binding (van Leeuwen *et*

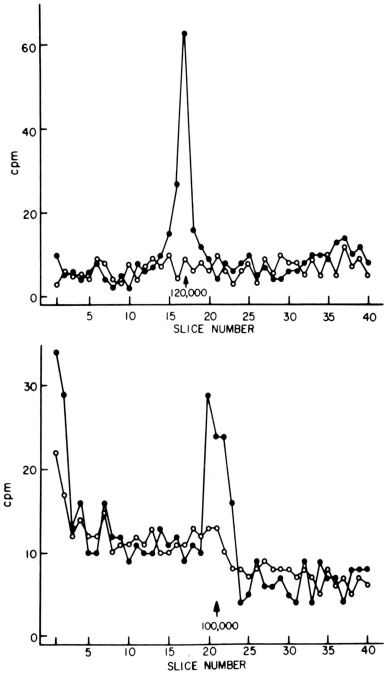

Figure 4. Profiles of iodinated material contained in precipitates formed by addition of serum containing antibody to Pl^{A1} antigen (●————●) or normal serum (○————○) to an NP-40 soluble extract of iodinated platelets, followed by addition of rabbit anti-human IgG. The washed precipitates were solubilized in SDS and subjected to SDS–PAGE on 7.5% polyacrylamide gels nonreduced (lower profile) or reduced with 2-mercaptoethanol (upper profile). Gels were cut into 2 mm slices and their ^{125}I content determined. From Kunicki and Aster (1979).

al., 1983). These different effects may result from either varying antibody concentrations, from different specificities for the Pl^A polypeptide determinants, or from different immunoglobulin subclasses composing the antibody.

4.1.2.   Pl^E

Complement-fixing antibodies reactive with Pl^{E1} and Pl^{E2} were found in a multitransfused patient and the mother of a child with neonatal thrombocytopenia, respectively (Shulman et al., 1964). The Pl^E determinant is almost universally distributed on human and primate platelets, but is not found on those of dog, rabbit, guinea pig, or rat.

4.1.3.   Ko

Antibodies recognizing Ko^a were reported by van der Weerdt et al. (1961) and Dausset and Berg (1963). They agglutinated platelets, but did not fix complement. Since Ko antigens are normally expressed on platelets from patients with Glanzmann's thrombasthenia (van Leeuwen et al., 1981), they are not associated with GP IIb or GP IIIa.

4.1.4.   DUZO

Apart from the initial report by Moulinier (1961), no further information on this antigen has been obtained.

4.1.5.   Bak

An antibody in the serum of a woman whose child had neonatal thrombocytopenia recognized determinants termed Bak^a, distinct from those described above (von dem Borne et al., 1980b). The antigen was absent on all 11 patients with Glanzmann's thrombasthenia studied by van Leeuwen et al. (1981), but there was no linkage between the genes coding for Pl^A and Bak^a. Anti-Bak^a immunoprecipitated GP IIb from ^{125}I-labeled platelets (Mulder et al., 1982).

4.1.6.   Lek

Another antigen, Lek^a, reduced or absent in platelets from seven patients with Glanzmann's thrombasthenia, has been described by Boizard and Wautier (1984). Analysis of platelet glycoproteins by Western blotting showed that anti-Lek^a, from a patient with posttransfusion thrombocytopenia, bound to GP IIb in nonreduced samples and to the α subunit of GP IIb after disulfide reduction (Kieffer et al., 1984). It did not bind to GP IIb of normal Lek^a-negative platelets. It remains to be determined if Bak and Lek are the same antigen.

4.2.   The Receptor for Drug-Dependent Antiplatelet Antibodies

A group of antibodies exists that are formed in response to drug ingestion and that attack platelets only in the presence of the drug. They are detected by the procedures described in Section 2 when the drug is added to the test system. The list of drugs that cause drug-dependent antiplatelet thrombocytopenia is seemingly endless. More than 50 drugs have been implicated; however, of these, apronal, quinine, and quinidine have received most attention. The serological findings and drugs involved in this type of immunological platelet damage have been the subject of two reviews (Hackett *et al.*, 1982; Shulman and Jordan, 1982).

4.2.1. Antibody Specificity

A high degree of specificity for the ingested drug is usually exhibited, i.e., antibodies for quinine do not react with quinidine (Shulman, 1964; van der Weerdt, 1967). In some instances, the antibodies may react with a drug metabolite rather than the ingested drug, e.g., quininone reacted better than quinine or quinidine with four antibodies reported by Shulman (1972), and an antibody arising after ingestion of *N*-acetyl-*p*-aminophenol reacted with a sulfate conjugate, but not to the drug itself (Eisner and Shahidi, 1972). Drug-dependent antibodies also show remarkable cell specificity. In the same serum, quinidine-dependent antibodies directed toward both granulocytes and platelets have been described (Chong *et al.*, 1983). When platelet-bound quinidine-dependent antibodies were eluted, they reacted with platelets and not with granulocytes and those eluted from granulocytes did not react with platelets. Platelet-directed antibodies are mostly IgG (van Leeuwen *et al.*, 1982a) and occasionally IgA (Eisner and Shahidi, 1972). Those of the IgM class may be directed toward red cells rather than platelets, producing hemolytic anemia (Shulman, 1964). Platelets from human and other primates, but not those from other laboratory animals, share the receptor for antibody attack.

4.2.2. The Platelet Receptor for Quinine- and Quinidine-Dependent Antibodies

4.2.2.a. Decreased Expression of the Receptor on Bernard-Soulier Syndrome Platelets. Platelets from patients with Bernard-Soulier syndrome (BSS) are less susceptible than those of normal individuals to immune lysis by quinine- or quinidine-dependent antibodies (Kunicki *et al.*, 1978, 1981b; George *et al.*, 1981). Parents of the patients showed intermediate susceptibility (George *et al.*, 1981). It was thus suggested that glycoproteins deficient in BSS (GP Ib, GP V, and GP IX, as discussed in Chapter 16) are involved in the platelet receptor. Subsequent studies have shown that at high drug concentrations, some of the antibodies could lyse BSS platelets (Shulman and Jordan, 1982) and that there is considerable variability both in the amount of IgG antibody bound by platelets from different BSS patients and in the degree of platelet lysis produced (van Leeuwen *et al.*, 1982a). It may be that the receptors of BSS platelets are altered, rather than absent, and that some antibodies may still bind to them, producing varying degrees of lysis.

4.2.2.b. Biochemical Characterization. Platelet membranes solubilized in Triton X-100 and subjected to affinity chromatography on wheat germ agglutinin–Sepharose yielded a fraction that inhibited platelet lysis by quinine- or quinidine-dependent antibodies (Kunicki *et al.*, 1981b). This fraction contained two glycoproteins: GP Ib and another with $M_r = 190,000–210,000$. Glycocalicin, the soluble glycopeptide derived by cleavage of GP Ib by the platelet's Ca^{2+}-activated protease, did not have receptor activity. Since receptor activity on intact platelets is resistant to proteolysis by trypsin or chymotrypsin (Kunicki *et al.*, 1978), it could be on the portion of GP Ib close to the membrane surface. Destruction of the receptor in intact platelets by reducing agents implies that disufide bonds are required for its correct configuration (Cimo and Gerber, 1979).

4.2.2.c. Immunoprecipitation. When solubilized platelets, in which surface carbohydrates had been labeled by the [^{3}H]borohydride procedure (see Chapter 3) were incubated with quinidine and a quinidine-dependent antibody and antigen–antibody complexes then adsorbed to *S. aureus* and subsequently eluted, two radiolabeled components were obtained. One component had the characteristics of GP Ib and the other was a glycoprotein of $M_r = 22,000$ (GP IX) (Chong *et al.*, 1983). The latter glycoprotein may be a structural analogue of GP Ib since monoclonal antibodies to GP Ib also reacted with a similar glycoprotein (Chong *et al.*, 1983, Coller *et al.*, 1983). Both glycoproteins were missing in BSS platelets (Berndt *et al.*, 1983). Since the purified preparation of receptor described in Section 4.2.2.b. contained no low-molecular-weight substances, it could be that GP Ib is the drug-dependent antibody receptor and that GP IX coprecipitates. Alternatively, GP IX may be the receptor and a complex of it with GP Ib may have formed the component of higher molecular weight.

4.2.3. Nature of the Drug-Dependent Antigen

Notwithstanding immunoprecipitation of platelet components by drug-dependent antibodies, the actual epitope has not been defined and different mechanisms for drug-dependent antibody reactions with platelets must be considered. Since the drugs are usually too small to be immunogenic alone, it is likely that they stimulate antibody formation when attached to a protein carrier. The antibodies may then recognize the drug as a hapten or recognize the drug-carrier complex (Figure 5).

4.2.3.a. The Drug as Hapten. Shulman (1958) proposed that the drug, either alone or associated with a plasma protein carrier, is the immunogen and that the antibodies, bound to the drug as hapten (Figure 5A1), react with the platelet Fc receptor (see Section 5.1.3). This is supported by a number of experimental findings. Quinine binding to platelets is weak except in the presence of specific antibody (Shulman *et al.*, 1958; Christie and Aster, 1982). Excess drug promotes antibody binding, an unlikely occurrence if the antibody were directed toward platelet-bound drug when hapten inhibition would occur. Since F(ab)$'_2$ fragments of a quinidine-dependent antibody did not bind to platelets (van Leeuwen *et al.*, 1982a), it could be that the Fc portion is required. It was not clear, however, whether the F(ab)$'_2$ fragments still bound the drug. A mechanism involving a complex of the drug and a plasma protein that has affinity for the platelet (Figure 5A2) would also be compatible with the above findings.

Ackroyd (1953) proposed that drug-dependent antibodies bind to platelet-bound drug (Figure 5A3). This is supported by the requirement for platelet, drug, and antibody together for complement fixation, the degree of platelet specificity, immunoprecipitation of GP Ib and GP IX, and the lack of evidence of a soluble antibody–drug complex.

A platelet–drug complex was directly observed when platelets pretreated with cotrimoxazole bound IgG from two separate cotrimoxazole-dependent antibodies

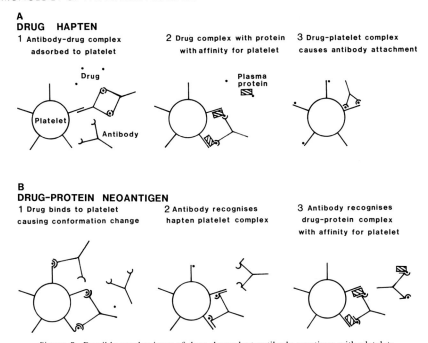

A
DRUG HAPTEN

1 Antibody-drug complex adsorbed to platelet

2 Drug complex with protein with affinity for platelet

3 Drug-platelet complex causes antibody attachment

B
DRUG-PROTEIN NEOANTIGEN

1 Drug binds to platelet causing conformation change

2 Antibody recognises hapten platelet complex

3 Antibody recognises drug-protein complex with affinity for platelet

Figure 5. Possible mechanisms of drug-dependent antibody reactions with platelets.

(Claas *et al.*, 1979). Since only 30% of the platelet donors tested reacted with the antibodies, a specific platelet antigen would seem to be involved (Claas *et al.*, 1981).

It has been suggested that a platelet–drug complex acts as immunogen since lymphocytes from patients who had produced drug-dependent antibodies proliferated in response to drug only in the presence of platelets (Hosseinzadeh *et al.*, 1980). Since this has not subsequently been confirmed (Pfueller and Firkin, 1984), it may be that these observations were the result of nonspecific immune complex-induced stimulation of lymphocytes that were still producing antibody.

4.2.3.b. Drug–Protein Interaction Producing a "Neoantigen" (Figure 5B). Several additional observations suggest that none of the above models are totally satisfactory. That drug–antibody complexes bind to the platelet Fc receptor is questionable since this receptor is expressed normally on Bernard-Soulier platelets (Pfueller *et al.*, 1984) and since heat-aggregated IgG, reactive with the Fc receptor, does not interfere with platelet lysis by drug-dependent antibody (Kunicki *et al.*, 1981b).

It is possible that a neoantigen is formed by either a drug-induced conformational change in the platelet "receptor" or by juxtaposition of platelet and drug determinants (Figure 5B1, B2). In both cases, weak association between drug and platelet must be postulated requiring excess drug for antibody binding to occur. Alternatively, reaction of the drug with a protein having affinity for the platelet may form a neoantigen (Figure 5B3). This mechanism was supported by the observation that a high-molecular-weight fraction of VIII-related antigen (von Willebrand factor), which binds to GP Ib (see Chapter 10), was required for platelet damage by quinine and quinidine-dependent antibodies (Pfueller *et al.*, 1981a). Subsequent studies (Pfueller and Firkin, 1984) have shown, however, that although platelet-rich plasma from patients with severe von Willebrand's disease, lacking both plasma and platelet von Willebrand factor (VWF), has decreased responsiveness to some of these antibodies—the defect is not corrected by purified VWF. Christie and Aster (1982) and Chong *et al.* (1983) have found that antibody binding was not affected by levels of VWF. Thus the role of VWF as part of a neoantigen requires further investigation.

The question of antibody heterogeneity must be considered when faced with the difficulty of reconciling the conflicting findings related to the nature of the drug-dependent antigen. It is possible that different antibodies are directed toward different platelet–drug or plasma–drug combinations and may thus have differing effects on platelets.

4.2.4. EDTA-Dependent Antibodies

Rare IgG antibodies occur that react with platelets only in the presence of the anticoagulant EDTA (Gowland *et al.*, 1969) or of the related EGTA. Since, for some of these antibodies, the calcium and magnesium salts of these chemicals were also effective (Pegels *et al.*, 1982), divalent metal ion chelation was not involved. These antibodies did not bind to platelets from patients with Glanzmann's thrombasthenia suggesting a role for GP IIb or GP IIIa as antigens. It may be that the antigenic sites are portions of these glycoproteins only exposed by EDTA or that they consist of an EDTA–glycoprotein complex. Other EDTA-dependent antibodies will react with platelets when the calcium concentration is below a critical level (Onder *et al.*, 1980).

4.3. Autoantigens

Autoantibodies to platelets are believed to cause idiopathic thrombocytopenia (ITP) (see Section 6.3). Immunoglobulin G and IgM antibodies from serum of patients with this disorder bind specifically to platelets (von dem Borne *et al.*, 1980a; Kelton *et al.*, 1981b). The failure of platelets from patients with Glanzmann's thrombasthenia to bind a majority of ITP antibodies (van Leeuwen *et al.*, 1982b) suggests that the antigens involved are located on GP IIb or GP IIIa. However, the antigens appear to be distinct from Pl[A1], Ko, or Bak[a]. The location of autoantigens of GP IIb-IIIa has been demonstrated directly in a microtiter IgG-binding assay and by immunoprecipitation (Woods *et al.*, 1984).

4.4. Isoantigens

Some patients with Glanzmann's thrombasthenia and Bernard-Soulier Syndrome have produced antibodies to GP IIb-IIIa that react with platelets from all normal donors and are not restricted to just the previously described alloantigenic determinants. These are discussed in Chapter 16.

5. PLATELET REACTIONS WITH IMMUNOGLOBULINS AT OTHER THAN THE ANTIGEN-SPECIFIC SITE

Immune complexes (IC), i.e., antigen–antibody complexes or other IgG aggregates, may mediate platelet stimulation by activating complement or by causing production of platelet-aggregating factor (PAF) from other cells, e.g., macrophages and basophils (reviewed by Henson and Ginsberg, 1981). They may also activate platelets directly in a reaction requiring the Fc portion of IgG (Pfueller and Lüscher, 1972a; Henson and Spiegelberg, 1973; Israels *et al.*, 1973; Pfueller and Cosgrove, 1980; Gross *et al.*, 1983). This reaction has not received the attention given to platelet activation by ADP and "conventional" aggregating agents, since, in humans, effects are usually observed only in washed platelet systems. However, recent work has revealed that IC binding to platelets may occur in plasma and has defined circumstances in which platelets may also become activated.

5.1. The Fc Receptor

5.1.1. Species Differences

Platelets from primates, pigs, sheep, goats, and cattle do not have receptors for the third component of complement (C3b), but possess Fc receptors. Most studies of these receptors have been conducted on human platelets. Platelets from dog, mouse, horse, and rabbit bear receptors for C3b, but not for Fc and therefore require complement for stimulation by IC. Activation of platelets by both mechanisms has been reviewed by Osler and Siraganian (1972), Pfueller and Lüscher (1972b), and Henson and Ginsberg (1981).

5.1.2. Types of Immune Complexes That Activate Human Platelets

Activation of platelets by IC requires that IgG molecules form macromolecular complexes, either bound to specific antigen (Bettex-Galland *et al.*, 1963; Mueller-Eckhardt and Lüscher, 1968a; Clark *et al.*, 1982), adsorbed to latex particles (Mueller-Eckhardt and Lüscher, 1968b; Glynn *et al.*, 1965) or glass (Packham *et al.*, 1969), aggregated noncovalently by heating or organic solvents, or linked covalently with bis-diazobenzidine (Pfueller and Lüscher, 1972a). Since chemical aggregates of IgG of defined size and type are readily prepared, they have often been studied as models of antigen–antibody complexes. All subclasses (1–4) of human IgG activate human platelets, whereas IgM, IgA, IgD, IgE, and bovine IgG do not (Pfueller and Lüscher, 1972a; Henson and Spiegelberg, 1973).

5.1.3. Characteristics of the Fc Receptor of Human Platelets

5.1.3.a. Quantitative Aspects. Binding studies using labeled IgG monomers or aggregates have shown that while there are between 400 and 2000 Fc receptors per platelet, binding to less than 100 is sufficient to activate platelets (Pfueller *et al.*, 1977b; Karas *et al.*, 1982). Immunoglobulin G monomers bind weakly and are not stimulatory; dimers have an affinity constant 0f 2.2×10^7 M^{-1} (Karas et al., 1982) and cause platelet activation (Ginsberg and Henson, 1978). In specificity and avidity, the platelet Fc receptor is similar to that on neutrophils, but the platelet has fewer per unit surface area. Since platelets from female blood donors are more responsive to IgG aggregates than those of males (Moore et al., 1981), they may have more Fc receptors or bind IC more strongly. Increased numbers of Fc receptors occur on platelets in myeloproliferative disease (Moore and Nachman, 1981).

5.1.3.b. Modulation by Plasma Components.   Immunoglobulin G and other plasma components inhibit both binding of soluble IgG aggregates and their stimulating ability (Pfueller et al., 1977b, Sugiura et al., 1981); however eight to ten times more IgG is needed to inhibit binding than to block platelet activation. Binding alone of IC to a small number of Fc receptors does not constitute a signal for platelet activation since the release reaction can be blocked by the subsequent addition of IgG (Pfueller et al., 1977b). Thus the signal requires bridging of a number of platelet surface molecules. Even though no release reaction occurs in platelet-rich plasma, soluble IC still bind, but at a slower rate than to washed platelets. Thus platelets in vivo may bind IC without becoming activated in the process. Insoluble DNA–antiDNA complexes will cause platelet stimulation even in the presence of plasma (Clark et al., 1982). This process can be inhibited by soluble complexes. Thus a factor determining the degree of response to IC binding is the solubility of the complex and therefore the antigen–antibody ratio.

Another possible determining factor may be the capacity of IC to react with fibrinogen. When latex particles coated with IgG were used as models for IC, they were unable to cause platelet aggregation and the release reaction in the presence of plasma unless fibrinogen was also present on the particle surface (Pfueller and Cosgrove, 1980). It is possible that binding of particles to both Fc and to fibrinogen receptors (see Chapter 9) on GP IIb and GP IIIa may form a ternary complex, providing sufficient bridging of surface molecules for an activation signal.

5.1.3.c. Modulation by Bacterial Lipids.   As well as having a generalized enhancing effect on platelet aggregation, lipid A-rich bacterial lipopolysaccharides specifically magnify platelet responses to IgG aggregates (Ginsberg and Henson, 1978). Since lipid A can bind both to platelets and to IgG, it is possible that the resulting ternary complex stimulates platelets by increasing bridging of surface molecules.

5.1.3.d. Reactions Involving Microorganisms.   When zymosan (a preparation of yeast cell walls) is incubated with plasma, it causes platelet aggregation and the release reaction (Pfueller and Lüscher, 1974b; Zucker and Grant, 1974; Zucker et al., 1974). The particles have complement components (activated by the alternative pathway) fibrinogen, and IgG on their surface (Pfueller and Lüscher, 1974b). This IgG, which is a ubiquitous naturally occurring antibody to zymosan (Martin et al., 1978) is the primary stimulus for platelet response (Nelson et al., 1980) and reacts with the platelet Fc receptor. Platelet stimulation by Histoplasma capsulatum occurs by a similar mechanism, except that complement is not required (Des Prez et al., 1980). Infection by this organism is often associated with thrombocytopenia. It is possible that platelet stimulation results, as described above, from multisite reactions involving both Fc and fibrinogen receptors.

5.1.3.e. Molecular Nature of the Fc Receptor. Since the ability of the Fc receptor to bind IC was not destroyed by treatment of platelets with proteases, it appeared that it was not related to major surface glycoproteins (Pfueller *et al.*, 1977a). However, since binding of complexes of keyhole-limpet hemocyanin and its specific antibody was inhibited by ristocetin–von Willebrand factor (VWF) and since platelet agglutination by ristocetin–VWF was inhibited by IgG and Fc fragments (Moore *et al.*, 1978), it was suggested that GP Ib may contain both the Fc receptor and that for ristocetin–VWF. Since binding of ovalbumin–antiovalbumin was not affected by ristocetin–VWF (Moore and Nachman, 1981), it could be that two types of Fc receptor exist.

That GP Ib is not the major Fc receptor is suggested by findings that Bernard-Soulier platelets were activated normally by latex particles coated with IgG (Pfueller *et al.*, 1983) and that chymotrypsin-treated platelets, which were depleted of surface glycoproteins, responded normally to IgG-coated latex, heat-aggregated IgG, and antigen–antibody complexes.

A glycoprotein with $M_r = 255,000$ and subunits of 50,000 has been implicated as the Fc receptor by Cheng and Hawiger (1979). It was extracted from platelets by 2 M KBr and bound to both soluble or immobilized Fc fragments. However, only 22% of the total was membrane located. In intact platelets, Steiner and Lüscher (1982) have shown that photoaffinity-labeled IgG binds specifically to GP IIa. The physiological roles of these glycoproteins as Fc receptors must yet be demonstrated.

5.1.3.f. Platelet-Associated IgG (PA-IgG). Washed platelets from normal subjects always have IgG associated with them; between 2 and 9 fg (8,000-25,000 molecules) per platelet have been measured on their surface and between 1.2 and 12 fg (5,000–50,000 molecules) per platelet in platelet lysates (reviewed by Kelton and Gibbons, 1982). Since this represents approximately 100 times as much as is found on normal red cells, which do not have Fc receptors, it is possible that this relatively high level of PA–IgG is due to binding of either plasma IgG or of immune complexes to the platelet Fc receptor.

5.2. Role of Complement in Activation of Human Platelets

Since the first component of complement inhibits IC-induced platelet activation (Pfueller and Lüscher, 1972a), it was proposed as the Fc receptor itself (Wautier *et al.*, 1974). However, since levels of platelet complement are low (Henson and Ginsberg, 1981) and since IgG4 activates platelets but not complement, it is unlikely to be a physiologically active receptor.

The requirement of complement for platelet activation by zymosan remains a puzzle since these platelets have no C3b receptor. However, involvement of complement is not without precedent. The rate of platelet aggregation by IC in diluted plasma was decreased by prior treatment of plasma with a cobra venom factor that inactivates complement (Pfueller and Lüscher, 1974a). It could be that these complement requirements are related to reactions described by Polley and Nachman (1975, 1979) and Polley *et al.* (1981), in which complement (C5–C9) enhanced platelet responses to thrombin and arachidonic acid. Macromolecular C5b–C9 complexes on the platelet surface could be seen by electron microscopy as lesions similar to those produced by complement action on red cells. The surface "receptors" for these complexes are unknown.

6. CLINICAL CONDITIONS ARISING FROM IMMUNOLOGICALLY MEDIATED PLATELET DAMAGE

6.1. Alloantibodies

6.1.1. Refractoriness to Transfusion

Patients requiring repeated platelet transfusions over a prolonged period may develop alloantibodies and become refractory to further transfusion. *In vivo* platelet recovery after transfusion is markedly reduced. HLA incompatibility is most frequently responsible (reviewed by Hackett *et al.*, 1982), whereas the effects of ABO incompatibility are not significant (Duquesnoy *et al.*, 1979). Even using HLA-matched platelets, however, approximately 20% of platelet transfusions are unsuccessful. The development of antibodies in these patients may be due either to as yet undefined HLA antigens or to platelet-specific alloantigens.

Adequate cross-matching techniques to detect granulocyte- and platelet-specific antibodies are thus likely to be important in managing refractoriness particularly in marrow-suppressed patients. Promising results have been obtained using antiglobulin consumption (Kickler *et al.*, 1983) and immunofluorescence procedures (Waters *et al.*, 1981).

6.1.2. Posttransfusion Purpura (PTP)

Because of the rarity and unusual characteristics of this disorder (first described by van Loghem *et al.*, 1959), its serology and pathophysiology have been the subject of a number of detailed discussions (Abramson *et al.*, 1974; Shulman *et al.*, 1964; Klein and Blajchman, 1982; Shulman and Jordan, 1982). Seven to ten days after transfusion, severe thrombocytopenia develops, sometimes accompanied by chills, fever, purpura, or mucosal bleeding. The thrombocytopenia is usually self-limiting, lasting from two to six weeks. Most patients are female with a history of transfusion or pregnancy that could have sensitized them to the antigen.

An antibody directed towards Pl^{A1} is most frequently responsible. It is mostly IgG, although IgM may occur as well (Pegels *et al.*, 1981). However, the patient's own platelets are Pl^{A1} negative. Neither Pl^{A1}-positive nor Pl^{A1}-negative platelets are effective in correcting the thrombocytopenia, and the patient's platelets, as well as those being transfused, are destroyed. Re-exposure to Pl^{A1}-positive platelets may cause a reoccurrence of PTP, but does not always do so.

Thus the mechanism of this disorder is not easily explainable. Several hypotheses have been proposed. Anti-Pl^{A1} may cross-react with another perhaps closely related antigen on Pl^{A1}-negative platelets. However, since the antibodies do not react with GP IIIa from Pl^{A1}-negative platelets (Section 4.1.1) and since recovery from PTP often occurs while the anti-Pl^{A1} titer is high, this seems unlikely. Shulman *et al.* (1964) have proposed that a Pl^{A1} antigen from transfused platelets may remain in circulation and that complexes of it with anti-Pl^{A1} may cause destruction of patient Pl^{A1}-negative platelets. However, attempts to detect either soluble Pl^{A1} antigen (Abramson et al., 1974) or effects of Pl^{A1} immune complexes on Pl^{A1}-negative platelets (Cimo and Aster, 1972) were unsuccessful. Alternatively, the Pl^{A1} antigen or membrane micro-particles containing it (George et al., 1982) may be released from the damaged donor platelets and then be absorbed onto the surface of the patient's platelets, making them reactive with the anti-Pl^{A1} antibody. The occurrence in PTP sera of antibodies that give weak reactions with Pl^{A1}-negative or Pl^{A1}-null platelets (Pegels et al., 1981) suggests that antibodies other than those directed toward Pl^{A1} also may be involved.

Although corticosteroids have occasionally been successfully used in treatment, exchange transfusion and plasmapheresis appear most useful (Cimo and Aster, 1972; Abramson et al., 1974).

6.1.3. Neonatal Alloimmune Thrombocytopenia

In this type of neonatal thrombocytopenia, IgG antibodies, produced by the mother, cross the placenta. They are directed toward paternal antigens on platelets of the fetus. Although this is a rare disorder, 12 to 14% of cases are fatal. Pl^{A1} is the antigen most commonly involved, but cases of Ko (Grenet et al., 1965), DUZO (Moulinier, 1961), Pl^{E2} (Shulman et al., 1964), Bak^a, HLA, and others that have not yet been characterized (von dem Borne, 1981) have been described. Immunoglobulin M anti-bodies as well as IgG may also circulate in the mother (von dem Borne, 1981). The mother has normal levels of platelet-associated IgG (PA–IgG), whereas it is increased in the infant since the antigens are missing on maternal platelets. This allows distinction between this type of thrombocytopenia of the newborn and that resulting from idiopathic thrombocytopenia in the mother in which autoantibodies, able to cross the placenta, may cause raised PA–IgG levels in both mother and infant (Cines et al., 1982).

Methods of treating this disorder have included steroids, exchange transfusion, or transfusions of platelets that do not contain the reactive antigen.

6.2. Thrombocytopenia as a Result of Drug Ingestion

6.2.1. Drug-Induced Immune Thrombocytopenia

Ingestion of a drug for more than seven days or even a single dose of a drug to which a patient has been previously exposed, may cause thrombocytopenia due to drug-dependent antiplatelet antibodies described in Section 4.2. Antibody-coated platelets are destroyed either by complement-mediated lysis or sequestration by the reticuloendothelial system.

The onset of thrombocytopenia is sudden and usually accompanied by petechiae and mucosal bleeding. Since the antibodies are usually IgG, levels of PA–IgG on the patient's platelets are raised, whereas IgG from patient serum, which is bindable to normal platelets, is raised only in the presence of the drug (Kelton et al., 1981a; Faig and Karpatkin, 1982). Recovery usually occurs 1–14 days after drug ingestion is discontinued, although laboratory tests may still show antibody in the patient's serum.

Readministration of the offending drug usually causes a further thrombocytopenic episode, but after longer periods (1–2 years), this response may not occur (Shulman and Jordan, 1982).

Since drugs are foreign chemicals, it has been suggested that they may always elicit antibody production; only in some people are enough antibodies of sufficiently high affinity ($> 10^5$ moles/liter) produced to reduce platelet numbers (Shulman and Jordan, 1982). In fact, 15% of 109 patients who were taking quinidine, but who had normal platelet numbers, had detectable quinidine-dependent antibodies (Okuno and Crockatt, 1976).

Other than cessation of drug therapy, treatment includes adrenocorticosteroids and platelet transfusions.

6.2.2. Drug-Associated Thrombocytopenia

Antibodies formed as a result of drug ingestion may cause thrombocytopenia in a manner distinct from that described above (reviewed by Hackett *et al.*, 1982; Shulman and Jordan, 1982).

6.2.2.a. Gold. Thrombocytopenia is one of the most frequent side-effects of gold therapy. Evidence for its immunologic nature includes shortened platelet survival, normal to increased numbers of megakaryocytes in the bone marrow, increased platelet-bindable IgG formed in cultures of patient splenic lymphocytes, and demonstration of gold-dependent binding of IgG from patient serum to normal platelets (Kelton *et al.*, 1981a).

6.2.2.b. Heparin. Although heparin-related thrombocytopenia may be a result of direct effects of heparin on platelets, heparin-dependent antibodies also occur causing increased levels of PA–IgG (Cines *et al.*, 1980). Platelet damage *in vitro* occurs in the presence of heparin and IgG from the patient's plasma (Green *et al.*, 1978; Chong *et al.*, 1981; Cines *et al.*, 1980) and is accompanied by C3 deposition on to platelets (Cines *et al.*, 1980). In some studies, evidence of specific IgG and heparin complexes was obtained (Cines *et al.*, 1980). Plasma from some nonthrombocytopenic patients receiving heparin also contained IgG that could mediate platelet damage at high heparin concentrations. It is possible that heparin–antiheparin antibody complexes cause platelet damage or that heparin binding to GP Ib and other surface proteins (Gogstad *et al.*, 1983) facilitates IgG binding to platelets.

6.3. Idiopathic Thrombocytopenic Purpura (ITP)

This is a term applied to thrombocytopenias of unknown etiology. Many investigators would, however, agree that ITP is synonymous with autoimmune thrombocytopenic purpura (reviewed by Mueller-Eckhardt, 1977; Karpatkin, 1980; McMillan, 1981; Kelton and Gibbons, 1982). There are two distinct varieties of the disease. Acute ITP is found most often in children 2–6 years old of either sex. The onset of symptoms is usually rapid, often occurring one to six weeks after a viral illness. Recovery usually occurs spontaneously within six months. Chronic ITP is seen most frequently in females in their third or fourth decade. Symptoms usually appear slowly and spontaneous remission is rare. Petechiae, purpura, and mucosal bleeding are observed in the absence of splenomegaly. The most serious complication is intracranial haemorrhage. Thrombocytopenia is usually the only abnormality of the peripheral blood film on which an increased percentage of large, dense platelets (megathrombocytes) may be seen. In the bone marrow, the number of megakaryocytes is normal to increased, suggesting that both synthesis and turnover of platelets must be increased. Platelet life span is usually less than 10% of the normal ten days due to platelet sequestration in the liver and spleen. Platelet function, e.g., aggregation or exposure of platelet factor 3, may also be impaired.

6.3.1. Immunological Basis of ITP

That ITP was an immune disorder was established by Harrington *et al.* (1956) and Shulman *et al.* (1975). Plasma from ITP patients contains a 7S gamma globulin that causes thrombocytopenia when infused into normal recipients. This "ITP factor" is absorbed by normal or ITP patient's platelets, is species specific, and behaves in the serologic tests outlined in Section 2 as an IgG antibody. Thus increased levels of PA–IgG and platelet-bindable IgG are thought to represent binding of this ITP factor to platelets. Raised levels of platelet-associated IgM (Nel *et al.*, 1983) and platelet-bindable IgM (van Leeuwen *et al.*, 1982a) have also been observed. However, raised PA–IgG in some thrombocytopenias not of immune origin (Mueller-Eckhardt *et al.*, 1980b; Pfueller *et al.*, 1981c; Kelton *et al.*, 1982b) and lack of correlation between platelet life span and PA–IgG in such thrombocytopenias (Mueller-Eckhardt *et al.*, 1982c) indicate that raised PA–IgG may not necessarily represent the factor responsible for platelet destruction. Levels of platelet-bindable IgG in serum may be a better indicator of the presence of antiplatelet factors, particularly in ITP occurring in pregnancy in which levels of maternal serum antibodies predict more accurately than PA–IgG the likelihood that the infant will also have ITP (Cines *et al.*, 1982).

6.3.2. Nature of the Antigen to Which ITP Antibodies Are Directed

In some sera, ITP antibodies directed toward platelet glycoproteins have been found (Section 4.3). However, the existence of the platelet Fc receptor (Section 5.1) that can bind antigen–antibody complexes raises the possibility that some ITP factors may be complexes of antibodies with foreign antigens. Techniques used to characterize the ITP factor as IgG would not distinguish between monomeric IgG and that bound to low-molecular-weight antigens. Indeed, raised levels of immune complexes have been found in patients with ITP (Lurhuma *et al.*, 1977; Clancy *et al.*, 1980) and with other forms of thrombocytopenia (Trent *et al.*, 1980), but a causal relationship has not been established. Although binding to platelets and megakaryocytes (McMillan *et al.*, 1978) of IgG from cultures of spleen cells from ITP patients does not require the Fc portion of the molecule (McMillan *et al.*, 1980), the relationship between splenic IgG and the serum ITP factor must be further investigated.

6.3.3. Treatment

Idiopathic thrombocytopenic purpura is usually treated initially with corticosteroids and if this is not effective, other treatments such as splenectomy or administration of vinca alkaloids or immunosuppressants may be undertaken. Transfusion or plasmapheresis have been used in emergencies. In childhood ITP, intravenous infusion of high doses of IgG has been recently suggested to be promising (reviewed by Imbach and Jungi, 1983). A small percentage of patients remain refractory to treatment.

6.4. Disease-Associated Immune Thrombocytopenia

Immune-mediated thrombocytopenia may develop in a range of other diseases such as systemic lupus erythematosus, septicemia, Graves' disease, Evans syndrome, Hashimoto's thyroiditis, lymphoproliferative disorders, Hodgkin's disease, and idiopathic glomerulonephritis (reviewed by Kelton and Gibbons, 1982; Shulman and Jordan, 1982). This may be a result of immune complexes (Kasai *et al.*, 1981) or of antibodies directed toward platelet antigens. In systemic lupus erythematosus and rheumatoid arthritis, antibodies to β_2 microglobulin (Falus *et al.*, 1982) and unidentified platelet antigens (Weissbarth *et al.*, 1982) have been found that can activate platelets.

Infection by virus (rubella, rubeola, mumps, Epstein-Barr), parasites, and bacteria may be associated with thrombocytopenia thought to be of immunologic origin. Antibodies to the infecting agent may cross-react with platelet membrane components or, when combined with antigen, may react with the platelet Fc receptor. In malaria, thrombocytopenia results when the platelet adsorbs the parasite and, subsequently, parasite-specific antibody (Kelton *et al.*, 1983).

7. CONCLUSION

Immunoglobulins thus react with the platelet in a variety of ways altering its reactivity and causing thrombocytopenia. It is becoming apparent that Fc receptors and antigens involved are located on platelet surface glycoproteins. The structure of the epitopes for alloantibodies, autoantibodies, and drug-dependent antibodies and of the Fc receptors are areas awaiting further investigation.

ACKNOWLEDGMENTS. Support from the National Health and Medical Research Council of Australia (Grant Numbers 820234 and 820233) and secretarial assistance by Miss Louise Fischer and Mrs. Marjorie Brown are gratefully acknowledged.

REFERENCES

Abramson, N., Eisenberg, P. D., and Aster, R. H., 1974, Post-transfusion purpura: Immunologic aspects and therapy, *N. Engl. J. Med.* **291**:1163–1166.

Ackroyd, J. F., 1953, Allergic purpura, including purpura due to foods, drugs and infections, *Am. J. Med.* **14**:605–632.

Aster, R. H., 1965, Effect of anticoagulant and ABO incompatibility on recovery of transfused human platelets, *Blood* **26**:732–743.

Aster, R. H. 1984, Platelet antigen systems, in: *Immunohaematology* (C. P. Engelfriet and A. E. G. von dem Borne, eds.), Elsevier Biomedical Press, Amsterdam, pp. 23–32.

Aster, R. H., Miskovich, B. H., and Rodey, G. E., 1973, Histocompatibility antigens of human plasma; Localization to the HLD-3 lipoprotein fraction, *Transplantation* **16**:205–210.

Berndt, M. C., Gregory, C., Chong, B. H., Zola, H., and Castaldi, P. A., 1983, Additional glycoprotein defects in Bernard-Soulier syndrome: Confirmation of genetic basis by parental analysis, *Blood* **62**:800–807.

Bernier, I., Dautigny, A., Colombani, J., and Jollès, P., 1974, Detergent-solubilized HL-A antigens from human platelets: A comparative study of various purification techniques, *Biochim. Biophys. Acta* **356**:82–90.

Bettex-Galland, M., Lüscher, E. F., Simon, G., and Vasali, P., 1963, Induction of viscous metamorphosis in human blood platelets by means other than by thrombin, *Nature (London)* **200**:1109–1110.

Boizard, B., and Wautier, J. L., 1984, Lek[a], A new platelet antigen absent in Glanzmann's thrombasthenia, *Vox Sang.* **46**:47–54.

Cartron, J. P., and Nurden, A. T., 1979, Galactosyltransferase and membrane glycoprotein abnormality in human platelets from Tn-syndrome donors, *Nature (London)* **282**:621–623.

Cartron, J. P., Nurden, A. T., Blanchard, D., Lee, H., Dupuis, D., and Salmon, C., 1980, The Tn receptors of human red cells and platelets, *Blood Transf. Haematol.* **23**:613–628.

Cheng, C. M., and Hawiger, J., 1979, Affinity isolation and characterisation of immunoglobulin G Fc fragment-binding glycoprotein from human blood platelets, *J. Biol. Chem.* **254**:2165–2167.

Chong, B. H., Grace, C. S., and Rozenberg, M. C., 1981, Heparin-induced thrombocytopenia: Effect of heparin platelet antibody on platelets, *Br. J. Haematol.* **49**:531–540.

Chong, B. H., Berndt, M. C., Koutts, J., and Castaldi, P. A., 1983, Quinidine-induced thrombocytopenia and leukopenia: Demonstration and characterization of distinct anti-platelet and anti-leukocyte antibodies, *Blood* **62**:1218–1223.

Christie, D. J., and Aster, R. H., 1982, Drug-antibody-platelet interaction in quinine and quinidine-induced thrombocytopenia, *J. Clin. Invest.* **70**:989–998.

Cimo, P. L., and Aster, R. H., 1972, Post transfusion purpura. Successful treatment by exchange transfusion, *N. Engl. J. Med.* **287**:290–292.

Cimo, P. L., and Gerber, S. A., 1979, AET-treated platelets: Their usefulness for platelet antibody detection and an examination of their altered sensitivity to immune lysis, *Blood* **54**:1101–1108.

Cines, D. B., and Schreiber, A. D., 1979, Effect of anti-Pl[A1] antibody on human platelets. I. The role of complement, *Blood* **53**:567–577.

Cines, D. B., Kaywin, P., Bina, M., Tomaski, A., and Schreiber, A. D., 1980, Heparin-associated thrombocytopenia, *N. Engl. J. Med.* **14**:788–795.

Cines, D. B., Dusak, B., Tomaski, A., Mennutti, M., and Schreiber, A. D., 1982, Immune thrombocytopenic purpura and pregnancy, *N. Engl. J. Med.* **14**:826–831.

Claas, F. H. J., van der Meer, J. W. M., and Langerak, J., 1979, Immunological effect of co-trimoxazole on platelets, *Br. Med. J.* **2**:898–899.

Claas, F. H. J., Langerak, J., and van Rood, J. J., 1981, Drug-induced antibodies with restricted specificity, *Immunol. Lett.* **2**:323–326.

Clancy, R., Trent, R. Davis, V., and Davidson, R., 1980, Autosensitisation and immune complexes in chronic idiopathic thrombocytopenic purpura, *Clin. Exp. Immunol.* **39**:170–175.

Clark, W. F., Tevaarwerk, G. J. M., and Reid, B. D., 1982, Human platelet-immune complex interaction in plasma, *J. Lab. Clin. Med.* **100**:917–931.

Coller, B. S., Peerschke, E. I., Scudder, L. E., and Sullivan, C. R., 1983, Studies with a murine monoclonal antibody that abolishes ristocetin-induced binding of von Willebrand factor to platelets: Additional evidence in support of GP Ib as a platelet receptor for von Willebrand factor, *Blood* **61**:99–110.

Coombs, R. R. A., and Bedford, D., 1955, The A and B antigen on human platelets demonstrated by means of mixed erythrocyte-platelet agglutination, *Vox Sang.* **5**:111–115.

Dausset, J., and Berg, P., 1963, Un nouvel exemple d'anticorps anti-plaquettaire Ko, *Vox Sang.* **8**:341–347.

Dautigny, A., Bernier, I., Colombani, J., and Jollès, P., 1975, Spontaneous release of soluble HL-A antigens from platelets during conservation, *Immunol. Commun.* **4**:443–451.

Des Prez, R. M., Steckley, S., Stroud, R. M. and Hawiger, J., 1980, Interaction of *Histoplasma capsulatum* with human platelets, *J. Infect. Dis.* **142**:32–39.

Duquesnoy, R. J., Anderson, A. J., Tomasulo, P. A., and Aster, R. H., 1979, ABO compatibility and platelet transfusion of alloimmunized thrombocytopenic patients, *Blood* **54**:595–599.

Eisner, E. V., and Shahidi, N. T., 1972, Immune thrombocytopenia due to a drug metabolite, *N. Engl. J. Med.* **287**:376–381.

Faig, D., and Karpatkin, S., 1982, Cumulative experience with a simplified solid-phase radioimmunoassay for the detection of bound antiplatelet IgG, serum auto-, allo-, and drug-dependent antibodies, *Blood* **60**:807–813.

Falus, A., Merétey, K., Bagdy, D., Diószegi, M., Böhm, U., Csák, E., and Bozsóky, S., 1982, β-2-microglobulin-specific autoantibodies cause platelet aggregation and interfere with ADP-induced aggregation, *Clin. Exp. Immunol.* **47**:103–109.

George, J. N., Reimann, T. A., Moake, J. L., Morgan, R. K., Cimo, P. L., and Sears, D. A., 1981, Bernard-Soulier disease: II. A study of four patients and their parents, *Br. J. Haematol.* **48**:459–467.

George, J. N., Thoi, L. L., McManus, L. M., and Reimann, T. A., 1982, Isolation of human platelet membrane microparticles from plasma and serum, *Blood* **60**:834–840.

Ginsberg, M. H., and Henson, P. M., 1978, Enhancement of platelet response to immune complexes and IgG aggregates by lipid A-rich bacterial lipopolysaccharides, *J. Exp. Med.* **147**:207–218.

Glöckner, W. M., Kaulen, H. D., and Uhlenbruch, G., 1978, Immunochemical detection of the Thomsen-Friedenreich antigen (T-antigen) on platelet plasma membranes, *Thromb. Haemostasis* **39**:186–192.

Glynn, M. F., Movat, H. Z., Murphy, E. A., and Mustard, J. F., 1965, Study of platelet adhesiveness and aggregation, with latex particles, *J. Lab. Clin. Med.* **65**:179–201.

Gockerman, J. P., and Jacob, W., 1979, Purification and characterization of papain-solubilised HLA antigens from human platelets, *Blood* **53**:838–850.

Gogstadt, G. O., Solum, N. O., and Krutnes, M. B., 1983, Heparin-binding platelet proteins demonstrated by crossed affinity immunoelectrophoresis, *Br. J. Haematol.* **53**:563–573.

Gowland, E., Kay, H. E. M., Spillman, J. C., and Williamson, J. R., 1969, Agglutination of platelets by a serum factor in the presence of EDTA, *J. Clin. Pathol.* **22**:460–464.

Green, D., Harris, K., Reynolds, N., Roberts, M., and Patterson, R., 1978, Heparin-immune thrombocytopenia: Evidence for a heparin-platelet complex as the antigenic determinant, *J. Lab. Clin. Med.* **91**:167–175.

Grenet, P., Dausset, J., Dugas, M., Petit, D., Badoual, J., and Tangun, Y., 1965, Purpura thrombocytopenique neonatal avec isoimmunisation foeto-maternelle anti-Ko[a], *Arch. Fr. Pediatr.* **22**:1165–1174.

Gross, B., Haessig, A., Lüscher, E. F., and Nydegger, U. E., 1983, Monomeric IgG preparations for intravenous use inhibit platelet stimulation by polymeric IgG, *Br. J. Haematol.* **53**:289–299.

Gudino, M., and Miller, W. V., 1981, Application of the enzyme linked immunospecific assay (ELISA) for the detection of platelet antibodies, *Blood* **57**:32–37.

Gurevitch, J., and Nelken, D., 1954, ABO groups in blood platelets, *J. Lab. Clin. Med.* **44**:562–570.

Hackett, T., Kelton, J. G., and Powers, P., 1982, Drug-induced platelet destruction, *Semin. Thromb. Hemost.* **8**:116–137.

Hakomori, S., 1981, Blood group ABH and Ii: Antigens of human erythrocytes: Chemistry, polymorphism, and their developmental change, *Semin. Hematol.* **18**:39–62.

Harrington, W. J., Sprague, C. C., Minnich, V., Moore, C. V., Aulvin, R. C., and Dubach, R., 1953, Immunologic mechanisms in idiopathic and neonatal thrombocytopenic purpura, *Ann. Intern. Med.* **38**:433–469.

Harrington, W. J., Minnick, V., and Arimura, G., 1956, The autoimmune thrombocytopenias, *Prog. Hematol.* **1**:166–192.

Heinrich, D., Stephinger, U., and Mueller-Eckhardt, C., 1977, Specific interaction of HLA antibodies (eluates) with washed platelets, *Br. J. Haematol.* **35**:441–452.

Henson, P. M., and Ginsberg, M. H., 1981, Immunological reactions of platelets, in: *Platelets in Biology and Pathology*, Volume 2 (J. L. Gordon, ed.), Elsevier/North Holland Biomedical Press, Amsterdam, pp. 265–308.

Henson, P. M., and Spiegelberg, H. L., 1973, Release of serotonin from human platelets induced by aggregated immunoglobulins of different classes and subclasses, *J. Clin. Invest.* **52**:1282–1288.

Hosseinzadeh, P. K., Firkin, B. G., and Pfueller, S. L., 1980, Study of the factors that cause specific transformation in cultures of lymphocytes from patients with quinine- and quinidine-induced immune thrombocytopenia, *J. Clin. Invest.* **66**:638–645.

Imbach, P., and Jungi, T. W., 1983, Possible mechanisms of intravenous immunoglobulin treatment in childhood idiopathic thrombocytopenic purpura (ITP), *Blut* **46**:117–124.

Israels, E. D., Nisli, G., Paraskevas, F., and Israels, L. G., 1973, Platelet Fc receptor as a mechanism for Ag-Ab complex-induced platelet injury, *Thromb. Diathes. Haemorrh.* **29**:434–444.

Kanz, L., Straub, G., Bross, K. G., and Fauser, A. A., 1982, Identification of human megakaryocytes derived from pure megakaryocytic colonies (CFU-M), megakaryocytic-erythroid colonies (CFU-M/E), and mixed hemopoietic colonies (CFU-GEMM) by antibodies against platelet associated antigens, *Blut* **45**:267–274.

Karas, S. P., Rosse, W. F., and Kurlander, R. J., 1982, Characterisation of the IgG-Fc receptor on human platelets, *Blood* **60**:1277–1282.

Karpatkin, S., 1980, Autoimmune thrombocytopenic purpura, *Blood* **56**:329–343.

Kasai, N., Parbtani, A., Cameron, J. S., Yewdall, V., Shepherd, P., and Verroust, P., 1981, Platelet-aggregating immune complexes and intraplatelet serotonin in idiopathic glomerulo-nephritis and systemic lupus, *Clin. Exp. Immunol.* **43**:64–72.

Kelton, J. G., and Gibbons, S., 1982, Autoimmune platelet destruction: Idiopathic thrombocytopenic purpura, *Semin. Thromb. Hemost.* **8**:83–104.

Kelton, J. G., Meltzer, D., Moore, J., Giles, A. R., Wilson, W. E., Barr, R., Hirsh, J., Neame, P. B., Powers, P. J., Walker, I., Bianchi, F., and Carter, C. J., 1981a, Drug-induced thrombocytopenia is associated with increased binding of IgG to platelets both *in vivo* and *in vitro*, *Blood* **58**:524–529.

Kelton, J. G., Moore, J., Gauldie, J., Neame, P. B., Hirsh, J., and Tozman, E., 1981b, The development and application of a serum assay for platelet-bindable IgG (S-PBIgG), *J. Lab. Clin. Med.* **98**:272–279.

Kelton, J. G., Hamid, C., Aker, S., and Blajchman, M. A., 1982a, The amount of blood group A substance on platelets is proportional to the amount in the plasma, *Blood* **59**:980–985.

Kelton, J. G., Powers, P. J., and Carter, C. J., 1982b, A prospective study of the usefulness of the measurement of platelet-associated IgG for the diagnosis of idiopathic thrombocytopenia purpura, *Blood* **60**:1050–1053.

Kelton, J. G., Keystone, J., Moore, J., Denomme, G., Tozman, E., Glynn, M., Neame, P. B., Gauldie, J., and Jensen, J., 1983, Immune-mediated thrombocytopenia of malaria, *J. Clin. Invest.* **71**:832–836.

Kickler, T. S., Braine, H. G., Ness, P. M., Koester, A., and Bias, W., 1983, A radiolabeled antiglobulin test for crossmatching platelet transfusions, *Blood* **61**:238–242.

Kieffer, N., Boizard, B., Didry, D., Wautier, J.-L., and Nurden, A. T., 1984, Immunochemical characterization of the platelet-specific alloantigen, Lek^a. A comparative study with the Pl^A1 antigen, *Blood,* **64**:1212–1219.

Klein, C. A., and Blajchman, M. A., 1982, Alloantibodies and platelet destruction, *Semin. Thromb. Hemost.* **8**:105–115.

Kunicki, T. J., and Aster, R. H., 1978, Deletion of the platelet-specific alloantigen Pl^A1 from platelets in Glanzmann's thrombasthenia, *J. Clin. Invest.* **61**:1225–1231.

Kunicki, T. J., and Aster, R. H., 1979, Isolation and Immunologic Characterisation of the Human Platelet Alloantigen Pl^A1, *Mol. Immunol.* **16**:353–360.

Kunicki, T. J., Johnson, M. M., and Aster, R. H., 1978, Absence of the platelet receptor for drug-dependent antibodies in the Bernard-Soulier syndrome, *J. Clin. Invest.* **62**:716–719.

Kunicki, T. J., Pidard, D., Casenave, J.-P., Nurden, A. T., and Caen, J. P., 1981a, Inheritance of the human platelet alloantigen Pl^A1, in type I Glanzmann's thrombasthemia, *J. Clin. Invest.* **67**:717–724.

Kunicki, T. J., Russell, N., Nurden, A. T., Aster, R. H., and Caen, J. P., 1981b, Further studies of the human platelet receptor for quinine- and quinidine-dependent antibodies, *J. Immunol.* **126**:398–402.

Lalezari, P., and Driscoll, A. M., 1982, Ability of thrombocytes to acquire HLA specificity from plasmas, *Blood* **59**:167–170.

Lane, J., Brown, M., Bernstein, I., Wilcox, P. K., Slichter, S., and Nowinski, R. C., 1982, Serological and biochemical analysis of the Pl^A1 alloantigen of human platelets, *Br. J. Haematol.* **50**:351–359.

Lewis, J. H., Draude, J., and Kuhns, W. J., 1960, Coating of "O" platelets with A and B blood group substances, *Vox Sang.* **5**:434–441.

Liebert, M., and Aster, R. H., 1977, Expression of HLA-B12 on platelets, on lymphocytes and in serum: A quantitative study, *Tissue Antigens* **9**:199–208.

Lurhuma, A. Z., Riccomi, H., and Masson, P. L., 1977, The occurrence of circulating platelet immune complexes and viral antigens in idiopathic thrombocytopenic purpura, *Clin. Exp. Immunol.* **28**:49–55.

Martin, S. E., Breckenridge, R. T., Rosenfeld, S. I., and Leddy, J. P., 1978, Responses of human platelets to immunologic stimuli: Independent roles for complement and IgG in zymosan activation, *J. Immunol.* **120**:9–14.

McMillan, R., 1981, Chronic idiopathic thrombocytopenic purpura, *N. Engl. J. Med.* **304**:1135–1147.

McMillan, R., Luiken, G. A., Levy, R., Yelenosky, R., and Longmire, R. L., 1978, Antibody against megakaryocytes in idiopathic thrombocytopenic purpura, *J. Am. Med. Ass.* **239**:2460–2462.

McMillan, R., Tani, P., and Mason, D., 1980, The demonstration of antibody binding to platelet-associated antigens with immune thrombocytopenic purpura, *Blood* **56**:993–995.

McMillan, R., Mason, D., Tani, P., and Schmidt, G. M., 1982, Evaluation of platelet surface antigens: Localization of the Pl^{A1} alloantigen, *Br. J. Haematol.* **51**:297–304.

Moore, A., and Nachman, R. L., 1981, Platelet Fc receptor. Increased expression in myeloproliferative disease, *J. Clin. Invest.* **67**:1064–1071.

Moore, A., Ross, G. D., and Nachman, R. L., 1978, Interaction of platelet membrane receptors with von Willebrand factor, ristocetin and the Fc region of immunoglobulin G, *J. Clin. Invest.* **62**:1053–1060.

Moore, A., Weksler, B. B., and Nachman, R. L., 1981, Platelet Fc IgG receptor: Increased expression in female platelets, *Thromb. Res.* **21**:469–474.

Moreau, P., and Andre, A., 1954, Blood groups of human blood platelets, *Nature (London)* **174**:88.

Moulinier, J., 1961, New observations on the platelet group Duzo (abstract), *Vox Sang.* **6**:242.

Mueller-Eckhardt, C., 1977, Idiopathic thrombocytopenic purpura (ITP): Clinical and immunologic considerations, *Semin. Thromb. Hemostas.* **3**:125–159.

Mueller-Eckhardt, C., and Lüscher, E. F., 1968a, Immune reactions of human blood platelets. I. A comparative study on the effects on platelets of heterologous antiplatelet antiserum, antigen–antibody complexes, aggregated gammaglobulin thrombin, *Thromb. Diathes. Haemorrh.* **20**:155–167.

Mueller-Eckhardt, C., and Lüscher, E. F., 1968b, Immune reactions of human blood platelets. II. The effect of latex particles coated with gammaglobulin in relation to complement activation, *Thromb. Diathes. Haemorrh.* **20**:168–179.

Mueller-Eckhardt, G., Hauck, M., Kayser, W., and Mueller-Eckhardt, C., 1980a, HLA-C antigens on platelets, *Tissue Antigens* **16**:91–94.

Mueller-Eckhardt, C., Kayser, W., Mersch-Baumert, K., Mueller-Eckhardt, G., Breidenbach, M., Kugel, H. G., and Graubner, M., 1980b, The clinical significance of platelet-associated IgG: A study on 298 patients with various disorders, *Br. J. Haematol.* **46**:123–131.

Mueller-Eckhardt, C., Kayser, N., Forster, C., Mueller-Eckhardt, G., and Ringenberg, C., 1982a, Improved assay for detection of platelet-specific Pl^{A1} antibodies in neonatal alloimmune thrombocytopenia, *Vox Sang.* **43**:76–81.

Mueller-Eckhardt, C., Marks, H. J., Baur, M. P., and Mueller-Eckhardt, G., 1982b, Immunogenetic studies of the platelet-specific antigen Pl^{A1} (Zw [a]), *Immunobiology* **160**:375–381.

Mueller-Eckhardt, C., Mueller-Eckhardt, G., Kayser, W., Voss, R. M., Wegner, J., and Kuenzlen, E., 1982c, Platelet-associated IgG, platelet survival, and platelet sequestration in thrombocytopenic states, *Br. J. Haematol.* **52**:49–58.

Mulder, A., Tetteroo, P., Veenboer, T., Lansdorp, P., and von dem Borne, A. E. G. K., 1982, Radioimmunoelectrophoretic analyses of platelet-specific antigens, *Proc. 19th Int. Soc. Thromb. Haematol. and Inta Soc. Blood Transfusion, Budapest* 291a (abstract).

Muller, J. Y., Patereau, C., and Soulier, J. P., 1978, Thrombasthenie de Glanzmann: Antigene Pl^{A1} et anticorps anti-Glanzmann, *Rev. Fr. Transfus. Immuno-hematol.* **21**:1069–1078.

Myers, T. J., Kim, B. K., Steiner, M., Bishop, J., and Baldini, M. G., 1981, Detection of Platelet Alloimmunity with a Platelet-associated IgG Assay, *J. Lab. Clin. Med.* **97**:855–863.

Nel, J. D., Stevens, K., Mouton, A., and Pretorius, F. J., 1983, Platelet-bound IgM in Autoimmune Thrombocytopenia, *Blood* **61**:119–124.

Nelson, L. A., Rosenfeld, S. I., and Leddy, J. P., 1980, Properties of immune complexes mediating aggregation of human platelets, *Arthritis Rheum.* **23**:725a.

Nurden, A. T., Dupuis, D., Pidard, D., Kieffer, N., Kunicki, T. J., and Cartron, J.-P., 1982, Surface modifications in the platelets of a patient with alpha-N-acetyl-D-galactosamine residues; the Tn-syndrome, *J. Clin. Invest.* **70**:1281–1291.

Okuno, T., and Crockatt, D., 1976, Anti-drug-related antibodies in nonthrombocytopenic cardiac patients, *Am. J. Clin. Pathol.* **65**:523–527.

Onder, O., Weinstein, A., and Hoyer, L. W., 1980, Pseudothrombocytopenia caused by platelet agglutinins that are reactive in blood anticoagulated with chelating agents, *Blood* **56**:177–182.

Osler, A. G., and Siraganian, R. P., 1972, Immunologic mechanisms of platelet damage, *Prog. Allergy* **16**:450–498.

Packham, M. A., Evans, G., Glynn, M. F., and Mustard, J. F., 1969, The effect of plasma proteins on the interaction of platelets with glass surfaces, *J. Lab. Clin. Med.* **73**:686–697.

Pegels, J. G., Bruynes, E. C. E., Engelfriet, C. P., and von dem Borne, A. E. G. K., 1981, Post-transfusion purpura: A serological and immunochemical study, *Br. J. Haematol.* **49**:521–530.

Pegels, J. G., Bruynes, E. C. E., Engelfriet, C. P., and von dem Borne, A. E. R. K., 1982, Pseudothrombocytopenia: An immunologic study on platelet antibodies dependent on ethylene diamine tetra-acetate, *Blood* **59**:157–161.

Perkins, H. A., 1979, Current status of the HLA system, *Am. J. Hematol.* **6**:285–292.

Pfueller, S. L., and Cosgrove, L. J., 1980, Activation of human platelets in PRP via their Fc-receptor by antigen–antibody complexes or immunoglobulin G: Requirement for particle-bound fibrinogen, *Thromb. Res.* **20**:97–108.

Pfueller, S. L., and Firkin, B. G., 1984, Retraction, *J. Clin. Invest.* **73**:1243.

Pfueller, S. L., and Lüscher, E. F., 1972a, The effects of aggregated immunoglobulins on human blood platelets in relation to their complement-fixing abilities. I. Studies of immunoglobulins of different types, *J. Immunol.* **109**:517–525.

Pfueller, S. L., and Lüscher, E. F., 1972b, The effects of immune complexes on blood platelets and their relationship to complement activation, *Immunochemistry* **9**:1151–1165.

Pfueller, S. L., and Lüscher, E. F., 1974a, Studies of the mechanisms of the human platelet release reaction induced by immunologic stimuli. I. Complement-dependent and complement-independent reactions, *J. Immunol.* **112**:1201–1210.

Pfueller, S. L., and Lüscher, E. F., 1974b, Studies of the mechanisms of the human platelet release reaction induced by immunologic stimuli. II. The effects of zymosan, *J. Immunol.* **112**:1211–1218.

Pfueller, S. F., Jenkins, C. S. P., and Lüscher, E. F., 1977a, A comparative study of the effect of modification of the surface of human platelets on the receptors for aggregated immunoglobulins and for ristocetin-von Willebrand factor, *Biochim. Biophys. Acta* **465**:614–626.

Pfueller, S. L., Weber, S., and Lüscher, E. F., 1977b, Studies of the mechanism of the human platelet release reaction induced by immunologic stimuli. III. Relationship between the binding of soluble IgG aggregates to the Fc receptor and cell response in the presence and absence of plasma, *J. Immunol.* **118**:514–524.

Pfueller, S. L., Hosseinzadeh, P. K., and Firkin, B. G., 1981a, Quinine- and quinidine-dependent antiplatelet antibodies. Requirement of factor VIII-related antigen for platelet damage and for *in vitro* transformation of lymphocytes from patients with drug-induced thrombocytopenia, *J. Clin. Invest.* **67**:907–910.

Pfueller, S. L., Tew, D., Cosgrove, L., and de Rosbo, N., 1981b, The relationship of platelet associated-IgG to anti-platelet antibodies (abstract), *Clin. Exp. Pharmacol. Physiol.* **8**:382.

Pfueller, S. L., Cosgrove, L., Firkin, B. G., and Tew, D., 1981c, Relationship of raised platelet IgG in thrombocytopenia to total platelet protein content, *Br. J. Haematol.* **49**:293–302.

Pfueller, S. L., Kerlero de Rosbo, N., and Bilston, R., 1984, Platelets deficient in glycoprotein I have normal Fc receptor expression, *Br. J. Haematol.* **56**:607–615.

Pincus, J. H., Kahan, B. D., and Mittal, K. K., 1976, A role for cAMP in the preparation of human platelets for the extraction of histocompatibility antigens, *Immunochemistry* **13**:565–570.

Polley, M. J., and Nachman, R. L., 1975, Ultrastructural lesions on the surface of platelets associated with either blood coagulation or with antibody-mediated immune injury, *J. Exp. Med.* **141**:1261–1268.

Polley, M. J., and Nachman, R., 1979, Human complement in thrombin-mediated platelet function. Uptake of the C5-9 complex, *J. Exp. Med.* **150**:633–645.

Polley, M. J., Nachman, R. L., and Webster, B. B., 1981, Human complement in the arachidonic acid transformation pathway in platelets, *J. Exp. Med.* **153**:257–268.

Ruan, C., Tobelem, G., McMichael, A. J., Drouet, L., Legrand, Y., Degos, L., Kieffer, N., Lee, H., and Caen, J. P., 1981, Monoclonal antibody to human platelet glycoprotein I. II. Effects on human platelet function, *Br. J. Haematol.* **499**:511–519.

Schiffer, C. A., and Young, V., 1983, Detection of platelet antibodies using a micro-enzyme-linked immunosorbent assay (ELISA), *Blood* **61**:311–317.

Shulman, N. R., 1958, Immunoreactions involving platelets. I. A steric and kinetic model for formation of a complex from a human antibody, quinidine as haptene, and platelets; and for fixation of complement by the complex, *J. Exp. Med.* **107**:665–690.

Shulman, N. R., 1964, A mechanism of cell destruction in individuals sensitised to foreign antigens and its implications in autoimmunity, *Ann. Intern. Med.* **60**:507–521.

Shulman, N. R., 1972, Immunologic reactions to drugs, *N. Engl. J. Med.* **287**:408–409.

Shulman, N. R., and Jordan, J. V., 1982, Platelet immunology, in *Hemostasis and Thrombosis, Basic Principles and Clinical Practice* (R. W. Colman, J. Hirsh, V. J. Marder, and E. W. Salzman, eds.), J. B. Lippincott Company, Philadelphia, Toronto, pp. 274–342.

Shulman, N. R., Aster, R. H., Leitner, A., and Hiller, M. C., 1961, Immunoreactions involving platelets. V. Post-transfusion purpura due to a complement-fixing antibody against a genetically controlled platelet antigen. A proposed mechanism for thrombocytopenia and its relevance in "autoimmunity," *J. Clin. Invest.* **40**:1597–1620.

Shulman, N. R., Marder, V. J., Aledort, L. M., and Hiller, M. C., 1962, Complement-fixing isoantibodies against antigens common to platelets and leukocytes, *Trans. Assoc. Am. Physicians* **75**:89–98.

Shulman, N. R., Marder, V. J., Hiller, M. C., and Collier, E. M., 1964, Platelet and leukocyte isoantigens and their antibodies: Serologic, physiologic and clinical studies, *Prog. Hematol.* **4**:222–304.

Shulman, N. R., Marder, V. J., and Weinrach, R. S., 1965, Similarities between known antiplatelet antibodies and the factor responsible for thrombocytopenia in idiopathic purpura: Physiologic, serologic and isotopic studies, *Ann. N.Y. Acad. Sci.* **124**:499–542.

Stefanini, M., Plitman, G. I., Damashek, W., Chatterjea, J. P., and Mednicoff, I. B., 1953, Studies on platelets XI, antigenicity of platelets and evidence of platelet groups and types in man, *J. Lab. Clin. Med.* **42**:723–738.

Steiner, M., and Lüscher, E. F., 1982, Photoaffinity labeling of IgG receptor on human platelets (abstract), *Blood* **60**:192.

Sugiura, K., Steiner, M., and Baldini, M., 1981, Physiological effects of nonimmune platelet associated immunoglobulin G, *Thromb. Haemostasis* **45**:27–33.

Svejgaard, A., 1969, Iso-antigenic systems of human blood platelets, *Ser. Haematol.* **2**(3):5–87.

Szatkowski, N. S., and Aster, R. H., 1980, HLA antigens of platelets. IV. Influence of "private" HLA-B locus specificities on the expression of BW4 and BW6 on human platelets, *Tissue Antigens* **15**:361–368.

Trägårdh, L., Klareskog, L., Curman, B., Rask, L., and Peterson, P. A., 1979, Isolation and properties of detergent-solubilized HLA antigens obtained from platelets, *Scand. J. Immunol.* **9**:303–314.

Trent, R. J., Clancy, R. L., Davis, V., and Basten, A., 1980, Immune complexes in thrombocytopenic patients, *Br. J. Haematol.* **44**:645–654.

Vainchenker, W., Testa, U., Deschamps, J. F., Henri, A., Titeux, M., Breton-Gorius, J., Rochant, H., Lee, D., and Cartron, J. P., 1982, Clonal expression of the Tn antigen in erythroid and granulocyte colonies and its application to determination of the clonality of the human megakaryocyte colony assay, *J. Clin. Invest.* **69**:1081–1091.

Van der Weerdt, C. M., 1967, Thrombocytopenia due to quinidine or quinine: Report on a series of 28 patients, *Vox Sang.* **12**:265–272.

Van der Weerdt, C. M., van de Wiel-Dorfmeyer, H., Engelfriet, C. P., and van Loghem, J. J., 1961, A new platelet antigen, in: *Proceedings of the Eighth Congress of the European Society of Haematology*, S. Karger, Basel, p. 379.

Van der Weerdt, C. M., Veenhoven-von Riesz, L. E., Nijenhuis, L. E., and van Loghem, J. J., 1963, The Zw blood group system in platelets, *Vox Sang.* **8**:513–530.

Van de Wiel, T. W. M., van de Wiel-Dorfmeyer, H., and van Loghem, J. J., 1961, Studies on platelet antibodies in man, *Vox Sang.* **6**:641–668.

Van Leeuwen, E. F., Zonneveld, G. T. E., von Riesz, L. E., Jenkins, C. S. P., van Mourik, J. A., and von dem Borne, A. E. G. K., 1979, Absence of the complete platelet-specific alloantigens Zw(Pl[A]) on platelets in Glanzmann's thrombasthenia and the effect of anti-Zw[a] antibody on platelet function, *Thromb. Haemostasis* **42**:422a.

Van Leeuwen, E. F., von dem Borne, A. E. G. K., von Riesz, L. E., Nijenhuis, L. E., and Engelfriet, C. P., 1981, Absence of platelet-specific alloantigens in Glanzmann's thrombasthenia, *Blood* **57**:49–54.

Van Leeuwen, E. F., Engelfriet, C. P., and von dem Borne, A. E. G. K., 1982a, Studies on quinine- and quinidine-dependent antibodies against platelets and their reaction with platelets in the Bernard-Soulier syndrome, *Br. J. Haematol.* **51**:551–560.

Van Leeuwen, E. F., van der Ven, J. T. M., Engelfriet, C. P., and von dem Borne, A. E. G. K., 1982b, Specificity of autoantibodies in autoimmune thrombocytopenia, *Blood* **59**:23–26.

Van Leeuwen, E. F., Leeksma, O. C., van Mourik, J. A., Engelfriet, C. P., and von dem Borne, A. E. G. K., 1983, The effect of the binding of anti-Zw^a antibodies on platelet function, *Vox Sang.* **47**:280–289.

Van Loghem, J. J., Dorfmeijer, H., van der Hart, M., and Schreuder, F., 1959, Serological and genetical studies on a platelet antigen (Zw), *Vox Sang.* **4**:161–169.

Von dem Borne, A. E. G. K., Helmerhorst, F. M., van Leeuwen, E. F., Pegels, H. G., von Riesz, E., and Engelfriet, C. P., 1980a, Autoimmune thrombocytopenia: Detection of platelet autoantibodies with the suspension immunofluorescence test, *Br. J. Haematol.* **45**:319–327.

Von dem Borne, A. E. G. K., von Riesz, E., Verheugt, F. W. A., Ten Cate, J. W., Koppe, J. G., Engelfriet, C. P., and Nijenhuis, L. E., 1980b, Bak^a, a new platelet-specific antigen involved in neonatal allo-immune thrombocytopenia, *Vox Sang.* **39**:113–120.

Von dem Borne, A. E. G. K., van Leeuwen, E. F., von Riesz, L. E., van Boxtel, C. J., and Engelfriet, C. P., 1981, Neonatal alloimmune thrombocytopenia: Detection and characterization of the responsible antibodies by the platelet immunofluorescence test, *Blood* **57**:649–656.

Walsh, F. S., and Crumpton, M. J., 1977, Orientation of cell surface antigens in the lipid bilayer of lymphocyte plasma membrane, *Nature (London)* **269**:307–311.

Waters, A. H., Minchinton, R. M., Bell, R., Ford, J. M., and Lister, T. A., 1981, A cross-matching procedure for the selection of platelet donors for alloimmunised patients, *Br. J. Haematol.* **48**:59–68.

Wautier, J. L., Tobelem, G. M., Peltier, A. P., and Caen, J. P., 1974, Evidence for C1 on human platelets, *Haemostasis* **2**:281–286.

Weissbarth, E., Baryth, B., Mielke, H., Liman, W., and Deicher, H., 1982, Platelets as target cells in rheumatoid arthritis and systemic lupus erythematosus: A platelet specific immunoglobulin inducing the release reaction, *Rheumatol. Int.* **2**:67–73.

Woods, V. L., Oh, E. H., Mason, D., and McMillan, R., 1984, Autoantibodies against the glycoprotein IIb/IIIa complex (GPIIb/GPIIIa) in patients with chronic idiopathic thrombocytopenic purpura (ITP), *Blood* **63**:368–375.

Zucker, M. B., and Grant, R. A., 1974, Aggregation and release reaction induced in human blood platelets by zymosan, *J. Immunol.* **112**:1219–1230.

Zucker, M. B., Grant, R. A., Alper, C. A., Goodkofsky, I., and Lepow, I. H., 1974, Requirement for complement components and fibrinogen in the zymosan-induced release reaction of human blood platelets, *J. Immunol.* **113**:1744–1751.

16

Glycoprotein Defects Responsible for Abnormal Platelet Function in Inherited Platelet Disorders

Alan T. Nurden

1. INTRODUCTION

The study of congenital bleeding syndromes has yielded much information on how normal platelets function. In particular, the characterization of the molecular defects in disorders of platelet adhesion and aggregation has resulted in specific roles being proposed for different membrane glycoproteins and α-granule proteins in these processes. In this chapter, we will highlight three major inherited disorders of platelet function: Bernard-Soulier syndrome (BSS), gray platelet syndrome (GPS), and Glanzmann's thrombasthenia (GT). Each disorder will be introduced by a brief description of the platelet function abnormalities present. This will be followed by a detailed discussion of the specific molecular defects that distinguish the platelets of each syndrome. Emphasis will be placed on reviewing recent data, although an attempt will be made to interpret previous studies in the context of the new advances. Where possible, the different molecular abnormalities will be assessed in terms of how their presence leads to modifications of platelet adhesion and aggregation mechanisms in hemostasis.

2. BERNARD-SOULIER SYNDROME

Bernard-Soulier syndrome is a congenital disorder of platelet adhesion that may cause severe, even fatal, hemorrhagic disease. It is rare and has an autosomal recessive

---

Alan T. Nurden • Unité 150 INSERM, Hôpital Lariboisière, Paris, Cedex 10, France.

inheritance. Bernard-Soulier syndrome is characterized by a moderate to severe thrombocytopenia (low circulating platelet count), the presence of large, even giant platelets on peripheral blood smears, a defective prothrombin consumption test, and a markedly prolonged bleeding time. Early studies and clinical aspects of this disorder have been discussed in detail by Bellucci *et al.* (1983) and George *et al.* (1984).

2.1. Platelet Function

The primary hemostatic lesion in BSS results from a decreased ability of the platelets to attach to exposed subendothelium. Bernard-Soulier syndrome platelets also aggregate less well with thrombin.

2.1.1. Ristocetin-Induced Platelet Agglutination

Initial evidence for the presence of a specific platelet function defect came from the findings that BSS platelets were not agglutinated by bovine factor VIII preparations (Bithell *et al.*, 1972; Weiss *et al.*, 1974) or by ristocetin in the presence of normal human plasma (Howard *et al.*, 1973; Weiss *et al.*, 1974). Current opinion suggests that these processes are mediated by the binding of bovine von Willebrand factor (VWF) (Kirby, 1982) or human VWF (Kao *et al.*, 1979) to specific receptors on normal human platelets (see Chapters 10 and 11). The lack of reactivity of BSS platelets suggested that they were missing, or were unable to express, these receptors.

2.1.2. Adhesion to Subendothelium

A fundamental observation was made by Weiss *et al.* (1974), who reported that BSS platelets showed a diminished ability to adhere to subendothelium. These authors used the *in vitro* system developed by Baumgartner (1973) in which citrated human blood was passed through a flow chamber containing annular segments of rabbit aorta denuded of endothelial cells. The adhesion defect was confirmed by Caen *et al.* (1976) and was also observed when studies were performed without anticoagulant (Weiss *et al.*, 1978). In all of these reports, the primary abnormality was said to concern the initial contact or attachment of platelets to the exposed subendothelium. Where BSS platelets did attach, platelet aggregation normally followed. Although a decreased adhesion was observed at all shear rates, the abnormality was more apparent at the higher shear rates tested. In this respect, the abnormality resembled that observed in classic von Willebrand's disease (VWD) where VWF is reduced or abnormal (see Meyer and Baumgartner, 1983; Chapter 10).

2.1.3. Defective Binding of VWF

Binding studies have now established that BSS platelets fail to bind significant amounts of VWF in the presence of ristocetin (Zucker *et al.*, 1977; Moake *et al.*, 1980). The abnormality is not to be confused with ADP- or thrombin-induced VWF binding that occurs through a different mechanism and is normal in BSS (Ruggeri *et al.*, 1983).

2.1.4. Thrombin-Induced Platelet Aggregation

An apparently normal and rapid aggregation is observed when BSS platelets are stimulated with ADP, epinephrine, or collagen (Howard *et al.*, 1973; Weiss *et al.*, 1974; Walsh *et al.*, 1975). Malmsten *et al.* (1977) also reported a normal dense granule release and thromboxane B_2 synthesis in response to collagen, ADP, arachidonic acid, or prostaglandin G_2. However, BSS platelets respond less well to stimulation by thrombin. Aggregation occurs, but at a reduced rate. In addition, increased threshold amounts of thrombin are required to initiate aggregation (Jamieson and Okumura, 1978; M. Jandrot-Perrus and A. T. Nurden, unpublished observations). In a preliminary report, Drouin *et al.* (1980) also described how thrombin-induced dense granule secretion is abnormal in BSS. The above studies suggest a specific defect in the thrombin–BSS platelet interaction. A markedly decreased density in both high- and low-affinity thrombin-binding sites on the BSS platelet surface has been reported and suggested to be related to the abnormal BSS platelet response to thrombin (Jamieson and Okumura, 1978).

2.2. Surface Charge Deficiency

Grottum and Solum (1969) made a fundamental discovery when they observed that the electrophoretic mobility of the platelets of three BSS patients was reduced to between 50% and 70% of that of normal human platelets. This finding was subsequently confirmed for the platelets of other patients (Evensen *et al.*, 1974; Coller, 1978). The defect was shown to be due to a platelet surface charge deficiency caused by a decreased sialic acid density (Grottum and Solum, 1969; Evensen *et al.*, 1974; Nurden *et al.*, 1981a).

2.3. Membrane Glycoprotein Ib Deficiency

The above observations suggested the presence of membrane glycoprotein (GP) abnormalities in BSS platelets and this has been shown to be the case.

2.3.1. Initial Studies

Nurden and Caen (1975) first reported a severe reduction in the periodate-Schiff (PAS) staining of the then-termed GP I band following sodium dodecyl sulfate–polyacrylamide gel electrophoresis (SDS–PAGE) of membrane fractions isolated from the platelets of two BSS patients. This abnormality was confirmed by the direct analysis of unfractionated BSS platelets by SDS–PAGE (Caen *et al.*, 1976) and independently described by Jenkins *et al.* (1976). These and other early studies reporting a ''GP I'' abnormality in BSS platelets were reviewed by Nurden and Caen (1979).

2.3.2. SDS–PAGE Followed by PAS Staining

As defined by Phillips and Poh Agin (1977a), GP Ib is a high-molecular-weight membrane glycoprotein that is composed of two subunits. Cleavage of intermolecular disulfides results in the separation of a small (*Mr* 22,000) β-subunit from a large M_r

148,000) α-subunit. Clemetson *et al.* (1981a) have summarized recent data and proposed a model for GP Ib in which both subunits are attached to the lipid bilayer with the disulfides linking the α- and β-subunits exposed at the external surface of the membrane (see also Chapter 3). The development of improved electrophoresis procedures allowed the separation of GP Ib from other closely migrating PAS-staining membrane glycoproteins. As illustrated by the "control" tracing in Figure 1, GP Ib

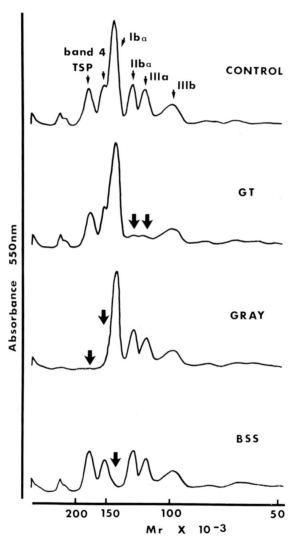

Figure 1. Initial screening of platelets for glycoprotein defects using single-dimension SDS–PAGE followed by PAS staining. Platelets isolated from a normal donor and from typical patients with Glanzmann's thrombasthenia (GT), gray platelet syndrome (gray), and the Bernard-Soulier syndrome (BSS) were solubilized with SDS, and after disulfide reduction, analyzed by SDS–PAGE on 6% acrylamide rod gels according to the procedures of Nurden *et al.* (1981a). The heavy arrows highlight the distinct and characteristic glycoprotein abnormalities present in the platelets of patients of each disorder.

stains intensely in the PAS reaction. Here, reduced samples have been analyzed and the GP Ib α-subunit predominates on the normal platelet glycoprotein pattern. A severe and specific reduction in the PAS stain of GP Ibα is apparent on analysis of BSS platelets. Other membrane glycoproteins were normally located as was the closely migrating "band 4" peak (discussed in Section 3.2.). Identical results to those illustrated in Figure 1 were reported for the platelets of four BSS patients by Nurden *et al.* (1981a). Analysis of nonreduced samples results in an even better separation of GP Ib from other membrane glycoproteins (Nurden *et al.*, 1981a) and is particularly recommended for a preliminary screening of a possible BSS platelet defect (illustrated by [125]I- and ^{3}H-radiolabeling procedures in Sections 2.4 and 2.7).

2.3.3. Glycocalicin

Early reports emphasized an apparent absence of glycocalicin (also termed GP Is or GP-S) from BSS platelet lysates (Caen *et al.*, 1976; Solum *et al.*, 1977; Jamieson *et al.*, 1979). It is now known that glycocalicin is derived from GP Ibα through the action of an endogenous platelet Ca^{2+}-activated protease (see Chapters 3 and 5). As the absence of glycocalicin from BSS platelet lysates is now recognized to be a consequence of the GP Ib deficiency, it will not be described further in this review.

2.4. ^{125}I-Labeling of the Bernard-Soulier Platelet Surface

Jenkins *et al.* (1976) described a highly abnormal ^{125}I-labeling pattern of the surface proteins of platelets isolated from a BSS patient. Using lactoperoxidase-catalyzed iodination followed by SDS–PAGE of the ^{125}I-labeled platelets, these authors located no radioactivity in the positions of "GP I" and GP IIb, while numerous low-molecular-weight peptides showed increased amounts of labeling. Subsequently, Hagen and Solum (1978) failed to locate differences when ^{125}I-labeled BSS platelets were analyzed and compared with ^{125}I-labeled normal platelets using two different SDS–PAGE procedures. Nurden *et al.* (1981a) clarified the situation when they studied the platelets of three BSS patients using the lactoperoxidase-catalyzed ^{125}I-labeling procedure. Here, the radiolabeled proteins were detected by autoradiography. This allowed the analysis of the less heavily labeled glycoproteins, which includes GP Ib when using lactoperoxidase-catalyzed iodination. Identical results were obtained for each patient. A typical result obtained using single-dimension SDS–PAGE and showing a specific GP Ib defect is to be seen in Figure 2. The advantage of nonreduced samples for investigating GP Ib is well illustrated in this figure. An even better resolution was obtained when the two-dimensional SDS–PAGE procedure of Phillips and Poh Agin (1977a) was used. Here, first-dimension electrophoresis was performed using nonreduced samples. Disulfide reduction was then performed in the gel prior to the second-dimension electrophoresis. A limited mapping of the radiolabeled surface proteins was achieved (Figure 3). An absence of ^{125}I-labeling in the position of GP Ibα is clearly revealed. Coomassie Blue staining also showed the apparent lack of GP Ibβ (Nurden *et al.*, 1981a). The results shown in Figure 2 and 3 emphasize that the bulk of the major membrane glycoproteins (other than Ib) are present in the BSS platelet surface membrane and are accessible for ^{125}I-labeling. The different result of Jenkins *et al.* (1976) will be discussed in Section 2.10.

NONREDUCED   REDUCED

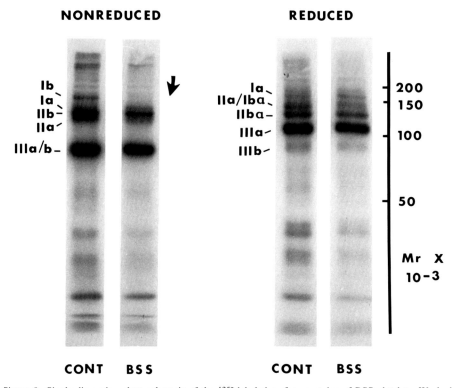

CONT   BSS   CONT   BSS

Figure 2. Single-dimension electrophoresis of the ^{125}I-labeled surface proteins of BSS platelets. Washed suspensions of normal human platelets and those isolated from a BSS patient were incubated in the presence of ^{125}I and the surface proteins labeled by the lactoperoxidase-catalyzed method. SDS-soluble samples were analyzed by SDS–PAGE on 7–12% gradient acrylamide slab gels both in the absence of (nonreduced) or following (reduced) disulfide reduction with 2-mercaptoethanol. The ^{125}I-labeled proteins were detected on dried gels by autoradiography. The absence of ^{125}I-labeling from the GP Ib position on the BSS profile is most clearly apparent following electrophoresis of nonreduced samples (heavy arrow). Full details of the methods employed are given by Nurden et al. (1981a) from where this figure is published by copyright permission of the American Society for Clinical Investigation.

2.5. Crossed Immunoelectrophoresis

Immunochemical evidence for the possible deletion of GP Ib from BSS platelets was first obtained by Hagen et al. (1980), who analyzed Triton X-100 extracts of the platelets of two patients by crossed immunoelectrophoresis (CIE). Using a rabbit antibody raised against normal human platelets, these authors observed that a specific immunoprecipitate, termed protein 13, was missing from the Coomassie Blue-stained pattern obtained for BSS platelets. Furthermore, when antiglycocalicin serum was incorporated into the intermediate gel prior to the second-dimension electrophoresis of BSS platelet proteins, no specific precipitation of glycocalicin or structurally related proteins was observed. Kunicki et al. (1981b) also failed to locate this precipitate on analysis of ^{125}I-labeled BSS platelets and confirmed that protein 13 was indeed the immunoprecipitate given by GP Ib. The analysis of Triton X-100-soluble proteins of

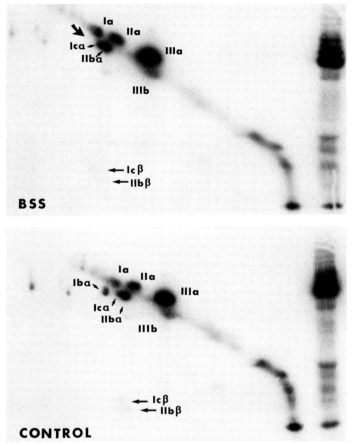

Figure 3. Two-dimensional SDS–PAGE analysis of the ^{125}I-labeled surface proteins of BSS platelets. First-dimension electrophoresis of nonreduced samples (100 μg protein) was performed using 7% acrylamide rod gels. Unstained gels were then incubated with 2-mercaptoethanol prior to a second dimension electrophoresis of the now reduced proteins on 7–12% gradient acrylamide slab gels. In addition, a single sample of reduced ^{125}I-labeled proteins (100 μg) of the sample being analyzed was added to the right-hand side of the slab gel. ^{125}I-labeled proteins were detected on dried Coomassie Blue-stained gels by autoradiography. Typical autoradiographs are illustrated. The absence of ^{125}I-labeling from the position of GP Ibα on the BSS profile is highlighted (heavy arrow). Full details of the methods employed are given by Nurden *et al.* (1981a) from where this figure is published by copyright permission of the American Society for Clinical Investigation.

BSS platelets by CIE is illustrated in Figure 4. The precipitates given by GP Ib and gly-cocalicin (difficult to distinguish on the figure) are absent from the Coomassie Blue-stained BSS platelet pattern. A faint precipitate usually masked by that containing GP Ib on the control platelet profile was normally present. Some of the precipitates given by α-granule proteins were augmented on the BSS platelet pattern. This might be a consequence of the increased size and protein content of the platelets (Nurden *et al.*, 1981a). Bernard-Soulier syndrome platelets have been previously shown to have an increased number of dense granules (Rendu *et al.*, 1981). The content of VWF in BSS platelets is also increased (Howard *et al.*, 1974). The abnormal position of the precipi-

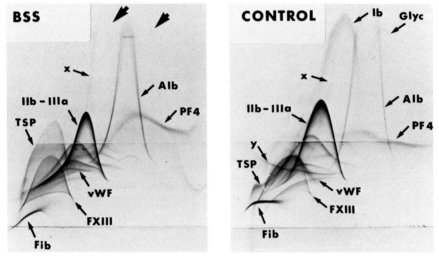

Figure 4. Analysis of Triton X-100 soluble proteins of normal human and BSS platelets by CIE. Samples (100 μg protein) were electrophoresed in the second-dimension agarose gel against a rabbit anti-human platelet antibody preparation. Precipitates were located by Coomassie Blue staining. Full details of the methods used are given by Hagen *et al.* (1980) and by Kunicki *et al.* (1981b). To be noted are the absence from the BSS pattern of the precipitates given by GP Ib and glycocalicin (Glyc). X = a precipitate migrating close to but separate from that containing GP Ib on the control platelet pattern, Y = an artifactual precipitate. The identification of the precipitates containing α-granule proteins was described by Nurden *et al.* (1982a).

tate containing VWF on the BSS platelet pattern could signify an altered multimer distribution. Full details of the CIE procedure, which is performed in agarose gels containing Triton X-100, are to be found in Chapters 4 and 5.

2.6. Use of High-Resolution Two-Dimensional Polyacrylamide Gel Electrophoresis

As described in detail by Clemetson in Chapter 3, separation of many of the minor (in terms of their ease of detection) membrane glycoproteins requires high-resolution PAGE, such as in achieved using the two-dimensional system developed by O'Farrell (1975). Here, isoelectric focusing (IF) is used in the first dimension; thus charge differences caused by major changes in glycosylation of the glycoproteins can also be detected. Nibu *et al.* (1982) studied platelets of four BSS patients using IF/SDS–PAGE and visualized the glycoproteins by PAS staining. These authors failed to detect GP Ib on analysis of platelets of each patient, whereas glycoproteins Ig (TSP), Ia, IIb, III (IIIa), and IV (IIIb) were all clearly located and possessed a normal isoelectric point (pI) and molecular weight. Peterson *et al.* (1982) reported a similar study. Again, a specific absence of GP Ib from the platelets of four BSS patients was described. Other glycoproteins that were normally located were identified as Ia, Ic, Id, IIa, IIb, III (IIIa), and IV (IIIb). Thus both of the above studies emphasized the presence of a distinct molecular defect in BSS platelets involving GP Ib.

2.7. ³H-Labeling of the Bernard-Soulier Platelet Surface

A major advance in our understanding of the nature of the BSS platelet lesion came from the application of ³H-labeling procedures to the study of their surface glycoproteins. The methods that have been used involve either (1) periodate oxidation followed by sodium [³H]borohydride reduction or (2) sequential treatment of platelets with neuraminidase, galactose oxidase, and sodium [³H]borohydride. In the former procedure, ³H is incorporated primarily into terminal sialic acid residues; in the latter method, ³H is incorporated into galactose or *N*-acetylgalactosamine moieties exposed following sialic acid removal. As detailed in Chapter 3, analysis of ³H-labeled platelets

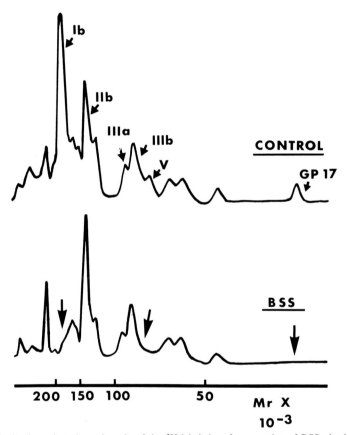

Figure 5. Single-dimension elecrophoresis of the ³H-labeled surface proteins of BSS platelets. Washed suspensions of normal human and BSS platelets were sequentially incubated with neuraminidase, galactose oxidase, and sodium [³H]borohydride according to the procedures of Nurden *et al.* (1982b). Samples (100 μg protein) of the SDS-solubilized platelets were analyzed without disulfide reduction on 7–12% gradient acrylamide slab gels. ³H-labeled proteins were detected in Coomassie Blue-stained gels by fluorography (Nurden *et al.,* 1982b). Densitometric scans of typical fluorographs are illustrated. The absence of ³H labeling from the positions of GP Ib, GP V, and a M_r 17,000 glycoprotein (GP 17) on the BSS profile is highlighted (heavy arrows). Published from Caen and Nurden (1983) with copyright permission.

by SDS–PAGE permits the detection of membrane glycoproteins not located by PAS staining or ^{125}I-labeling procedures. It was of interest, therefore, when two preliminary reports suggested additional glycoprotein abnormalities in BSS platelets analyzed in this way (Clemetson *et al.*, 1981b; Nurden and Dupuis, 1981). These studies suggested abnormalities of GP V and a M_r 17,000 glycoprotein (GP$_{17}$) in addition to a GP Ib deficiency. Figure 5 illustrates the analysis of BSS platelets treated with neuraminidase, galactose oxidase, and sodium [^{3}H]borohydride. Samples have been analyzed by single-dimension SDS–PAGE in the absence of disulfide reduction. A severe diminution or absence of labeling in the positions normally occupied by GP Ib, GP V, and GP$_{17}$ is to be observed on the BSS platelet pattern. The analysis of ^{3}H-labeled BSS platelets by single-dimension or two-dimensional electrophoresis has now been the subject of several reports. In an elegant study, Clemetson *et al.* (1982) used a high-resolution IF/SDS–PAGE procedure to separate the labeled glycoproteins. Both periodate and neuraminidase plus galactose oxidase-treated platelets were analyzed. Severe deficiencies of GP Ibα, GP Ibβ, GP V, and GP$_{17}$ were observed for the platelets of two BSS patients. Platelets of a third patient possessed intermediate levels of these glycoproteins (see Section 2.9). A feature of this study was again the normal pI of the remaining major glycoproteins of BSS platelets. Berndt *et al.* (1983) used two-dimensional nonreduced/reduced SDS–PAGE to analyze platelets labeled using the periodate and sodium [^{3}H]borohydride procedure. Three BSS patients were studied. The platelets of two patients had no detectable GP Ibα, GP Ibβ, GP V, and GP$_{17}$ (termed GP IX and given a M_r of 22,000 by these authors). Furthermore, the apparent absence of an additional minor glycoprotein of M_r 100,000 was noted.

2.8. Glycoprotein V Deficiency

As described by Berndt and Phillips (1981) (see also Chapter 7), GP V is a substrate for thrombin on the platelet surface and its hydrolysis is accompanied by the release of a M_r 69,500 hydrolytic fragment (GP V$_{f1}$) into the supernatant. Figure 6 illustrates the incubation of ^{3}H-labeled normal human and BSS platelets with human α-thrombin. No hydrolytic fragments of GP V were located in the supernant after incubation of the BSS platelets with the enzyme. The lack of release of GP V$_{f1}$ is further evidence for a structural defect or deficiency of GP V in the platelets of this patient. During the incubation of the control platelets at 37 °C, some hydrolysis of GP Ib and glycocalicin release occurred. This hydrolysis was also seen without added thrombin (not illustrated). No glycocalicin release was observed from the BSS platelets in the absence of GP Ib.

2.9. Heterogeneity within Bernard-Soulier Syndrome

The platelets from eight BSS patients have now been examined in the authors laboratory. A simple screening test involving PAS staining following SDS–PAGE revealed the apparent absence of GP Ib from the platelets of seven of these patients. Both PAS staining and ^{3}H-labeling procedures revealed low, but detectable, amounts of GP Ib in the platelets of one patient (A.C.) (Nurden *et al.*, 1983a). Densitometric scanning suggested that GP Ib was reduced to about 10% of the normal platelet

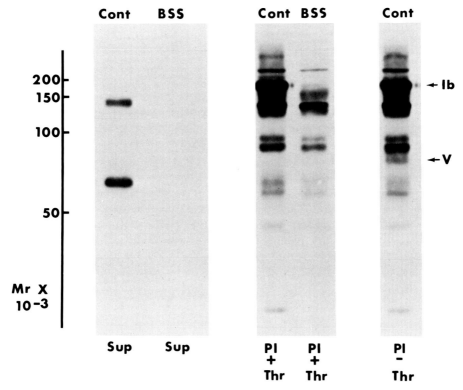

Figure 6. Incubation of thrombin with ³H-labeled BSS platelet membrane glycoproteins. Washed suspensions of normal human and BSS platelets were incubated with neuraminidase, galactose oxidase, and sodium [³H]borohydride prior to their resuspension at 1×10^9 platelets/ml (Cont) or 5×10^8 platelets/ml (BSS). The lower concentration of BSS platelets was chosen to compensate for their increased protein content and surface area (as discussed by Nurden *et al.*, 1981a). Human α-thrombin (0.5 NIH U/ml) was added and the suspensions incubated for 5 min at 37 °C. Samples of the supernatants (80 μl) and of platelets (100 μg protein) incubated with (+) or without (−) thrombin (Thr) were analyzed by SDS–PAGE on 7–12% gradient acrylamide slab gels in the absence of disulfide reduction. ³H-labeled glycoproteins were detected by fluorography. Analysis of the supernatant obtained from thrombin-treated normal platelets revealed the hydrolytic fragment of GP V, GP V_{f1} (lower band), and glycocalicin (upper band). Neither band was detected in the supernatant obtained from thrombin-treated BSS platelets.

concentration. Severe deficiencies of GP V and GP_{17} were also observed, but not quantified (Nurden *et al.*, 1983a). This patient resembles patient (P.R.) described by Berndt *et al.* (1983), where approximately 7% of the normal platelet level of GP Ibα and severe deficiencies of GP V and GP_{17} were detected. In contrast, Clemetson *et al.* (1982) described a patient (E.H.) with an approximate 40% level of platelet GP Ib. Significantly, GP V and GP_{17} were similarly decreased. The platelets of this patient showed a much decreased agglutination response with ristocetin suggesting that the GP Ib was functionally defective. Heterozygotes for BSS with 50% levels of GP Ib, as detected by PAS staining, react normally with ristocetin (George *et al.*, 1981). Another unusual case (M.S.) was reported by McGregor *et al.* (1980) who described a decreased ^{3}H-labeling of both GP Ia and Ib. Since other studies have failed to show a GP Ia defect in BSS platelets, it would appear that patient (M.S.) may represent yet another variant of this disorder.

2.10.   Problems of Analyzing Bernard-Soulier Platelet Glycoproteins

A distinct characteristic of BSS platelets is their increased size and protein content (McGill *et al.*, 1980; Nurden *et al.*, 1981a; Rendu *et al.*, 1981). This leads to problems in separating the platelets from erythrocytes and white cells. Special procedures are recommended for their isolation (see Solum *et al.*, 1977; Nurden *et al.*, 1981a). The presence of large numbers of white cells in isolated BSS platelet suspensions may result in proteolytic modifications of the platelet glycoproteins. Shulman *et al.* (1983) have recently shown that elastase-like enzymes in granulocyte membranes are able to degrade GP IIb and GP IIIa. These authors concluded that such a degradation was the cause of their initial report highlighting the absence of GP IIb-IIIa from BSS platelet membranes (Shulman and Karpatkin, 1980). Similar modifications could also account for the observations of Jenkins *et al.* (1976), who located an increased number of ^{125}I-labeled low-molecular-weight peptides on analysis of ^{125}I-labeled BSS platelets by SDS–PAGE.

2.11.   Monoclonal Antibodies

Diagnosis of BSS is now rendered simpler by the availability of monoclonal antibodies specific for GP Ib. The first such antibody to be described, AN51, bound to 1.6×10^4 sites per normal human platelet, but failed to bind to platelets of three BSS patients (McMichael *et al.*, 1981). The antibody reacted with GP Ib and glycocalicin in indirect immunoprecipitation procedures, therefore its epitope was presumed to be carried by the GP Ibα chain (Ruan *et al.*, 1981). AP-1, 6D1, and FMC 25 are other monoclonal antibodies directed against determinants carried by GP Ib (Montgomery *et al.*, 1983; Coller *et al.*, 1983a; Berndt *et al.*, 1983). None of these antibodies bound to BSS platelets, either by immunofluorescence or in binding assays using radiolabeled antibody. For example, AP-1 binding to the platelets of three BSS patients was assessed as being less than 2% by Montgomery *et al.* (1983). These authors employed a rapid whole blood assay requiring as little as 0.2 ml blood, thus emphasizing the potential of such procedures.

2.12.   Basic Genetic Lesion in Bernard-Soulier Syndrome

In view of the presence of multiple defects the question arises as to the nature of the primary genetic abnormality in BSS. An obvious possibility is that the deficiencies are due to a defect in a step common to the biosynthesis of each of the affected glycoproteins or to an abnormality of a common precursor. However, there is no evidence to suggest that GP Ib, GP V, and GP$_{17}$ are structurally related glycoproteins and at least GP Ib and GP V have been shown to have distinct structural characteristics (see Chapters 3 and 7). A common abnormality of glycosylation cannot be discounted, although BSS platelets normally contain sialyltransferase and galactosyltransferase activities (Bauvois *et al.*, 1981; Cartron and Nurden, 1979). Although there is no evidence for an unusual protease activity in BSS blood (Clemetson *et al.*, 1982), the possibility that the deficiencies arise through a degradation of normally synthesized components must also be considered. The involvement of a posttranslational event concerning a regulator gene or a degradative process could also explain the presence of small amounts of GP Ib in the platelets of some patients. A fundamental advance to our understanding of the basic lesion in BSS may date from the observations of Coller (1983a) and Berndt *et al.* (1983), who showed that the monoclonal antibodies 6D1, FMC 25, and AN51 immunoprecipitated both GP Ib and GP_{17} from detergent extracts of ^{3}H-labeled normal platelets. These results suggest that GP Ib and GP_{17} may exist as a complex in the platelet plasma membrane. The presence in platelets of heterozygotes of intermediate levels of GP Ib (George *et al.*, 1981) and of GP Ib, GP V, and GP_{17} (Berndt *et al.*, 1983) suggests a link between the glycoprotein deficiencies and the inheritance of BSS. Intermediate levels of GP Ib, GP V, and GP_{17} were also present in the variant described by Clemetson *et al.* (1982). Thus the genetic lesion appears to affect these components equally. One possibility is that the primary lesion affects the biosynthesis of GP Ib or GP_{17}, and that in the absence (or reduced concentration) of one component, the other glycoprotein is unable to form a complex and is not incorporated into the membrane. Glycoprotein V, as a peripheral and loosely attached surface glycoprotein (see Chapter 3), could depend on the presence of GP Ib and/or GP_{17} for its attachment to the membrane. Further studies, probably on megakaryocytes, will be required to prove or disprove these theories.

2.13. Conclusions on Bernard-Soulier Syndrome

Bernard-Soulier syndrome platelets are characterized by their inability to adhere to exposed subendothelial tissue as a result of the absence or structural modification of a specific population of receptors for VWF. Evidence establishing a direct role for GP Ib in the binding of VWF to human platelets is detailed in Chapters 10 and 11 and will not be discussed further here. Bernard-Soulier syndrome is important in studies of platelet function for it represents the only example of an inherited platelet disorder with a membrane defect causing abnormal platelet adhesion. It should be compared with classic VWD where VWF, the plasma cofactor of the initial platelet attachment to the vessel wall, is affected (reviewed by George *et al.*, 1984). Furthermore, BSS contrasts strongly with pseudo-VWD where high-molecular-weight VWF multitimers have an increased reactivity with the platelet surface (see George *et al.*, 1984; and Chapter 10).

Other defects in the interaction of ligands with BSS platelets are more difficult to assess. Both GP Ib and GP V have been proposed as initial sites for thrombin interaction with the normal platelet membrane (see Chapter 7). Their absence from BSS platelets correlates with a decreased aggregation response of the abnormal platelets with thrombin. This would be compatible with one or other (or both) of these components playing a role in thrombin-induced platelet aggregation. However, with increased amounts of thrombin, some platelet aggregation does occur in BSS (Jamieson and Okumura, 1978). At least one pathway for platelet activation by thrombin independent of GP Ib and V would therefore appear to be present. Further studies are required to define the interaction between thrombin and the surface components of normal and

BSS platelets. Also to be explained is the defective prothrombin consumption test (Bellucci *et al.*, 1983; George *et al.*, 1984). A decreased binding of coagulation factor XI to BSS platelets has been noted (Walsh *et al.*, 1975) and may be related to this functional defect.

The deletion of GP Ib, and possibly also of GP V and GP_{17}, has consequences for the immunology of BSS platelets. Glycoprotein Ib has been implicated in the expression of drug-dependent antibody receptors by normal platelets and these receptors are lacking or are abnormal in BSS (discussed in Chapter 15). In the absence of GP Ib, there is always the risk that patients receiving multiple transfusions to arrest bleeding episodes may develop an immune response and form antibodies against the additional antigens present on the transfused platelets. The IgG P is such an antibody discovered in the serum of a BSS patient (Tobelem *et al.*, 1976). The antibody inhibited the adhesion of normal human platelets to subendothelium and generally induced a BSS platelet functional activity on normal human platelets (Tobelem *et al.*, 1976). Indirect-immunoprecipitation tests confirmed that the antibody was directed against GP Ib (Degos *et al.*, 1977). Studies with this antibody provided the first direct correlation between the molecular defects of the BSS platelet surface and the altered platelet function in this disorder.

3. GRAY PLATELET SYNDROME

Whereas BSS is a disorder of platelet adhesion, the gray platelet syndrome (GPS) is essentially a disorder of platelet secretion. Gray platelet syndrome is a special example of what are more generally described as ''storage pool diseases'' (Holmsen and Weiss, 1979; Bellucci *et al.*, 1983) where secretion of the contents of dense granules and/or α-granules is defective. Evaluation of the ultrastructure of gray platelets by electron microscopy combined with cytochemical procedures showed that they specifically lack α-granules (White, 1979; Gerrard *et al.*, 1980; Levy-Toledano *et al.*, 1981). As a consequence, gray platelets are an important model for understanding the role of secreted α-granule proteins in platelet function. Gray platelet syndrome is a rare congenital disorder, although five such patients have now been studied in the author's laboratory. A characteristic is the typical lack of cytoplasmic granulation on stained blood smears. This abnormality gives the platelet a gray appearance. Early studies and clinical aspects of this disorder have been reviewed by Bellucci *et al.* (1983) and George *et al.* (1984).

3.1. Platelet Function

Platelet adhesion to collagen fibrils was found to be quantitatively normal for the platelets of two patients (Levy-Toledano *et al.*, 1981). Platelet aggregation in GPS has been tested by two groups of investigators. Gerrard *et al.* (1980) found deficiencies in ADP-, collagen-, and thrombin-induced aggregation. However, no consistent abnormalities were observed for platelet aggregation induced by arachidonic acid or epinephrine, or for ristocetin-induced platelet agglutination. Levy-Toledano *et al.* (1981) reported a normal aggregation of gray platelets with ADP, arachidonic acid, and

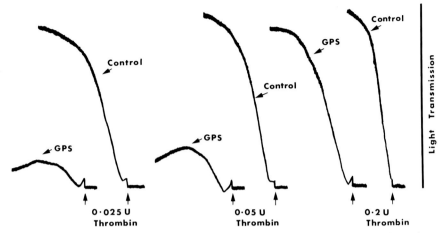

Figure 7. Abnormal aggregation response of gray platelets to α-thrombin. Washed suspensions of normal human platelets and those isolated from a GPS patient were prepared according to the method of Lee *et al.* (1981). Platelet aggregation was studied by standard procedures by recording light transmission changes in a stirred platelet suspension (3 × 10⁸ platelets/ml) incubated at 37 °C in a platelet aggregometer. Note the reduced response of gray platelets to all but the highest dose of thrombin.

ionophore A23187, but obtained little or no aggregation with collagen, while thrombin-induced platelet aggregation was also severely abnormal. An important additional observation from this study was a much decreased dense granule secretion induced by collagen and thrombin (Levy-Toledano *et al.*, 1981). The abnormal aggregation response of gray platelets to thrombin is illustrated in Figure 7.

3.2. Protein and Glycoprotein Deficiencies

The major surface-membrane glycoproteins of gray platelets were typically present when located by PAS staining following SDS–PAGE and were normally labeled with ^{125}I during lactoperoxidase-catalyzed iodination (Gerrard *et al.*, 1980; Nurden *et al.*, 1982a). However, strikingly absent from the PAS-stained glycoprotein profile was the band given by thrombospondin (TSP) (also termed GP-G or GP Ig). The analysis of gray platelet glycoproteins is illustrated for reduced samples in Figure 1. Nurden *et al.* (1982a) also described decreases in minor high-molecular-weight bands and the absence of a peak migrating close to and just behind GP Ibα after reduction. Termed band 4, the components of this peak were not identified, but may be of intracellular origin since no differences were to be observed in the same region of the gel following SDS–PAGE of surface-labeled (^3H)gray and control platelets (A. T. Nurden, unpublished observation).

The protein composition of gray platelets has been analyzed by both single and two-dimensional SDS–PAGE (Gerrard *et al.*, 1980; Nurden *et al.*, 1982a). Figure 8 illustrates the severely reduced levels of TSP and fibrinogen in the platelets of a typical GPS patient. At the same time, GP IIb and GP IIIa are confirmed as being normally present. Analysis of the gray platelet polypeptides on 7–20% gradient acrylamide gels allowed the additional detection of deficiencies of β-thromboglobulin (β-TG) and

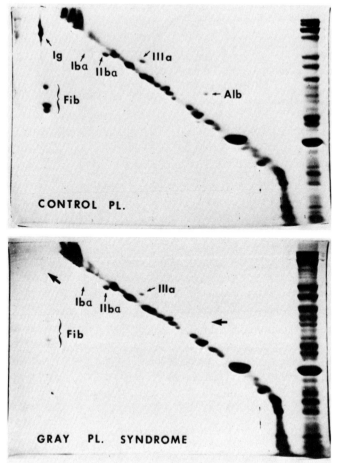

Figure 8. Major α-granule protein deficiencies of gray platelets. Two-dimensional nonreduced/reduced SDS–PAGE was performed according to the procedures described in the legend to Figure 3. Samples (100 μg protein) of SDS-solubilized normal human platelets and those of a GPS donor were analyzed. Proteins were detected by Coomassie Blue staining. Severe deficiencies of TSP (GP Ig), fibrinogen, and albumin in gray platelets were shown by this procedure. Published from Nurden *et al.* (1982a) with copyright permission.

platelet factor 4 (PF4) (not illustrated). All of the above abnormalities were confirmed on examination of the supernatant fractions obtained following the incubation of washed gray platelets with thrombin (Gerrard *et al.*, 1980; Nurden *et al.*, 1982a). In fact, the SDS–PAGE polypeptide profiles of gray platelets resembled greatly those of thrombin-treated control platelets. Further studies performed by CIE using a poly-specific rabbit anti-human platelet antibody or by rocket immunoelectrophoresis using monospecific antisera showed that a moderate deficiency of albumin (50% of normal platelet levels) and severe deficiencies of VWF and fibronectin were also present (Nurden *et al.*, 1982a). Finally, quantitation using a biological assay also revealed a severe reduction in the platelet levels of platelet-derived growth factor (Gerrard *et al.*, 1980).

3.3. Abnormalities in Megakaryocytes

Ultrastructural studies suggest a basic abnormality of α-granule formation in GPS megakaryocytes (White, 1979; Breton-Gorius *et al.*, 1981). Although megakaryocytes have been shown to synthesize VWF (Nachman *et al.*, 1977) and PF4 (Ryo *et al.*, 1983), direct proof for their synthesis of the other α-granule proteins has yet to be obtained. These proteins are presumed either to be synthesized by megakaryocytes or to be taken up from the environment before or after megakaryocyte maturation and platelet release. The presence of elevated amounts of β-TG and PF4 in the plasma of GPS patients (Levy-Toledano *et al.*, 1981) would be compatible with either an early release from the marrow or an abnormal uptake. The assumption that synthesis of α-granule proteins occurs in the megakaryocyte leads to the question as to why all of the proteins are deficient in gray platelets. Electron microscope studies on the mega-karyocytes of two patients led Breton-Gorius *et al.* (1981) to suggest that predestined α-granule proteins were extended into the marrow prior to megakaryocyte maturation. Thus the primary lesion in GPS could relate to the packaging of normally synthesized α-granule proteins and may even involve an α-granule membrane defect. The abnor-mal secretion from the megakaryocytes of the α-granule proteins, including platelet-derived growth factor, may be related to the development of a myelofibrosis in the marrow of some patients (Breton-Gorius *et al.*, 1981; Levy-Toledano *et al.*, 1981).

3.4. Significance of the Protein and Glycoprotein Deficiencies

The GPS is a model for studying the importance of secreted α-granule proteins in platelet function. Recent studies have shown that secreted fibronectin and TSP proba-bly play a role during platelet adhesion to collagen (Lahav *et al.*, 1982). Although the initial attachment of gray platelets to fibrillar collagen is quantitatively normal (Levy-Toledano *et al.*, 1981), little is known about the reactions of the gray platelets subse-quent to the adhesion (e.g., platelet spreading, thrombus formation). Unfortunately, platelet interaction with the vessel wall in a flowing blood system has yet to be investigated in GPS. Thus in several respects the GPS still represents a new model for future investigations.

Although studies remain at a preliminary stage, it appears that secretion-depen-dent platelet aggregation in GPS is highly modified. Gartner (see Chapter 12) has proposed that thrombin-induced platelet aggregation is mediated, at least in part, by an agglutinin secreted from the α-granules. This agglutinin, which was lacking from gray platelets (Gartner *et al.*, 1981), was subsequently shown to be TSP (Jaffe *et al.*, 1982). Thus, the abnormal aggregation response of gray platelets to thrombin may be due to their severely decreased TSP content. The situation is not entirely clear cut, however, for as illustrated in Figure 7, some aggregation of gray platelets can occur in a washed platelet suspension on addition of high doses of thrombin. Does this represent aggrega-tion that is mediated by trace amounts of secreted α-granule proteins or does it repre-sent a secretion-independent mechanism? Furthermore, preliminary experiments in the author's laboratory suggest that addition of fibrinogen alone may correct, at least in part, the abnormal aggregation response of gray platelets to low-dose thrombin (A. T. Nurden and M. Hasitz, unpublished observations). Finally, it should be recalled that not only aggregation but also dense body secretion may be abnormal in GPS. Binding of proteins such as TSP or fibrinogen to their memrane receptors may have a role in signal generation or amplification as part of the overall process of platelet activation.

4. GLANZMANN'S THROMBASTHENIA

Glanzmann's thrombasthenia (GT) is a bleeding disorder whose chief diagnostic feature is an absence of macroscopic platelet aggregation in response to all physiologic stimuli. Patients with GT have a notable hemorrhagic tendency and *in vitro* tests show a prolonged bleeding time, although the platelet count is nearly always normal. Although rare, 42 patients have now been examined in the author's laboratory. Although the bulk of the evidence suggests an autosomal recessive inheritance, recent studies point to GT being a heterogeneous disorder. The first comprehensive report on this disease was made by Caen *et al.* (1966), who assessed clinical data and the results of platelet function testing on samples taken from 15 patients. More recent reviews include those by Bellucci *et al.* (1983) and George *et al.* (1984).

4.1. Platelet Function

The primary hemostatic lesion in GT results from an inability of the platelets to adhere one with another after activation with the result that thrombus formation *in vivo* does not occur. In the majority of patients, platelets are also unable to support clot retraction.

4.1.1. Platelet Adhesion to Subendothelium

Thrombasthenic platelets adhere normally to collagen fibers (Caen *et al.,* 1966; Zucker *et al.,* 1966) and, in contrast to BSS, the initial platelet attachment to exposed subendothelial tissue occurs normally in flowing blood (Tschopp *et al.,* 1975; Baumgartner *et al.,* 1977). Morphological examinations performed in the above studies suggested that both platelet spreading on the subendothelial surface and α-granule release had usually occurred. A striking feature, however, was the complete absence of thrombus formation. Although not a physiologic test system, an abnormal surface reactivity of GT platelets is shown by their inability to adhere to glass. Recently, Rosenstein *et al.* (1981) reinvestigated this phenomenon and showed that the abnormal platelets exhibited both an initial reluctance to attach to the glass surface, and, when adherence did occur, the platelets showed a reduced ability to spread.

4.1.2. Platelet Aggregation

When tested in a platelet aggregometer, no aggregation of GT platelets is observed with physiologic stimuli such as ADP, thrombin, epinephrine, or collagen (Caen *et al.,* 1966). Although significant changes in light transmission do not occur, Caen *et al.* (1966) visually noted the formation of small clusters or microaggregates in the platelet-rich plasma of some patients following addition of ADP or collagen. This finding is recalled for Heptinstall *et al.* (1983) have noted in a preliminary communication that the percentage of single platelets fell when aggregation-inducing agents were added to citrated GT blood. Despite such findings, all studies point to a defective mechanism of platelet aggregation in thrombasthenia as it relates to thrombus formation and primary hemostasis.

Thrombasthenic platelets normally express primary receptors for ADP and thrombin, and binding of these agonists follows the normal kinetic parameters. The initial stimulus–receptor interaction is followed by pseudopod formation and a usual shape change (see Nurden and Caen, 1979). Primary receptor occupancy may also lead to a normal contractile wave and centralization of granules (Gerrard and White, 1976), therefore signal transmission may be normally accomplished. Secretion from both dense granules and α-granules can occur normally in the presence of high doses of thrombin or ionophore A23187 (Levy-Toledano et al., 1979; Gartner et al., 1981). However, in citrated platelet-rich plasma, both ADP and collagen-induced dense granule secretion were significantly low, as were the amounts of thromboxane B_2 formed (Malmsten et al., 1977). This result may be secondary to the aggregation abnormality and suggests that secretion induced by some agonists is dependent on platelet surface contact interactions.

4.1.3. Clot Retraction

Early studies established that clot retraction was either absent or reduced in thrombasthenia (Caen et al., 1966; Caen, 1972). The demonstration that addition of Mg^{2+} partially corrected the clot retraction defect, but did not restore the ability of GT platelets to aggregate, distinguished the abnormalities. It is important to note that a deficient fibrin clot retraction by fibroblasts has also been noted in thrombasthenia (Donati et al., 1977). This suggests that cell defects in thrombasthenia may not be confined to platelets. The recent observation by Rosenstein et al. (1981) that GT platelets fail to bind to fibrin strands is confirmation of the involvement of a platelet surface abnormality in the clot retraction defect.

4.2. Initial Studies Reporting Deficiencies of Glycoprotein IIb and Glycoprotein IIIa

The first demonstration of a membrane abnormality in GT platelets came from Nurden and Caen (1974), who described an altered glycoprotein profile when membrane fractions isolated from the platelets of three patients were analyzed by SDS–PAGE. The presence of molecular differences in thrombasthenic platelet membranes was confirmed by Phillips et al. (1975). These and other early studies have been reviewed by Nurden and Caen (1979). It gradually became clear that a decreased PAS staining of two membrane glycoproteins was a characteristic finding for GT platelets. Originally these were termed GP II and GP III, but as the glycoprotein composition of normal platelet membranes became more precisely defined, it was recognized that GP IIb and GP IIIa (still termed GP III by some authors) were the affected glycoproteins. The analysis of GT platelets by single-dimension SDS–PAGE followed by PAS staining is illustrated in Figure 1. This remains a viable method for the preliminary screening of a thrombasthenialike defect in platelets, but it is not quantitative. Other minor PAS-staining glycoproteins that migrate close to or with GP IIb and GP IIIa may contribute to any PAS staining observed in this region of the gel. The presence of these components (e.g., GP Ic, GP IIa) is well illustrated by PAS staining following the analysis of normal and GT platelet glycoproteins by high-resolution two-dimensional electrophoresis procedures (Clemetson *et al.*, 1980; Holahan and White, 1981).

4.3. Surface-Labeling Procedures

A comprehensive study involving the use of single- and two-dimensional SDS–PAGE procedures was reported by Phillips and Poh Agin (1977b), who analyzed GT platelets whose surface proteins had been radiolabeled either with ^{125}I during lactoperoxidase–catalyzed iodination or ^{3}H following sequential treatment of the platelets with neuraminidase, galactose oxidase, and sodium [^{3}H]borohydride. Glycoproteins were located by Coomassie Blue staining, autoradiography (^{125}I), or fluorography (^{3}H). Five patients were examined in this study. The combined results strongly suggested the presence of severe molecular deficiencies of GP IIb and GP IIIa in the platelets of each patient.

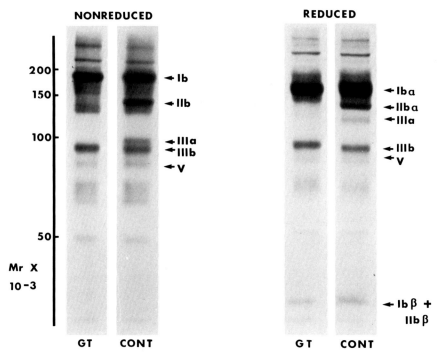

Figure 9. Single-dimension electrophoresis of the ^{3}H-labeled surface proteins of GT platelets. Washed suspensions of normal human platelets and those of a type I GT patient were sequentially incubated with neuraminidase, galactose oxidase, and sodium [^{3}H]borohydride according to the procedure of Nurden *et al.* (1982b). Samples (100 μg protein) of the SDS-solubilized platelets were analyzed by SDS–PAGE on 7–12% gradient acrylamide slab gels both following (reduced) and in the absence of (nonreduced) disulfide reduction with 2-mercaptoethanol. ^{3}H-labeled proteins were detected by fluorography. No ^{3}H labeling was located in the positions of GP IIb, IIbα, or IIIa on the GT platelet pattern. ^{3}H labeling of GP IIbβ was also reduced, but this abnormality was masked by the comigration of GP Ibβ and IIbβ on the 7–12% acrylamide gradient gel. Published from Caen and Nurden (1983) with copyright permission.

The analysis of ³H-labeled GT platelets by single-dimension SDS–PAGE is illustrated in Figure 9. No labeling was located in the positions normally occupied by GP IIb and GP IIIa. In contrast, the other major glycoproteins of the platelets of this patient were of a normal migration and labeling intensity. The characteristic changes in migration of GP IIb and GP IIIa with or without disulfide reduction are illustrated in Figure 9. It should be emphasized that IIb and IIIa are structurally different glycoproteins. Their individual structural characteristics are detailed by Clemetson (Chapter 3). Glycoprotein IIb consists of two subunits linked by disulfides. The larger, more heavily glycosylated, α-subunit is clearly absent from the reduced GT platelet glycoprotein pattern illustrated in Figure 9. Although also decreased, GP IIbβ comigrated with GP Ibβ in the system used and thus its absence could not be accurately determined.

Figure 10 illustrates the separation of the major ¹²⁵I-labeled surface proteins of GT platelets by IF/SDS–PAGE. Again, the apparent absence of GP IIbα and GP IIIa from the thrombasthenic platelet pattern is to be noted, together with the normal presence of other less intensely ¹²⁵I-labeled membrane glycoproteins in the same molecular weight range. High-resolution two-dimensional electrophoresis procedures have been used in several studies to analyze GT platelet glycoproteins. Two reports

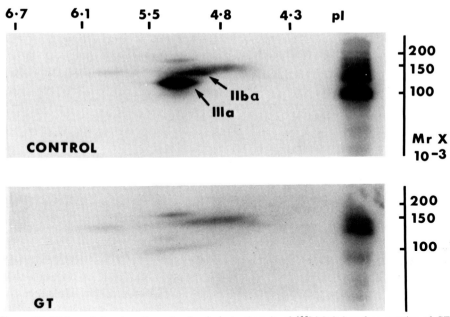

Figure 10. High-resolution two-dimensional gel electrophoresis of ¹²⁵I-labeled surface proteins of GT platelets. ¹²⁵I-Labeling of the surface proteins of washed normal human platelets and those of a type I GT patient was by the lactoperoxidase-catalyzed method. Washed, ¹²⁵I-labeled platelets were solubilized by 2% SDS in the presence of 5 mM N-ethylmaleimide. Prior to a first-dimension separation by isoelectric focusing (IF), the samples (200 μg protein) were made 9 M with urea and Triton X-100 added to give a 4:1 ratio with respect to the SDS. IF was performed according to the procedure of Ames and Nikaido (1976). After IF, the gels were incubated with 2% SDS and 5% 2-mercaptoethanol and second-dimension electrophoresis of the reduced proteins performed on 7–12% gradient acrylamide slab gels. ¹²⁵I-labeled proteins were located in dried Coomassie Blue-stained gels by autoradiography. Typical autoradiographs are illustrated. The absence of GP IIbα and GP IIIa from the GT platelet pattern is clearly shown.

have suggested that the major glycoproteins actually present in the GT platelet membrane have an altered pI due to unspecified changes in their carbohydrate composition (Clemetson et al., 1980, McGregor et al., 1981). However, this does not appear to be a consistent finding for others have failed to detect pI changes of thrombasthenic platelet membrane glycoproteins (Holahan and White, 1981; Peterson and Wehring, 1981; Herrmann et al., 1982). Also, no obvious modifications in pI are to be observed for the ^{125}I-labeled glycoproteins of the GT platelets analyzed in Figure 10.

The highly sensitive methodology used by McGregor et al. (1981) enabled them to detect glycoproteins not previously identified by other workers. As a consequence, additional deficiencies of low-molecular-weight glycoproteins (termed IVa, IVb, and VII) and one high-molecular-weight glycoprotein in GT platelets were reported. These results were interpreted as providing additional evidence for the presence of a generalized structural perturbation of the GT platelet membrane. However, the possibility remains that the additional glycoproteins in fact represent trace degradation or cross-linked products of GP IIb and/or GP IIIa in the normal platelet samples.

4.4. Type I and Type II Thrombasthenia

It has been known for some time that GT platelets may be deficient in fibrinogen (Jackson et al., 1963; Castaldi and Caen, 1965). Increasing evidence of heterogeneity led Caen (1972) to distinguish two subgroups of patients: type I thrombasthenia in which lack of aggregation was associated with an apparent absence of platelet fibrinogen and a profound defect in clot retraction, and the rare type II thrombasthenia in which a lack of aggregation was accompanied by subnormal but clearly detectable levels of platelet fibrinogen and only a moderately defective clot retraction. A major advance in our understanding of the molecular defects in the platelets of the two subgroups came from the application of CIE to the analysis of GT platelets.

4.5. Crossed Immunoelectrophoresis

Crossed immunoelectrophoresis of Triton X-100 extracts of ^{125}I-labeled normal human platelets using a polyspecific rabbit anti-human platelet antiserum revealed a prominent ^{125}I-labeled precipitate, initially termed band 16, that was shown to contain both GP IIb and GP IIIa (Hagen et al., 1980; Kunicki et al., 1981a). It is now known that these glycoproteins are present as Ca^{2+}-dependent heterodimers (GP IIb-IIIa) under the conditions of the assay. Evidence for the presence of both GP IIb and GP IIIa in the band 16 precipitate and for the heterodimer nature of the GP IIb-IIIa complex is detailed in Chapters 4 and 5. Analysis of platelets from several patients with type I thrombasthenia revealed no traces of the GP IIb-IIIa precipitate and confirmed that the platelets were severely deficient in fibrinogen (Hagen et al., 1980; Kunicki et al., 1981b,c). Crossed immunoelectrophoresis of the ^{125}I-labeled surface proteins of the platelets of a typical type I GT patient is illustrated in Figure 11. The complete absence of the GP IIb-IIIa precipitate and the normal presence of the precipitates given by the other major membrane glycoproteins (Ia, Ib, IIa, IIIb) is to be noted. A similar analysis of the platelets of a second type I GT patient is illustrated in Chapter 4 (Figure 6). In contrast, analysis of platelets from two type II GT patients revealed the presence of low

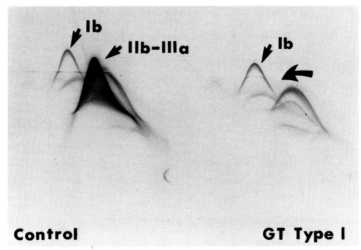

Figure 11. Analysis of Triton X-100 extracts of ^{125}I-labeled type I GT platelets by CIE. Methods were as detailed in the legend to Figure 4. In this experiment first-dimension electrophoresis was from right to left. ^{125}I-labeled proteins were detected by autoradiography. The apparent absence of the GP IIb-IIIa complex from the GT platelet sample is highlighted (heavy arrow). Published from Nurden *et al.* (1981b) with copyright permission.

levels of the GP IIb-IIIa complex (Hagen *et al.*, 1980; Kunicki *et al.*, 1981b). For these patients, the amounts of GP IIb-IIIa were estimated as being 13% and 15% of that present in normal platelets. For the same patients, platelet fibrinogen levels were 30% and 50% of that of normal platelets (see Lee *et al.*, 1981; A. T. Nurden, unpublished data). An advantage of CIE is that quantitative studies can be performed. A direct link between the inheritance of the thrombasthenic trait and the glycoprotein abnormalities was strongly suggested by the finding that platelets from obligate heterozygotes for type I GT contained approximately 50% levels of GP IIb-IIIa (Kunicki *et al.*, 1981c; Stormorken *et al.*, 1982).

4.6. Patient Heterogeneity

The presence or absence of fibrinogen in GT platelets can also be readily established by SDS–PAGE of nonreduced samples followed by Coomassie Blue staining. Such a preliminary screening of the platelets of a series of patients is illustrated in Figure 12. Here, protein patterns of platelets from three type I GT patients, one type II GT patient, and one BSS patient are compared. The fibrinogen band is clearly lacking or is severely reduced from the profiles obtained for the platelets of the type I patients (GT 1–3), but is present in significant amounts in the type II platelets (GT 4). Using 7–12% gradient acrylamide gels and nonreduced samples, GP IIb is well separated from other polypeptides and can also be distinguished. In contrast, GP IIIa is not separated and comigrates with more strongly Coomassie Blue-staining polypeptides. Glycoprotein IIb was not detected on analysis of platelets from all of the GT patients. Only a faint band corresponding to GP Ib was missing from the BSS platelet protein profile.

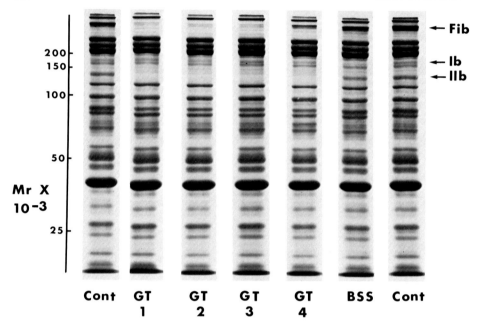

Figure 12. Screening of the polypeptide composition of abnormal platelets by single-dimension SDS–PAGE. Washed platelets from four GT, one BSS, and two control donors were solubilized with SDS and samples (100 μg protein) analyzed on 7–12% gradient acrylamide slab gels according to the procedure of Nurden *et al.* (1981a). No disulfide reduction was performed. Proteins were detected by Coomassie Blue staining. Levels of GP IIb in the platelets of all GT patients were below the limits of detection with this procedure. Fibrinogen was severely reduced in the platelets of GT 1–3, but present in significant amounts in the platelets of GT 4. Heterogeneity in the platelet lesion in thrombasthenia is therefore illustrated. Only a faint band representing GP Ib is missing from the BSS platelets.

The question arose as to whether GP IIb (and GP IIIa) were present in the platelets of the GT patients (especially GT 4) but at concentrations too low to be revealed by Coomassie Blue staining.

4.6.1. "Western" Blot

A sensitive approach to detecting trace amounts of glycoproteins is to use a "Western" blot procedure combined with antibodies that are monospecific for the glycoproteins. Figure 13 illustrates such an analysis in which small amounts of GP IIb and GP IIIa are revealed in platelets from a type II GT patient using a mixture of an alloantibody directed against GP IIb (anti-Lek[a]) and an alloantibody against GP IIIa (anti-Pl[A1]). Further details on these antibodies are given in Chapter 15. It is now clear from such studies that the platelets of some type II patients (such as GT 4 in Figure 12) contain lower amounts of GP IIb and GP IIIa than do the platelets of the two type II patients originally studied in Paris and discussed in Section 4.5. Furthermore, the

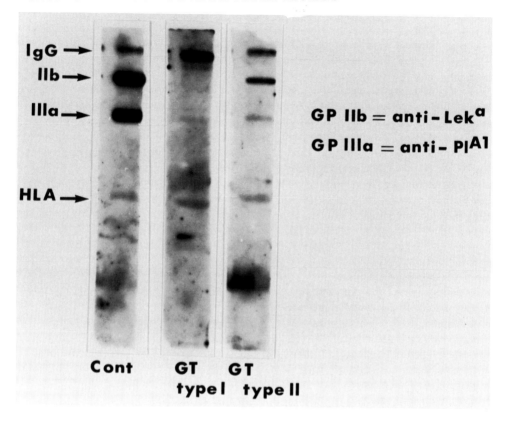

Figure 13. Detection of trace amounts of GP IIb and IIIa in GT platelets using a Western blot procedure. Washed platelets were isolated from a control donor and from two GT patients classified by platelet function testing as being type I and type II GT. SDS-soluble samples (50 μg protein) were electrophoresed on 7–12% gradient acrylamide slab gels after which the proteins were electrophoretically transferred to nitrocellulose membrane. Individual membrane strips were incubated with a mixture of an alloantibody reacting with GP IIb (anti-Lek[a]) and an alloantibody that binds to GP IIIa (anti-Pl[A1]). Bound antibody was located using [125I]protein A followed by autoradiography. The methods employed were based on those of Bennett *et al.* (1982). The autoradiographs were overexposed to allow detection of minor components. Note the presence of detectable GP IIb and GP IIIa in the type II platelets and the possible presence of trace amounts of GP IIIa in the type I platelets. The type II patient was probably a heterozygote for the Pl[A1] antigen.

analysis of type I platelets using the "Western" blot procedure has revealed the possible presence of trace amounts of GP IIIa in the platelets of some patients (Figure 13). It is probable, therefore, that classic GT represents a range of patients from those containing no detectable GP IIb or GP IIIa to those with upward of 20% or more of these glycoproteins. Notwithstanding this heterogeneity, it is still apparent from our studies that only those patients with detectable GP IIb *and* GP IIIa contain platelet fibrinogen in significant amounts.

4.6.2. Surface Labeling of Type II Glanzmann's Thrombasthenic Platelets

There is evidence to suggest that GP IIb and/or GP IIIa in the platelets of some type II patients may have qualitative defects. Peterson and Wehring (1981) located small amounts ($<$ 11% of normal) of both glycoproteins in the platelets of each of four GT patients studied. Platelets of each patient were labeled with ^{125}I using the lactoperoxidase-catalyzed method. After separation of GP IIb and GP IIIa by IF/SDS–PAGE, the spots corresponding to the individual glycoproteins were excised from the gel and the amount of radioactivity associated with each glycoprotein determined. Increased amounts of ^{125}I were incorporated in GP IIb relative to GP IIIa as compared with the results obtained for normal platelets. This finding suggests an altered structure or organization of GP IIb and/or GP IIIa in the platelets of these patients. McGregor *et al.* (1981) also obtained evidence for an altered structure of GP IIb and GP IIIa of type II GT platelets using ^{3}H-labeling procedures.

4.7. Platelet-Specific Alloantigens

Because GT platelets lack or have a reduced expression of GP IIb and GP IIIa, they also have a reduced expression of those platelet-specific alloantigens carried by GP IIb (Bak^a, Lek^a) or GP IIIa (Pl^{A1}/Pl^{A2}) (Kunicki and Aster, 1978; van Leuwen *et al.*, 1981) (see Chapter 16 and George *et al.*, 1984). The heredity of GP IIb-IIIa and the Pl^{A1} determinant was studied in detail among 20 kindred of two large gypsy families in the Strasbourg region of France (Kunicki *et al.*, 1981c). Heterozygotes containing approximately 50% levels of GP IIb-IIIa were found to have either 54% or 28% of the normal platelet amounts of Pl^{A1} antigen. Thus heterozygotes for the thrombasthenia trait showed a normal inheritance of the Pl^{A1} antigen. This finding confirms that the thombasthenic platelet glycoprotein deficiencies and the inheritance of the Pl^{A1} antigen are controlled by different genes. Patients with type II GT express low concentrations of Pl^{A1} antigen (Kunicki *et al.*, 1981c).

4.8. Monoclonal Antibodies

Diagnosis of GT is now possible using monoclonal antibodies directed against determinants carried by the GP IIb-IIIa complex. The first such antibody to be described, Tab, bound to 3.9×10^4 sites per normal human platelet. Platelets from four GT patients bound less than 5% of the normal amount of Tab, whereas platelets from obligate heterozygotes for GT bound intermediate amounts (McEver *et al.*, 1980). The Tab antibody was subsequently shown to react exclusively with GP IIb (McEver *et al.*, 1983). Other monoclonal antibodies not reacting with GT platelets include 10E5 (Coller *et al.*, 1983b) and AP-2 (Montgomery *et al.*, 1983). Using AP-2 in a whole blood binding assay, Montgomery *et al.* (1983) reported that they were able to distinguish between type I and type II thrombasthenia. Both 10E5 (Coller *et al.*, 1983c) and AP-2 (Pidard *et al.*, 1983) bind only to the GP IIb-IIIa complex. These antibodies do not bind to dissociated GP IIb or IIIa and their binding sites may be conformation dependent.

4.9. Newly Described Variants with Glanzmann's Thrombasthenia

Recent evidence suggests the detection of patients where GP IIb and GP IIIa are present in normal concentrations in the platelet membrane, but which are functionally inactive due to defects of their structure or organization. The first such report was by Lightsey et al. (1981), who studied three siblings from Guam. Both GP IIb and GP IIIa were normally radiolabeled with ^{125}I during lactoperoxidase-catalyzed iodination and exhibited a normal pI during isoelectric focusing. However, the platelets did not aggregate with ADP and exhibited all of the functional characteristics of GT. Patient M. was referred to us in Paris as a type I GT patient. There was no ADP-induced platelet aggregation and clot retraction was defective. As with the Guam siblings, GP IIb and GP IIIa were present and were normally radiolabeled with ^{125}I during lactoperoxidase-catalyzed iodination (Nurden et al., 1983b). The platelet fibrinogen content was, however, severely decreased. When M.'s platelets were prepared for CIE using standard methods, the precipitate given by the GP IIb-IIIa complex was unusually small and clearly abnormal (Nurden et al., 1983b). The initial impression was that the platelets of this patient contained a high proportion of uncomplexed GP IIb and IIIa. However, our recent results suggest that the platelets of this patient have a structural, or organizational, defect of GP IIb-IIIa that results in the complex being more susceptible to dissociation with EDTA (A. T. Nurden, and J-P. Rosa, unpublished findings). This explains why M.'s platelets that were isolated at pH 7.4 in the presence of EDTA, conditions which do not normally result in complex dissociation (see Chapter 4), failed to bind monoclonal antibodies (AP-2, 10E5) specific for the GP IIb-IIIa complex (Nurden et al., 1983b). In future studies, this patient will be referred to as variant Lille I. A second variant under study in our laboratory, patient P. (variant Paris I), resembles a heterozygote in that CIE revealed the presence of 50% of the normal level of GP IIb-IIIa in his platelets (Caen et al., 1983). Unlike patient M., platelet fibrinogen levels were not decreased, although ^{125}I-fibrinogen binding to ADP-stimulated platelets did not occur. No abnormalities of GP IIb-IIIa complex dissociation have been detected (A. T. Nurden, unpublished data). Although some platelet function abnormalities may occur in heterozygotes, platelet aggregation is not absent (Stormorken et al., 1982). These newly described variants of thrombasthenia, with possible structural defects of the GP IIb-IIIa complex, offer a unique chance to locate the functional zones of the complex.

4.10. Altered Surface Reactivity of Glanzmann's Thrombasthenic Platelets

An important observation was made when Bennett and Vilaire (1979) and Mustard et al. (1979) independently showed that no specific fibrinogen binding occurred to ADP-stimulated thrombasthenic platelets. This was a crucial finding since fibrinogen had long been known as an essential cofactor for platelet aggregation (see Chapter 9).

4.10.1. Fibrinogen Receptor Deficiency

The above results suggested that either GT platelets lacked the fibrinogen receptor or that the stimulus initiated by ADP-receptor occupancy was not transmitted within the thrombasthenic platelet membrane. Additional patients were studied by Peerschke *et al.* (1980a) and by Lee *et al.* (1981), who confirmed the defective GT platelet–fibrinogen interaction. In France, Lee *et al.* (1981) studied four type I patients and two type II patients. Although platelets of the type I patients failed to bind fibrinogen, platelets of the French type II patients bound low amounts. There seemed to be a correlation between the reduced level of GP IIb-IIIa in the membrane and the amounts of fibrinogen bound. Such a result argues in favor of a direct role for GP IIb-IIIa in fibrinogen receptor expression.

4.10.2. Fibrinogen Binding to Chymotrypsin-Treated Glanzmann's Thrombasthenic Platelets

Kornecki *et al.* (1981) challenged the theory that the primary defect of GT platelets was the absence of fibrinogen receptors by showing that, after chymotrypsin treatment, the abnormal platelets bound fibrinogen to a discrete population of high-affinity binding sites. Nonetheless, the total amount of fibrinogen bound and the extent of aggregation was much less than reported for chymotrypsin-treated normal platelets by the same authors (Kornecki *et al.*, 1981; Niewiarowski *et al.*, 1981). The GT patients in this study were subsequently revealed to have measurable GP IIb-IIIa in their membranes and to be of the type II subtype (Kornecki *et al.*, 1983). This is especially important for these authors also reported that fibrinogen-induced aggregation of chymotrypsin-treated normal human platelets is mediated by a M_r 66,000 fragment of GP IIIa (Kornecki *et al.*, 1983). Thus residual GP IIb-IIIa may have mediated the fibrinogen binding to the type II GT platelets used in the original study. The interest in studies on chymotrypsin-treated platelets stems from the fact that the protease digestion bypasses the ADP-mediated step and directly exposes fibrinogen receptors.

4.10.3. Reduced Binding of Fibronectin, VWF, and Thrombospondin

It has recently been shown that thrombin-stimulated GT platelets are also deficient in their ability to bind VWF (Ruggeri *et al.*, 1982) and fibronectin (Ginsberg *et al.*, 1983). Data from immunocytochemical studies indicate that they also bind much reduced amounts of TSP (Hourdillé *et al.*, 1984). Binding of the bulk of these proteins to normal platelets requires both platelet activation (or activation plus secretion) and the presence of divalent cations. The above findings are fundamental observations for they mean that a defect in fibrinogen receptor expression is not the only altered surface reactivity of GT platelets.

4.10.4. Defective Ca²⁺ Binding

Peerschke *et al.* (1980b) also made an important discovery when they showed that GT platelets took up 50% less calcium than normal platelets when incubated in a medium containing ^{45}Ca. It was postulated that thrombasthenic platelets lacked surface binding sites for divalent cations and this has been confirmed (Johnston *et al.*, 1983; Brass and Shattil, 1984). Gogstad *et al.* (1983) have also demonstrated that $^{45}Ca^{2+}$ binds to both GP IIb and IIIa. Thus, the results of Peerschke *et al.* (1980b) may be at least partially explained by the GP IIb-IIIa deficiency of thrombasthenic platelets.

4.10.5. Ristocetin-Induced VWF Binding

Although GT platelets do not bind VWF after ADP- or thrombin-induced platelet activation, they do bind this protein in the presence of ristocetin (Moake *et al.*, 1980). This shows that the receptor mechanism for GP Ib-mediated VWF binding (see Chapter 11) functions normally in GT platelets. However, it should be pointed out that ristocetin-induced platelet agglutination occurs in cycles in thrombasthenia and, unlike with normal platelets, ADP inhibits the agglutination (Chediak *et al.*, 1979).

4.11. Basic Genetic Lesion in Thrombasthenia

The question arises as to why two apparently different membrane glycoproteins, GP IIb and GP IIIa, are either absent or severely deficient from GT platelets. Despite major differences in the structure of GP IIb and GP IIIa (see Chapter 3), a single gene deletion responsible for a lack of synthesis of both glycoproteins remains a possibility. A peptide or oligosaccharide sequence common to both glycoproteins cannot yet be discounted. Alternatively, the abnormality could occur as a posttranslational event involving either a regulator gene or a degradative process. The occurrence of small amounts of GP IIb and GP IIIa in the platelets of some patients favors this possibility. It is interesting to note that in platelets of at least some type II GT patients, GP IIb and GP IIIa are present as the GP IIb-IIIa complex (Hagen *et al.*, 1980; Kunicki *et al.*, 1981b). Thus the proportion of GP IIb to GP IIIa has been at least approximately maintained. This could suggest either a parallel decrease in synthesis or a primary lesion of one glycoprotein accompanied by the normal synthesis and subsequent degradation of the other. The formation of the GP IIb-IIIa complex could be important for the stabilization of the glycoproteins within the membrane and provide a protection against proteolytic degradation.

A second question concerns the absence or reduction of fibrinogen from GT platelets. A recent preliminary report appears to confirm that normal human megakaryocytes synthesize fibrinogen (Belloc *et al.*, 1983). The detection of significant levels of fibrinogen in the platelets of type II GT patients strongly suggests that this is related to the presence of small amounts of GP IIb-IIIa. A recent finding that the GP IIb-IIIa complex is a component of the platelet α-granule membrane (Gogstad *et al.*, 1981) allows speculation that the complex plays a role in fibrinogen incorporation into the α-granules. Thus, in the absence of GP IIb-IIIa, no fibrinogen would be incorporated. Studies using megakaryocytes from GT patients are now required for a further development of our understanding of the primary genetic lesion(s) in this disorder.

4.12. Conclusions on Thrombasthenia

The basic picture is of a heterogeneous disorder characterized by a lack of platelet aggregation due to the inability of thrombasthenic platelets to normally express receptors for fibrinogen. A lack of binding of secreted α-granule proteins such as fibronectin, VWF and TSP has also been noted, but the physiologic relevance of these findings has yet to be evaluated. In contrast, abundant evidence points to a key role for fibrinogen in the aggregation mechanism and the GP IIb-IIIa complex in fibrinogen receptor expression by normal platelets (detailed in Chapter 9). The simplest model for aggregation is where fibrinogen cross-links GP IIb-IIIa complexes on adjacent platelets. However, it has not been clearly established that fibrinogen once attached to its receptor then binds directly, or through other fibrinogen molecules, to the same receptor on adjacent platelets. An interesting experiment to perform would be to see if GT platelets bind to normal platelets activated in the presence of fibrinogen. In fact, Gerrard et al. (1979) have reported preliminary data to suggest that thrombasthenic platelets are incorporated into platelet aggregates when mixed with platelets containing GP IIb and GP IIIa. This suggests that the primary defect in thrombasthenia is in the initial binding of the aggregation cofactor to its platelet receptor. Perhaps secreted "adhesive" α-granule proteins can supplement or replace fibrinogen in this process. A recent finding suggests that fibrinogen, VWF, and fibronectin are recognized by a common receptor site on activated platelets (Plow et al., 1984). This is indirect evidence that the GP IIb-IIIa defect may be responsible for the lack of binding of each of these proteins.

Finally, it should be pointed out that the absence of GP IIb-IIIa from the platelets of GT patients may have an important side-effect, for it raises the possibility of antibody formation as a consequence of blood transfusion during treatment for the bleeding episodes that the patients occasionally encounter. One such antibody, the IgG L., has been isolated and characterized. It inhibits platelet aggregation by preventing fibrinogen binding (Lee et al., 1981) and reacts with GP IIb-IIIa in CIE (Hagen et al., 1980; Kunicki et al., 1981a). After GP IIb-IIIa dissociation by EDTA only a weak reactivity with GP IIb remained (see Figure 1, Chapter 4). Thus the bulk of the IgG appears to react with complex-dependent determinants (Rosa et al., 1984). Studies with this antibody provided the first direct correlation between the molecular defects of the GT platelet surface and the altered platelet function in this disorder.

5. GENERAL CONCLUSIONS

It has not been the aim of this chapter to provide a comprehensive review of inherited platelet diseases. Recent reviews of this nature include those by Bellucci et al. (1983) and George et al. (1984). Instead, those disorders have been highlighted which have provided most insight into the roles of platelet glycoproteins in normal human platelet function. Each disorder reflects platelet involvement in a different step in the initial arrest of bleeding with the primary lesions involving platelet adhesion (BSS), secretion (GPS), and aggregation (GT). Elucidation of the molecular defects responsible for the bleeding syndromes has been of major importance in the identification of the principal functional groups of the platelet surface. Furthermore, it has emphasized how the platelet membrane is specialized to meet the functional requirements placed upon it.

ACKNOWLEDGMENTS. The technical skills of Jean-Philippe Rosa, Dominique Pidard, Nelly Kieffer, and Martine Jandrot-Perrus are gratefully acknowledged. Especially, I would like to express my thanks to Dominique Didry for her essential role in the platelet glycoprotein studies performed in Paris over many years. I also thank Mrs. Judi Skinner for her preparation of this manuscript.

REFERENCES

Ames, G. F., and Nikaido, K., 1976, Two-dimensional gel electrophoresis of membrane proteins, *Biochemistry* **15**:616–623.

Baumgartner, H. R., 1973, The role of blood flow in platelet adhesion, fibrin deposition and formation of mural thrombi, *Microvasc. Res.* **5**:167–179.

Baumgartner, H. R., Tschopp, T. B., and Weiss, H. J., 1977, Platelet interaction with collagen fibrils in flowing blood. II. Impaired adhesion-aggregation in bleeding disorders. A comparison with subendothelium, *Thromb. Haemostasis.* **37**:17–28.

Bauvois, B., Cartron, J. P., Nurden, A., and Caen, J. P., 1981, Glycoprotein-sialyltransferase activity of normal human, thrombasthenic, and Bernard-Soulier platelets, *Vox Sang.* **40**:71–78.

Belloc, F., Hourdille, P., Fialon, P., Boisseau, M. R., and Soria, J., 1983, Fibrinogen synthesis by isolated human megakaryocytes, *Thromb. Haemostasis* **50**:18a.

Bellucci, S., Tobelem, G., and Caen, J. P., 1983, Inherited platelet disorders, *Prog. Hematol.* **13**:223–264.

Bennett, J. S., and Vilaire, G., 1979, Exposure of platelet fibrinogen receptors by ADP and epinephrine, *J. Clin. Invest.* **64**:1393–1401.

Bennett, J. S., Vilaire, G., and Cines, D. B., 1982, Identification of the fibrinogen receptor on human platelets by photoaffinity labeling, *J. Biol. Chem.* **257**:8049–8054.

Berndt, M. C., and Phillips, D. R., 1981, Purification and preliminary physicochemical characterization of human platelet membrane glycoprotein V, *J. Biol. Chem.* **256**:59–65.

Berndt, M. C., Gregory, C., Chong, B. H., Zola, H., and Castaldi, P. 1983, Additional glycoprotein defects in Bernard-Soulier syndrome: Confirmation of genetic basis by parental analysis, *Blood* **62**:800–807.

Bithell, T. C., Parekh, S. J., and Strong, R. R., 1972, Platelet-function studies in the Bernard-Soulier syndrome, *Ann. N.Y. Acad. Sci.* **201**:145–160.

Brass, L. F., and Shattil, S. J., 1984, Identification and function of the high affinity binding sites for Ca^{2+} on the surface of platelets, *J. Clin. Invest.* **73**:626–632.

Breton-Gorius, J., Vainchenker, W., Nurden, A., Levy-Toledano, S., and Caen, J., 1981, Defective α-granule production in megakaryocytes from gray platelet syndrome. Ultrastructural studies of bone marrow cells and megakaryocytes growing in culture from blood precursors, *Am. J. Pathol.* **102**:10–19.

Caen, J. P., 1972, Glanzmann's thrombasthenia, *Clin. Haematol.* **1**:383–392.

Caen, J. P., and Nurden, A. T., 1983, Inherited abnormalities of platelet glycoproteins, *Surv. Synth. Pathol. Res.* **1**:274–281.

Caen, J. P., Castaldi, P. A., Leclerc, J. C., Inceman, S., Larrieu, M. J., Probst, M., and Bernard, J., 1966, Congenital bleeding disorders with long bleeding time and normal platelet count. I. Glanzmann's thrombasthenia (report of fifteen patients), *Am. J. Med.* **41**:4–26.

Caen, J. P., Nurden, A. T., Jeanneau, C., Michel, H., Tobelem, G., Levy-Toledano, S., Sultan, Y., Valensi, F., and Bernard, J., 1976, Bernard-Soulier syndrome. A new platelet glycoprotein abnormality. Its relationship with platelet adhesion to subendothelium and with the factor VIII von Willebrand protein, *J. Lab. Clin. Med.* **87**:586–596.

Caen, J. P., Rosa, J-P., Boizard, B., and Nurden, A. T., 1983, Thrombasthenia Paris I variant, a model for the study of the platelet (GP) IIb-IIIa complex, *Blood* **62**:25a.

Cartron, J. P., and Nurden, A. T., 1979, Galactosyltransferase and membrane glycoprotein abnormality in human platelets from Tn-syndrome donors, *Nature (London)* **282**:621–623.

Castaldi, P., and Caen, J. P., 1965, Platelet fibrinogen, *J. Clin. Pathol.* **18**:579–585.

Chediak, J., Telfer, M. C., van der Laan, B., Maxey, B., and Cohen, I., 1979, Cycles of agglutination-disagglutination induced by ristocetin in thrombasthenic platelets, *Br. J. Haematol.* **43**:113–126.

Clemetson, K. J., Capitanio, A., Pareti, F. I., McGregor, J. L., and Luscher, E. F., 1980, Additional platelet membrane glycoprotein abnormalities in Glanzmann's thrombasthenia: A comparison with normals by high resolution two-dimensional polyacrylamide gel electrophoresis, *Thromb. Res.* **18:**797–806.

Clemetson, K. J., Naim, H. Y., and Luscher, E. F., 1981a, Relationship between glycocalicin and glycoprotein Ib of human platelets, *Proc. Natl. Acad. Sci. U.S.A.* **78:**2712–2716.

Clemetson, K. J., McGregor, J. L., James, E., Dechavanne, M., and Luscher, E. F., 1981b, Membrane glycoprotein defects in Bernard-Soulier syndrome platelets, *Thromb. Haemostasis* **46:**108a.

Clemetson, K. J., McGregor, J. L., James, E., Dechavanne, M., and Luscher, E. F., 1982, Characterization of the platelet membrane glycoprotein abnormalities in Bernard-Soulier syndrome and comparison with normal by surface-labeling techniques and high-resolution two-dimensional gel electrophoresis, *J. Clin. Invest.* **70:**304–311.

Coller, B. S., 1978, The effects of ristocetin and von Willebrand factor on platelet electrophoretic mobility, *J. Clin. Invest.* **61:**1168–1175.

Coller, B. S., Peerschke, E. I., Scudder, L. E., and Sullivan, C. A., 1983a, Studies with a murine monoclonal antibody that abolishes ristocetin-induced binding of von Willebrand factor to platelets: Additional evidence in support of GP Ib as a platelet receptor for von Willebrand factor, *Blood* **61:**99–110.

Coller, B. S., Peerschke, E. I., Scudder, L. E., and Sullivan, C. A., 1983b, A murine monoclonal antibody that completely blocks the binding of fibrinogen to platelets produces a thrombasthenic-like state in normal platelets and binds to glycoproteins IIb and/or IIIa, *J. Clin. Invest.* **72:**325–338.

Coller, B. S., Peerschke, E. I., Nurden, A. T., and Rosa, J-P., 1983c, The epitope of a monoclonal antibody that inhibits fibrinogen binding to platelets is EDTA-sensitive: Implications for the presence of the GP IIb/IIIa complex on unactivated platelets, *Thromb. Haemostasis* **50:**314a.

Degos, L., Tobelem, G., Lethielleux, P., Levy-Toledano, S., Caen, J., and Colombani, J., 1977, Molecular defect in platelets from patients with Bernard-Soulier syndrome, *Blood* **50:**899–903.

Donati, M. B., Balconi, G., Remuzzi, G., Borgia, R., Morasco, L., and de Gaetano, G., 1977, Skin fibroblasts from a patient with Glanzmann's thrombasthenia do not induce fibrin clot retraction, *Thromb. Res.* **10:**173–174.

Drouin, J., Wong, S. C., Jamieson, G. A., and Rock, G. A., 1980, Release of ATP and serotonin from Bernard-Soulier platelets in response to various aggregating agents, in: *18th Congress of the International Society of Hematology* (Abstracts), Montreal, p. 219a.

Evensen, S. A., Solum, N. O., Grottum, K. A., and Hovig, T., 1974, Familial bleeding disorder with a moderate thrombocytopenia and giant platelets, *Scand. J. Haematol.* **13:**203–214.

Gartner, T. K., Gerrard, J. M., White, J. G., and Williams, D. C., 1981, The endogenous lectin of human platelets is an α-granule component, *Blood* **58:**153–157.

George, J. N., Reimann, T. A., Moake, J. L., Morgan, R. K., Cimo, P. L., and Sears, D. A., 1981, Bernard-Soulier disease: A study of four patients and their parents, *Br. J. Haematol.* **48:**459–467.

George, J. N., Nurden, A. T., and Phillips, D. R., 1984, Molecular defects in interactions of platelets with the vessel wall, *New Engl. J. Med.* **311:**1084–1098.

Gerrard, J. M., and White, J. G., 1976, The influence of aspirin and indomethacin on the platelet contractile wave, *Am. J. Pathol.* **82:**513–526.

Gerrard, J. M., Schollmeyer, J. V., Phillips, D. R., and White, J. G., 1979, α-Actinin deficiency in thrombasthenia, *Am. J. Pathol.* **94:**509–528.

Gerrard, J. M., Phillips, D. R., Rao, G. H. R., Plow, E. F., Waltz, D. A., Ross, R., Harker, L. A., and White, J. G., 1980, Biochemical studies of two patients with the gray platelet syndrome. Selective deficiency of platelet alpha granules, *J. Clin. Invest.* **66:**102–109.

Ginsberg, M. H., Forsyth, J., Lightsey, A., Chediak, J., and Plow, E. F., 1983, Reduced surface expression and binding of fibronectin by thrombin-stimulated thrombasthenic platelets, *J. Clin. Invest.* **71:**619–624.

Gogstad, G. O., Hagen, I., Korsmo, R., and Solum, N. O., 1981, Characterization of the proteins of isolated human platelet α-granules. Evidence for a separate α-granule pool of the glycoproteins IIb and IIIa, *Biochim. Biophys. Acta* **670:**150–162.

Gogstad, G. O., Krutnes, M-B., and Solum, N. O., 1983, Calcium-binding proteins from human platelets. A study using crossed immunoelectrophoresis and $^{45}Ca^{2+}$, *Eur. J. Biochem.* **133:**193–199.

Grottum, K. A., and Solum, N. O., 1969, Congenital thrombocytopenia with giant platelets: A defect in the platelet membrane, *Br. J. Haematol.* **16:**277–290.

Hagen, I., and Solum, N. O., 1978, Further studies on the protein composition and surface structure of normal platelets and platelets from patients with Glanzmann's thrombasthenia and Bernard-Soulier syndrome, *Thromb. Res.* **13:**845–855.

Hagen, I., Nurden, A., Bjerrum, O. J., Solum, N. O., and Caen, J. P., 1980, Immunochemical evidence for protein abnormalities in platelets from patients with Glanzmann's thrombasthenia and Bernard-Soulier syndrome, *J. Clin. Invest.* **65:**722–731.

Heptinstall, S., Burgess-Wilson, M., Cockbill, S. R., Fox, S. C., and Sills, T., 1983, Platelet aggregation in whole blood. Studies in normal subjects and in Glanzmann's thrombasthenia, *Thromb. Haemostasis* **50:**216a.

Herrmann, F. H., Meyer, M., and Ihle, E., 1982, Protein and glycoprotein abnormalities in an unusual subtype of Glanzmann's thrombasthenia, *Haemostasis* **12:**337–344.

Holahan, J. R., and White, G. C., 1981, Heterogeneity of membrane surface proteins in Glanzmann's thrombasthenia, *Blood* **57:**174–181.

Holmsen, H., and Weiss, H. J., 1979, Secretable storage pools in platelets, *Annu. Rev. Med.* **30:**119–134.

Hourdillé, P., Hasitz, M., Belloc, F., and Nurden, A. T., 1984, Immunocytochemical study of the binding of fibrinogen and thrombospondin to ADP- and thrombin-stimulated human platelets, *Blood* (in press).

Howard, M. A., Hutton, R. A., and Hardisty, R. M., 1973, Hereditary giant platelet syndrome: A disorder of a new aspect of platelet function, *Br. Med. J.* **4:**586–588.

Howard, M. A., Montgomery, D. C., and Hardisty, R. M., 1974, Factor-VIII-related antigen in platelets, *Thromb. Res.* **4:**617–624.

Jackson, D. P., Morse, E. E., Zieve, P. D., and Conley, C. L., 1963, Thrombocytopenic purpura associated with defective clot retraction and absence of platelet fibrinogen, *Blood* **22:**827a.

Jaffe, E. A., Leung, L. L. K., Nachman, R. L., Levin, R. I., and Mosher, D. F., 1982, Thrombospondin is the endogenous lectin of human platelets, *Nature (London)* **295:**246–248.

Jamieson, G. A., and Okumura, T., 1978, Reduced thrombin binding and aggregation in Bernard-Soulier platelets, *J. Clin. Invest.* **61:**861–864.

Jamieson, G. A., Okumura, T., Fishback, B., Johnson, M. M., Egan, J. J., and Weiss H. J., 1979, Platelet membrane glycoproteins in thrombasthenia, Bernard-Soulier syndrome, and storage pool disease, *J. Lab. Clin. Med.* **93:**652–660.

Jenkins, C. S. P., Phillips, D. R., Clemetson, K. J., Meyer, D., Larrieu M-J., and Luscher, E. F., 1976, Platelet membrane glycoproteins implicated in ristocetin-induced aggregation. Studies of the proteins on platelets from patients with Bernard-Soulier syndrome and von Willebrand's disease, *J. Clin. Invest.* **57:**112–124.

Johnston, G. I., Taylor, P. M., and Heptinstall, S., 1983, Studies on platelets from patients with Glanzmann's thrombasthenia suggest that the surface located Ca-binding sites that are relevant to platelet aggregation are associated with glycoproteins IIb and IIIa, *Thromb. Haemostasis* **50:**215a.

Kao, K. J., Pizzo, S. V., and McKee, P. A., 1979, Demonstration and characterization of specific binding sites for Factor VIII/von Willebrand factor on human platelets, *J. Clin. Invest.* **63:**656–664.

Kirby, E. P., 1982, The agglutination of human platelets by bovine factor VIII:R, *J. Lab. Clin. Med.* **100:**963–976.

Kornecki, E., Niewiarowski, S., Morinelli, T. A., and Kloczewiak, M., 1981, Effects of chymotrypsin and adenosine diphosphate on the exposure of fibrinogen receptors on normal human and Glanzmann's thrombasthenic platelets, *J. Biol. Chem.* **256:**5696–5701.

Kornecki, E., Tuszynski, G. P., and Niewiarowski, S., 1983, Inhibition of fibrinogen receptor-mediated platelet aggregation by heterologous anti-human platelet membrane antibody. Significance of an Mr = 66,000 protein derived from glycoprotein IIIa, *J. Biol. Chem.* **258:**9349–9356.

Kunicki, T. J., and Aster, R. H., 1978, Deletion of the platelet-specific alloantigen Pl[A1] from platelets in Glanzmann's thrombasthenia, *J. Clin. Invest.* **61:**1225–1231.

Kunicki, T. J., Pidard, D., Rosa, J-P., and Nurden, A. T., 1981a, The formation of Ca^{2+}-dependent complexes of platelet membrane glycoproteins IIb and IIIa in solution as determined by crossed immunoelectrophoresis, *Blood* **58:**268–278.

Kunicki, T. J., Nurden, A. T., Pidard, D., Russell, N. R., and Caen, J. P., 1981b, Characterization of human platelet glycoprotein antigens giving rise to individual immunoprecipitates in crossed immunoelectrophoresis, *Blood* **58**:1190–1197.

Kunicki, T. J., Pidard, D., Cazenave, J-P., Nurden, A. T., and Caen, J. P., 1981c, Inheritance of the human platelet alloantigen, Pl^{A1}, in type I Glanzmann's thrombasthenia, *J. Clin. Invest.* **67**:717–724.

Lahav, J., Schwartz, M. A., Hynes, R. O., 1982, Analysis of platelet adhesion with a radioactive chemical crosslinking reagent: Interaction of thrombospondin with fibronectin and collagen, *Cell* **31**:253–262.

Lee, H., Nurden, A. T., Thomaidis, A., and Caen, J. P., 1981, Relationship between fibrinogen binding and the platelet glycoprotein deficiencies in Glanzmann's thrombasthenia type I and type II, *Br. J. Haematol.* **48**:47–57.

Levy-Toledano, S., Maclouf, J., Rendu, F., Rigaud, M., and Caen, J. P., 1979, Ionophore A23187 and thrombasthenic platelets: A model for dissociating serotonin release and thromboxane formation from true aggregation, *Thromb. Res.* **16**:453–462.

Levy-Toledano, S., Caen, J. P., Breton-Gorius, J., Rendu, F., Cywiner-Golenzer, C., Dupuy, E., Legrand, Y., and Maclouf, J., 1981, Gray platelet syndrome: α-granule deficiency. Its influence on platelet function, *J. Lab. Clin. Med.* **98**:831–848.

Lightsey, A. L., Plow, E. F., McMillan, R., and Ginsberg, M. H., 1981, Glanzmann's thrombasthenia in the absence of GP IIb and III deficiency, *Blood* **58**(suppl):199a.

Malmsten, C., Kindahl, H., Samuelsson, B., Levy-Toledano, S., Tobelem, G., and Caen, J. P., 1977, Thromboxane synthesis and the platelet release reaction in Bernard-Soulier, Glanzmann's thrombasthenia and Hermansky-Pudlak syndrome, *Br. J. Haematol.* **35**:511–520.

McEver, R. P., Baenziger, N. L., and Majerus, P. W., 1980, Isolation and quantitation of the platelet membrane glycoprotein deficient in thrombasthenia using a monoclonal hybridoma antibody, *J. Clin Invest.* **66**:1311–1318.

McEver, R. P., Bennett, E. M., and Martin, M. N., 1983, Identification of two structurally and functionally distinct sites on human platelet membrane glycoprotein IIb—IIIa using monoclonal antibodies, *J. Biol. Chem.* **258**:5269–5275.

McGill, M., Jamieson, G. A., Drouin, J., and Rock, G. A., 1980, Size and shape of Bernard-Soulier platelets in native and anticoagulated blood, in: *18th Congress of the International Society of Hematology* (Abstracts), Montreal, p. 219a.

McGregor, J. L., Clemetson, K. J., James, E., Luscher, E. F., and Dechavanne, M., 1980, A comparison of the major platelet membrane glycoproteins from Bernard-Soulier syndrome with normals after radiolabelling of sialic acid or terminal galactose/N-acetylgalactosamine residues, *Thromb. Res.* **17**:713–718.

McGregor, J. L., Clemetson, K. J., James E., Capitanio, A., Greenland, T., Luscher, E. F., and Dechavanne, M., 1981, Glycoproteins of platelet membranes from Glanzmann's thrombasthenia. A comparison with normal using carbohydrate-specific or protein-specific labelling techniques and high-resolution two-dimensional gel electrophoresis, *Eur. J. Biochem.* **116**:379–388.

McMichael, A. J., Rust, N. A., Pilch, J. R., Sochynsky, R., Morton, J., Mason, D. Y., Ruan, C., Tobelem, G., and Caen, J., 1981, Monoclonal antibody to human platelet glycoprotein I. I. Immunological studies, *Br. J. Haematol.* **49**:501–509.

Meyer, D., and Baumgartner, H. R., 1983, Role of von Willebrand factor in platelet adhesion to the subendothelium, *Br. J. Haematol.* **54**:1–9.

Moake, J. L., Olson, J. D., Troll, J. H., Tang, S. S., Funicella, T., and Peterson, D. M., 1980, Binding of radioiodinated human von Willebrand factor to Bernard-Soulier, thrombasthenic and von Willebrand's disease platelets, *Thromb. Res.* **19**:21–27.

Montgomery, R. R., Kunicki, T. J., Taves, C., Pidard, D., and Corcoran, M., 1983, Diagnosis of Bernard-Soulier syndrome and Glanzmann's thrombasthenia with a monoclonal assay on whole blood, *J. Clin. Invest.* **71**:385–389.

Mustard, J. F., Kinlough-Rathbone, R. L., Packham, M. A., Perry, D. W., Harfenist, E. J., and Pai, K. R. M., 1979, Comparison of fibrinogen association with normal and thrombasthenic platelets on exposure to ADP or chymotrypsin, *Blood* **54**:987–993.

Nachman, R. L., Levine, R. F., and Jaffe, E. A., 1977, Synthesis of factor VIII antigen by cultured guinea pig megakaryocytes, *J. Clin. Invest.* **60**:914–921.

Nibu, K., Ideguchi, H., and Hamasaki, N., 1982, Platelet glycoproteins of Bernard-Soulier syndrome:

Analysis by isoelectric focusing and sodium dodecyl sulfate polyacrylamide gel electrophoresis in a two-dimensional technique, *Thromb. Res.* **28**:67–74.

Niewiarowski, S., Budzynski, A. Z., Morinelli, T. A., Brudzynski, T. M., and Stewart, G. J., 1981, Exposure of fibrinogen receptors on human platelets by proteolytic enzymes, *J. Biol. Chem.* **256**:917–925.

Nurden, A. T., and Caen, J. P., 1974, An abnormal platelet glycoprotein pattern in three cases of Glanzmann's thrombasthenia, *Br. J. Haematol.* **28**:253–260.

Nurden, A. T., and Caen, J. P., 1975, Specific roles for platelet surface glycoproteins in platelet function, *Nature (London)* **255**:720–722.

Nurden, A. T., and Caen, J. P., 1979, The different glycoprotein abnormalities in thrombasthenic and Bernard-Soulier platelets, *Semin. Hematol.* **16**:234–250.

Nurden, A. T., and Dupuis, D., 1981, The reduced aggregation response of Bernard-Soulier platelets to thrombin may be related to an abnormal glycoprotein V, *Thromb. Haemostasis* **46**:216a.

Nurden, A. T., Dupuis, D., Kunicki, T. J., and Caen, J. P., 1981a, Analysis of the glycoprotein and protein composition of Bernard-Soulier platelets by single and two-dimensional sodium dodecyl sulfate-polyacrylamide gel electrophoresis, *J. Clin. Invest.* **67**:1431–1440.

Nurden, A. T., Dupuis, D., Pidard, D., Kunicki, T., and Caen, J. P., 1981b, Biochemistry and immunology of platelet membranes with reference to glycoprotein composition, *Ann. N.Y. Acad. Sci.* **370**:72–86.

Nurden, A. T., Kunicki, T. J., Dupuis, D., Soria, C., and Caen, J. P., 1982a, Specific protein and glycoprotein deficiencies in platelets isolated from two patients with the gray platelet syndrome, *Blood* **59**:709–718.

Nurden, A. T., Dupuis, D., Pidard, D., Kieffer, N., Kunicki, T. J., and Cartron, J-P., 1982b, Surface modifications in the platelets of a patient with α-N-Acetyl-D-Galactosamine residues, the Tn-syndrome, *J. Clin. Invest.* **70**:1281–1291.

Nurden, A. T., Didry, D., and Rosa, J-P., 1983a, Molecular defects of platelets in the Bernard-Soulier syndrome, *Blood Cells* **9**:333–358.

Nurden, A. T., Rosa, J-P., Boizard, B., Didry, D., Legrand, C. and Parquet, A., 1983b, Evidence that GP IIb-IIIa complexes are required for ADP-induced platelet aggregation. Studies on the platelets of a patient with a new type of Glanzmann's thrombasthenia, *Thromb. Haemostasis* **50**:216a.

O'Farrell, P. H., 1975, High resolution two-dimensional electrophoresis of proteins, *J. Biol. Chem.* **250**:4007–4021.

Peerschke, E. I., Zucker, M. B., Grant, R. A., Egan, J. J., and Johnson, M. M., 1980a, Correlation between fibrinogen binding to human platelets and platelet aggregability. *Blood* **55**:841–847.

Peerschke, E. I., Grant, R. A., and Zucker, M. B., 1980b, Decreased association of ^{45}calcium with platelets unable to aggregate due to thrombasthenia or prolonged calcium deprivation, *Br. J. Haematol.* **46**:247–256.

Peterson, D. M., and Wehring, B., 1981, Isoelectric characteristics and surface radioiodination of normal and thrombasthenic platelet membrane glycoproteins, *Thromb. Res.* **22**:53–65.

Peterson, D. M., Hirst, A., and Wehring, B., 1982, Comparison of normal and Bernard-Soulier platelet membrane glycoproteins. Isoelectric characteristics and surface radiolabel, *J. Lab. Clin. Med.* **100**:26–36.

Phillips, D. R., and Poh Agin, P., 1977a, Platelet plasma membrane glycoproteins. Evidence for the presence of nonequivalent disulfide bonds using nonreduced-reduced two-dimensional gel electrophoresis, *J. Biol. Chem.* **252**:2120–2126.

Phillips, D. R., and Poh Agin, P., 1977b, Platelet membrane defects in Glanzmann's thrombasthenia, *J. Clin. Invest.* **60**:535–545.

Phillips, D. R., Jenkins, C. S. P., Luscher, E. F., and Larrieu, M-J., 1975, Molecular differences of exposed surface proteins on thrombasthenic platelet plasma membranes, *Nature (London)* **257**:599–600.

Pidard, D., Montgomery, R. R., Bennett, J. S., and Kunicki, T. J., 1983, Interaction of AP-2, a monoclonal antibody specific for the human platelet glycoprotein IIb-IIIa complex, with intact platelets, *J. Biol. Chem.* **258**:12582–12586.

Plow, E. F., Srouji, A. H., Meyer, D., Marguerie, G., and Ginsberg, M. H., 1984, Evidence that three adhesive proteins interact with a common recognition site on activated platelets, *J. Biol. Chem.* **259:**5388–5391.

Rendu, F., Nurden, A. T., Lebret, M., and Caen, J. P., 1981, Further investigations on Bernard-Soulier platelet abnormalities. A study of 5-hydroxytryptamine uptake and mepacrine fluorescence, *J. Lab. Clin. Med.* **97:**689–699.

Rosa, J.-P., Kieffer, N., Didry, D., Pidard, D., Kunicki, T. J., and Nurden, A. T., 1984, The human platelet membrane glycoprotein complex GP IIb-IIIa expresses antigenic sites not exposed on the dissociated glycoproteins, *Blood* **64:**1246–1253.

Rosenstein, R., Zacharski, L. R., and Allen, R. D., 1981, Quantitation of human platelet transformation on siliconized glass: Comparison of "normal" and "abnormal" platelets, *Thromb. Haemostas,* **46:**521–524.

Ruan, C., Tobelem, G., McMichael, A. J., Drouet, L., Legrand, Y., Degos, L., Kieffer, N., Lee, H., and Caen, J. P., 1981, Monoclonal antibody to human platelet glycoprotein I. II. Effects on human platelet function, *Br. J. Haematol.* **49:**511–519.

Ruggeri, Z. M., Bader, R., and De Marco, L., 1982, Glanzmann's thrombasthenia: Deficient binding of von Willebrand factor to thrombin-stimulated platelets, *Proc. Natl. Acad. Sci. U.S.A.* **79:**6038–6041.

Ruggeri, Z. M., De Marco, L., Gatti, L., Bader, R., and Montgomery, R. R., 1983, Platelets have more than one binding site for von Willebrand factor, *J. Clin, Invest.* **72:**1–12.

Ryo, R., Nakeff, A., Huang, S. S., Ginsberg, M., and Deuel, T. F., 1983, New synthesis of a platelet-specific protein: Platelet factor 4 synthesis in a megakaryocyte-enriched rabbit bone marrow culture system, *J. Cell. Biol.* **96:**515–520.

Shulman, S., and Karpatkin, S., 1980, Crossed immunoelectrophoresis of human platelet membranes. Diminished major antigen in Glanzmann's thrombasthenia and Bernard-Soulier syndrome, *J. Biol. Chem.* **255:**4320–4327.

Shulman, S., Wiesner, R., Troll, W., and Karpatkin, S., 1983, Re-evaluation of the presence of the major antigen Ca^{++} complex in Bernard-Soulier syndrome platelets. Elastase degradation of the complex in Bernard-Soulier syndrome platelet preparations, *Thromb. Res.* **30:**61–69.

Solum, N. O., Hagen, I., and Gjemdal, T., 1977, Platelet membrane glycoproteins and the interaction between bovine factor VIII related protein and human platelets, *Throm. Res.* **38:**914–923.

Stormorken, H., Gogstad, G. O., Solum, N. O., and Pande, H., 1982, Diagnosis of heterozygotes in Glanzmann's thrombasthenia, *Thromb. Haemostasis* **48:**217–221.

Tobelem, G., Levy-Toledano, S., Bredoux, R., Michel, H., Nurden, A., Caen, J. P., and Degos, L., 1976, New approach to determination of specific functions of platelet membrane sites, *Nature (London)* **263:**427–429.

Tschopp, T. B., Weiss, H. J., and Baumgartner, H. R., 1975, Interaction of thrombasthenic platelets with subendothelium, normal adhesion, absent aggregation, *Experientia* **31:**113–117.

Van Leeuwen, E. F., van dem Borne, A. E. G. K., von Riesz, L. E., Nijenhuis, L. E., and Engelfriet, C. P., 1981, Absence of platelet-specific alloantigens in Glanzmann's thrombasthenia, *Blood* **57:**49–59.

Walsh, P. N., Mills, D. C. B., Pareti, F. I., Stewart, G. J., Macfarlane, D. E., Johnson, M. M., and Egan, J. J., 1975, Hereditary giant-platelet syndrome. Absence of collagen-induced coagulant activity and deficiency of factor XI binding to platelets, *Br. J. Haematol.* **29:**639–655.

Weiss, H. J., Tschopp, T. B., Baumgartner, H. R., Sussman, I. I., Johnson, M. M., and Egan, J. J., 1974, Decreased adhesion of giant (Bernard-Soulier) platelets to subendothelium—further implications on the role of the von Willebrand factor in hemostasis, *Am. J. Med.* **57:**920–925.

Weiss, H. J., Turitto, V. T., and Baumgartner, H. R., 1978, Effect of shear rate on platelet interaction with subendothelium in citrated and native blood. I. Shear rate-dependent decrease of adhesion in von Willebrand's disease and the Bernard-Soulier syndrome, *J. Lab. Clin. Med.* **92:**750–764.

White, J. G., 1979, Ultrastructural studies of the gray platelet syndrome, *Am. J. Pathol.* **95:**445–462.

Zucker, M. B., Pert, J. H., and Hilgartner, M. W., 1966, Platelet function in a patient with thrombasthenia, *Blood* **28:**524–534.

Zucker, M. B., Kim, S. J., McPherson, J., and Grant, R. A., 1977, Binding of factor VIII to platelets in the presence of ristocetin, *Br. J. Haematol.* **35:**535–549.

VI

Conclusion

17

The Role of Membrane Glycoproteins in Platelet Formation, Circulation, and Senescence
Review and Hypotheses

James N. George

1. INTRODUCTION

Circulating blood platelets are required to form an immediate hemostatic plug at the site of a vascular endothelial gap. The properties of the platelet membrane glycoproteins related to the specific functions of adhesion and aggregation during this hemostatic reaction have been the focus of the preceding chapters of this book. This chapter will review current knowledge and hypotheses on the broader and less defined phenomena of platelet production from megakaryocytes and their appearance in the circulation, their lifespan and senescence, and finally their ultimate removal from the circulation. Each of these events seems to involve membrane contact phenomena that are unique to the platelet: (1) within the bone marrow the megakaryocyte occupies a special position adjacent to the sinusoidal wall; (2) platelets are released from fragmenting megakaryocyte cytoplasm either within the marrow sinuses or later within the pulmonary capillaries; (3) circulating platelets undergo reversible adhesion encounters during their lifespan, losing fragments of their surface membrane in the process; and (4) finally an unknown senescent change on the surface membrane signals the ultimate sequestration of the platelet. This entire process may be viewed as a sequence of

JAMES N. GEORGE • Division of Hematology, Department of Medicine, University of Texas Health Science Center at San Antonio, San Antonio, Texas 78284.

reversible contact interactions among platelets as well as between platelets and the vessel wall resulting in surface glycoprotein changes and membrane fragmentation.

2. MEGAKARYOCYTE DEVELOPMENT IN THE MARROW

2.1. The Origin of Megakaryocytes from Pluripotent Stem Cells

Studies in cell culture and clincial observations in man have documented that megakaryocytes are derived from a hematopoietic stem cell in common with the other cells produced in the marrow: red cells, granulocytes, monocytes-macrophages, and B lymphocytes (Quesenberry and Levitt, 1979; Williams and Levine, 1982; Suda *et al.*, 1983). The existence of such a pluripotent stem cell is confirmed by the pattern of cell proliferation in various hematopoietic malignant disorders (Vainchenker *et al.*, 1982) and by the simultaneous presence of erythroid and megakaryocyte-specific proteins in a single cell line derived from erythroleukemia cells (Tabilio *et al.*, 1984). Also, *in vitro* mixed cell colonies can develop from single cells, containing granulocytic, erythroid, monocyte-macrophage, and megakaryocytic elements (Ash *et al.*, 1981a). The pluripotent stem cell from which these colonies develop is present in both the peripheral blood and marrow of normal human subjects. The morphologic features of this stem cell are unknown, but the earliest recognizable megakaryocyte precursor has been identified as a small (9–12 μm) mononuclear cell using antibodies specific for the membrane glycoproteins of mature platelets. Rabellino *et al.* (1981) demonstrated reactivity of megakaryocytes and mononuclear cells in human marrow using mono-specific polyclonal rabbit antisera prepared against human platelet membrane glyco-proteins Ib, IIb, and IIIa. Ash *et al.* (1981b) identified mature megakaryocytes and their mononuclear precursors in *in vitro* cultures of human marrow cells using the mouse monoclonal antibody to platelet membrane glycoprotein IIb, "Tab" (McEver *et al.*, 1980, 1983). Figure 1 illustrates these data, showing strong immunofluorescent reactions of the Tab antibody with both a morphologically identifiable megakaryocyte

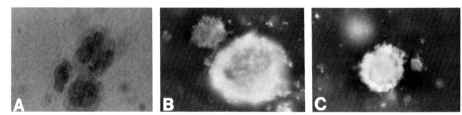

Figure 1. Identification of the platelet membrane GP IIb on megakaryocytes and their mononuclear precur-sor cells. Human marrow mononuclear cells cultured with erythropoietin and media conditioned by human leukocytes in the presence of phytohemagglutinin formed colonies of pure megakaryocytes and larger mixed colonies containing granulocytic, erythroid, monocyte-macrophage, and megakaryocytic elements. (Left) Megakaryocyte, hematoxylin stain. (Center) Megakaryocyte and small mononuclear cell, immunofluores-cent reaction with fluoroscein isothiocyanate-conjugated (FITC)-"Tab." Tab is a mouse monoclonal anti-body directed against the GP IIb of mature platelets. (Right) Mononuclear cell reacting with FITC-Tab. Magnification of all photographs is 400X. Data are from G. D. Roodman and R. P. McEver, unpublished observations, with their permission. Reproduced at 75%.

and a smaller mononuclear cell isolated from *in vitro* culture colonies. Megakaryocytes not only synthesize platelet-specific membrane glycoproteins, but also at least two secretable platelet α-granule proteins, von Willebrand factor (VWF), and platelet factor 4 (Nachman *et al.*, 1977; Ryo *et al.*, 1983).

2.2. The Development of Megakaryocyte Membrane Systems

As the megakaryocyte matures, intracytoplasmic membranes become more complex (Tavassoli, 1980). These new membranes are termed the "demarcation membrane system" because they delineate the zones of eventual individual platelets. There is evidence that the demarcation membranes originate by invagination at multiple sites from the megakaryocyte surface plasma membrane (Behnke, 1968; Tavassoli, 1980). A variety of extracellular tracers that do not penetrate intact membranes diffuse into the cysternae of the demarcation system, demonstrating communication with the surrounding medium. Following intravenous infusion of thorotrast and horseradish peroxidase into rats, these particles appeared within the cavities of the megakaryocyte demarcation membrane system. Staining of fixed megakaryocytes with ruthenium red and lanthanum nitrate demonstrated a continuity of the carbohydrate-rich surface coat of the plasma membrane with the external face of the demarcation membranes (Behnke, 1968). Also freeze-fracture studies have demonstrated the same intramembranous particles in plasma membranes and demarcation membranes, and no similarities with other intracellular membranes (Shaklai and Tavassoli, 1978). However other data indicate differences between the megakaryocyte demarcation membrane and the platelet plasma membrane, suggesting a unique intracellular origin for the demarcation membrane system (Zucker-Franklin and Petursson, 1984). Following the development and maturation of the megakaryocyte membrane systems, the cytoplasm of the mature megakaryocyte is like a nest of nascent platelets. Each immature platelet is surrounded by its eventual plasma membrane and packaged within this area are the organelles required for the platelet's function and survival. The residual deeply invaginated plasma membranes of circulating, individual platelets forming the open canalicular system may be a legacy of the megakaryocyte's demarcation membranes.

3.   PLATELET PRODUCTION FROM MEGAKARYOCYTES

3.1. Megakaryocyte Release from the Marrow and Platelet Release in the Lungs

Within the marrow, the megakaryocyte occupies a specific location adjacent to the sinus wall (Figure 2) (Lichtman *et al.*, 1978; Tavassoli and Aoki, 1981). In these areas, the megakaryocyte replaces the subendothelial reticular cell layer and provides an adventitial supporting function. The development of megakaryocytes as part of the marrow sinus-supporting structure is consistent with the synthesis and sharing of many proteins that are common to megakaryocytes, platelets, and the vessel wall (George *et al.*, 1984). Von Willebrand factor is synthesized by both megakaryocytes and endothelial cells (Nachman *et al.*, 1977; Jaffe *et al.*, 1974). Fibronectin and thrombospondin are found both as secretable proteins in platelet α-granules and in the extracellular

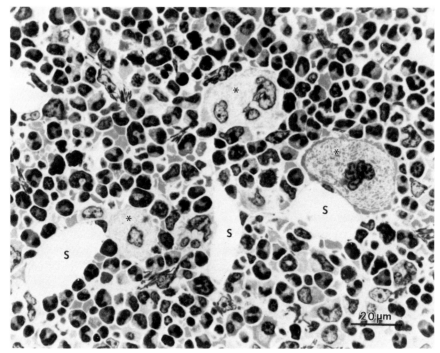

Figure 2. Demonstration of the parasinusoidal location of megakaryocytes in the marrow. The photograph illustrates a thick (1 μ) section of mouse femoral marrow stained with toluidine blue. Three megakaryocytes are identified by asterisks. Two are immediately adjacent to a sinus (s) and a third is close in this section. Reproduced from Lichtman *et al.* (1978) with permission. Reproduced at 80%.

connective tissue matrix of vessel walls (Mosher, 1980; Jaffe *et al.*, 1983). Even an intrinsic plasma membrane glycoprotein, GP IIIa, may be shared by megakaryocytes, platelets, and endothelial cells (Thiagarajan *et al.*, 1983). This nonrandom position of megakaryocytes, compared with the other progeny of the marrow stem cells, is the initial indication of a special relationship between megakaryocytes-platelets and the vessel wall, and implies that the sinus wall provides a favored site for megakaryocyte development from the pluripotent stem cell (Fedorko and Lichtman, 1982). From this location, cytoplasmic projections of the megakaryocyte flow through gaps in the sinusoidal basement membrane. These may separate from the body of the megakaryocyte, forming cytoplasmic fragments that are capable of producing several or up to 1200 platelets; alternatively the intact megakaryocyte itself may negotiate endothelial apertures and enter the marrow sinusoidal circulation (Tavassoli, 1980; Tavassoli and Aoki, 1981; Fedorko and Lichtman, 1982). It has been estimated from studies of right heart blood in humans that 20–50% of mature megakaryocytes migrate from the marrow to the lungs, and about one third of these have a full load of cytoplasm (Kaufman *et al.*, 1965a,b; Pederson, 1978). Approximately 80% of these megakaryocytes and large fragments are stopped within the pulmonary circulation where they release their platelets, accounting for the higher platelet count in the pulmonary vein than in the pulmonary artery (Kallinikos-Maniatis, 1969; Pederson,

1978). The force of right ventricular ejection provides the impact to initiate cytoplasmic fragmentation when the megakaryocytes or their cytoplasmic pieces hit the small pulmonary vessels. Estimates of the platelet production in the lungs range from 7% to 100% of total platelet production (Kaufman *et al.*, 1965b; Pederson, 1978; Tavassoli, 1980; Trowbridge *et al.*, 1982). It is conceivable that the thrombocytopenia noted in patients with cyanotic congenital heart disease (Goldschmidt *et al.*, 1974), who shunt their returning venous blood past the lungs and directly into the left heart chambers or aorta, may be due to diminished platelet release in the pulmonary circulation.

3.2. The Analogy between Platelet Separation from Megakaryocytes and Reversible Aggregation of Mature Platelets

The adhesive force that holds the demarcation membrane surfaces together but then allows them to separate upon exposure to circulatory shear stress and form individual platelets is unknown. This sequence is similar, however, to the ability of mature platelets to associate into transient aggregates and then be broken apart again by the shear forces of flowing blood and return to the circulation with no apparent membrane surface alteration. In fact, the electron microscopic appearance of the cytoplasm of a mature platelet-producing megakaryocyte (for example, in Bessis, 1973) is very similar to the appearance of a reversible aggregate of platelets attached to a vessel wall thrombus (Figure 3) (Baumgartner and Muggli, 1976). In experimental animals, the phenomenon of transient aggregation of mature, individual platelets can be induced by an intravenous bolus of ADP: the peripheral venous platelet count decreases by two thirds within 1 min and by 3 min the platelets have all returned to the circulation with no loss of their membrane proteins (see Section 4.4 and Figure 5A) (George, 1978). This is analogous to the ability of ADP to induce reversible fibrinogen binding to the platelet surface and reversible platelet aggregation *in vitro* with no change of the membrane glycoproteins (Mustard *et al.*, 1978; George *et al.*, 1980a,b; Kinlough-Rathbone *et al.*, 1983a,b). Close contact among platelets can itself have a stimulatory effect on secretion and further aggregation. This can be most simply demonstrated by the secretion resulting from centrifugation of platelet-rich plasma at an elevated pH (Massini and Lüscher, 1971). Close platelet contact can also augment the activation initiated by a variety of other agonists: ADP, epinephrine, thrombin, and calcium ionophore (Massini and Lüscher, 1971; Charo *et al.*, 1977; Holmsen and Dangelmaier, 1981). Therefore, simply the origin of the demarcation membranes with their external faces in close opposition may provide the stimulus for the cohesive force among the nascent platelets, similar to the phenomenon of "aggregation-mediated activation" of mature platelets. But clearly the cohesion cannot be intimate enough to stimulate secretion and the further results of activation.

3.3. The Origins of Density and Size Heterogeneity of Circulating Platelets

Many studies have addressed the cause of variable size and density among circulating platelets. There are reports that platelet density decreases with age in the circulation (Karpatkin, 1972; Rand *et al.*, 1983), possibly due to reversible activation

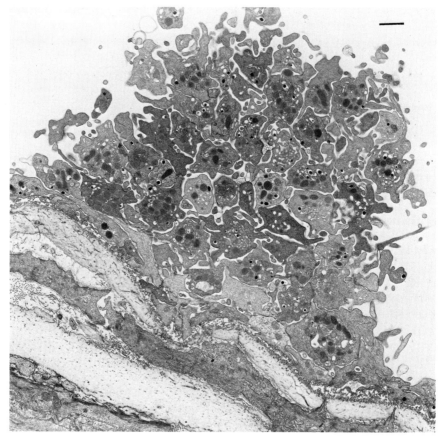

Figure 3. Electron micrograph of a mural platelet thrombus formed *in vivo* 10 min after removal of the endothelium from an iliac artery in a rabbit. The platelets close to the subendothelial surface contain their mitochondria, but have lost most of their α-granules and dense granules. The more peripheral platelets retain their normal morphology and most soon separate from the thrombus and disaggregate in the blood stream. The black bar indicates 1 μm (magnification 9800×). Reproduced from Baumgartner and Muggli (1976) with permission. Reproduced at 70%.

and secretion of α-granule contents (van Oost *et al.*, 1984). Other reports provide evidence supporting the opposite conclusion, that platelet density increases with age (Mezzano *et al.*, 1981; Savage *et al.*, 1983). This latter observation could be the result of membrane loss from circulating platelets (George *et al.*, 1976), similar to the mechanism of decreasing density of aging red blood cells (Luthra *et al.*, 1979). Other data suggest that platelet volume and density are independent variables resulting from the normal production process and that platelet size is not related to platelet age (Trowbridge *et al.*, 1982; Martin *et al.*, 1983; Thompson *et al.*, 1983; Levin and Bessman, 1983).

The factors regulating the heterogeneous size of the released platelets are unknown. The log-normal distribution of platelet size has been proposed to be a function of the pattern of megakaryocyte demarcation membranes determining the new platelet zones (Paulus, 1975) and a result of the fragmentation process of platelet release

(Trowbridge *et al.*, 1982). When thrombocytopenia occurs due to increased peripheral destruction of platelets or when the marrow is recovering from transient suppression, thrombopoiesis can be increased up to fivefold to deliver more platelets to the circulation (Harker and Finch, 1969). These platelets are larger than normal (Paulus, 1975) and may be considered to be analogous to the large "stress" polychromatophilic reticulocytes released from the marrow following extreme erythropoietin stimulation (Hillman, 1969). The regulation of thrombopoiesis may involve two stimuli: one, termed "megakaryocytic-colony-stimulating activity," enhances the differentiation and proliferation of the megakaryocyte-precursor stem cells; a second, termed "thrombopoietin," may act on recognizable megakaryocytes to accelerate maturation and platelet release (Hoffman *et al.*, 1981; Mazur *et al.*, 1981). Accelerated thrombopoiesis is associated with an increased number, size, and nuclear ploidy of marrow megakaryocytes (Harker and Finch, 1969; Martin *et al.*, 1982). Somehow the accelerated maturation of the megakaryocytes alters the formation of the demarcation membrane system, resulting in larger platelets.

3.4. Platelet Size and Platelet Membrane Glycoproteins

More striking giant platelets are seen in a hereditary disorder of platelet production, the Bernard-Soulier syndrome (see Chapter 16). In these patients, the size and frequency of giant platelets seem to be more consistent within a family group than related to the degree of thrombocytopenia (George *et al.*, 1981b). The characteristic feature of Bernard-Soulier platelets is their severe deficiency of the normal plasma membrane glycoproteins (GP) Ib, V, and IX. This is associated with a decreased membrane density of sialic acid causing a reduced net negative charge, and the decreased charge may result in less electrostatic repulsion and allow more cohesion among the surface membranes (Evensen *et al.*, 1974). Alteration of the megakaryocyte demarcation zone formation due to this plasma membrane glycoprotein abnormality is a tempting hypothesis, but it remains untested. Subjects heterozygous for the Bernard-Soulier syndrome have normal platelet function associated with a half-normal concentration of the membrane GPs Ib and IX (GP V has not yet been quantified in these subjects) (George *et al.*, 1981b; Berndt *et al.*, 1983). The observation of large platelets in these heterozygous subjects suggests that the development of megakaryocyte demarcation membranes may be more sensitive to the concentration of one or more of these glycoproteins than is the function of platelet adhesion. However, a simple relationship between demonstrable abnormalities of platelet membrane glycoproteins and platelet size does not exist. Glanzmann's thrombasthenic platelets lack a major membrane glycoprotein required for platelet aggregation, GP IIb-IIIa, but are of normal size. Conversely, the May-Hegglin anomaly is associated with large platelets, but the membrane proteins are normal (Coller and Zarrabi, 1981)

4.   THE FUNCTION OF CIRCULATING PLATELETS

4.1.   Transient-Sequestration of Platelets in the Spleen

After their release from megakaryocytes, circulating platelets continue to have a unique association with vessel wall structures. This is most apparent in the spleen, where blood cells must pass through gaps in the endothelial cell-basement membrane barrier to move from the red pulp cords into the sinusoids and then back into the venous system (Crosby, 1963; Weiss, 1974). Normally, about thirty % of circulating platelets are reversibly detained in the spleen (Aster and Jandl, 1964). This fraction can increase to ninety % in patients with portal vein hypertension and dilated, congested splenic cords, but as in normal subjects, the sequestration is reversible and the platelet lifespan remains normal (Aster, 1966). That this fraction of detained platelets may not be random is suggested by evidence that the spleen preferentially retains the youngest and most hemostatically effective platelets (Shulman et al., 1968). This observation is consistent with numerous reports that younger platelets are more hemostatically competent and also have a greater surface concentration of membrane glycoproteins to facilitate interaction with vessel wall structures and other platelets (Wright, 1942; Hirsh et al., 1968; Blajchman et al., 1981). Perhaps these younger platelets more readily adhere to the supporting connective tissue as they pass through the splenic sinus wall. If the slow splenic circulation allows sufficient cell-to-cell contact, then perhaps younger platelets also preferentially associate in transient aggregates. There is some evidence to suggest that this transient splenic sequestration of new platelets results in diminished hemostatic ability (Shulman et al., 1968), which could occur due to loss of surface membrane fragments analogous to the splenic removal of redundant surface membrane of reticulocytes (Crosby, 1963).

4.2. The Hypothesis of Continual Endothelial Support by Circulating Platelets

Other than their obvious accumulation at the site of blood vessel trauma, platelets may also continually support the integrity of normal vessels. This vascular support function hypothesis is supported by the clinical observation that an acute, severe decrease in the number of circulating platelets results in the prompt appearance of innumerable petechiae. The petechial hemorrhages are not randomly distributed throughout the body, but are concentrated in the lower extremeties where increased hydrostatic pressure would be expected to cause the greatest challenge to the integrity of the vessel wall. Experimental thrombocytopenia in rabbits is associated with morphologic changes in the capillary wall: endothelial cells are thinner than normal, with discrete thin spots and fenestrations that are not present in normal capillaries (Kitchens, 1977). This observation has been questioned, but the conflicting data are not a valid comparison because they were obtained with a different animal model that had neither such severe nor such prolonged thrombocytopenia (Shepro et al., 1980). The beneficial effect of viable platelets in maintaining normal capillary structure and function has also been demonstrated in a variety of in vitro organ perfusion systems (Danielli, 1940; Gimbrone et al., 1969; Dodds et al., 1973). The role of platelets in vessel wall support is consistent with their diffusion to a peripheral location in circulating blood, adjacent to the endothelial cells (Baumgartner and Muggli, 1976; Turitto, 1982). Being less dense, platelets will be displaced by red cells to the lateral margin of the flowing column of blood and therefore will be in a position to readily adhere to endothelial gaps. Extremely large proteins, such as VWF, may be similarly displaced toward the vessel wall (Turitto, 1982). Vascular damage disrupts the laminar flow pattern and the resulting turbulence serves to trap platelets within vortices, allowing even greater platelet–vessel wall contact. The clinical result of this red blood cell effect can be seen with the increased platelet adhesion on subendothelium with an increasing hematocrit (between 10 and 40%) (Turitto and Weiss, 1980) and the prolonged bleeding time in severely anemic patients (Hellem *et al.*, 1961).

4.3. Platelet Membrane Fragmentation during Reversible Adhesion Encounters

Platelet involvement in an adhesion encounter with the vessel wall, or in an aggregate with other platelets, may be reversible (Figure 3). Since younger platelets have greater hemostatic ability, these platelets may be preferentially involved in endothelial support and then return to the circulation with an alteration of surface structure that decreases their hemostatic effectiveness (Buchanan *et al.*, 1979). Platelet survival studies in rabbits and baboons using a radioisotope label of surface membrane glycoproteins (DD^{125}ISA) simultaneous with ^{51}Cr as a label of internal cytoplasm demonstrated a more rapid disappearance of the membrane label (Figure 4) (George *et al.*, 1976; Hanson and Harker, 1981; Rand, 1982). The label is lost equally from all major glycoproteins as if intact pieces of the platelet surface membrane are separating from the platelets. These experiments suggest continuing membrane fragmentation of circulating platelets. Platelets may form transient contacts with their reversibly extended pseudopods (Massini *et al.*, 1982) and these pseudopods may be vulnerable to fragmentation during platelet separation from an aggregate. Membrane loss may be facilitated by fusions within the invaginated plasma membrane network of the open canalicular system in a manner analogous to the invaginated demarcation plasma membrane of the megakaryocyte. In fact, the open canalicular system itself may begin, or at least be extended, by the fusion of α-granule membranes with the surface membranes during the secretion process (Stenberg *et al.*, 1984). The surface glycoproteins may be especially vulnerable to loss during membrane fragmentation if they cluster (Polley *et al.*, 1981) at the points of contact with the vessel wall or other platelets. The fragmentation of a thin pseudopod or filament of membrane may therefore account for the glycoprotein loss without significantly affecting platelet size or density since normal platelets contain excess plasma membrane capable of extending their surface area fourfold (Frojmovic and Milton, 1982). Repeated loss of membrane fragments may reduce hemostatic effectiveness since the availability of critical surface glycoproteins would be progressively diminished.

4.4. Platelet Membrane Microparticles

A corollary event to the loss of glycoproteins from the platelet during hemostasis is the presence of platelet membrane microparticles in normal human plasma and their increased concentration in serum after blood coagulation (George *et al.*, 1982). The concentration of platelet membrane microparticles in carefully collected plasma obtained by centrifugation at 12,000*g* for 10 min (Levine and Krentz, 1977) is equivalent to 0.1% of circulating platelets (George, unpublished observations). The membrane glycoprotein composition of the plasma microparticles is similar to intact platelets. In

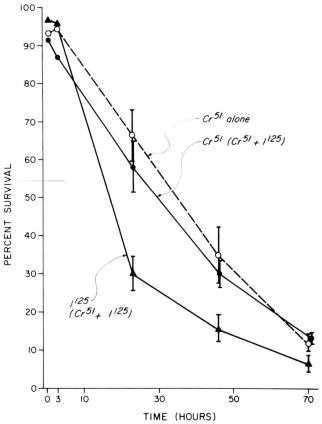

Figure 4. Platelet survival in rabbits. Platelets were labeled either with ^{51}Cr alone, which binds to cytoplasmic constituents, or simultaneously with ^{51}Cr plus DD^{125}ISA, which binds to the exposed surface platelet membrane glycoproteins. The point of maximal *in vivo* recovery of each isotope, which occurred at 30 min or 3 hr after injection, was assigned a value of 100%. Mean values ± S.E. are plotted. There was no difference among any of the values at 30 min and 3 hr and no difference between the rate of subsequent disappearance of ^{51}Cr of singly labeled platelets (○) and doubly labeled platelets (●) ($P > 0.2$). The differences between the ^{51}Cr (●) and DD^{125}ISA (▲) disappearance from the doubly labeled platelets were significant at 23 hr ($P < 0.002$), 46 hr ($P < 0.02$), and 70 hr ($P < 0.05$). Reproduced from George *et al.* (1976) with permission.

serum, microparticles are increased over 40-fold and their membrane glycoprotein composition reflects that of thrombin-activated intact platelets. It may be assumed that these circulating microparticles are even more effectively displaced toward the vessel wall than intact platelets, and it is possible that they play a role in hemostasis, similar to the adhesion-promoting function of membrane glycoprotein-containing microparticles (termed ''adherons'') shed from cultured retinal cells (Cole and Glaser, 1984). The presence of soluble glycocalicin, a catabolic product of platelet membrane GP Ib, in normal human plasma may also reflect membrane alterations of circulating platelets (Coller *et al.*, 1984).

The cohesion forces within a platelet aggregate become stronger following secretion (Kinlough-Rathbone *et al.*, 1983a,b). This is apparent in Figure 3 by the contrast between the permanently adherent platelets adjacent to the subendothelium that have lost their secretion granules and the peripheral, reversibly adherent platelets that have retained their secretion granules (Baumgartner and Muggli, 1976). Some reversible platelet aggregates, such as those induced by ADP *in vitro* and *in vivo* and that are unaccompanied by secretion, result in no membrane surface changes (Figure 5A) (George, 1978; George *et al.*, 1980a,b). Disruption of the stronger cohesive bonds of aggregates formed in association with platelet secretion, such as caused by a thrombin infusion into rabbits, is associated with membrane loss from the circulating platelets (Figure 5B) (George and Lewis, 1978). These stronger cohesive forces may be related to the firm association of cross-linked fibrin with the platelet cytoskeleton (Tuszynski *et al.*, 1984). However, rabbit platelets recovered from *in vitro* thrombin-induced aggregates associated with secretion and disappearance of visible granules can still circulate normally (Reimers *et al.*, 1976). The absence of platelet granules indicating prior secretory activity may also be seen in circulating human platelets in diseases with intravascular thrombi or following extracorporeal circulation (Harker *et al.*, 1980; van Oost *et al.*, 1983). Therefore, platelets that probably have a decreased surface density of glycoproteins can continue to circulate normally, even if they may be less involved in hemostatic interactions (Buchanan *et al.*, 1979). Some additional membrane change must occur that signals the platelet's removal from the circulation.

5. PLATELET SENESCENCE AND REMOVAL FROM THE CIRCULATION

5.1. The Occurrence of Senescent Antigens

The signal for the final sequestration of platelets is unknown. Perhaps the recurrent loss of membrane fragments and alteration of membrane proteins is a process that begins with the platelet's separation from the megakaryocyte, continues through its detention in the spleen and time in the circulation, and finally results in a senescent cell to be sequestered and destroyed. A postulated mechanism for senescence and removal of blood cells is the progressive loss of surface sialic acid causing a diminished net negative surface charge. Evidence has been presented that this occurs in human red blood cells by loss of the major surface glycoprotein, glycophorin (Lutz and Fehr, 1979). Also in red cells, a "senescent" antigen has been demonstrated to be asialoglycophorin, and this antigen reacts with antibodies present in normal human serum (Alderman *et al.*, 1981). This antigen–antibody interaction may then promote the final red cell sequestration and phagocytosis. A senescent antigen that interacts with the same naturally occurring antibody has also been to demonstrated occur on human platelets (Kay, 1981). Nugent *et al.* (1984) have isolated a monoclonal antibody from spleen cells of a patient with accelerated platelet destruction that reacts selectively with senescent platelets.

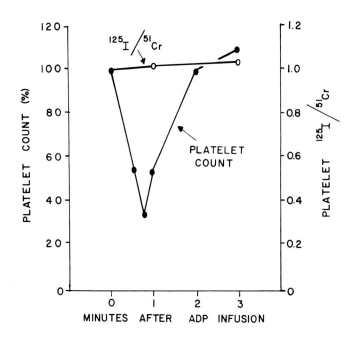

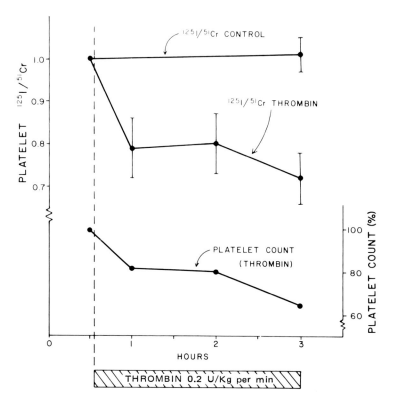

5.2. T and Tn Determinants on Glycoprotein Ib as Possible Platelet Senescent Antigens

The senescent antigen on erythrocyte asialoglycophorin may be the "T" determinant, which is present on normal human red cells after treatment with neuraminidase to remove sialic acid and expose galactose residues. If the penultimate galactose residue is also removed from the cell surface glycoprotein along with the terminal sialic acid, exposing *N*-acetylgalactosamine, then the neoantigen is termed "Tn." Anti-T and anti-Tn agglutinins are present in all human sera, except those from newborn infants (Mollison, 1983). Tn agglutinins are important in an acquired clonal abnormality of the stem cell that produces red cells, granulocytes, and platelets, termed the "Tn syndrome," in which there is a deficiency of β-3-D-galactosyltransferase activity (Cartron *et al.*, 1978; Nurden *et al.*, 1982). This disorder is associated with the cell surface exposure of *N*-acetylgalactosamine residues within the abnormal membrane glycoproteins. These residues react with anti-Tn agglutinins, causing accelerated destruction of red cells, granulocytes, and platelets. In platelets, the affected membrane glycoprotein has been demonstrated to be GP Ib by its altered electrophoretic mobility and its lectin-binding properties (Nurden *et al.*, 1982).

Therefore, a postulated mechanism for platelet senescence could involve the alteration of GP Ib to expose neoantigens on the carbohydrate moiety, and these antigens could interact with naturally occurring antibodies, such as anti-T and anti-Tn. An interaction of aging platelets with such antibodies may be one of the mechanisms for the accumulation of large amounts of IgG on circulating platelets (Kelton and Denomme, 1982). A critical amount of platelet antibody may then eventually cause platelet sequestration and removal from the circulation. Also, the loss of surface sialic acid, either from the carbohydrate side chain or from loss of a larger portion of the protein molecule, would diminish the platelet surface negative charge allowing closer contact with other cells and facilitating phagocytosis. Glycoprotein Ib would be a likely site for these platelet membrane senescent changes since it is normally the major sialic acid-containing glycoprotein and the major contributor to the platelet surface negative charge (see Chapter 3). Glycoprotein Ib may be particularly vulnerable to alteration or loss from the platelet membrane during circulation since it can form into large clusters after thrombin stimulation (Polley *et al.*, 1981) and it is the contact point

←———————————————————————————————

Figure 5. Effect of *in vivo* platelet aggregation by ADP and thrombin on the membrane glycoproteins of circulating rabbit platelets. In each experiment, platelets were labeled simultaneously with ^{51}Cr (internal cytoplasm) and DD^{125}ISA (surface membrane glycoproteins) and infused into recipient rabbits. After 30 min, samples for platelet counts and radioactivity were obtained and assigned the values of 100% for the platelet count and 1.0 for the ratio of platelet DD^{125}ISA/platelet ^{51}Cr. Then the agonist, either ADP or thrombin, was infused. (A) ADP (5 mg/kg) was administered as a rapid intravenous infusion (<1 sec) in 0.2 ml of saline. Despite the extreme but transient thrombocytopenia, no membrane glycoproteins were lost from the platelets, indicated by the constant DD^{125}ISA/^{51}Cr ratio. Data are the mean values for 2 rabbits. (B) Bovine thrombin was dissolved in saline for a constant intravenous infusion of 0.2 U/kg body weight per min as 0.2 ml/min for 150 min. The thrombin infusion caused progressive thrombocytopenia, a decrease in plasma fibrinogen, and an increase of the plasma prothrombin time. Associated with this was a loss of membrane glycoproteins from the circulating platelets, indicated by the decreased ratio of DD^{125}ISA/^{51}Cr. Data are the mean values (± S.E.) for 6 experimental and 6 control rabbits. Reproduced from George and Lewis (1978), with permission.

for interaction with subendothelium via VWF (see Chapter 10). Also, GP Ib is uniquely sensitive to the platelet's endogenous calcium-activated protease (Solum *et al.*, 1983), which may contribute to surface membrane changes during reversible platelet aggregation if the protease gains access to platelet surfaces within the clump. Such a phenomenon is suggested by the *in vitro* observation of membrane GP Ib loss that occurs with the centrifugation and resuspension of platelet washing (George *et al.*, 1981a) and the *in vivo* occurrence of glycocalicin in human plasma (Coller *et al.*, 1984).

In summary, the membrane glycoproteins that are critical to platelet function may also be important to the unique manner in which platelets develop and are released into the circulation, and to their ultimate sequestration and removal from the circulation.

ACKNOWLEDGMENTS.   I thank Dr. Rodger McEver (San Antonio) for his advice and criticisms, and Mrs. Judi Skinner for the preparation of this manuscript. Supported by research grants from the NHLBI (HL 19996) and NASA (NAG 9-5).

REFERENCES

Alderman, E. M., Fudenberg, H. H., and Lovins, R. E., 1981, Isolation and characterization of an age-related antigen present on senescent human red blood cells, *Blood* **58**:341–349.

Ash, R. C., Detrick, R. A., and Zanjani, E. D., 1981a, Studies of human pluripotential hematopoietic stem cells (CFU-GEMM) in vitro, *Blood* **58**:309–316.

Ash, R. C., McEver, R. P., McGinnis, M., Lindquist, D., and Zanjani, E. D., 1981b, Growth of pure and mixed human megakaryocytic colonies identified by immunofluorescent labeling with monoclonal antibody, *Tab, Blood* **58**:117a.

Aster, R. H., 1966, Pooling of platelets in the spleen: Role in the pathogenesis of "hypersplenic" thrombocytopenia, *J. Clin. Invest.* **45**:645–657.

Aster, R. H., and Jandl, J. H., 1964, Platelet sequestration in man. I. Methods, *J. Clin. Invest.* **43**:843–855.

Baumgartner, H. R., and Muggli, R., 1976, Adhesion and aggregation: Morphological demonstration and quantitation *in vivo* and *in vitro*, in: *Platelets in Biology and Pathology*, Volume 1 (J. L. Gordon, ed.), Elsevier/North-Holland, Amsterdam, pp. 23–60.

Behnke, O., 1968, An electron microscope study of the megakaryocyte of the rat bone marrow. I. The development of the demarcation membrane system and the platelet surface coat, *J. Ultrastruct. Res.* **24**:412–433.

Berndt, M. C., Gregory, C., Chong, B. H., Zola, H., and Castaldi, P. A., 1983, Additional glycoprotein defects in Bernard-Soulier syndrome: Confirmation of genetic basis by parental analysis, *Blood* **62**:800–807.

Bessis, M., 1973, *Living Blood Cells and Their Ultrastructure*, Springer-Verlag, New York.

Blajchman, M. A., Senyi, A. F., Hirsh, J., Genton, E., and George, J. N., 1981, Hemostatic function, survival, and membrane glycoprotein changes in young versus old rabbit platelets, *J. Clin. Invest.* **68**:1289–1294.

Buchanan, M. R., Carter, C. J., and Hirsh, J., 1979, Decreased platelet thrombogenicity in association with increased platelet turnover and vascular damage, *Blood* **54**:1369–1375.

Cartron, J.-P., Andreu, G., Cartron, J., Bird, G. W. G., Salmon, C., and Gerbal, A., 1978, Demonstration of T-transferase deficiency in Tn-polyagglutinable blood samples, *Eur. J. Biochem.* **92**:111–119.

Charo, I. F., Feinman, R. D., and Detwiler, T. C., 1977, Interrelations of platelet aggregation and secretion, *J. Clin. Invest.* **60**:866–873.

Cole, G. J., and Glaser, L., 1984, Inhibition of embryonic neural retinal cell-substratum adhesion with a monoclonal antibody, *J. Biol. Chem.* **259**:4031–4034.

Coller, B. S., and Zarrabi, M. H., 1981, Platelet membrane studies in the May-Hegglin anomaly, *Blood* **58**:279–284.

Coller, B. S., Kalomiris, E., Steinberg, M., and Scudder, L. E., 1984, Evidence that glycocalicin circulates in normal plasma, *J. Clin. Invest.* **73:**794–799.

Crosby, W. H., 1963, Hyposplenism: an inquiry into normal functions of the spleen, *Annu. Rev. Med.* **14:**349–370.

Danielli, J. F., 1940, Capillary permeability and oedema in the perfused frog, *J. Physiol.* **98:**109–129.

Dodds, W. J., Raymond, S. L., and Pert, J. H., 1973, Isolated kidney perfusion: A model for testing platelet function, *Proc. Soc. Exp. Biol. Med.* **144:**189–194.

Evensen, S. A., Solum, N. O., Grottum, K. A., and Hovig, T., 1974, Familial bleeding disorder with a moderate thrombocytopenia and giant blood platelets, *Scand. J. Haematol.* **13:**203–214.

Fedorko, M. E., and Lichtman, M. A., 1982, Megakaryocyte structure, maturation, and ecology, in: *Hemostasis and Thrombosis* (R. W. Colman, J. Hirsh, V. J. Marder, and E. W. Salzman, eds.), J. B. Lippincott, Philadelphia, pp. 210–224.

Frojmovic, M. M., and Milton, J. G., 1982, Human platelet size, shape, and related functions in health and disease, *Physiol. Rev.* **62:**185–261.

George, J. N., 1978, Platelet behavior and aging in the circulation, in: *The Blood Platelet in Transfusion Therapy* (T. J. Greenwalt and G. A. Jamieson, eds.), Alan R. Liss, New York, pp. 39–64.

George, J. N., and Lewis, P. C., 1978, Studies on platelet plasma membranes. III. Membrane glycoprotein loss from circulating platelets in rabbits: Inhibition by aspirin-dipyridamore and acceleration by thrombin, *J. Lab. Clin. Med.* **92:**301–306.

George, J. N., Lewis, P. C., and Sears, D. A., 1976, Studies on platelet plasma membranes. II. Characterization of surface proteins of rabbit platelets *in vitro* and during circulation *in vivo* using diazotized (^{125}I)-diiodosulfanilic acid as a label, *J. Lab. Clin. Med.* **88:**247–260.

George, J. N., Lyons, R. M., and Morgan, R. K., 1980a, Membrane changes associated with platelet activation, *J. Clin. Invest.* **66:**1–9.

George, J. N., Lyons, R. M., and Morgan, R. K., 1980b, Membrane alterations caused by platelet aggregation and secretion, in: *Platelets: Cellular Response Mechanisms and their Biological Significance* (A. Rotman, F. A. Meyer, C. Gitler, and A. Silberberg, eds.), John Wiley and Sons, Ltd., Chichester, pp. 81–93.

George, J. N., Thoi, L. L., and Morgan, R. K., 1981a, Quantitative analysis of platelet membrane glycoproteins: Effect of washing procedures and isolation of platelet density subpopulations, *Thromb. Res.* **23:**69–77.

George, J. N., Reimann, T. A., Moake, J. L., Morgan, R. K., Cimo, P. L., and Sears, D. A., 1981b, Bernard-Soulier disease: A study of four patients and their parents, *Br. J. Haematol.* **48:**459–467.

George, J. N., Thoi, L. L., McManus, L. M., and Reimann, T. A., 1982, Isolation of human platelet membrane microparticles from plasma and serum, *Blood* **60:**834–840.

George, J. N., Nurden, A. T., and Phillips, D. R., 1984, Molecular defects that cause abnormalities of platelet-vessel wall interactions, *N. Engl. J. Med.* **311:**1084–1098.

Gimbrone, M. A., Aster, R. H., Cotran, R. S., Corkery, J., Jandl, J. H., and Folkman, J., 1969, Preservation of vascular integrity in organs perfused *in vitro* with a platelet-rich medium, *Nature (London)* **222:**33–36.

Goldschmidt, B., Sarkadi, B., Gardos, G., and Matlary, A., 1974, Platelet production and survival in cyanotic congenital heart disease, *Scand. J. Haematol.* **13:**110–115.

Hanson, S. R., and Harker, L. A., 1981, Survival of baboon platelets labeled with diazotized (^{125}I)-iodosulfanilic acid: No effect of drugs that modify platelet behavior, *Thromb. Res.* **23:**133–143.

Harker, L. A., and Finch, C. A., 1969, Thrombokinetics in man, *J. Clin. Invest.* **48:**963–974.

Harker, L. A., Malpass, T. W., Branson, H. E., Hessel, E. A. II, and Slichter, S. J., 1980, Mechanism of abnormal bleeding in patients undergoing cardiopulmonary bypass: Acquired transient platelet dysfunction associated with selective α-granule release, *Blood* **56:**824–834.

Hellem, A. J., Borchgrevink, C. F., and Ames, S. B., 1961, The role of red cells in haemostasis: The relation between hematocrit, bleeding time, and platelet adhesiveness, *Br. J. Haematol.* **7:**42–50.

Hillman, R. S., 1969, Characteristics of marrow production and reticulocyte maturation in normal man in response to anemia, *J. Clin. Invest.* **48:**443.

Hirsh, J., Glynn, M. F., and Mustard, J. F., 1968, The effect of platelet age on platelet adherence to collagen, *J. Clin. Invest.* **47:**466–473.

Hoffman, R., Mazur, E., Bruno, E., and Floyd, V., 1981, Assay of an activity in the serum of patients with disorders of thrombopoiesis that stimulates formation of megakaryocytic colonies, *N. Engl. J. Med.* **305:**533–538.

Holmsen, H., and Dangelmaier, C. A., 1981, Evidence that the platelet plasma membrane is impermeable to calcium and magnesium complexes of A23187. A23187-induced secretion is inhibited by Mg^{2+} and Ca^{2+}, and requires aggregation and active cyclooxygenase, *J. Biol. Chem.* **256:**10449–10452.

Jaffe, E. A., Hoyer, L. W., and Nachman, R. L., 1974, Synthesis of von Willebrand factor by cultured human endothelial cells, *Proc. Natl. Acad. Sci. U.S.A.* **71:**1906–1909.

Jaffe, E. A., Ruggiero, J. T., Leung, L. L. K., Doyle, M. J., McKeown-Longo, P. J., and Mosher, D. F., 1983, Cultured human fibroblasts synthesize and secrete thrombospondin and incorporate it into extra-cellular matrix, *Proc. Natl. Acad. Sci. U.S.A.* **80:**998–1002.

Kallinikos-Maniatis, A., 1969, Megakaryocytes and platelets in central venous and arterial blood, *Acta Haematol.* **42:**330–335.

Karpatkin, S., 1972, Platelet senescence, *Annu. Rev. Med.* **23:**101–128.

Kaufman, R. M., Airo, R., Pollack, S., Crosby, W. H., and Doberneck, R., 1965a, Origin of pulmonary megakaryocytes, *Blood* **25:**767–775.

Kaufman, R. M., Airo, R., Pollack, S., and Crosby, W. H., 1965b, Circulating megakaryocytes and platelet release in the lung, *Blood* **26:**720–731.

Kay, M. M. B., 1981, Isolation of the phagocytosis-inducing IgG-binding antigen on senescent somatic cells, *Nature (London)* **289:**491–494.

Kelton, J. G., and Denomme, G., 1982, The quantitation of platelet-associated IgG on cohorts of platelets separated from healthy individuals by bouyant density centrifugation, *Blood* **60:**136–139.

Kinlough-Rathbone, R. L., Mustard, J. F., Perry, D. W., Dejana, E., Cazenave, J.-P., Packham, M. A., and Harfenist, E. J., 1983a, Factors influencing the deaggregation of human and rabbit platelets, *Thromb. Haemostasis* **49:**162–167.

Kinlough-Rathbone, R. L., Mustard, J. F., Packham, M. A., and Harfenist, E. F., 1983b, Factors influencing the deaggregation of chymotrypsin-treated human platelets aggregated by fibrinogen, *Thromb. Haemostasis* **49:**196–198.

Kitchens, C. S., 1977, Amelioration of endothelial abnormalities by prednisone in experimental thrombocytopenia in the rabbit, *J. Clin. Invest.* **60:**1129–1134.

Levin, J., and Bessman, J. D., 1983, The inverse relation between platelet volume and platelet number. Abnormalities in hematologic disease and evidence that platelet size does not correlate with platelet age, *J. Lab. Clin. Med.* **101:**295–307.

Levine, S. P., and Krentz, L. S., 1977, Development of a radioimmunoassay for human platelet factor 4, *Thromb. Res.* **11:**673–686.

Lichtman, M. A., Chamberlain, J. K., Simon, W., and Santillo, P. A., 1978, Parasinusoidal location of megakaryocytes in the marrow: A determinant of platelet release, *Am. J. Hermatol.* **4:**303–312.

Luthra, M. G., Friedman, J. M., and Sears, D. A., 1979, Studies of density fractions of normal human erythrocytes labeled with iron-59 *in vivo*, *J. Lab. Clin. Med.* **94:**879–895.

Lutz, H. U., and Fehr, J., 1979, Total sialic content of glycophorins during senescence of human red blood cells, *J. Biol. Chem.* **254:**11177–11180.

Martin, J. F., Trowbridge, E. A., Salmon, G. L., and Salter, D. N., 1982, The relationship between platelet and megakaryocyte volumes, *Thromb. Res.* **28:**447–459.

Martin, J. F., Shaw, T., Heggie, J., and Penington, D. G., 1983, Measurement of the density of human platelets and its relationship to volume, *Br. J. Haematol.* **54:**337–352.

Massini, P., and Lüscher, E. F., 1971, The induction of the release reaction in human platelets by close cell contact, *Thromb. Diath. Haemorrh.* **25:**13–20.

Massini, P., Naf, U., and Lüscher, E. F., 1982, Clot retraction does not require calcium ions and depends on continuous contractile activity, *Thromb. Res.* **27:**751–756.

Mazur, E., Hoffman, R., and Bruno, E., 1981, Regulation of human megakaryocytopoiesis, *J. Clin. Invest.* **68:**733–741.

McEver, R. P., Baenziger, N. L., and Majerus, P. W., 1980, Isolation and quantitation of the platelet membrane glycoprotein deficient in thrombasthenia using a monoclonal hybridoma antibody, *J. Clin. Invest.* **66:**1311–1318.

McEver, R. P., Bennett, E. M., and Martin, M. N., 1983, Identification of two structurally and functionally distinct sites on human platelet membrane glycoprotein IIb-IIIa using monoclonal antibodies, *J. Biol. Chem.* **258**:5269–5275.

Mezzano, D., Hwang, K., Catalano, P., and Aster, R. H., 1981, Evidence that platelet buoyant density, but not size, correlates with platelet age in man, *Am. J. Hematol.* **11**:61–76.

Mollison, P. L., 1983, *Blood Transfusion in Clinical Medicine,* 7th ed., Blackwell, Oxford, pp. 468–473.

Mosher, D. F., 1980, Fibronectin, *Prog. Hemost. Thromb.* **5**:111–151.

Mustard, J. F., Packham, M. A., Kinlough-Rathbone, R. L., Perry, D. W., and Regoeczi, E., 1978, Fibrinogen and ADP-induced platelet aggregation, *Blood* **52**:453–466.

Nachman, R. L., Levine, R. F., Jaffe, E. A., 1977, Synthesis of factor VIII antigen by cultured guinea pig megakaryocytes, *J. Clin. Invest.* **60**:914–921.

Nurden, A. T., Dupuis, D., Pidard, D., Kieffer, N., Kunicki, T. J., and Cartron, J. P., 1982, Surface modifications in the platelets of a patient with β-N-acetyl-D-galactosamine residues, the Tn syndrome, *J. Clin. Invest.* **70**:1281–1291.

Nugent, D., Berglund, L., and Bernstein, I., 1984, Platelet senescence antigen recognized by human monoclonal autoantibody, *Blood* **64**:89a.

Paulus, J. M., 1975, Platelet size in man, *Blood* **46**:321–336.

Pederson, N. T., 1978, Occurrence of megakaryocytes in various vessels and their retention in the pulmonary capillaries in man, *Scand. J. Haematol.* **21**:369–375.

Polley, M. J., Leung, L. L. K., Clark, F. Y., and Nachman, R. L., 1981, Thrombin-induced platelet membrane glycoprotein IIb and IIIa complex formation: An electron microscope study, *J. Exp. Med.* **154**:1058–1068.

Quesenberry, P., and Levitt, L., 1979, Hematopoietic stem cells, *N. Engl. J. Med.* **301**:755–760, 819, 868–872.

Rabellino, E. M., Levene, R. B., Leung, L. L. K., and Nachman, R. L., 1981, Human megakaryocytes. II. Expression of platelet proteins in early marrow megakaryocytes, *J. Exp. Med.* **154**:88–100.

Rand, M. L., 1982, Studies of changes in rabbit platelets as they age *in vivo,* Ph.D. Thesis, University of Toronto.

Rand, M. L., Packham, M. A., and Mustard, J. F., 1983, Survival of density subpopulations of rabbit platelets: Use of ⁵¹Cr- or ᴵᴵᴵIn-labeled platelets to measure survival of least dense and most dense platelets concurrently, *Blood* **61**:362–367.

Reimers, H. J., Kinlough-Rathbone, R. L., Cazenave, J. P., Senyi, A. F., Hirsh, J., Packham, M. A., and Mustard, J. F., 1976, *In vitro* and *in vivo* function of thrombin-treated platelets, *Thrombos. Haemostasis* **35**:151–166.

Ryo, R., Nakeff, A., Huang, S. S., Ginsberg, M., and Deuel, T. F., 1983, New synthesis of a platelet specific protein: Platelet factor 4 synthesis in a megakaryocyte-enriched rabbit bone marrow culture system, *J. Cell Biol.* **96**:515–520.

Savage, B., Kotze, H. F., Hanson, S. R., and Harker, L. A., 1983, Platelet density increases with platelet aging in baboons, *Circulation* **68**(Suppl III):318a.

Shaklai, M., and Tavassoli, M., 1978, Demarcation membrane system in the rat megakaryocyte and the mechanism of platelet formation: A membrane reorganization process, *J. Ultrastruct. Res.* **62**:270–285.

Shepro, D., Sweetman, H. E., and Hechtman, H. B., 1980, Experimental thrombocytopenia and capillary ultrastructure, *Blood* **56**:937–939.

Shulman, N. R., Watkins, S. P., Jr., Itscoitz, S. B., and Students, A. B., 1968, Evidence that the spleen retains the youngest and most hemostatically effective platelets, *Trans. Assoc. Am. Physicians* **81**:302–313.

Solum, N. O., Olsen, T. M., Gogstad, G. O., Hagen, I., and Brosstad, F., 1983, Demonstration of a new glycoprotein Ib-related component in platelet extracts prepared in the presence of leupeptin, *Biochim. Biophys. Acta* **729**:53–61.

Stenberg, P. E., Shuman, M. A., Levine, S. P., and Bainton, D. F., 1984, Redistribution of alpha-granules and their contents in thrombin-stimulated platelets, *J. Cell Biol.* **98**:748–760.

Suda, T., Suda, J., and Ogawa, M., 1983, Single-cell origin of mouse hematopoietic colonies expressing multiple lineages in various combinations, *Proc. Soc. Natl. Acad. Sci. U.S.A.* **80**:6689–6693.

Tabilio, A., Rosa, J. P., Testa, U., Kieffer, N., Nurden, A. T., Del Carizo, M. C., Breton-Gorius, J., and

Vainchenker, W., 1984, Expression platelet membrane glycoproteins and α-granule proteins by a human erythroleukemia cell line (HEL), *EMBO J.* **3:**453–459.

Tavassoli, M., 1980, Megakaryocyte-platelet axis and the process of platelet formation and release, *Blood* **55:**537–545.

Tavassoli, M., and Aoki, M., 1981, Migration of entire megakaryocytes through the marrow-blood barrier, *Br. J. Haematol.* **48:**25–29.

Thiagarajan, P., Shapiro, S. S., De Marco, L., Levine, E., and Yalcin, A., 1983, Monoclonal and polyclonal antibodies to human platelet membrane glycoprotein IIIa detect a related protein in human endothelial cells, *Clin. Res.* **31:**485A.

Thompson, C. B., Love, D. G., Quinn, P. G., and Valeri, C. R., 1983, Platelet size does not correlate with platelet age, *Blood* **62:**487–494.

Trowbridge, E. A., Martin, J. F., and Slater, D. N., 1982, Evidence for a theory of physical fragmentation of megakaryocytes, implying that all platelets are produced in the pulmonary circulation, *Thromb. Res.* **28:**461–475.

Turitto, V. T., 1982, Blood viscosity, mass transport, and thrombogenesis, *Prog. Hemost. Thromb.* **6:**139–177.

Turitto, V. T., and Weiss, H. J., 1980, Red blood cells: Their dual role in thrombus formation, *Science* **207:**541–543.

Tuszynski, G. P., Kornecki, E., Cierniewski, C., Knight, L. C., Koshy, A., Srivastava, S., Newiarowski, S., and Walsh, P. N., 1984, Association of fibrin with the platelet cytoskeleton, *J. Biol. Chem.* **259:**5247–5254.

Vainchenker, W., Guichard, J., Deschamps, J. F., Bouguet, J., Titeux, M., Chapman, J., McMichael, A. J., and Breton-Gorius, J., 1982, Megakaryocyte cultures in the chronic phase and in the blast crisis of chronic myeloid leukaemia: Studies on the differentiation of the megakaryocte progenitors and on the maturation of megakaryocytes *in vitro*, *Br. J. Haematol.* **51:**131–146.

Van Oost, B. A., van Hien-Hagg, I., Timmermans, A. P. M., and Sixma, J. J., 1983, The effect of thrombin on the density distribution of blood platelets: Detection of activated platelets in the circulation, *Blood* **62:**433–438.

Van Oost, B. A., Timmermans, A. P. M., and Suxma, J. J., 1984, Evidence that platelet density depends on the α-granule content in platelets, *Blood* **63:**482–485.

Weiss, L., 1974, A scanning electron microscopic study of the spleen, *Blood* **43:**665–691.

Williams, N., and Levine, R. F., 1982, The origin, development and regulation of megakaryocytes, *Br. J. Haematol.* **52:**173–180.

Wright, H. P., 1942, Changes in the adhesiveness of blood platelets following parturition and surgical operations, *J. Pathol. Bacteriol.* **54:**461–468.

Zucker-Franklin, D., and Petursson, S., 1984, Thrombocytopoiesis–analysis by membrane tracer and freeze-fracture studies on fresh human and cultured mouse megakaryocytes, *J. Cell Biol.* **99:**390–402.

Index

A23187, 36, 131, 202, 262, 284
N-Acetylgalactosamine, 69, 112, 260, 330, 407
N-Acetylglucosamine, 53, 65, 69, 72, 112, 260
N-Acetylmannosamine, 252, 260
Acid phosphatase, 27
Acrasis rosea, 245–246
Actin, 273–274
  filament crosslinking by actin-binding protein,
    288, 315
  filament ultrastructure, 279
  measurement of polymerized state, 278
  membrane-associated, 17–20, 36–38, 290, 312
  regulation of filament content, 279, 283, 315
Actin-binding protein, 20, 38, 277
  association with GP Ib, 99, 118
  association with cytoskeleton, 282
  association with the plasma membrane, 36–37
  crosslinking actin filaments, 288, 315
  hydrolysis by Ca^{2+}-dependent protease, 288
  regulation of actin polymerization, 285
α-Actinin, 276
  association with the plasma membrane, 284,
    290, 313
  regulation of actin polymerization, 315
Adenosine, 23, 153
Adenosine diphosphate (ADP)
  binding to platelet proteins, 154
  dense granule content, 181
  platelet aggregation, 6, 399, 406
  receptors on platelets, 152–155
  regulation of adenylate cyclase activity, 153
  role in fibrinogen binding, 5, 197–201
Adenosine nucleotides, *see* Adenosine diphosphate
  (ADP); Adenosene triphosphate (ATP)

Adenosine triphosphate (ATP)
  antagonist of ADP stimulation, 153
  dense granule content, 181
Adenylate cyclase
  regulation by ADP stimulation, 153
  regulation of platelet–VWF interaction, 8
  surface membrane marker, 26, 27, 34, 35
Adrenergic receptors on platelets, 159–162, 200
Afibrinogenemia, 193, 200, 265
Albumin
  endogenous, platelet-associated, 88, 108, 123
    position on CIE, 88, 106
  exogenous, effect on platelets, 23
Alcian blue, 55
Alkaline phosphatase, 26
Alpha granules, 14, 17, 25, 120, 171–186
Amiloride, 131
Amphiphilic proteins, 111–114
Ankyrin, 290
Antibodies
  alloantibodies
    anti-Bak, 345, 382
    anti-Lek, 380, 382
    anti-Pl[A1], 94, 344, 380, 382
    IgG-L, 70, 90, 91, 106, 108, 110
    IgG-P, 70, 229
  drug-dependent, 94, 330
  EDTA-dependent, 340
  methods for detection, 327
  monoclonal antibodies
    A_2A_9, 207–208
    An-51, 229, 232, 330, 368
    AP-1, 91, 93, 94, 368
    AP-2, 91, 93–96, 97, 207, 382